6,701 BUILDING FIRES
Survive in the Flow-path

A Manual for Fire Commanders, Fire Design Engineers, Architects, Students and Building Regulators.

Euro **Firefighter 2**
Firefighting Tactics and Fire Engineer's Handbook

Paul Grimwood PhD, FIFireE
Kent Fire and Rescue Service

Published by D&M Heritage Press

Unit 7, Park Valley
Meltham Road
Huddersfield
West Yorkshire
HD4 7BH

www.dandmheritage.co.uk
www.firebookshop.co.uk

First published 2017
Text and images © Paul Grimwood, unless otherwise attributed

The moral right of Paul Grimwood to be identified as
 the author of this work has been asserted.

All rights reserved. No part of this book may be
reproduced in any form or by any means without
prior permission in writing from the publisher.

ISBN: 978-1-911148-10-4 (paperback)
ISBN: 978-1-911148-12-8 (hardback)

Contents

Foreword by Battalion Commander Jerry Tracy FDNY (retired)		i
The Author		iii
Acknowledgements		vi
Introduction		viii
European to USA Conversion Factors		xvii
Chapter 1	Firefighting – Knowledge Gaps	1
	1.1 Common Design Approaches in the USA	2
	1.2 Common Design Approaches in the UK and Europe	3
	1.3 Knowledge Gaps that exist	4
	1.4 Research by NIST, USA	5
	• Coordinating Tactical Ventilation with Fire Attack	6
	• Transitional Attack	6
	• Positive Pressure Ventilation (PPV)	7
	• Wind driven Fires and Ventilation Flow-paths	9
	• NIST Lessons learned	10
	• 'Smart Firefighting' Project (Future and ongoing Research by NIST)	10
	1.5 Research by UL, USA	11
	• Firefighter exposure to Smoke Particulates wearing PPE (Contaminants)	11
	• The Performance of Engineered Floor Systems in Domestic Dwellings	14
	• UL Study of the effectiveness of the Fire Service Positive Pressure Ventilation during Fire Attack in Single Family Homes incorporating modern Construction Practices	15
	1.6 Hazards that should be considered when using PPA	22
	1.7 20 key tactical considerations from UL Research	23
	1.8 Negative Pressure Attack (NPA) Flow-path control	24
	1.9 RECEO-REVAS-RICE and SLICERS	26

	1.10	Staffing issues and limited crewing arrangements	30
		• Firefighting or Search and Rescue first?	30
		• Tactical Objectives	31
		• Search and Rescue	33
	1.11	Equipment trends	34
	1.12	Corridor Fires	34
Chapter 2		**The Building – what the Firefighter needs to know**	**36**
	2.1	Firefighting Access and Facilities (UK Fire Codes)	36
		• Firefighting Access	36
		• Firefighting Shaft	37
		• Firefighting Lift	38
		• Firefighting Stairs	46
		• Rising Fire Mains (Standpipes)	47
	2.2	Smoke Control to assist Firefighting	47
		• System Categories	47
		• System Types	48
		• System Configurations	48
		• Possible impacts on Fire Behaviour	49
	2.3	Horizontal Fire Mains	50
	2.4	Identifying Residential Construction Types	52
		• Traditional Construction	52
		• Non-traditional Construction	52
		• Modern methods of Construction	52
	2.5	Behaviour of Materials and general Structural Forms in Fire	53
	2.6	Lightweight Construction	62
		• Lightweight timber Frame	62
		• Lightweight steel modular Construction	63
		• Firefighting Tactics in lightweight Construction	66
	2.7	Heavyweight engineered Wood Products (EWPs)	67
		• Structural timber composites (STC)	67
	2.8	Fire Dynamics in EWP mass timber Buildings	73
	2.9	Recent Research into the Fire performance of CLT timber Walls and Floors	77
		• Fire Resistance	79
		• 18-storey EWP High-rise in Canada 2016	79
	2.10	Encapsulation of mass timber with Gypsum – how effective?	84
		• Fire Resistance	84

Chapter 3		Practical Fire Dynamics & extreme Fire Behaviour	87
	3.1	Fuel-controlled Fire	88
	3.2	Ventilation-controlled fire	89
	3.3	Under-ventilated fire	89
	3.4	Fire Load	89
	3.5	The Heat Profile	90
	3.6	Heat, Temperature and Heat flux	90
	3.7	Fire Growth-rate (time to Flashover)	91
	3.8	Fire Spread-rate (time taken for Fire to travel vertically or horizontally)	91
	3.9	Smoke layer 'rate of drop'	92
		• The Hot Gas Layer and Neutral Plane	92
	3.10	Heat Release-rate (HRR) (Fire intensity)	95
	3.11	Energy and Power (Fire intensity)	95
	3.12	Tactical actions (control measures) by Firefighters	96
	3.13	'Adequate' Firefighting Water	97
	3.14	Ventilation Profile	99
	3.15	Compressed Air Foam (CAF)	100
	3.16	A changing Environment may require new Tactics	102
	3.17	Design Fire Loads (Fire Engineering) and 'Design Fires'	102
		• The impact of including 'Permanent Fire Load' in Total Fire Load Calculations	106
	3.18	Travelling Fires in large open-plan Compartments	107
		• Time Temperature curves	108
	3.19	Do modern Compartment Fires burn 'faster' and 'hotter'?	111
		• Maximum HRR as a function of available Oxygen	111
		• The role of Thermal inertia	112
		• The six main Parameters that have the greatest impact on Compartment Fire intensity	115
		• Thermal runaway	118
	3.20	Phase change Materials (PCMs)	119
	3.21	Extreme Fire Behaviour	120
		• Flashover	121
		• Backdraft	121
		• Fire Gas Ignitions (FGIs)	122
		• The differences between a Backdraft and a Smoke Explosion	124
	3.22	Flashover or Backdraft – can one follow the other?	127
	3.23	The Kings Cross Underground Railway Fire – London 1987	130

Chapter 4	The 'House Fire'		**133**
	4.1	'Go in together – come out together'	133
	4.2	Working in Smoke	134
	4.3	The Risk Profile of a House Fire	135
	4.4	Changing Levels	135
Chapter 5	The 'Apartment (Flat) Fire'		**136**
	5.1	RICE (Rescue – Intervention – Containment – Evacuation)	136
	5.2	Smoke Blocker – Door Control Curtain	137
	5.3	NHBC Foundation Fire Research	137
		• The Fire Performance of new lightweight steel and timber-framed residential Buildings	141
	5.4	Stair-shaft fires	142
		• Un-vented Stair-shaft Fire	142
	5.5 Corridor fires		142
Chapter 6	The 'Office Fire'		**145**
	6.1	Office Fire (CCAB) Chicago 2004	146
Chapter 7	The Town Centre 'Shop Fire'		**151**
	7.1	The Town Centre Shop Problem	151
	7.2	Firefighting Tactics at Town Centre Shop Fires	151
Chapter 8	The 'Warehouse or Industrial Unit Fire'		**153**
	8.1	Sprinkler experience in US Warehouses	154
	8.2	The Life Safety Issue	155
	8.3	Removal of Sprinklers (Fire Engineer's Table-top Exercise)	155
		• The Firefighter's Perspective	161
	8.4	Firefighting Operations in large Industrial Buildings and warehouses	164
		• The most hazardous Environment	164
		• Structural Stability and interior Fire Service Operations	164
		• Firefighting Access	165
		• Firefighting Water	166
		• Firefighting Deployments	166
		• Firefighter Tenability	166
		• Firefighter Physiology	166
		• Crew relief Cycles	166
		• Crew briefing and debriefing	168
		• National guidance on the deployment of Firefighters wearing BA	168

		• 2014 guidance provided by the Fire Brigades Union for extended relief Firefighting Operations	168
		• Research into Firefighter Body Core Temperatures	169
Chapter 9	**The 'Car Park Fire'**		**172**
	9.1	Car Park Fire Statistics (UK)	173
	9.2	Ventilation of Underground Car Parks	174
		• Design for Ventilation by Code – BS 7346	175
		• Clause 9 Systems: impulse Ventilation to achieve Smoke Clearance (Post-fire)	175
		• Clause 10 systems: impulse Ventilation to assist Firefighting Access (during Firefighting)	175
		• Clause 11 systems: impulse Ventilation to protect means of escape)	176
	9.3	Sprinklers in Underground Car Parks	176
		• Some of the findings of the BRE Research on Fire Spread in Car Parks	176
Chapter 10	**The 'High-rise Fire'**		**178**
	10.1	Report by Fire Officer Karel Lambert (Brussels Fire Service – Belgium)	179
	10.2	Rising Fire Mains (wet and dry Pipes) – Design Specifications	179
	10.3	Nozzle reaction force	182
	10.4	BDAG research 2004 – minimum K-Factor Nozzle requirements	184
		• 51mm Hose – optimised for High-rise Firefighting	186
	10.5	Telstar House Office Fire – London 2003	188
	10.6	Hose-line and Nozzle configuration for High-rise	188
Chapter 11	**Compartment Fire Behaviour Training (CFBT)**		**191**
	11.1	Andy Fredericks FDNY Squad 18 New York City	192
	11.2	Little Drops of Water – Lloyd Layman – Royer/Nelson – UK FRS	194
	11.3	Fire Brigade Intervention Model (FBIM) Australia-New Zealand	197
	11.4	London, October 1984	199
	11.5	The first use of Swedish Water-fog Tactics in the UK	200
	11.6	A Decade of Fire-ground experience using Swedish Water-fog Tactics	201
	11.7	'100 Fires' in London 1989	202
	11.8	The Catalyst for Tactical Changes in the UK – February 1996	205

11.9		Is the Knowledge Gap met by Fire Behaviour Training being effectively transferred onto the Fire-ground in its current Format?	208
11.10		Analysis of 3D Water-fog's Gas and Flame cooling capabilities	210
11.11		Leeds University (UK) Research into 3D Water-fog Tactics	212
11.12		Thermal Classes	217
	•	The accuracy of 'Thermal Classes' and Temperature monitoring against Firefighter Thermal Exposures in live Fire Training Scenarios	218
	•	Firefighter Tenability – Personal Alarms	221
11.13		Knowledge Gaps in Training and Tactics	221
	•	Communication	221
	•	Tactical Ventilation and Flow-path Control	221
	•	Strategic Placement of Hose-lines	222
	•	Door Entry Procedures	222
	•	Advancing the right Nozzle – using the right Techniques – for the Job at Hand	223
	•	Fire Behaviour in 'Big-box' Buildings	224
	•	Limited Staffing on-scene and the impact on Tactical decision-making	224
11.4		Recent Research from Sweden into Gas cooling	225
11.5		The efficiency of Gas cooling in Compartments with high Ceilings	227

Chapter 12 Adequate 'Firefighting Water' – *you need this!* 228

12.1	There were 5,401 Building Fires – resulting in guidance in BS PD 7974-5-2014	230
12.2	How much Water is 'adequate' for tackling Building Fires?	233
12.3	The theory of Fire suppression using Water Cooling Ratios	239
	• Paul Grimwood's Research (based on 5,401 'working' Building fires 2009–2012)	245
12.4	The amount of Firefighting Water used by Firefighters	249
12.5	Building Fire damage and Firefighting Water Flow-rate – the Link	251
12.6	Other most recent Research into Firefighting Water Flow-rates	254
	• Fire Protection Research Foundation (NFPA)	254
	• Stefan Sardqvist's Research 1998	256
	• Hadjisophocleous & Richardson 2005	257

	12.7	Fire Protection Engineer's Design guidance	258
		• Prescriptive guidance in the UK	258
		• Modern performance based Codes in the UK	259
		• Obtaining Sprinkler compensations	263
	12.8	Firefighting Water Calculator Tool (based on BS 7974-5:2014)	267
	12.9	Travelling Fires	267
	12.10	The impact of inadequate Fire Main capacity on Fire Service intervention	276
	12.11	Then there were 249 more Fires … 2015–16 (Post-script)	278
	12.12	19mm Diameter Hose-reels	282
	12.13	A sliding-scale Flow-rate analysis for Fire Commanders	284
Chapter 13	**Prescriptive & Performance based Building Designs**		**286**
	13.1	Prescriptive Codes and Regulations	286
	13.2	Performance based Fire Engineering	286
	13.3	Structural Fire Engineering	287
Chapter 14	**Fire Safety Regulation**		**289**
	14.1	The functional Requirements of ADB (England and Wales)	289
	14.2	The Housing Act 2004	291
	14.3	The Regulatory Reform (Fire Safety) Order (RRO) 2005	291
	14.4	The Fire Risk Assessment	293
	14.5	The Licensing Act	293
Chapter 15	**Structural stability under Fire Attack**		**295**
	15.1	A 'reasonable' period	296
	15.2	Disproportionate collapse	296
		• SFARP	298
		• Structural Fire Engineering	298
	15.3	Total 'Burnout' Compartments	298
		• Section 2.4 Full Burnout Design Fires NZ C/VM2	299
		• Time equivalence	299
	15.4	Furnace Testing does not replicate 'Real Fire' conditions	302
	15.5	Cooling phase and residual capacity	305
	15.6	Failure criteria in composite frame structures	305
	15.7	Granary Warehouse wall collapse, London 1978	306
		• Granary Warehouse Construction	309

Chapter 16	Wind Driven building fires	311
16.1	The 360° reconnaissance and size-up	312
16.2	Over-pressurizing the Fire Compartment	312
16.3	Working against a Head-wind	312
16.4	The Beacon Street Fire, Boston 2014	313
16.5	Wind Driven Fire Research (NIST USA)	316
	• The Fires	318

Chapter 17	Rules of Engagement (interior deployment guidance)	322
17.1	Rules of engagement	324
17.2	Firefighter injuries caused by structural collapse	325
17.3	Firefighter Training in Building Construction	326
17.4	National operational guidance (NOG)	327
17.5	Rules of Engagement (ROE) – B-SAHF	328
17.6	Fire resistance ratings (FRRS) as a 'Rule of Engagement'	330
	• British Standard 9999:2017 (UK Code of Practice)	331
	• Cardington full-scale Office Fire Test 1997	336
	• Cardington full-scale Fire Test 2003	336
	• Time equivalence	337
	• Consistency and 15 minute increments in FRRs	337
17.7	Rules of Engagement – Thermal Imaging	339

Chapter 18	Tactical Ventilation – survive in the 'Flow-path'	345
18.1	Tactical Ventilation – Flow-path Control	346
	• Key points in the Tac-Vent strategy	349
	• Tactical Horizontal Ventilation	350
	• Tactical Vertical Ventilation	350
	• Venting 'Hot-wall' or 'Cold-wall' Fire Compartments	350
18.2	Tactical (Vertical) Ventilation	351
18.3	The 'Four Tenets' and 'Six Commandments' of Tactical Ventilation	352
	• The Four Tenets of Tactical Ventilation	353
	• The Six Commandments of Ventilation	359
18.4	Vent – Enter – Isolate – Search (VE(I)S)	360
18.5	Positive Pressure Ventilation (Attack) – PPA	362
	• Air-flow – practical objectives	363
	• Wind effects	365
	• Air-flow versus fire pressure	365
	• NIST Test Fire data compared to author's Test House (non-fire) air-flow data	367

	18.6	'Air-track' or 'Flow-path'?	374
	18.7	Smoke Blocking Devices	376
	18.8	Flow-paths are driven by pressure differentials	376
		• Wind Driven flow-paths	376
		• Flow-path creation	377
	18.9	Flow-path reversals	377
		• Watts Street Fire, New York City, 1994	379
		• Cherry Road Fire, Washington DC, 1999	380
		• Shirley Towers Fire, Southampton UK, 2010	382
	18.10	Door Control and reversing the Flow-path to Tactical Advantage	382
		• Advantages of Fire isolation	384
		• Disadvantages of Fire isolation	384
	18.11	Door Control Assignment	384
	18.12	Hazardous conditions in the Flow-path	386
		• Bidirectional Flow-path	387
		• Unidirectional Flow-path	387
	18.13	Taking Control of the Flow-path	388
	18.14	Basement Flow-path reversals and Tactics	391
Chapter 19		**Smoke Control – System Categories and Configurations**	**394**
	19.1	Mechanical Smoke Ventilation Systems (MSVS)	394
		• 'EuroFire' Fire Engineering Conference 2011 – Paris	394
		• Author's Research Paper – Data inputs	396
		• Fire Behaviour	397
		• Study Findings	399
	19.2	Smoke Control in a London Residential Tower	402
	19.3	Extended length Corridors in UK Buildings	402
	19.4	Smoke Control using Pressure Differentials	409
		• Two basic requirements for a Pressure Differential System	409
		• BS EN 12101 – Design Objectives for Pressurisation Systems	410
	19.5	Pressurisation of escape routes, Stairs and stack effect	411
		• Stack effect and Neutral Pressure Plane (NPP)	411

	19.6	Pressure Differential Systems in very tall buildings >30 metres	412
		• Pressurization Systems	413
		• Effect of stack and wind effect on a Vertical Shaft	414
		• Pressurization of a Shaft in accordance with BS EN 12101-6:2005	421
		• Limitations	422
		• Performance based Design of a Pressurization Shaft	425
		• Depressurization Systems	426
		• Lobby Ventilation System	427
		• Lobby Ventilation System with insufficient capacity	430
		• Design Tools	431
	19.7	Smoke-flushing (Dilution) Systems	432
		• Key Design Specifications	435
	19.8	Fire Engineer's Smoke Control Calculations (simplified approach)	436
Chapter 20		**1-5-12 Fire Command Concept**	**439**
	20.1	Decisions made in the first sixty seconds can save occupants' Lives	440
	20.2	Decisions made in the first five minutes can save Property	441
	20.3	Decisions made in the first twelve minutes can save Firefighter Lives	441
	20.4	Bill Gough writes on Safety Culture	442
		• But what is the 'cultural' dimension of habit?	444
	20.5	'Describe yourself in one word'	446
		• Probabilistic analysis – ('Flapper's' story)	446
Postscript by Group Commander Rob Lawson KFRS (retired)			**448**

Foreword

By Battalion Commander Jerry Tracy
New York City Fire Department (ret)

'The most important attitude that can be formed is that of a desire to go on learning.'

John Dewey

Photo courtesy of Tony Greco – FDIC International 2016

I have been a student of Paul Grimwood for decades and find his principles and paradigms of the fire service engaging. As a fellow firefighter, I believe 'we are made of the same cloth and character'. I have always considered Paul a *Seeker of Truth* to appreciating the efficiency of water, effective techniques of application and appliances used to control and extinguish fire in all types of construction design, with various fuel loads. The experience Paul has acquired in his career assigns credence to his work and study of this craft. The fact that his work is recognized and respected in code standards speaks volumes. Paul has not only been a voice of reason but an advocate to integrate appropriate extinguishment and ventilation tactics and the application of effective water flows (GPM / LPM) to be safe and successful. He has personally witnessed the efficiency and application of proper placement of hose streams and flows during his detachment to the FDNY during its years of heavy fire duty in the South Bronx. His observation of the ventilation practice of the FDNY and those of other departments throughout the world was the foundation of his concept and submission of Tactical Ventilation to the Fire Service. Paul has assimilated his experience from many departments throughout the world to embrace best practice and keen understanding of our collective challenge to provide life safety to the populations we serve.

Paul has been influential to introduce and stimulate our understanding of fire dynamics some refer to as fire behaviour. Paul investigated the events of the flash fire that engulfed an old wooden escalator at the King's Cross underground station back in 1987. Thirty-one people perished in that disaster including a firefighter – Colin Townsley, a station officer from the Soho Fire Station in central London. That investigation revealed and elucidated the dynamics of a fire propagating from a trench like configuration (escalator) in a rising slope within a shaft like tunnel. Paul has now inoculated the fire service to integrate expected fire dynamics into our pre-fire planning, Fire department Standard Operating Procedures, Architectural designs of structures, Fire protection plans, Codes and Standards assessment and finally the critiques and assessments of our success or failures on the fire ground.

Our equipment and application in extinguishment may be different on this side of the pond yet our ideology is the same. Paul has the ability to present a logical 'proactive' approach, meeting the challenge to reduce the 'time-lag' or as we here in North America use the term, 'reflex time' to place water on the fire, from the moment we are notified to our intervention. Paul presents a methodology proven by experience and promotes professionalism in every aspect of our service. I wish to thank Paul for his invaluable contributions to educate the fire service and all those associated with fire safety.

Gerald 'Jerry' Tracy

Battalion Commander
Fire Department New York City (FDNY) (ret)

The Author

'Let today be the start of something new'

I feel like I'm the luckiest man alive! Being able to spend such a long career in the fire service and work with such an inspirational and courageous group of men and women makes me feel extremely fortunate. Having just turned eighteen years old, I was accepted into the London Fire Brigade (LFB) in August 1971 as London's youngest ever firefighter. I had no idea back then of how the next five decades would impact my life as I walked into LFB's Southwark training centre for the hardest twelve weeks of my life. There was a mix of excitement, elation, loyalty and close bonding within the firefighting teams I worked with, leading to lifelong friendships like no other. The daily adrenalin rush was on another level and was clearly addictive. I desperately wanted to become the best I could be, doing the best I could do for those unfortunates who needed us at the worst possible times in their life. I served at the busiest of fire stations wherever I went and my strongminded curiosity ensured I learned the job of a firefighter pretty quickly. The LFB were just emerging from the 1960s when I first climbed a wooden ladder, wearing oxygen breathing apparatus, heading up to search for people trapped in several of London's hotels or public building fires. At this point in time the 1971 Fire Precautions Act was still to take effect and the fires and rescues were almost weekly.

Looking back across those five decades, I have also been fortunate and honoured to contribute in excess of 200 trade journal articles and academic papers on innovative firefighting strategies and fire engineering developments; four books on firefighting tactics; delivered presentations and technical papers at conference in over a dozen countries to an international audience of fire chiefs, firefighters and fire engineers; whilst introducing and developing a broad range of firefighting strategies within the UK fire service that I have worked with and learned on my travels.

Tactical Ventilation – Swedish 3D Water-fog tactics and training – Positive Pressure Ventilation

'Paul Grimwood is a pioneer who has introduced and developed a number of innovative fire-fighting techniques in the UK Fire Service over the past decade'.

Fire Protection Association's review of 'Fog Attack' 1992

London's youngest ever firefighter Paul Grimwood (2nd from left) at an eight-pump fire in central London, 1972

The background research for introducing and developing a range of new firefighting techniques in the UK and Europe began in 1975 when I first visited the New York City Fire Department (FDNY), with subsequent detachments from LFB to fire departments in the USA occurring in 1976-77 and again in 1990. Whilst in the USA I worked on assignment with ten big city fire departments over several months and also as a volunteer firefighter on Long Island NY, gaining an in-depth understanding of North American firefighting strategy.

I have spent years comparing systems, analysing methods and techniques and studying response and deployment models. In some situations, I used my own experience of fires and compared international strategies with our own. Would New York City firefighters, with their predicated methods and systems, have provided a better outcome at any particular fire than our own local response? Would Swedish firefighting methods have been more effective at a specific building fire or perhaps using a North American positive pressure venting attack as opposed to a water-fog insertion approach may have produced better outcomes in comparison to our own? One thing is certain, there are so many different ways to approach a building fire using a wide range of strategic and tactical options, we should always ask ourselves if and how we might have achieved the best outcome and in what way we could improve or optimise our tactical approaches in future.

I am now retired from operational firefighting but remaining in uniformed service with Kent Fire and Rescue Service (KFRS) as a fire engineer, attached to the Technical Fire Safety department. I inspect building plans and liaise with architects and designers over large and complex constructions that include a range of fire safety measures. I also undertake command training and provide additional support to operational research and development.

This fourth book on firefighting tactics and fire engineering investigates how the ever-changing built environment is shaping current and future firefighting tactics and how creative building design may assist firefighters in their work.

In memory of **Assistant Chief Fire Officer Roy Baldwin** *and* **Station Officer Tom Stanton** *(London Fire Brigade) whom, without their help and encouragement, none of this work would have been possible.*

100 percent of author's royalties from this book are donated to the Katie Piper (Burns) Foundation (Registered charity: 1133313). The Katie Piper Foundation provides support to those suffering severe disfigurement caused by burns and scalds and offer a range of surgical and non-surgical assistance to enhance appearance.

Acknowledgements

I would wish to thank all the following for their sincere friendship, support, technical guidance and contributions in the production of this book.

To my wonderful wife Lorraine, for her never ending support in endless hours of research, and my two sons, Paul Jnr and Richard, for their belief and encouragement

Assistant Chief Officer Roy Baldwin

Station Officer Tom Stanton

Station Officer Tim Marcham

Sub Officer Robin Taylor

Leading Firefighter Andy Green

London Fire Brigade

Deputy Chief William Bohner

Deputy Chief Vincent Dunn

Battalion Chief Gerald Tracy

Firefighter William McLoughlin E222

Capt. John Ceriello

New York City Fire Department

Chief Executive Ann Millington

Director of Operations Sean Bone-Knell

AD Paul Flaherty

GM Rob Lawson

GM Mark Richards

WM Mick South

WM Phil Bailey

WM Dave Harris

WM Liam Hudson

WM Mark Duly

WM Barry Healey

WM Royston (Roy) Ingram

Kent Fire and Rescue Service

For their contributions in this book

Bill Gough

Anders Dagneryd

Nick Papa

Andrew Starnes

Karel Lambert

Jim Dave

Russell Acton (Acton Ostry)

Graham Couchman (SCI)

Nick Mellor (LEIA)

Dr. Stefan Svensson

Dr. Iain Sanderson

Dr. Stefan Sardqvist

Cover graphics by:
Christine Engert
Graphic Design, Brighton UK
David Myers Photography
Brighton UK

John Chubb

Juan Carlos Campaña

Stephane Morizot

Michel Persoglio

Szymon Kokot-Gora

Martin Arrowsmith

Mark Fishlock

Dennis LeGear

Franck Gaviot-Blanc

Christoph Albert

Dogan Gurer

Idan Braun

PJ Norwood

Christoph Gruber

Colin Hopkins

Pilou Angeli

Siemco Baaji

Hans Nieling

Dietmar Kuhn

Christopher Naum

John McDonough

Chief Ed Hartin

Julian Gifford

Michael Reick

Scott Corrigan

Darren Masini

Jose Miguel Basset

Jan Sudmersen

Steve Kerber

Adrian Ridder

Introduction

'6,701 Building Fires'

'It was a warm summer evening in 1975 at a tenement fire around 179th Street in the south Bronx when Tony Bonnano (Engine 45) handed me the nozzle – I was fighting fire in the big apple and this all felt like another world, compared to London's west-end'!'

This book is about 6,701 UK building fires. As firefighters and fire design professionals these fires have taught us a lot. They provided benchmark data that enabled us to gauge how successful we are in what we do, providing unique opportunities to explore how we might further develop and optimise our strategic approaches, whilst also reducing the hazard levels to best effect. So, what was special about these fires and why did they represent the largest data analyses of firefighting operations (working fires**) ever researched in the UK?

This book will explain more.

What did these 6701 fires teach us and what is discussed throughout this book in greater depth?

- Less water applied on a building fire during the early stages of firefighting intervention (general) results in increased building fire damage and heavier demands on resources
- More ventilation created or occurring demands higher flow-rates applied
- Where water supply is unavailable or inadequate, or not continuous, reduce ventilation
- Control and use tactical ventilation to advantage, not disadvantage
- Swedish fog-tactics are there to complement direct fire attack, not replace it
- Very few firefighters or senior fire officers understand the basic principles of transporting and applying adequate firefighting water most effectively
- The term 'adequate water' is little understood and was not defined in any building codes
- The building codes are rapidly becoming outdated in many respects and performance based fire engineering is taking over many important design issues
- The term 'reasonable period' for aligning fire resistance to structural stability in line with firefighting operations has no definition
- The built environment is developing greater fire-ground hazards and firefighters exposure to risk is therefore constantly increasing

1984	A decade of inner city research of the operational use of Swedish firefighting tactics at building fires in London 1984–1994.	549 fires	Grimwood's research into 3D gas cooling and water-fog gas-phase fire suppression by firefighters.	Became a source of prime operational reference for European guidance in relation to the firefighting tactics that originated in Sweden.
1989	A study of 100 'working' building fires in London in 1989.	100 fires	Grimwood's firefighting water analysis to develop a rough fire-ground formula for use by fire commanders' on-scene to estimate likely water requirements.	The sole real fire data inputs used for Barnett's fire flow computer model that became a New Zealand design standard.
1998	A study of 307 building fires in London, 1994–1997.	360 fires	Sardqvist's firefighting water analysis of 307 fires in non-residential building fires in London.	Became a fundamental input to the FBIM (Australia) firefighter intervention model
2015	A study of 5,401 'working' building fires occurring in a County and a Metro fire service from 2009–2012.	5,401 fires	Grimwood's firefighting water analysis compared against resulting building fire damage.	A PhD research thesis by the author that provided several pages of input to the BS PD 7974-5 Fire Engineering standard in 2015.
2016	A study of 291 'working' building fires in a County fire service 2015–16.	291 fires	Kent Fire & Rescue Service's firefighting water analysis compared against resulting building fire damage, utilising alternate tactical approaches as a means of reducing building fire damage	For information and analysis for the development of a new firefighting response and intervention model.

** 'working fires' have been defined as any occupied building fire where firefighters wore breathing apparatus and used a flow-rate in excess of 100 Litres/minute to gain control of the fire. In effect, the vast majority of these fires caused between 1 and 500 square metres of fire damage, with almost half the fires spreading beyond the compartment of origin and a third spreading to involve other floors.

- Exposure to firefighter risk in lightweight buildings is creating a need for new tactical approaches and firefighting equipment – firefighter life losses in lightweight buildings is on the increase, pro rata
- The introduction of fire behaviour training in the UK since 1999/2000 has not reduced firefighter life losses – they've increased!
- Greater attention is needed to teach firefighters the importance of 'flow-path' creation, reversal and control – a key fire behaviour module
- Compartment fires are now burning with greater intensity and higher temperatures than they did thirty years ago and modern firefighter clothing is creating greater risks in some ways
- Firefighters must find ways, and resort to modern technology, to assess the compartment fire environment more effectively
- Building design is creating larger window openings and long corridors – the wind-driven fire has suddenly become an even greater hazard to firefighters
- In 1950 the average time to flashover was 15 minutes but now we can consider 5 minutes as more appropriate – this may affect both firefighter injuries as well as occupants
- Critical primary command time-windows of 1; 5 and 12 minutes have been identified by the author as key to a successful and safe outcome on the fire-ground
- 'Rules of engagement' (ROE) protocols should be introduced and clearly defined, allowing operational discretion where justified and reviewed
- Identifying risk control measures and implementing them through fire-ground protocols and SOPs have never been so important
- The design and configuration of smoke control systems are little understood and design practitioners must recognise that such designs must take into account firefighter tenability and safety, as well as human behaviour impacts on means of escape

Between 1975 and 1977 the author visited and responded alongside city of New York (FDNY) firefighters for several months on funded research detachments, studying the firefighting tactics that were so dissimilar from those practiced by the London Fire Brigade. The starkly varied approaches were brought about mainly due to differing philosophies in how to control the interior firefighting environment of fire-involved buildings, as well as certain construction issues. Releasing combustion products from the building and using high water flow-rates to achieve control was the American approach whereas the London tactics were to isolate the fire where possible from any air supply and use lower flows to suppress lesser amounts of fire. However, the author noted distinct advantages and disadvantages in both approaches and spent the next decade developing and publishing a 'middle-ground' strategy he termed *tactical ventilation*. This tactical approach recognised the benefits of vertical ventilation to control smoke movements but also established clear protocols where isolating the fire compartment and containing the fire was a better strategy.

Then in 1984 the author began to research some innovative Swedish firefighting techniques that appeared to complement his tactical ventilation strategy. The idea was simply to (a) control the openings within the fire building and reduce the amounts of air feeding the fire; (b) utilise controlled amounts of water-fog to inert and cool hot fire gas layers, and then remove them from the building; and (c) attempt to use the Swedish water-fog tactics to extinguish compartment fires from the interior. At this point in time from 1984 onwards, he began passing on this knowledge to a group of firefighters in London's west-end district, where he worked and over a ten-year period the tactics were used operationally **at over 500 'working' building fires**. In some instances, the methods worked very well but on occasions it became clear that the water flow-rate that appeared optimal was closely linked to the venting parameters, some of which the firefighters had no control over as fires self-vented.

In 1989 the author began more research into the flow-rates required to extinguish fires *safely and effectively* and as a London Fire Brigade fire investigator undertook a study of **100 'working' building fires** in a wide range of occupancies in order to provide some data on needed flow-rates. It became obvious that the low-flows that appeared optimal when using the Swedish firefighting tactics would sometimes reach their limits of suppressive capacity and cause firefighters to retreat to safety before additional hose-lines were laid in support.

In 1997 the fire service college at Moreton-in-Marsh (UK) took on the author's work and started to introduce fire behaviour training and ventilation tactics, based on his 1992 book Fog Attack and earlier published papers. At the same time, following the authors work with Warrington Research, the Home Office (UK) produced firefighting manuals based on the *tactical ventilation* approach.

However, just like the variance between US-UK ventilation tactics, the difference between the Swedish-UK approaches required some compromise in order to meet on middle ground. The author then created the *3D firefighting strategy*[1] (2002) that harnessed Swedish fog tactics using higher water flow-rates, encouraging firefighters to take greater control over the buildings air-track (flow-path) by utilising positive pressure ventilation as part of the overall attack strategy. At this point the Swedish and UK strategies were beginning to spread around the globe.

The between 2009 and 2012 the author undertook further research into the amounts of water used by firefighters in two UK fire services; a county brigade and a city metro brigade. This was the most detailed research into fire flow-rates ever undertaken **(5,401 working building fires)** and it provided benchmark data of how optimising the water-flow could reduce building fire damage. This research demonstrated that the methods and quantities used to apply firefighting water could impact building fire damage (and life safety) almost as much as response times and firefighting resources – 'adequate' water in was of extreme importance to the firefighter!

Then in 2016 the author carried out the latest analysis (**291 working building fires**) of the county fire service data for the 12 months 2015–2016. A comparison was made with the 2009-2012 data as since then, the Kent Fire and Rescue Service (county) had undergone a detailed capability review of response, deployment, equipment and firefighting tactics. They had also introduced higher flow smooth-bore nozzles to complement the lower flow fog nozzles and increased the diameter of primary attack hose-reels from 19mm bore to 22mm (doubling the flow-rate) on many of the new fire engines. The impact in reducing building fire damage was quite substantial at a time when the entire response model in Kent was being redeveloped (reduced in staffing and resources). The 291 'working' building fires were contained within the compartment of origin on 62 percent of occasions compared to just 51 percent in 2012 and only 20 percent of fires spread to involve multi-levels, compared to 35 percent in 2009.

The Fire-Ground

When there's a building fire, firefighters don't have time for street science or complex engineering calculations. They just want to pull levers, turn switches, throw up ladders,

---

1 Grimwood. P and Desmet. K; Tactical Firefighting CEMAC http://www.olerdola.org/documentos/cemac-kd-pg-2003.pdf

squirt water and rescue trapped occupants. This active approach is generally based around a pre-determined framework of supporting guidance notes and operating procedures. If the fire is a big one, then they'll use 'big water'. If it's getting bigger then they'll call for assistance and lay in additional hose-lines. The fire-ground is fast moving and dynamic enough without making it any more complex than it already is. In a wide range of firefighting settings from the busy inner-city fire stations, to the ghettos, to high-class neighbourhoods or rural havens; the firefighter's exposure to risk is driven by various social factors, work-load, construction standards, tradition, culture, risk based tactics and staffing/resource levels. However, if left to its own means, without appreciation and input from supporting background research into everything associated with equipment and tactics, design methodology and constructions, the fire-ground (or the training scenario) can become a far more dangerous place than it already is. It's essential that as the building fire environment changes and advances in equipment technology and systems continue to evolve, firefighters must learn to adapt their strategy and tactics if they are to maintain or increase performance levels and advance safety at fires.

The design of energy efficient insulated buildings (modern methods of construction) creates new challenges for the fire service. The Swedish SRSA (Swedish Rescue Services Agency) in a report on fire prevention states that: 'In 1950 the average time from ignition of a fire to flashover was 15 minutes. Then, in 1980, that time was down to 5 minutes and now fatal conditions can occur after 3 minutes. This change has come about because of the increase of plastics in our homes, nothing else'. During the period 1994-2014, the UK department of communities and local government (DCLG) reported that average UK fire service attendance times to general building fires have increased from 6 to 8 minutes. It is clear to see that a combination of an ever-increasing plastic, and adhesive impregnated wood, content in the fire load, coupled with energy efficient highly insulated buildings or pre-engineered mass timber, will result in faster fire growth rates; earlier flashovers and more fires spreading beyond the compartment or floor of origin. In effect, firefighters are being exposed to an increased risk of higher compartment fire temperatures and possibly earlier structural collapse.

Firefighter's Exposure to Risk in Lightweight Construction

It is concerning that a continuing trend towards lightweight energy efficient modular construction in the U.S. and more recently the UK and Europe, may have unwittingly exposed firefighters to greater risk, resulting in an increased firefighter life loss ratio (per 100,000 building fires) during a transitional period across three decades (from 1979) that saw the introduction of these fast-build modern methods of construction (MMC). It was established in a U.S government sponsored 'structural collapse' study related to firefighter life losses in the USA during interior firefighting operations (reported on later), that whilst a general reduction (by a third per 100,000 fires) in traumatic or fire related firefighter fatalities was achieved over a 23-year period, the life loss ratio in cases of 'structural collapse' remained steady at 1.5 per 100,000 building fires. Additionally, it was shown that the percentage of firefighters killed by structural collapse in residential apartment buildings and private houses had actually increased during this time period.

The Importance of 'Flow-Path' Control

This book does not represent a complex fire engineering text but aims to provide fire engineers and students, fire commanders, firefighters, building inspectors and regulators with basic but essential information in relation to adequate means of fire service access and installed fire protection facilities necessary to support safe and effective firefighting operations. The core of the text is directed to the firefighting tactics now being used by firefighters to control the natural movements of smoke, heat and combustible fire gases in the ventilation 'flow-path'. It also reviews the various categories of natural and mechanical smoke control systems commonly being encountered in buildings and discusses how their configurations might have both positive and negative impacts in the flow-paths. In some instances, the direction of an existing flow-path can be used to tactical advantage whilst in other situations, firefighters will need to take positive actions to reverse the flow-path to gain any advantage. In some smoke system designs in extended corridors, firefighter safety may be severely compromised by inappropriate system configurations.

Critical command time-scales
There is a sixty second period following firefighters arriving on-scene at a building fire where their primary actions may well determine the final outcome in terms of lives and property saved. This is then followed by a further five-minute period where secondary command decisions can influence the next five hours on-scene! Incident response data has demonstrated that a large percentage (25%) of fires become worse following fire service arrival, before control is achieved. Therefore, that initial sixty second period following arrival, followed by an additional five-minute command time-frame are critical and it's what happens here that can have major impact in saving lives and property. It should be a strategic objective of every fire service to train and prepare for those first few minutes of primary response in an effort to take control of the fire environment immediately on arrival.

The fire service in general also has legal responsibility to ensure buildings are safe for occupation following construction and completion. This book will therefore appeal to the architect who strives to achieve the very best outcome in any project. It is for the fire protection engineer or student who wishes to know more about the fire service experience in regulating buildings following occupation and tackling fires when a building is subjected to its greatest ever imposed loads. The text will offer some insight to those responsible for regulating building design through to construction, offering guidance on the importance of structural safety and fire resistance, where code definitions and design strategies may appear stretched.

Firefighter risk 'Savvy'
Captain Sir Eyre Massey Shaw led the London Fire Brigade from 1866 to 1891 and spent many weeks visiting the New York Fire Department (FDNY) in 1865 (*Metropolitan Fire District of New York* and the *Metropolitan Fire Department of Brooklyn (MFD)*, to compare firefighting equipment and tactics. Shaw was renowned for his simple creed by which he instructed his firefighters in the basics –

'A fireman to be successful, must enter buildings; He must get in below, above, on every side, from opposite houses, over brick walls, over side walls, through panels of doors, through windows, through loopholes cut by himself in the gates, the walls, the roof; he must know how to reach the attic from the basement by ladders placed on half burned stairs, and the basement from the attic by rope made fast on a chimney; His whole success depends on his getting in and remaining there, and he must always carry his appliances with him, as without them he is of no use'.

So, what did Shaw actually mean by his elucidation of a successful firefighter? On a recent visit to New York's FDNY training centre (The 'Rock') in 2016 I spoke with several retired and serving FDNY commanders and a discussion surrounding local knowledge raised the term 'firefighter savvy'. This is basically *an awareness, a thought process, perception or understanding* of critical local knowledge of their buildings and assigned response areas.

There was a distinct view from the older generation of fire commanders that firefighters in their day may have been more astute as to the layout and construction of buildings and their inter-connection with adjacent properties. However, the point was made that current firefighters may not be as 'firefighter savvy' and possess a generic and limited knowledge of their buildings. It may be that far more regeneration and redevelopment has since occurred, making local knowledge more difficult to obtain or recall. The author in fact remembers, for example, having a distinct personal knowledge of how building roofs interconnected in central London, making it possible to reach certain points above the fire and even rescue trapped occupants without ladders, simply by entering unaffected (by fire) buildings and accessing the entire flat roofing environment at some great height above the streets.

This is what I believe Shaw was alluding to – being aware, able and 'savvy' enough to be able to access and remain, to tactical advantage, any point of a building on fire using the personal equipment carried. This information wasn't learned through structured teaching, it was knowledge handed down. It is certain that the level of knowledge and awareness needed of local risks and buildings is dependent on the quantity of risk and the importance or relevance of information. Some station areas may have over 400 high-rise buildings and the quantity of risk may be disproportionate to the resources available to visit and record information. However, is your crew local risk 'savvy'? Are you personally as aware as you should be of important local information? Is your premises risk information recorded and used effectively, if at all! New risks in the built environment are evolving each day and it's important we locate them and familiarise ourselves with relevant information.

> **What does this mean to the Firefighter?**
>
> For firefighters wishing to advance quickly and speed-read some key points throughout this book, go to the tactical paragraph boxes provided in each chapter, indicated by the above motif headings. This will enable the important points to be covered more quickly without sifting through extensive background information and pages of engineering calculations. However, where a more in-depth understanding is required, you are advised to read the surrounding information provided in support.

THE FIREFIGHTER, FIRE COMMANDER AND TRAINING INSTRUCTOR

Firefighter training

Over the past five decades of international fire service, and as a firefighter serving on both sides of the Atlantic, I have seen many positive changes and numerous innovations. In general, our homes and workplaces have become far safer from fire and our firefighters have been supported with modern technology and equipment that aids the tactical approach. However, as we reduce the number of building fires through safer home and workplace technologies and community safety education, it means there is less experience from which to learn our trade. In turn, this places an increased onus on realistic but safe training, to make up for lesser exposures to real fire incidents. Add to this the changing role of the 21st century firefighter whose rescue and humanitarian aid role is becoming very diverse, far beyond that of just fighting fires, where training time for all disciplines becomes expensive and time restricted.

The fire environment is also changing and becoming more dangerous for the firefighter, with more rapid fire spread rates and hotter fires in less stable constructions. This means the logical answer is to simplify the firefighting approach, supported by innovating tactics and methodologies that may be applied with optimum precision and not whilst working at close quarters with the fire, as has been the traditional approach until now. This is not to say that interior firefighting should not take place, but rather that utilising methods to reduce the heat release and heat flux exposure by lowering compartment temperatures, should pave the way for safer tactical approaches in supporting search and rescue as a primary objective.

These strategies and innovations are now being seen amongst others as forward looking:

- External stream applications ahead of, and in support of, an interior deployment
- External fog insertion tools to reduce the heat release and 're-set' fire conditions
- External fog insertion tools to cut off void fires before opening them to air-inflow
- Taking control of *air flow-paths* to prevent, reduce or reverse their effects
- Creating *flow-paths* that support firefighting entry, utilising well timed and correctly sited horizontal ventilation points and PPV applied in an attack mode (PPA)
- Considering tools that can control *flow-paths* using negative pressure (NPV), such as 'HydroVent' – these tools may be deployed more quickly with less staffing on-scene
- Recognising how building design is changing and countering early and more rapid fire spread with higher water flow-rates (nothing new to some of us!)
- Using CAFS or water additives to increase the suppressive capability of water where adequate flow-rate is not immediately available
- The use of Thermal Image Cameras by all internal attack and search teams to locate the fire and trapped occupants
- The use of overhead drones to provide image scanning of how fires are spreading
- The provision and use of light and portable personal heat-flux monitors at firefighter locations (future innovation) to prevent firefighters working too long in dangerous environments.

THE FIRE ENGINEER, ARCHITECT, FIRE SAFETY OFFICER AND BUILDING CONTROL OFFICER

> **What does this mean to the Fire Engineer/Architect?**
>
> For fire engineers and architects wishing to advance quickly and speed-read some key points throughout this book, go to the tactical paragraph boxes provided, indicated by the above motif headings. This will enable the important points to be covered more quickly without sifting through extensive background information. However, where a more in-depth understanding is required, you are advised to read the surrounding information provided in support.

Fire Engineering

Current building codes may be classified as *prescriptive or performance-based* in nature. **Prescriptive codes** obtain their names from the fact that they prescribe specifically (in general) what to do in a given case. **Performance codes** express the desired objective to be accomplished and allow the architect, or fire engineer, to use any other *equivalent* and *acceptable* approach (to regulators) to achieve the required results. Prescriptive codes have evolved over many decades and have been used in specifying fire protection systems in buildings. These codes have developed, with newer requirements and updates often being imposed over existing ones. As a result, prescriptive codes have become complex and are often difficult to apply in the modern built environment. Prescriptive building codes are used in most countries and many people are now arguing the benefits in using alternative approaches. The reason behind the move in design towards the *fire engineered* performance approach is the expected advantages that the performance-based fire safety design can offer over the prescriptive design.

- Establishing clear fire safety goals and leaving the means of achieving those goals to the designer;
- Permitting innovative design solutions that meet the established performance requirements;
- Eliminating technical barriers to trade for a smooth flow of industrial products;
- Allowing international harmonization of regulation systems;
- Permitting the use of new knowledge as it becomes available;
- Allowing cost-effectiveness and flexibility in design;
- Enabling the prompt introduction of new technologies to the marketplace;
- Eliminating the complexity of the existing prescriptive regulations.

It is the intention of this book to look at modern approaches to design and construction principles and explore ways that firefighters may need to adapt their firefighting tactics in fire engineered performance-based constructions, in maintaining existing levels of fire-ground effectiveness and safety whilst optimising the resources available to them. It is also the intention to introduce firefighters to the basic principles of fire engineered buildings whilst at the same time, providing fire engineers and architects with a greater understanding of why firefighters will rely on *even higher standards* (than prescriptive codes might provide) of compartmentation, active suppression systems, correctly configured smoke control and ventilation, firefighting access (both externally and internally) and adequate firefighting water provisions, in such engineered buildings.

To design and construct on performance based principles but then only provide the minimum prescriptive requirements to assist firefighters, is failing to appreciate how the modern firefighting environment is also changing and the firefighter's job is becoming ever more hazardous. As fire loads and ventilation factors increase and lightweight heavily insulated buildings become more common, compartment fires will now burn faster and hotter and if fire protection features are inadequate, structural elements may collapse far sooner than they did thirty years ago, when existing prescriptive codes were introduced.

European to USA Conversion Factors

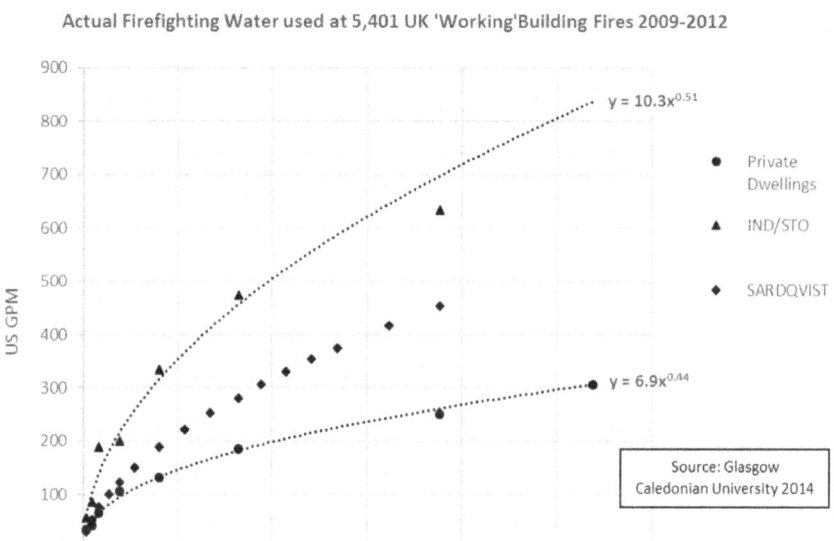

A quick 'rule of thumb' rough fire-ground flow calculation derived from 5,400 working UK fires is:

Square feet of fire (take the last digit off) and then add 100 = required gpm, so; For fires to 5,500 square feet of fire involvement (based on UK construction)		
200 Sq.ft of fire	= 20 + 100	= 120 gpm
3500 Sq.ft of fire	= 350 + 100	= 450 gpm

The author's PhD research included analysis of the quantities of firefighting water used for suppression at 5,401 'working' building fires in the UK between 2009 and 2012. The lower line represents houses and apartments with an upper line representing industrial units and warehouse fires. All other fires fall between these two lines, as represented by a median line of data provided by the Sardqvist research into non-residential premises. It should be noted that construction in the UK is widely solid masonry and structure fires are in general, only tactically ventilated at the point when fire is under control, or at least is 'surrounded'. However, lightweight building construction is now becoming more widely predominant in both the UK and Europe in general.

Estimating Firefighting Water Needs – The US National Fire Academy (NFA) Formula

Courtesy of Chief Ed Hartin MS, EFO, MIFireE
CFO *(Whidbey Island FD)*

In the mid-1980s the development team for the National Fire Academy Field course Preparing for Incident Command developed a formula to provide a simple method for estimating the flow requirements for offensive, interior operations where a direct attack was used to control and extinguish the fire. Interestingly enough the NFA Fire Flow Formula is not based on science (at least not physical science). The developers tapped into another valid source of information, being the knowledge of experienced fire officers.

The course developers designed a number of plot and floor plans showing different sizes of building with different configurations (e.g., rooms, doors, windows) with varied levels of involvement. These drawings were distributed to students attending the academy and they were asked how their fire department would control the fire (with the emphasis on the number, placement, and flow rate of hose-lines).

There are three major parameters used for the scenarios based on these plot and floor plans.

- All scenarios were designed to involve offensive, interior firefighting operations and as such, fire involvement was limited to 50% or less of the total floor area of the building.
- Operations were to be conducted as they normally would, with initial operations started by the first arriving company and additional tactics implemented as resources arrive.
- Primary search and ventilation tactics would be performed concurrently with fire control operations.

The student's responses were collected and analysed. For each scenario, when the floor area of the involved area in square feet (ft2) was divided by the total flow rate in US gallons per minute (GPM) for <u>all hose-lines used for attack, backup, and exposure protection</u>; the average result was three. Turning this around, flow rate in GPM can be determined by dividing the area of involvement in ft2 by three. In that the exterior of the building can be determined more easily than the area of involvement, the formula was adapted to determine the flow rate based on building size and approximate percentage of involvement as illustrated below:

$$\text{Needed flow (GPM)} = \left(\frac{\text{length } ft \times \text{width } ft}{3} \right) \times \% \text{ fire involvement}$$

Note: *This method does not translate easily into standard international (SI), simply converted the formula would be L/min = m^2 / 0.07.*

The course development team extended the application of this formula to include estimated flow required for exposure protection by adding 25% of the flow rate required for fire control (as determined by the basic formula) for each exposure. The full formula as used in pre-plan development is as follows:

$$\text{Needed flow (GPM)} = \left(\frac{\text{length ft} \times \text{width ft}}{3} + \text{Exposure charge} \right) \times \% \text{ fire involvement}$$

Note: Exposure charge = 25% of the basic fire flow, per exposure.

The development team believed that this formula would also be applicable to defensive attack for levels of involvement above 50%. However, this was not validated using the same type of methodology as used to develop the base fire flow formula.

Limitations

It is important to remember the limitations of this fire flow estimation method:

- The NFA Fire Flow Formula is designed for offensive, interior operations involving direct attack.
- The formula becomes increasingly inaccurate if the level of involvement exceeds 50% or the resulting flow is greater than 1,000 GPM.
- This method is not designed for defensive, master stream operations (even though the developers believed that it would provide a reasonable estimate of required flow rate for defense.
- The formula is based on area, not volume. If the ceiling height exceeds 10ft, the flow rate may be underestimated.
- The NFA Formula does not take into account the potential heat release rate of the fuel. Fuel with extremely high heat release rate may require a higher flow rate
- The developers of the NFA Formula made the assumption that the building was well ventilated (tactically). Increased ventilation can (if the fire is initially ventilation controlled) result in increased heat release rate.
- It may be tough to do the math at 0200 hours when faced with a rapidly developing fire! This method is best used in advance of the fire when developing pre-plans or working on tactical problems.

Convert from	Convert to	Calculate	Result
Litres/minute to	US Gallons/minute	Multiply by 0.26417	200 L/min = 53 GPM
Litres/minute to	US Gallons/minute	Divide by 3.78	200 L/min = 53 GPM
US Gallons/min to	Litres/minute	Divide by 0.26417	200 GPM = 757 L/min
US Gallons/min to	Litres/minute	Multiply by 3.78	200 GPM = 756 L/min
Square metres	Square feet	Multiply by 10.76	250 m^2 = 2690 ft^2
Square feet	Square metres	Divide by 10.76	2690 ft^2 = 253 m^2

Chapter 1

Firefighting – Knowledge Gaps

'The single biggest problem in communication is the illusion that it has taken place.'
George Bernard Shaw

It's an indisputable fact; we urgently need more housing for growing populations and both residential and commercial floor space in our cities is going vertical, where 'fast-build' methods of construction are now becoming the primary way of achieving this. At a time when building construction methods and modern structural designs are evolving through the 21st century as our economies face financial downturns, it is inevitable that we will undertake much research that influences changes and optimisation in the way we design, construct and protect our buildings and their occupants from fire, floods and other natural disasters. At the same time, we are directed to take a close look at our firefighting response and deployment models. If external pressures force us to decrease our capacity for intervention, then we should counter this with an increase in building protection and fire prevention activity. Unfortunately, this rarely occurs during global economic downturns. However, one thing is certain and that is the basic principles of firefighting intervention must further evolve, for failing to do so may expose firefighters to increased hazards. Just as the dynamics of compartment fires have changed over recent years due to the increasing contribution of plastics in the fire load coupled with an increasing energy efficiency incorporated in the boundary walls, lightweight buildings are becoming less stable under fire attack than the traditional constructions of past experience, so we must make sure our tactical approaches ensure the working environment is understood and any unnecessary exposure to risk to firefighters is countered.

A modern lightweight timber framed dwelling, or apartment block, should incorporate sufficient structural protection to enable an adequate time frame for occupants to escape unaided. However, many buildings of this type still employ the principle of 'defend occupants in place', or 'stay put', where occupants are encouraged to remain within their apartments whilst the fire service extinguish the fire. In a steel or concrete building that might be a viable strategy but the strategy may come into question in lightweight buildings. Recent research[1] has demonstrated that 18 different code compliant compartment wall lining and insulation configurations, attached to timber or steel frames, showed wall (barrier) failure times ranging between 23 and 60 minutes during unsuppressed fires

---

1   Xiao Li, Modeling of Barrier Failure and Fire Spread in 'CUrisk', Carleton University Ottawa, Ontario 2015

(eight barriers did not fail at all). The author himself has witnessed 30-minute fire resisting ceiling barriers fail in less than 15 minutes at fires where tall wooden wardrobes, fully laden with clothes, were involved in the fire at an early stage. At this point the lightweight structure becomes part of the involved the fire load. As our built environment changes the firefighting challenges we face in future will rely on 'fast water' applied to the fire from the exterior, to prevent fire spreading to adjacent buildings. In some cases, this will also mean 'heavy' water or foam applied from a fire appliance (engine) roof mounted monitor.

1.1 COMMON DESIGN APPROACHES IN THE USA

Statistics within the DCLG (UK Government) monitoring system repeatedly confirm the insurer experience that fire damage in lightweight timber frame (LTF) construction is disproportionately more than for conventional build methods. What the data did not inform was that, subject to the accuracy of the LTF building market share estimates, a fire is apparently more than twice as likely to occur in a LTF building. A report by the RISC authority, a group of industry insurers in the UK and published by the Fire Protection Association[2], provided detail of some USA experience, suggesting such problems are now spreading across Europe:

- Unprotected wood frame buildings under construction are the most frequent large loss fires in the US (over $5–10 million);
- Fires in unprotected wood frame residential buildings are the third most frequent large loss fires in the US;
- Fires in unprotected wood frame buildings account for the most firefighter injuries in the US;
- Within the US, unprotected wood frame construction dataset, fires in residential buildings accounted for the most firefighter injuries;
- Fires in unprotected wood frame buildings account for over half of all fires that result in catastrophic multiple fire deaths in the US;
- Fires in unprotected wood frame buildings account for over half of all fire fatalities in the US; and
- It is estimated that, on average, for every 100,000 populaces, there are 1.10 fire related deaths in the home in the US compared to 0.63 currently in the UK.

The report went on to say, 'from an insurance perspective, it is very interesting to note that the number of dwelling fires in England and Wales as a percentage of all building fires, compared with the US, is very similar: 71% UK versus 76% US. Where stark differences occur is in the associated financial loss. In England and Wales, these fires account for 35% of the total financial loss attributable to fire, whereas in the US it is 58% – a figure that may be relevant when considered against the 1.9% (UK) to 90% (US) LTF domestic housing stock difference.

Recent National Fire Protection Association (USA) data[3] demonstrates a rising trend in financial property losses despite the number of fires in private homes reducing. While the number of home fires that occur each year has fallen by nearly 50 percent since 1977, the total amount of resulting property damage (in dollars) is increasing. The graph (below)

---

2   BDM14, Fire in timber frame buildings: A review of fire statistics from the UK and the USA, Fire Protection Association (FPA) 2011
3   Fire Loss in the United States 2013, Michael J. Karter, Jr., NFPA, September 2014 and previous reports in the series. USA

is financially adjusted for inflation using the consumer price index. The transition over several decades towards lightweight construction methods is having a clear effect.

Figure 1.1: Financially adjusted (for inflation) residential fire Loss in the United States 1977 – 2013, Michael J. Karter, Jr., NFPA (Increasing property loss against a reducing number of home fires)

1.2 COMMON DESIGN APPROACHES IN THE UK AND EUROPE

When building out of bricks and mortar and other non-combustible materials, the resulting building may be quite tolerant of minor deviations and imperfections in build. UK Building Regulations (Fire Safety ADB) have been developed largely in the context of non-combustible building methods and materials and this may explain its looseness and lax process of inspection. LTF and other 'modern methods of construction' (MMC) processes involving combustible materials can be highly intolerant to any deviation in design, construction and alteration. The guidance in ADB needs to respond to this to ensure the fire safety ambition is achieved and maintained at completion, with an ongoing and increasing likelihood of being maintained over the life of the building'.

It is also factual that compartment fires are becoming more intense with faster growth rates that impact upon the structural elements much earlier in the fire development. However, the building codes and fire engineered design strategies may not effectively account for this.

- Fire loads increase as plastics become more widely used
- Fires develop faster in larger compartments than existing codes and strategies might suggest
- Open-plan office fires are now presenting as a 'fast' t-squared fire with a spread-rate of around 22m^2/minute
- Most building codes only apply a 'medium' t-squared rate of fire growth to both cellular and open-plan offices
- Vertical fire loads in storage warehouses are likely to burn out of control before firefighters are able to apply adequate firefighting water
- Energy efficient compartments create a faster build-up of flammable fire gases
- Larger window openings enable a greater heat release to be achieved

- Larger window openings exacerbate wind driven fire conditions where structural elements are subject to earlier failures
- The average time to flashover[4] has reduced over previous decades from fifteen to five minutes
- The average UK fire service attendance time to general building fires has increased from six to eight minutes over the past twenty years[5]
- A delay in applying *adequate* firefighting water leads to an increase in building fire damage[6]
- Research has shown that lightweight timber-frame elements can add to the fire load and increase the heat energy release in a building fire, requiring even more firefighting water

Since 1980 the fire service in Sweden have been developing firefighting tactics that place greater emphasis on dealing with the fire gases accumulating near the ceiling, not by ventilating them to the exterior but by cooling and inerting them with very short bursts of a water-fog stream. This method was seen to assert greater control over the hostile environment of a fire compartment and by 1997 these fire control techniques were spreading throughout Europe and across the UK. At the start of the new millennium the method, termed '3D Firefighting', had spread almost across the entire globe. However, to ensure operational effectiveness these techniques required regular use or extensive training and the low-flows (100–200 L/min) used to gain optimum effectiveness in gas cooling were not considered safe where the intensity of a fire was fast increasing due to increased ventilation caused by the failure of a window. More recent research[7] has shown that an adequate flow-rate (L/min) at the firefighting nozzle should correspond more appropriately against the potential heat energy (MJ/m^2) in the fire load (kg/m^2), likely to become involved at flashover.

1.3 KNOWLEDGE GAPS THAT EXIST

As modern methods of construction (MMC) using lightweight timber and steel frames become more common, it is essential that firefighters grasp the differences between MMC and older more traditional constructions. Perhaps more importantly, they must create amongst themselves a practical situational awareness of how either forms of construction are likely to behave in a fire. It is true that all methods of construction are designed to codes and standards that recommend minimum periods of fire resistance (minutes) to the structural elements, whatever the form of construction used. However, experience has shown that occupied MMC buildings under fire attack are more likely to collapse much earlier than the vast majority of more solid traditional constructions. Where these buildings are under construction or redevelopment, the hazards are greater still. What can be done to make these buildings less likely to suffer such damage in fire is a knowledge gap

---

4 'Flashover' is a scientifically derived term describing when a fire reaches total compartment involvement.
5 Average Response Times by location, 1994–95 to 2014–15, England DCLG Data
6 Grimwood. P and Sanderson. I; A performance based approach to defining and calculating adequate firefighting water using section .8.5 of the design guide BSPD7974:5:2014 (fire service intervention) Fire Safety Journal 78 (2015) 155–167
7 Grimwood. P, A study of 5401 UK building fires 2009-2012 comparing firefighting water deployments against resulting building fire damage; PhD Thesis, School of Engineering, Glasgow Caledonian University, 2015

on its own and the National Fire Protection Association (NFP is undertaking a five-year research project to help meet these challenges, in tall buildings particularly[8].

Similarly, there is greater situational awareness needed amongst firefighters of how fires grow, develop and spread in new energy efficient buildings and how flammable fire gases form and transport through ventilation flow-paths that occur naturally or are created by firefighters. The effects of these flow-paths in compartments, voids, stair-shafts and hallways/corridors are little understood by firefighters in general and an understanding in practical 'street' fire dynamics is now being encouraged.

Experience has also demonstrated that there are identifiable knowledge gaps and areas of concern on the fire-ground involving basic *firefighting tactics; face to face communication with interaction and reciprocal responses; crew deployment strategies; pre-planning (risk intelligence) and tactical command decision making.* Therefore, any fire service training officers would do well to concentrate on these key strategic learning needs.

1.4 RESEARCH BY NIST, USA

The National Institute of Standards and Technology (NIST) is a US is a non-regulatory federal agency within the U.S. Department of Commerce. The mission of the building and fire research programs at NIST is to anticipate and meet the measurement, science, standards, and technology needs of the U.S. construction and fire safety industries in areas of critical national need. Strategic goals include net-zero energy high-performance buildings, advancing infrastructure delivery and improving construction productivity through information integration and automation technologies, sustainable infrastructure materials, innovative fire protection, and disaster-resilient structures and communities, which includes work on hurricanes, earthquakes, and fires. NIST has specific statutory responsibilities for fire prevention and control, earthquake hazards reduction, windstorm impact reduction, and building and fire safety investigations. In respect of the fire service NIST has undertaken a vast amount of research and reported on some critical aspects of fire behaviour and firefighting tactics. In particular, their research areas extend to the following:

- Fire Detection
- Fire Fighting Technologies
- Fire Materials Research
- Fire Measurements
- Fire Modelling
- Fire Spread
- Fire Risk Reduction
- Performance-Based Design Methodologies for Structures in Fire

A NIST project into 'the enhanced effectiveness of firefighting tactics'[9] is aimed at demonstrating through the use of measurement science, the dynamics of fire behaviour in a structure and provide guidance on non-traditional means to mitigate the fire hazard in the structure in a manner that provides optimum safety and effectiveness for the fire fighter. The project has focused primarily on understanding the impact of *tactical or non-tactical ventilation* at structure fires. Further research focus diverted towards fire suppression and technology transfer to the fire service. NIST has been extremely focused

---

8 Fire Safety Challenges of Tall Wood Buildings; National Fire Protection Association, Phase One 2013; Phase Two 2017
9 https://www.nist.gov/programs-projects/enhanced-effectiveness-firefighting-tactics-project 2010

on tactical ventilation and since 2010 has shown the relationship between ventilation control and limiting fire growth, based on much of the earlier work done by this author.

Traditionally, the global fire service has broadly defined ventilation by firefighters *as the removal of heat and smoke from a burning structure while introducing cooler, cleaner air.* Traditionally this would be done with vertical ventilation – by making an exhaust vent in the roof and providing air inlets by opening doors or breaking windows. Another method is horizontal ventilation, where only windows are vented. Given the fuel rich environment found in structures, these actions may actually enable the fire to reach flashover at a much faster rate, increasing the hazard range further. Therefore, a study documenting the fundamentals of natural ventilation in a real scale residential structure was required. The results of the NIST ventilation studies, as well as recent studies by UL, have clearly shown the importance of coordinating ventilation and suppression tactics and have provided a strong scientific foundation in support of this notion. The experimental results coupled with investigations of fire incidents leads to the conclusion that with present day residential fuel loads, using ventilation alone is likely to cause an increase in the hazardous conditions from the structure fire. This is opposite of the desired effect. As a result, suppression tactics needed to be examined closely to determine a means to improve the effectiveness of using suppression and ventilation in combination.

Coordinating Tactical Ventilation with Fire Attack

Traditionally, suppression operations have been conducted from the interior of the structure as a means to reduce water damage, manage the 'thermal balance', and limit fire damage to structures. In current training manuals, it is stated in firefighting training texts and SOPs that these operations must be closely coordinated with the ventilation operations. However scientific guidance on compartment fire control has been limited. Previous research and examinations of line of duty deaths have shown that uncoordinated ventilation events occurring with fire fighters in the building prior to suppression taking place have led to tragic results. One means of reducing the possibilities of this occurrence would be an attempt to extinguish the fire prior to entering the building. A golden rule has often been to apply 'early water', that is to get water onto the fire as quickly as possible. However, there is often conflict from a moral perspective with the need for urgency and prioritisation in search and rescue operations, particularly for confirmed occupants trapped within.

Transitional Attack

As staffing issues and weight of attack in terms of the number of fire engines initially responding to building fires may be reduced, as optimisation of response and deployment models occurs, it is prudent to analyse such firefighting tactics to ensure safe operational working practice for firefighters but at the same time, maintain and prioritise occupant safety. In effect and when viable an offensive exterior fire attack, in which water from a straight stream pattern is directed at the ceiling of the fire compartment from the exterior, cooling the fire gases and reducing the heat release rate of the fire prior to the fire fighters entering the building. This is 'early water' acting like an internal sprinkler system can save lives and research has shown that a 'fire reset' (immediately reducing the heat release prior to an interior attack) can give additional 'survival seconds' to trapped victims.

As far back as 1830 CFO James Braidwood (London Fire Brigade) provided the first known guidance for 'transitional attack'[10] – 'when entrance by the door is impracticable …. the director (nozzle) should be pointed at the window in a perpendicular direction; the water striking the lintel and falling all around inside the window, will soon extinguish the fire at that point, sufficiently to render an entrance practicable'.

However, even in the USA transitional attack is certainly not new. The 'blitz' attack, as it was then called, became a popular strategy when in 1976 Captain William Olsen of the Chicago Fire Department [Engine 39] began experimenting with a pre-connected portable monitor (deck-gun) mounted on top of a newly assigned engine. Using a 28.5mm nozzle, large amounts of fire were knocked down from the street using just tank water, just prior to an interior deployment. This early interpretation of transitional attack used in Chicago became popular elsewhere and was used in the inner-city areas of New York, Detroit and other major cities. In fact, the author witnessed the use of roof mounted 'Stang' nozzles in New York City's South Bronx in 1975. The key to success was seen in using a 30-second water blitz to darken down a heavy body of fire and was particularly effective when combined with a follow-up portable 45mm hose-line in later years.

During 2012, measurements of temperature, total heat flux and other fire-ground data gathered during live fire experiments by FDNY, supported by NIST and UL, later provided some street science to the modern fire-suppression tactics associated with 'transitional fire attack'. Likened to the military concept of 'softening the target,' a transitional fire attack begins by applying water as soon as possible from the exterior of a burning house, before firefighters enter the structure, and then proceeds immediately into the interior having applied a 30–60 second exterior application.

An alternative action is to maintain a 'holding' position from the exterior, awaiting further assistance to arrive on-scene, prior to undertaking an interior attack against a reduced heat release fire. However, in some situations such a strategy can fail to 'hold' fire spread if the interior fire is too deep within the building to reach with an exterior stream. Water pressure, flowrate, timing and quantity are all critical variables to ensure fire fighter safety, yet they are not measured. As a first step towards 'Smart Firefighting', work is being undertaken by NIST and UL to explore the effective use of sensors to measure, record and transmit water flow through hoses, providing real-time data on critical parameters. Thus, measurements are being conducted to document the changes of the thermal environment within a structure. This work mirrors the author's UK research through Glasgow Caledonian University (GCU), reported later in chapter 12. As this new information is analysed and documented, a significant task undertaken by NIST is to share the information with the fire service.

Positive Pressure Ventilation (PPV)

NIST has also been at the forefront of recent research[11] into the use of PPV or positive pressure attack (PPA), where portable mechanical fans are used by the fire service to create a smoke and heat free flow-path for firefighters between the entry doorway to the building and the window opening/s serving the fire compartment. In effect, the directed airflow into the building from the street causes a controlled pressure differential between the inlet and outlets, forcing combustion products out and away from the building. This normally creates a major tactical advantage for firefighters searching for occupants and advancing

---

10 Braidwood. J; On the Construction of Fire Engines and Apparatus, the Training of Firemen and the Methods of Proceeding in Cases of Fire (Memoirs dated 1830)
11 http://www.nist.gov/fire/ppv.cfm

on the fire but at the same time, brings some potential hazards. There are concerns over the possibility of causing flames to 'flash-back' to the point of entry, possibly injuring or trapping firefighters and occupants. This is tactically countered by ensuring effective inlet/outlet opening size ratios in alignment with fan air-flow (m³/s) (see Chapter 18). There is also a risk that fire can be forced into unburned sections of a structure, or even the roof space. However, as a firefighting tool, PPV in *UK phase 3 attack mode* is certainly one of the most advantageous firefighting tools particularly in small to mid-sized buildings (<200m²), providing clear protocols are established and followed and firefighters are well trained in its use.

UK Phase 3 PPV –

Once crews are competent in the application of PPV while the fire is still burning but under control, Brigades can consider the introduction of PPV in the initial stages of an incident, on first arrival, <u>prior to committing BA crews.</u> This will create the best possible environment for crews and casualties.

This phase requires a greater degree of experience from the Incident Commander, as they must be able to estimate the effects of PPV with little knowledge of the internal layout, and no crews inside the building to provide them with feedback. It is also the point at which the greatest benefits to BA wearers is realised.

The NIST research into PPV included one room fires and multiple occupancy fires right through to high-rise stairwells. Full-scale experiments were conducted to characterise a Positive Pressure Ventilation (PPV) fan, in terms of velocity. Experiments were performed in an open atmosphere and in simple room geometry. The results of the experiments were compared with Fire Dynamic Simulator (FDS) output (computational fluid dynamics). The measurements of all experiments compared favourably with the FDS model results. The fire service may use ventilation blowers or fans to pressurise a structure prior to suppressing a fire. This pressurization or positive pressure ventilation (PPV) tactic can assist in the venting of smoke and high temperature combustion products and make attacking the fire easier than without PPV. However, this tactic also provides additional oxygen to the fire and can increase the rate of heat and energy being released. The NIST study examined gas temperatures, gas velocities and total heat release rate in a series of fires in a furnished room. The use of the PPV fan created slightly lower gas temperatures in the fire room and significantly lower gas temperatures in the adjacent corridor. The gas velocities at the window plane were much higher in the PPV case than in the naturally ventilated scenario. This higher velocity improved visibility significantly. PPV caused an increase in heat release rate for 200 seconds following initiation of ventilation but the heat release rate then declined at a faster rate than that of the naturally ventilated experiment.

The NIST analysis of 160 experiments conducted with small and large fans in the stairwell of a vacant 30-story office building in Toledo, Ohio, showed that the positive pressure ventilation (PPV) techniques used in small homes also worked in tall structures for keeping stairs smoke free. The NIST engineers found that portable PPV fans, if used correctly, can both limit the amount of smoke and heat entering the stairway, and push smoke and deadly gases out of the structure. Data were collected with training smoke and pressure measurement devices. Pressurisation smoke control systems, which usually consist of mounted wall fans, have been incorporated into high-rise buildings since the

1970s. The NIST experiments, however, represented the first scientific evaluation of positive pressure ventilation technology using portable fans for buildings without built-in pressurisation systems. The NIST researchers developed several guidelines for the most effective use and positioning of portable PPV fans. The researchers also noted that the noise level near the working fans can go as high as 110 decibels – comparable to the level of a chainsaw – making communication difficult, and recommended that the fire command post be sited well away from the fans. The Department of Homeland Security's Science and Technology Directorate sponsored the Toledo PPV experiments. The Toledo Fire and Rescue Department, the Fire Department of New York and the Chicago Fire Department provided guidance and assistance during the experiments.

Controlled fires on the third, 10th and 15th floors of a Chicago high-rise represented the next stage experiments to study the effectiveness in multi-storey buildings of positive pressure ventilation (PPV). All 16 floors were equipped with temperature and pressure monitors while the three burn floors also included cameras, heat flux gauges and typical apartment furnishings. The entire setup was connected to the data acquisition centre by seven miles of cable. Once the fires were under way, a variety of ventilation tests were conducted. For example, in one test, a large fan was placed at the front door to force cool air up through the building. In another test, two smaller fans – one on the first floor and one two floors below the fire floor, both forcing air into the stairwell – were used to achieve the same PPV effect. Results from both scenarios show that PPV significantly reduced the temperature and the amount of smoke in the corridors and stairwells outside the burn rooms. In one case, the temperature very quickly dropped from 316 degrees Celsius to 16 degrees.

Wind driven Fires and Ventilation Flow-paths

The dangers of wind driven fires exist in all buildings but are increased as buildings rise in height. The increasing fire loads coupled with larger window openings seen in modern buildings exacerbates this problem and multiples of firefighters have recently been killed or severely injured under circumstances where strong winds, or gusting conditions, have forced smoke, heat and flaming combustion directly into the path of searching or advancing firefighters. One particular phenomenon exemplified by exterior winds is the creation of a ventilation flow-path, driven by pressure differentials existing between the inlet and outlet vents, or openings. This effect has probably resulted in more firefighter deaths and injuries than any other and the outcome is more often seen as a fire behavioural event, such as flashover, backdraft or smoke explosion. Ventilation flow-paths may also be driven by temperature differences between the inside and outside of fire buildings. Therefore, where openings are made by firefighters or by the fire itself at high level, then a flow path may be created between the entry point lower down and the new opening. This causes a rapid movement of intensifying fire gases, heat and possibly flame along this route and if firefighters or occupants are located in this pathway, their chances of injury are dramatically increased.

The work of NIST[12] was to explore these phenomena and concluded that, based on the analysis of results from a range of studies, adjusting firefighting tactics to account for wind conditions in structural firefighting is critical to enhancing the safety and the effectiveness of fire fighters and remaining occupants.

12 http://www.nist.gov/manuscript-publication-search.cfm?pub_id=902177

NIST Lessons learned

- **Impact of PPV.** PPV fans alone could not overcome the effects of a wind driven condition. However, when used in conjunction with door control, WCDs, and floor below nozzles (FBNs) the PPV fans were able to maintain tenable and clear conditions in the stairwell.
- **Impact of Wind Control Devices (WCDs)** *(Exterior hung fire curtains located across window openings).* In these experiments, the WCDs reduced the temperatures in the corridor and the stairwell by more than 50% within 120 seconds of deployment. The WCDs also completely mitigated any velocity due to the external wind. The WCDs were exposed to a variety of extended thermal conditions without failure.
- **Impact of externally applied water.** In these experiments, the externally applied water streams were implemented in different ways; a fog stream inserted into the fire room window, a fog stream** flowed from the floor below into the fire room window opening, and a solid water stream flowed from the floor below into the fire room window opening. In all cases, the water flows suppressed the fires, thereby causing reductions in temperature in the corridor and the stairwell of at least 50%. The water flow rates used in these experiments were between 600 L/min and 750 L/min, demonstrating that a relatively small amount of water applied directly into the fire compartment can have a significant impact.
  ** Note UL research on exterior fog patterns

Previous studies demonstrated that applying water from the exterior, into the windward side of the structure can have a significant impact on controlling the fire prior to beginning interior operations. It should be made clear that in a wind-driven fire, it is most important to use the wind to your advantage and attack the fire from the windward side of the structure. Interior operations need to be aware of potentially rapidly changing conditions.

'Smart Firefighting' Project (Future and ongoing Research by NIST)

New opportunities to fuse emerging sensor and computing technologies with building control systems and firefighting equipment and apparatus are emerging. The resulting cyber-physical systems will revolutionize firefighting by collecting data globally, centrally processing the information, and distributing the results locally. Engineering, developing, and deploying these systems will require new measurement tools and standards. This project will focus on the needed tools and standards in three areas: smart building and robotic sensor technologies, smart firefighter equipment and robotic mapping technologies, and smart fire service apparatus and equipment. The results of this project will (1) mitigate total social costs of fires at both the community and the building scales, and (2) integrate cyber-physical systems into innovative fire protection technologies. It is intended that the work will fuse cyber-physical capabilities into a multi-dimensional integrated system that enables smart firefighting at three distinct levels: the individual fire fighter level, the firefighting team level, and the incident commander level; demonstrating how computer technologies can be used to augment existing fire models with real-time sensor data to provide powerful decision-making tools.

NIST have also undertaken a vast amount of research into structural safety and building collapse at fires, reported elsewhere in this book.

1.5 RESEARCH BY UL, USA

Underwriters Laboratories (UL) is a global independent safety science company with headquarters based in the US. UL's mission is to certify, validate, test, verify, inspect, audit, advise and educate. As part of UL, the Firefighter Safety Research Institute (FSRI) is dedicated to increasing firefighter knowledge to reduce injuries and deaths in the fire service and in the communities they serve. Over recent years UL has provided a wealth of research and information to increase firefighter awareness and develop fire-ground strategy and tactics. Much of this work is a follow on from previous research by NIST. Topics such as structural collapse, the evolution of the fire environment, fire suppression, ventilation, basement fires and many more have been studied in detail to examine fire dynamics and firefighting tactics. These studies generate information, data and observations that UL technical panels—composed of firefighters from the United States and around the world—develop into tactical considerations. The tactical considerations are tools that can be used to share science in such a way that firefighters can connect it with their experience and then integrate it into their department's operations.

A comparison of modern and legacy fire loads demonstrated how fires can reach flashover in a much shorter time-scale as plastics become more common in building contents and construction. Other work by UL has demonstrated how vertical or horizontal ventilation by firefighters is likely to impact of fire growth and development in single and multi-level residential buildings and suggested that greater attention to detail concerning how, when and where to tactically ventilate fire buildings is needed.

Firefighter exposure to Smoke Particulates wearing PPE (Contaminants)

A key research area partnered UL with the Chicago Fire Department and the University of Cincinnati, college of medicine, to further investigate the causal relationship between sub-micron smoke particles and the risk of cardiovascular problems[13]. During the study, data was collected on the smoke and gas effluents to which firefighters are exposed during routine firefighting operations, as well as contact exposure from contaminated personal protective equipment. The project included investigations on three fire scales: (1) fires in the Chicago metropolitan area, (2) residential room content and automobile fires, and (3) material-level fire tests. Detected effluent gases, airborne chemicals and smoke particulates were assessed by the University of Cincinnati, College of Medicine for their potential adverse health effects on fire service personnel.

Key findings from this research were:

- Concentrations of combustion products were found to vary tremendously from fire to fire, depending on the size, the chemistry of materials involved, and the ventilation conditions of the fire.
- The type and quantity of smoke particles and gases generated depended on the chemistry and physical form of the materials being burned. However, synthetic materials produced more smoke than natural materials.
- Combustion of the materials generated asphyxiates, irritants, and airborne carcinogenic by-products that could be potentially debilitating.

---

13 http://www.ul.com/global/documents/offerings/industries/buildingmaterials/fireservice/WEBDOCUMENTS/EMW-2007-FP-02093.pdf

- Multiple asphyxiates, irritants and carcinogenic materials were found in smoke during both suppression and overhaul phases. Carcinogenic chemicals may act topically, following inhalation, or subsequently through dermal absorption from contaminated personal protective (PPE) and other equipment.
- Long-term repeated exposure may accelerate cardiovascular mortality and the initiation and/or progression of atherosclerosis.
- All of the construction materials tested resulted in the production of water, carbon dioxide and carbon monoxide.
- Styrene based materials led to formation of benzene, phenols, and styrene.
- Vinyl compounds led to formation of acid gases (HCl and HCN) and benzene.
- Wood based products led to formation of formaldehyde, formic acid, HCN, and phenols.
- Roofing materials (US) led to formation of sulphur gas compounds such as sulphur dioxide and hydrogen sulphide.
- The actual field tests found a range of additional contaminants, including multiple heavy metals including arsenic, cobalt, chromium, lead, and phosphorous.
- Chemical composition of the smoke deposited and soot accumulated on firefighter gloves and hoods was virtually the same except concentrations on the gloves were 100 times greater than the hoods.

Jeffery and Grace Stull (Globe USA) inform us[14] that contamination occurs when some unintended substance gets on or in clothing or equipment. It may be minor such as light soiling or intrinsically dangerous like a splash of a strong acid. Contamination implies substances that are unwanted because of their potential health effects. It is also important to recognize that contamination may be transient, either leaving the clothing on its own with time or through cleaning. The greatest concern is for contamination that is persistent and continues to expose the firefighter to the hazard. The process of contamination itself varies greatly as does the environments in which the contamination occurs. During a building fire, much of the contamination is in the form of smoke particulate and fire gases. This form of contamination envelopes the firefighter, contaminating all exterior surfaces of the clothing ensemble. The contamination also penetrates various portions of the ensemble, most notably through joints and gaps between different clothing items or closures such as on the front of the coat or pants fly. Gases and fine particles easily get into interior spaces, so even underlying clothing can become contaminated. The penetration of contamination in this fashion will be more localized where there are easier pathways for entry. Ensemble areas such as the hood, pant legs and coat-to-pant overlap are common penetration pathways. Places where there are continuous films of flexible material retard this penetration, but some substances can still get through.

In addition to penetration, there is permeation. This is when contamination moves across a material on a molecular level as opposed to the bulk passage that occurs during penetration. Permeation is more insidious because it goes on unseen, and can occur at relatively high rates causing contamination to spread throughout the clothing. Large bulk substances such as smoke particles tend to coat surfaces in the form of soot. These products of combustion are particularly hazardous because the individual particles adsorb and retain the chemical gases created by the fire, causing a longer-lasting form of contamination. The particle sizes in smoke can be a hundredth of a micron, making the particles invisible except in the concentrated aggregate of the smoke cloud. Thus, the

---

14 https://www.firerecruit.com/articles/130280018-Firefighter-PPE-contamination-What-you-need-to-know

particles easily penetrate textile materials, but any material with a semi-solid surface, such as rubber or plastic, will generally stop this penetration. On the other hand, leather, which is porous, will allow some penetration. Gases are more ubiquitous. Given their form at a molecular level, these substances will penetrate textile and leather materials but not plastics and solid films. Nevertheless, depending on the type of the gas, primarily the size of the chemical molecule and the nature of the material, chemical gases will permeate most materials. Very dense and thick materials will reduce permeation and some materials such as metals will completely resist any chemical permeation.

Firefighters are also exposed to various types of liquids, whether through hose spray that has picked up contaminants or broken containers of liquids. Liquids can also penetrate clothing and well as permeate into materials. The ease of liquid contamination depends greatly on the amount of liquid, the characteristics of the liquid, the force behind that liquid and the characteristics of material. Materials may be able stop a small volume of liquid, but when in large quantities, many porous materials are likely to become wetted and permit penetration. Some liquids have greater contamination potential because they move more easily over material surface or degrade material due to corrosiveness or their solubility in the material. Many textile materials used in firefighter protective clothing are treated with repellent finishes that help. However, some liquids with low surface tensions enable the liquid to easily spread on surfaces (instead of beading like water). This allows the liquid to penetrate small openings. Unfortunately, already soiled and contaminated surfaces allow easier penetration. If there is pressure behind the liquid such as from kneeling in a puddle or the force of hose spray bouncing off a surface and back onto the individual, that pressure will push the liquid farther into clothing, particularly porous textiles and leather. As with gases, liquids can also permeate by dissolving in the material, even plastic and rubber materials. In some cases, some of these more solid materials can actually retain the chemical longer than can the porous materials.

A firefighter helmet best represents the full range of materials and illustrates how contamination can occur. The hard surface of the shell can be coated with solid soot. This soot coating may even absorb and hold fire-ground gases. The same soot will deposit on all exterior surfaces including the face shield or goggles, reflective trim, ear flaps and exposed portions of the suspension that include a range of plastic, textile and leather materials. While fire gases will not penetrate or even permeate a hard-thermoplastic shell, the gases will enter the helmet through any opening between the shell and suspension, be absorbed by the textile or leather materials and permeate plastics. Similarly, liquids that contact the helmet will be absorbed by any exposed textile or leather components. The retention of contaminants in clothing and other protective equipment depends mainly on the type of substance and how the contamination took place. Many fire gases are volatile and, while absorbed by soot or directly into materials, will off gas over time, usually at very low levels. Thus, removing the soot not only gets rid of the solid contaminant, but also takes away some of the trapped gases. Chemicals that are less volatile are more likely to remain in place, particularly if directly absorbed into materials. The increased affinity of the substance for the material can mean that some contaminants can stay for very long periods unless the item is thoroughly cleaned. The hardened veteran firefighter who wears a smoke blackened helmet may be one of those who are most at risk!

One important factor other than substance volatility is the degree to which the contamination is water-soluble. Not all combustion products dissolve in water and therefore rinsing may not be effective. Using detergents and soaps improves non-water soluble contaminants' solubility in water, particularly for oil-based chemicals.

Cleaning turnout gear properly takes expertise in removing contaminants. Many independent service providers have gained this experience and have found various solutions for removing both general and specific contaminants. Whilst on the subject, the UK Fire Brigades Union (FBU) recognizes that there is a serious threat to firefighters in relation to cancer in the workplace. They have produced an initial 10-point guidance document[15] (2016) which highlights the basic principles that firefighters should follow to try and prevent unnecessary contamination before, during and after incidents. This is undoubtedly an area of interest to firefighters where the knowledge gaps are gradually emerging and the future is likely to see dramatic changes in the way we decontaminate ourselves on-scene, prior to mounting the fire engine to return to station.

The Performance of Engineered Floor Systems in Domestic Dwellings

US Firefighters have expressed concern about the rate of engineered timber floor deflections prior to collapse when reporting on experiences upon entering a fire scene and performing life safety and fire extinguishment activities. The firefighter's reports indicate that *unprotected* lightweight wood construction collapses at a quicker rate as compared to traditional floors supported by 50mm x 250mm timber joists. The main objective of this research by UL was to improve firefighter safety by increasing the level of knowledge on the response to fire by residential flooring systems. Several types (or series) of experiments were conducted and analysed to expand the body of knowledge on the impact of fire on residential flooring systems. The results of the study[16] provide tactical considerations for the fire service to enable improved decision-making on the fire scene.

Experiments were conducted to examine several types of floor joists including, dimensional lumber, engineered I-joists, metal plate connected wood trusses, steel C-joists, castellated I-joists and hybrid trusses. They were performed at multiple scales to examine single floor system joists in a laboratory up through a full floor system in an acquired structure. Applied load, ventilation, fuel load, span and protection methods were altered to provide important information about the impact of these variables to structural stability and firefighter safety. Whilst the majority of lightweight wood flooring construction is protected in the UK, caution in firefighting and situational awareness of construction types is called for.

Tactical considerations firefighters can use immediately to improve their understanding, safety and decision-making when sizing-up a fire in a domestic timber framed dwelling include:

- Collapse times of all unprotected wood floor systems are within the operational time frame of the fire service regardless of response time.
- Size-up should include the location of the basement fire as well as the amount of ventilation. Collapse always originated above the fire and the more ventilation available the faster the time to floor collapse.
- When possible the floor should be inspected from below prior to operating on top of it. Signs of collapse vary by floor system; Dimensional lumber should be inspected for joist rupture or complete burn through, Engineered I-joists should be inspected for web burn through and separation from subflooring, Parallel Chord Trusses should be inspected for connection failure, and Metal C-joists should be inspected for deformation and subfloor connection failure.

15 https://www.fbu.org.uk/contaminants-guidance
16 http://www.ul.com/global/documents/offerings/industries/buildingmaterials/fireservice/basementfires/2009%20NIST%20ARRA%20Compilation%20Report.pdf

- Sounding the floor (a method used by firefighters using a tool to 'hear' the floor's tone on striking) for stability is not reliable and therefore should be combined with other tactics to increase safety.
- Thermal imagers may help indicate there is a basement fire but can't be used to assess structural integrity from above.
- Attacking a basement fire from a stairway places firefighters in a high-risk location due to being in the flow path of hot gases flowing up the stairs and working over the fire on a flooring system which has the potential to collapse due to fire exposure.
- It has been thought that if a firefighter quickly descended the stairs cooler temperatures would be found at the bottom of the basement stairs. The experiments in this study showed that temperatures at the bottom of the basement stairs where often worse than the temperatures at the top of the stairs.
- Coordinating ventilation is extremely important. Ventilating the basement created a flow path up the stairs and out through the front door of the structure, almost doubling the speed of the hot gases and increasing temperatures of the gases to levels that could cause injury or death to a fully protected firefighter.
- Floor sag is a poor indicator of floor collapse, as it may be very difficult to determine the amount of deflection while moving through a structure.
- Gas temperatures in the room above the fire can be a poor indicator of both the fire conditions below and the structural integrity of the flooring system.
- Charged hose lines should be available when opening up void spaces to expose wood floor systems.

During all of these experiments where the variables were systematically controlled, there were no reliable and repeatable warning signs of collapse. In the real world, the fire service will never respond to two fires that are exactly the same. On the fire-ground there are many variables to consider and most of the parameters being considered are often unknown which makes decision making that much more difficult. Information such as how long the fire has been burning, what type of floor system, was it built to code or altered at any point, is it protected with gypsum board, what is the loading on the floor and how long is the span are all unknown to the responding firefighters. There are also no collapse indicators that guarantee the floor system is safe to operate on top of.

UL Study of the effectiveness of Fire Service Positive Pressure Ventilation during Fire Attack in Single Family Homes incorporating modern Construction Practices

The test two main houses in this research were designed by a residential architectural company to meant to be representative of a one or two-storey home constructed in the late-twentieth to early twenty-first century. The experiments aimed to examine the fire dynamics in structures of this type and to further understand the impact of positive pressure attack on tenability throughout the structure. The buildings demonstrated air leakage paths in excess of those acceptable in newly constructed houses but were similar to earlier constructions pre-2009. The thermal insulation found in the boundary construction of new energy efficient homes was not installed and this may have some major impact on fire development, compartment temperatures and overall test outputs. Even so, the research is of great interest to the fire service.

A project Technical Panel was established of fire service personnel from all over the world who assisted the technical process of live burns in purpose built test structures. The information from the full-scale tests was reviewed to develop tactical considerations for the use of PPV fans in residential single family structures. A brief summary of these tactical considerations are as follows:

1. **Understanding the Basics of Positive Pressure Ventilation/Attack –**

   An understanding of pressure, how pressure creates flow, and how flow is associated with ventilation is essential to fully understanding if PPA or PPV will be either effective or ineffective in ventilating a residential single family structure.

2. **Horizontal, Vertical and Positive Pressure Attack are different tactics –**

   No one tactic will work in every scenario. Understanding the fire environment with emphasis on ventilation limited fire dynamics and how fire department operations impact those will ensure the tactic chosen is most effective. Water must be applied in coordination with each of these ventilation tactics for successful outcomes. For example, in a compartmentalized structure, PPA will transition a ventilation limited compartment fire to flashover faster, however, temperatures will be lower in adjacent spaces and return to ambient in those spaces faster with water application.

3. **The setback of the fan or development of a cone of air is not as important as the exhaust size –**

   In the application of PPA a great deal of emphasis has been placed on the flow occurring at the front door. Ensuring the 'cone of air' does not equate to most effective flow. An adequate size exhaust is more important for creating the intended flow. The intent of PPA is to increase the pressure in adjacent compartments higher than the fire compartment to prevent the flow of products of combustion from the fire compartment to the remainder of the structure.

4. **During PPA, an ongoing assessment of inlet and exhaust flow is imperative to understanding whether or not a fan flow path has been established and if conditions are improving –**

   The fire attack entrance cannot tell you the conditions at the exhaust location(s). Assessing both the inlet, exhaust locations and interior conditions together provide the best assessment of PPA effectiveness. When assessing the exhaust locations/s, the impact of PPA will be noticeable within seconds. When the exhaust is first created the buoyant flows will result in a neutral plane located somewhere in the window depending on the location of the fire and stage of fire growth. The high pressure hot gases (smoke) will flow out the top of the window above the neutral plane. A gravity flow of cooler ambient air will flow in the bottom, below the neutral plane. Once the fan is turned on, the neutral plane should drop to the windowsill and the exhaust should become a unidirectional (one-way) flow indicating the fan flow path has been established. A neutral plane above the window sill on the exhaust opening while conducting PPA indicates more flow is required or an obstruction exists between the inlet and the exhaust. This indicates additional actions such as increasing the fan flow by adding a fan or increasing fan throttle are necessary while ensuring that no obstruction exists in the intended fan flow path. If increased exhaust vent flow cannot be established within a short period of time, crews should stop the fan and consider implementing a different tactic.

In addition to monitoring exterior conditions, interior crews must be monitoring interior conditions. Ineffective PPA has the potential to cause conditions to deteriorate faster than would be noted in horizontal or vertical ventilation. Identifying and reacting to deteriorating conditions becomes even more essential during PPA. With the fan introduced, interior crews should notice a decrease in temperature and increased visibility over time. If this is not the case, the structure should be evacuated until the elements required for a successful PPA are re-evaluated and the fire location or the extent of fire spread is re-assessed.

> **What does this mean to the Firefighter?**
>
> **When using PPV to assist firefighters in gaining entry to the fire compartment, the exhaust vent should be opened first and firefighters should not enter the building ahead of the fan air-flow being deployed. In fact, it is sometimes safer to allow for a 30 second delay following the air-flow being directed into the building to allow the pressure to impact favourably on fire conditions. During this 30 second period the position of any neutral plane at the exhaust vent should be observed and the air-flow adjusted as necessary. A neutral plane above the window sill on the exhaust opening while conducting PPA indicates more flow is required or an obstruction exists between the inlet and the exhaust. This indicates additional actions such as increasing the fan flow by adding a fan or increasing fan throttle are necessary while ensuring that no obstruction exists in the intended fan flow path.**
>
> **Where a smoke control system is installed in large buildings, the effects of this should be closely reviewed whilst in operation, before any decision to deactivate the system and deploy PPV is made.**

5. Positive Pressure Attack is Exhaust Dependent –

   For PPA to be effective the pressure created by the fan must be greater than the pressure created by the fire. Although fan size does play a role in the effectiveness of PPA, exhaust size plays a greater role. Providing enough exhaust to reduce the pressure in the fire room below what the fan is capable of producing in the remainder of the structure is essential for safe PPA operations. A fire in post flashover state, venting to the exterior was seen to produce between 9Pa and 11Pa of pressure in the upper layer close to the ceiling. This means for the fan to prevent flow from the fire compartment to an adjacent compartment, the adjacent compartment needs to be at least 9Pa, preferably 11Pa or higher.

   If not enough exhaust is provided, the fire room pressure created by the fire and the fan [combined], will overwhelm the pressure created by the fan alone, causing the fire to grow back towards the inlet vent. The solution is to lower the pressure in the fire compartment by **(1)** creating more exhaust openings in the fire compartment or **(2)** applying water to reduce the pressure created by the fire.

   Testing has demonstrated a 2:1 exhaust to inlet ratio is much more effective than a ratio of 1:1 or less.

6. **An outlet of sufficient size must be provided, in the fire room to allow for effective PPA –**

    PPA effectiveness is directly dependent on the ability of the fan to exhaust products of combustion to the exterior. Any exhaust opening created in conjunction with PPA should be located in the fire compartment. When a vent is opened in the fire room, most of the heat and smoke will be vented out the window resulting in temperatures remaining low in the adjacent rooms. <u>When the exhaust vent is created in an adjacent room, it results in a temperature increase in that room. Even when an exhaust opening is located in the fire compartment, if it is not of sufficient size, the effect will be similar to creating an exhaust vent in an adjacent compartment.</u> The pressure created by the fire, combined with the pressure increase in the room due to the fan, will exceed the pressure created by the fan in the adjacent space, <u>resulting in flow from the fire compartment into adjacent compartments</u>. This results in increased temperatures, decreased visibility, and decreased survivability.

    As with any ventilation tactic, extension to exposures needs to be a consideration when utilizing PPA. The use of the fan greatly intensifies the volume of fire venting from the exhaust window. At 3.7 metres the transfer of heat from a window emitting fire increased by around 150% once the PPA fan was introduced, and by 300% at 2.4 metres from the window (in the UL research tests).

7. **During PPA, creating additional openings not in the fire room will create additional flow paths making PPA ineffective with the potential to draw the fire into all flow paths –**

    Additional openings not in the fire compartment will lower the pressure in the adjacent compartments, allowing for more flow from the fire compartment to the remainder of the structure. With four times the additional ventilation openings (*in the research*), the fan could not maintain the same amount of pressure in the living room. The reduction in pressure in the living room resulted in additional fire gas flow from the fire room to surrounding compartments.

8. **The safety of PPA is decreased when the location and extent of the fire is not known with a high degree of certainty –**

    To ensure the exhaust is provided in the most effective location it is essential to identify the location of the fire. Several indicators are available to aid firefighters in this identification such as heat signatures identified via thermal imaging cameras and smoke/neutral plane conditions. It is important to remember that venting a window is not a temporary action. Once glass is broken, it cannot be replaced. Whenever possible, a doorway should be used to identify the location of the fire. <u>Closing the door after inspecting the neutral plane will limit the heat release rate of the fire by limiting the available oxygen until crews are in position to implement PPA.</u>

9. **PPA will not be effective on a fire located in an open concept floor plan or any floor plan with high ceilings –**

    In order for positive pressure attack to be effective, the fan must be capable of increasing pressure in the adjacent compartments. This forces the products of combustion out of the structure rather than into adjacent compartments. This pressure increase is only possible where the fire is located within a [*non-open plan*] compartment. Without a doorway to separate the fire compartment from the remainder of the structure, the pressure increase in adjacent areas cannot be achieved. This causes the fan to create a churning or mixing of the fire gases. High ceilings compound the problem as buoyant gases are carried vertically instead of out of the structure, further mixing the interior environment.

10. **The application of water, as quickly as possible, whether from the interior or exterior prior to initiating PPA will increase the likelihood of a successful outcome –**

    The application of water onto a compartment fire has been shown to slow the growth rate, increasing firefighter and occupant safety while decreasing property loss. This makes rapid hose line deployment a top priority for first arriving crews. Although positive pressure attack can improve the efficiency of a hose stretch, is not a substitute for the application of water on the seat of the fire. This early application of water will aid in the effectiveness of PPA. For fire departments that choose to utilize PPA as their primary fire attack method, training should focus on minimizing the time lapse between the start of the fan and the application of water to the fire (but consider the 30 second rule provided above). If an interior attack is chosen, interior hose advancement should take advantage of the fan's ability to influence the flow path and reduce temperatures by following the intended fan flow towards the seat of the fire.

11. **PPA is not a replacement for using the reach of your hose stream –**

    Although PPA can reduce temperatures as crews approach a fire, it is not a replacement for the reach of a hose stream. Applying water as you approach the fire reduces the heat release rate making PPA most effective. However, if the flow from the fan is cut off or reduced for any reason, including being obstructed by firefighters located in the inlet or along the flow path, the result may be rapid temperature rise and onset of zero visibility. The energy, which was previously being exhausted, is now split between the exhaust and the inlet as if the fan was not in place. Firefighters should appreciate the fact that they are a potential impediment to the air flow they are using, especially as they approach the door to a fire room. The same is true for a crew advancing down a narrow hallway.

    Using the reach of your stream as you approach the fire from a safe distance will reduce the energy released by the fire. Should the fan become obstructed or shut off while fire fighters are approaching the fire compartment, upper layer gas cooling will decrease the thermal assault on the advancing attack crew. Putting water into smoke cools the space, which lowers the temperature, lowering the pressure in turn and making PPA more effective.

12. **During PPA, extension into void spaces is directly related to the exhaust capabilities of the void space –**

    In order for fire extension into structural void spaces there must be an entrance (penetration) for the fire and an exit (exhaust) for the products of combustion to leave. An inlet, only allows the products of combustion to accumulate, limiting oxygen, slowing the ignition of objects in that space, and thus limiting fire spread. To transfer enough heat into the void space to extend fire, there must be an exhaust opening provided. If the exhaust size is increased, as is the case in balloon construction (*timber-framed construction*), then combustion can occur in the stud space. There was charring of studs and flames observed from the exhaust point, even before PPA was applied. Temperatures in the stud cavity with balloon frame construction began approaching temperatures in the adjacent compartment.

13. **PPA does not negatively affect the survivability of occupants behind a closed door –**

Prior research shows the importance of having a closed door between occupants and the fire if they are unable to escape. These PPA experiments reinforced this assessment as temperatures and gas concentrations in the closed or isolated rooms remained tenable while conditions in open compartments exceeded tenability thresholds.

14. **When PPV is used post fire control, in single story residential structures, the more openings made in the structure during PPV the more effective it is at ventilating the structure –**

The 450mm petrol fan used was capable of moving so much air it could exhaust through more than five times the inlet size to most efficiently remove products of combustion. As the number of exhaust points increased the exhaust flow continued to increase. The greater the exhaust flow, the faster the structure was ventilated. With today's fans flow rates, systematically exhausting smoke one room at a time may not be not as effective as exhausting several rooms at once.

15. **When PPV is used post fire control, it is important to assess for fire extension –**

While the fan provides additional visibility after fire control by exhausting products of combustion out of the structure faster, it also has the potential to hide extension in void spaces. Directing attention to these spaces immediately following knockdown will limit the possibility of extension.

16. **When PPV is used post fire control, starting or turning on the fan immediately after fire control will provide the most benefit –**

Once water is on the fire and the attack crew has the upper hand, fans will assist with increasing visibility and reducing temperatures to ambient to allow for other fire-ground operations like search, rescue and overhaul to happen faster and more efficiently. The use of the fan must be coordinated with interior crews and incident command to ensure fire control has been achieved.

What does this mean to the Firefighter?

The research undertaken by UL in the USA provides some key information for firefighters in all countries upon which they can increase their knowledge and understanding of the practical Implementation of positive pressure attack (PPA) and post fire ventilation, using fans (PPV). The data from the research is applicable in general to most types of structure, when establishing some base protocols upon which SOPs are formed. Although there is nothing new here in terms of tactical guidance, what the UL research does is put some science to the strategy of using forced ventilation at building fires and their work should be studied.

It has always been known by experienced practitioners that PPA saves lives and improves working conditions for firefighters, making their approach safer. However, it is also a concern that many firefighters do not implement the strategy safely or effectively at fires, either through inadequate training or misunderstanding of the core principles.

Here are some of the key protocols and requirements for using PPA that are already well known, practiced and supported by the 2016 UL research, but may not be fully understood:

- Have an SOP that provides clear protocols as to when, where and how PPV/PPA can be used. This may particularly refer to minimum crew size, roles and responsibilities, triggers for use, type of buildings PPA can be used, fire behaviour observations and their impact on when and if to use the strategy.
- The time between an exterior fire stream being applied and entering the building to complete suppression should be around 30 seconds.
- The time between the full fan flow being directed towards the open entry door to pressurise the building and entry into the building to attack the fire should include a pause for observation of fan effects of around 30 seconds. Allow it to take effect.
- Provide practical and refresher training for firefighters in how such protocols impact on the fire ground use of PPV/PPA. Base this on information provided by the UL research reports that are feely available and practical fire dynamics input.
- Ensure sufficient PPA trained personnel and support firefighters are on-scene.
- Reduce air flowing into the fire (close the building entry door/s if viable).
- Locate the fire. If the fire has transitioned beyond a single compartment fire then structural stability and fire spread factors must be further considered.
- Determine if building type and compartment size are suited to a PPA strategy
- Position and start the fan at idle speed, pointing away from the entry door.
- Ensure crews are ready with a charged hose-line and SCBA started on air. Create an exhaust vent in the fire room, if not already existing.
- Consider an exterior fire stream application just prior to entering the structure to reduce the heat release and temperature in the fire compartment.

1.6 HAZARDS THAT SHOULD BE CONSIDERED WHEN USING PPA

1	Forcing air into an under-ventilated fire may cause the fire to escalate rapidly and possibly even backdraft.	Whenever directing a PPA airflow into a structure, there is always a possibility that fire conditions may worsen for a few seconds. The potential for a backdraft occurring also exists and if signs and symptoms are present then alternative tactics might precede PPA, such as fog insertion tools deployed from external or adjacent locations. It is important to have located the fire room, confirmed the fire is unlikely to have spread into other rooms and then wait for a 30 second period after the PPA airflow is directed into the building before deploying with a hose-line.
2	Fire may spread into unseen voids, the roof space or adjacent compartments.	The UL research advises this is unlikely to occur unless the void or roof space is open elsewhere. However, this still remains a possibility and should be anticipated and the airflow immediately directed away from the air inlet should it occur.
3	Fire may burn back towards the entry point, trapping or injuring firefighters.	This has occurred on occasions and is usually the result of an inlet/outlet ratio that favours a larger inlet area. This may also occur where the ventilator is too powerful for the space being vented. Again, a 30 second period for pause before deployment will allow the fan's airflow to take effect. If fire conditions appear to worsen then the fan can be directed away from the inlet point.
4	With fire suddenly burning back to the entry point with firefighters still located inside, a decision must be made whether to turn the airflow away from the inlet point or not?	This is always an opportunity for debate and serves as a useful tactical learning session. If the airflow is interrupted the fire conditions may improve and tenability levels may become viable to assist an escape. However, the fans continued airflow may be the only way that firefighters inside can cool through cool air to reach the entry door. If protocols are followed correctly and control measures are implemented, this scenario should not occur.
5	The air-flow from the fan will increase the burning rate by up to 15 percent in the fire compartment.	This will most likely occur and temperatures at the ceiling in the fire compartment and hallway serving the room will increase. However, the cooling effect of the fan at floor level will reduce heat flux and temperatures experienced at firefighter and victim locations. Consideration might be given to adding 15 percent to your nozzle flow-rate where PPA is used but this is not really necessary due to a greater percentage of the gaseous fire load being directed out of the exhaust vent or forced to burn off externally.
6	Temperatures in the hallway and fire room will also increase.	This may well occur for a short period of a few seconds as the airflow builds up but providing the air inlet/outlet ratio is adequate and airflow is not disrupted by firefighters blocking the doorways, within the 30 second pause and deploy period, this should begin to even itself out.
7	Fire may force trapped occupants to exit, or jump, from windows.	Unfortunately, this has occurred on occasions. It is therefore essential that an observer should be in visual contact of the exhaust vent and direct communication with the fan operator to coordinate the PPA operation safely and effectively. This external assignment may also be equipped with a hose-line to prevent fire spreading into the roof eaves

| 8 | Firefighters inside may not have located the fire correctly and unknowingly get into a position between the fire and the outlet. | This is again an event that has led to life loss and injury. Wherever PPA is used, firefighters must never go above the fire floor or change levels without first ensuring the fire has been located and water is being applied effectively to the fire. Similarly, they should not enter into a room where the fire is believed to exist until visibility has started to improve from the PPV airflow. If the room is geometrically shaped to allow any possibility that the firefighters could be situated between the fire and the outlet they should not enter the room without clear visibility and access to the fire and equipped with adequate water. |

Table 1.1: Potential hazards when using PPV in pre-attack mode (PPA)

These are clear and relevant concerns that should be addressed through effective training and generic risk assessment, ensuring adequate resource deployments and control measures are implemented prior to using PPV against developing ventilation controlled fires. This is why we write standard operating procedures (SOPs) and formulate strategic protocols that inform firefighters and commanders how to risk assess each situation effectively and implement PPA to best effect. The safe and effective use of PPA as a fireground strategy will rapidly improve conditions for firefighters, increase their safety and enable them to locate the fire and rescue trapped occupants faster. Wherever staffing is limited, PPA is one tool that will enable firefighters to enter house and apartment fires, according to protocol, getting water onto the fire quickly whilst undertaking rapid search and rescue operations in a more comfortable and safer working environment.

1.7 20 KEY TACTICAL CONSIDERATIONS FROM UL RESEARCH

A panel of four fire officers from UL's Firefighter Safety Research Advisory Board offered these key learning points[17] that came out of the broad range of UL research to date.

1. There's no substitute for knowledge
2. Your workplace has changed; you need to evolve
3. Follow the rules of live-fire training
4. Understand how heat transfers through turnout gear
5. Fire development changes when a fire becomes ventilation-limited
6. Fire flows from high pressure to low pressure
7. 'Nothing showing' means nothing
8. Keep the wind at your back
9. Flow path and suppression must be considered together
10. Water does not push fire (but a continuous fog pattern might)!
11. Initiate your firefight on the level the fire is on
12. Rapidly get water in the eaves for attic fires
13. The door closest to the apparatus should not dictate line/stream placement
14. Forcing the front door must be thought of as ventilation
15. Controlling the door limits the air and size of the fire
16. Never get stuck between the fire and where it wants to go without water or a door to close

17 http://brevity.next.firehouse.com/issue/561fcf363bab46bf780c3752

17. Well-timed and coordinated ventilation leads to improved conditions
18. Coordination of vertical ventilation with fire attack must occur just like with horizontal ventilation
19. Thermal image cameras can't assess structural integrity from above
20. Basement fires: Don't fall through or get caught in the flow path

If you take each of these tactical considerations, having researched the entire UL and NIST research series, they generate some critical learning points for firefighters and each one leads to great opportunities for discussion. These points may be so critical that a firefighter's life might depend on each and every one. In fact, if you review previous firefighter fatality reports, you may see a failure to address some of these points time and again.

1.8 NEGATIVE PRESSURE ATTACK (NPA) FLOW-PATH CONTROL

When deploying PPA as an attack strategy, one of the hindering factors is staffing. It is considered that six firefighters are required to safely deploy PPA and this normally means two engines in the UK are needed on-scene. In a few situations where one engine is on-scene for the first minutes without immediate support, PPA could not be deployed.

1. Incident Commander
2. Pump operator
3. PPV unit operator
4. External outlet vent – hose-line operator
5. SCBA wearer # 1
6. SCBA wearer # 2

The US system, termed HydroVent, offers a means to deploy a *negative pressure attack* (NPA) fire stream at the outlet vent using the 'venturi' principle, where tactically viable. What this tool does is extract smoke and heat out of the fire compartment and initiates a strong flow-path heading away from the interior hallways and adjacent rooms. In this case a bi-directional flow path may exist at the outlet vent but also consider creating an inlet vent, probably at the main entrance doorway, to enable air to replace the smoke being extracted. This is almost PPA in reverse.

The key factor for firefighters in using HydroVent would be to enable a single engine on-scene to have the option of taking control of the flow-path in a fire building where the outlet vent is viable and accessible and fire conditions support this action. Even with limited staffing this is certainly an option worth considering for it will create normally create cooler conditions inside the building and cause an untenable atmosphere move towards tenability, far sooner. Even without an internal SCBA deployment, the application should, in most cases, allow greater control over the fire.

1. Incident Commander
2. Pump operator
3. HydroVent firefighter
4. SCBA wearer # 1 (If staffing allows)
5. SCBA wearer # 2 (If staffing allows)
6. BA Control (or Rapid Deployment – If staffing allows)

Figures 1.2 and 1.3: The Hydrovent negative pressure tool used on a training container fire to extract smoke and heat and suppress fire in the room and at the outlet vent. **Courtesy of Hydrovent.US**

Where a building fire has spread beyond the compartment of origin, or where the fire is located further into the building, perhaps in a basement, the act of creating the negative pressure will draw fire to the vent outlet point. In general, this will not be a bad thing but if fire conditions appear to be worsening then just like PPA, the venting action of HydroVent should be curtailed and alternative tactics employed.

The HydroVent tool can be used both with and without a firefighting stream discharging into the room from the rear of the head.

Figure 1-4: Hydrovent on trial in France Photo courtesy of LADEPECHE.fr

Manufacturer's Specifications for HydroVent:

- Weight (dry) – 9.5 kg (21 lbs.)
- Length (when assembled) – 2.44 m (96 inches)
- Tool diameter -> 50mm (2 inches)
- Designed to operate at 413 kPa or 4.14 bar (60 psi) at the tip
- At 60psi HydroVent delivers 12 litres per second (190 gpm) or 720 L/min with a 360 L/min straight stream directed into the room and a 360 L/min fog ventilation nozzle
- Tested to 1379 kPa or 13.8 bar (200 psi)
- www.hydrovent.us

1.9 RECEO-REVAS-RICE AND SLICERS

There have been several acronyms used over many years previous that provided aid-memoires for firefighters, particularly in the USA, guiding those first on-scene as to what critical tasks required prioritization. In Eurofirefighter there was a chapter on *'guiding principles and managing risk at fires'* that discussed some of these acronyms in more detail. Some key principles were emphasised as '10 fire-ground golden rules' to follow, and these remain as primary considerations.

1. Remove those victims in the greatest danger first (visible at windows and balconies) (Don't hestitate)!
2. Always complete an initial 360) walk around, or get a view of as much of the fire building as possible.
3. If staffing is unable to achieve a coordinated fire intervention and interior search and rescue, place the first hose-line between possible victims and the fire, as this may save more lives.
4. Consider also the benefits of an immediate exterior attack off tank water to slow fire spread.

5. When adequate staffing is on-scene implement both firefighting and search and rescue in a coordinated attack.
6. Any ventilation should be delayed unless pre-documented SOP or direct on-scene communication from the incident commander (IC) permits this.
7. When there is no threat to occupants, the lives of firefighters should not be unduly endangered.
8. When making entry, consider the wind conditions and direction and any impact this may have on selecting the best point of entry.
9. Stabilise the fire conditions by isolating (confining) the fire. This one critical action can save both occupant and firefighter lives. If you pass or see an open door close it!
10. Identify and anticipate the ventilation flow-paths and their impact on fire spread and smoke (fire gases) transport.

USA Acronym	USA Acronym	UK (origin Kent FRS) Acronym
R – Rescue E – Exposures C – Confinement E – Extinguish O – Overhaul V – Ventilation S – Salvage	R – Rescue E – Evacuate V – Ventilation A – Attack S – Salvage	R – Rescue I – Intervention (Fire) C – Confinement E – Evacuation

Table 1.2: Useful Acronyms to remember (Critical tactical tasks to action, not necessarily in any order)

However, the most recent introduction of an acronym based on NIST and UL research is termed **SLICERS**[18]. This tactical approach SLICE-RS was a method created to help firefighters develop a mind-set that incorporates the fire dynamic lessons from the research into the actions of the first due engine. It is seen as less aggressive in relation to 'risk versus gain' and has caused some to voice concerns and others to demonstrate support for a guidance strategy that is based on a solid foundation of scientific research. Some say it prioritizes firefighter safety over occupant safety and others see it as a means of justifying staffing cuts on the fire-ground.

Either way, **SLICERS** does provide a range of risk-based tactical options that can be used effectively at fires by all fire services and departments in a range of occupancies and different situations. In truth, it can serve all from the busiest inner city fire services to the most rural and inactive departments. It recognises some of the common causal factors associated with life losses of firefighters at building fires and incorporates guidance that should not necessarily be implemented in the order listed, but applied in a way that prioritises necessary actions, according to available staffing and resources.

The 'size-up' (360^0 view of the building) is something not included in previous acronyms but is seen as a critical function of command. Although it is probably something that experienced commanders do without thinking, it still remains as one of the reasons that key fire behaviour indicators, alternative access points and external window rescues were not seen early enough prior to crews deploying internally. Locating the fire is another

---

18 Eddie Buchanan; Division Chief from Hanover Fire & EMS in Richmond, VA. USA

urgent tactical requirement that should come prior to deployment wherever possible and may result following a prompt '360 walk around'. Previous research has suggested that 25 percent of building fires continue to spread following fire service arrival and a targeted objective should be to reduce this statistic by containment actions. Taking control of the ventilation flow-paths immediately on arrival by simply closing the entrance door, if accessible and open, should assist this cause. Also, early water on the fire to take the heat release rate back down and reduce fire spread can sometimes be applied initially from an exterior position. In SLICERS this is termed 'Cooling' from a safe distance.

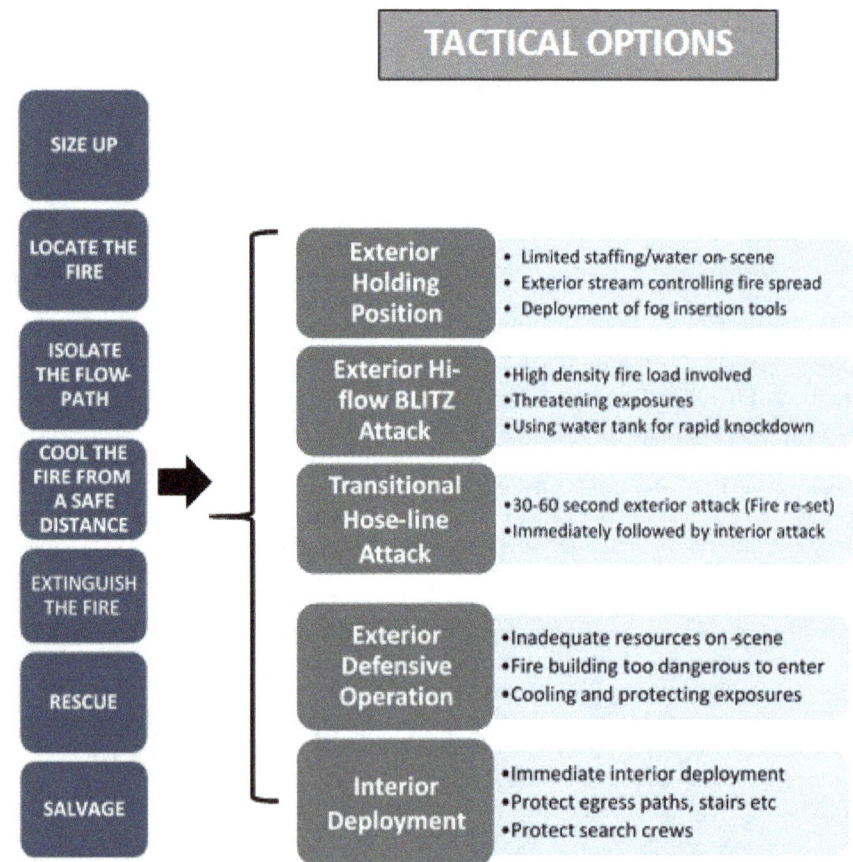

Figure 1.5: The SLICERS risk-based tactical option guide for primary fire response

This is where some may argue that 'cooling the fire' from a safe distance is very much a defensive stance that supports fighting a fire from the street, when an interior deployment of a hose-line will save more lives and property. However, the flow chart above shows five different tactical options that feed off into sub levels at 'cooling the fire' from a safe distance. Three of these options are actually 'offensive' options and each should be taken in context.

In certain inner-city fire services it is commonplace to arrive on-scene only to see heavy fire spreading to upper levels or adjacent properties. Some fire engines have roof-mounted monitors (we used to call them 'Stangs' – a manufacturer's term) whilst others carry portable high-flow monitors to mount an immediate 'blitz' attack through openings spewing fire, actually 'dumping' the entire contents of the on-board water-tank (1500–3000 litres). This strategy has been used since the 1950s and even earlier, with the objective to rapidly reduce the fire's heat release and burning rate, taking it back down to manageable levels prior to an interior attack taking place. This is nothing new! What firefighters should realise is that if applied correctly, this exterior application of water or CAFS, followed by an immediate interior deployment, can be very effective. The 'transitional attack', as it is now being termed, involves a similar strategy using a hand-line from the exterior to 're-set' the fire back down to a lower rate of burning. This strategy has been viewed in some circles as a delaying tactic for getting inside and fighting the fire from the inside. In fact it is another variation of the above approach using a high-flow monitor. There are distinct advantages in applying water this way prior to transitioning to an interior attack, but such a strategy demands effective hose management and adequate staffing for if the interior hose-line is not advancing into the fire compartment within a few short minutes, the fire may well have re-established itself into a steady state of post-flashover burning with little gained.

When determining which mode of attack a first response crew are in, of the first four of five cooling options given above, it is important to define objectives and prepare to deploy effectively from the outset.

- Are we going to discharge the entire contents of the water tank into an opening showing heavy fire to prevent rapid fire spread?
- Do we have a nearby water supply?
- Do we have a second engine in back-up?
- Are we going to transition to the interior after 30 seconds of water applied into a window spewing fire or are we in a 'holding' situation?
- Are crews ready and on air for the transition or will time be lost masking up?
- Will the line being used from the exterior be taken into the building or do we have a second line for that?
- Will the line be long enough to reach the fire room from inside?
- Are we reaching the fire's base or is the fire located in another part of the structure?
- Are we deploying in behind a PPA airflow and if so, will this cause the fire to re-establish itself to full development too quickly, or is this a viable strategy under controlled conditions?

All of these are important points for crew and station discussion and departmental pre-planning.

These are the questions we should be asking (and acting upon) at a very early stage for there can be much confusion as to which actual mode of attack the firefighters are in. A clear tactical objective should be communicated to all members on-scene.

> **What does this mean to the Firefighter?**
>
> Those key bullet points above are very relevant. So many times, firefighters arrive on-scene with a clear intention to optimise staffing availability in accordance with exterior fire showing from windows. Are we going to soften that fire and then deploy immediately? Or are we going to continue with an external fire attack until adequate staffing is on-scene for interior work? Or, are we going to use brief bursts of a firefighting stream to keep knocking the fire back whilst conserving tank water? Or are we going to deploy to the interior AND have someone hitting fire from the street at the same time? These points are CRITICAL and firefighters must know immediately what the strategy is and be ready with BA donned and started, for any interior deployment following some exterior knockdown must come within 30-60 seconds, or the fire is likely to redevelop back to where it was on arrival.
>
> Note: Any external use of a fog pattern or wide cone is likely to 'push' fire and heat further into the structure! A solid stream bounced off the ceiling, creating a sprinkler like effect, appears to offer optimum results.

1.10 STAFFING ISSUES AND LIMITED CREWING ARRANGEMENTS

As the increase in building fire protection and community based safety education programmes is seen to decrease the number of building fires, the likelihood is that the levels of fire cover will come under increasing review. In many areas the firefighter's workload is diversifying and increasing into special services, flood and natural disaster response and pre-hospital emergency medical care. However, the staffing levels on fire stations and fire engines is demonstrating a general global reduction and when building fires do occur, even though less frequently, staffing issues can impact on service delivery ad firefighter safety.

As discussed earlier, fires are now burning with greater intensity than 40 years ago as energy efficient buildings containing a greater plastic content in the fire load, coupled with larger window openings and increased ventilation mean that fires are less likely to be contained. As firefighters arrive the likelihood of flashover having already occurred is increasing and new lightweight construction methods mean that the structural frame can become the fire load as fire response-times increase.

It used to be the case that 8-12 firefighters were needed on scene within the first five minutes of a building fire to implement full interior search and rescue operations, supported by firefighting actions. It may soon be the case that the weight of attack is reduced to levels where only firefighting operations can be effected on arrival and interior search and rescue will become a secondary strategy, often taking place after the fire has been extinguished. In some situations, 'taking the fire first' is the best life-saving strategy there is!

Firefighting or Search and Rescue first?

Arriving on-scene at a working fire, firefighters may not have support for 10–15 minutes or more. This means that due to limited staffing on-scene they may have to prioritise either search and rescue or firefighting actions over each other. There are straightforward tactical considerations to be made here.

- Don't deploy too quickly – take a few seconds and deploy according to the IC's brief
- Are there persons <u>known</u> to be trapped or missing?
- Are there persons <u>believed</u> to be trapped or missing?
- Are there persons requiring rescue at windows or on balconies? Then take them now!
- What is the wind speed and direction? Is it relevant to the fire?
- Is there fire showing from a window/s?
- Is air feeding into the fire at existing external door openings?
- At what stage in fire development is the fire?
- Can the fire compartment be isolated and the fire confined?
- Would an immediate firefighting action save more lives or is a 'snatch rescue' viable?
- Do we have enough resources on-scene to undertake interior search and rescue?
- What is the status of our water supply? Tank only, or augmented from source?
- Has the fire compartment been located? Is it obvious?
- Is the fire spreading out of and beyond the original fire compartment?
- Is this single storey or multi-level?
- Think SLICERS!

There are a number of issues here that require more detail and there is a need to discuss a range of alternative tactics that will serve to enhance a limited staffed response.

> **What does this mean to the Firefighter?**
>
> **A firefighting action can be the most effective life-saving strategy there is. By placing a hose-line directly on the fire to achieve rapid extinguishment may solve all your problems. Second to this, the fire room should be isolated before firefighters advance above or beyond the fire. If this cannot be done a hose-line crew should be sited at the base of a stair, or adjacent to the fire compartment, or in a corridor/hallway, to protect any crew searching beyond or above the hose-line to maintain their path of egress.**

Tactical Objectives
1. Take control of the fire
2. Reduce the heat release rate of the fire
3. Take control of the flow-path

1. Taking control of the fire

Attempting to 'take control of the fire' as a primary action may, in itself, save lives. Depending on the stages of fire development, we can immediately do a number of things to control a developing fire. We can close exterior doors, preventing air feeding the fire with much needed air, until we are ready to deploy internally. We can consider an external attack on the fire where flames are issuing from a window, providing we are certain that the fire is not travelling from the floor below to an upstairs window vent. Any such attempt from the street should encompass a straight stream (not wide cone or fog pattern) directed at the ceiling of the fire compartment and held static (not sweeping across the ceiling) to represent a room sprinkler effect. There are several approaches we might consider here – are we attempting to 're-set' fire conditions (heat release rate) down to a level where we might then immediately deploy internally to finish extinguishing the fire? If that is the case, we need to get into the fire compartment within 30-60 seconds of the initial external attack to ensure the fire does not redevelop back to its original state or we have gained nothing. Is this possible? Are we prepared and ready to do this with firefighters already under air (BA) and is it viable for them to relocate and deploy the hose-line used for exterior attack or are we to deploy using another line? With limited staffing these are serious considerations. Of course, we may be using two hose-lines with one person remaining with the primary external line whilst a second line is taken to the fire compartment. In such circumstances, will the initial external knockdown of a heavy fire enable the internal deployment of a smaller fast attack UHP hose-reel? What is the status of our water supply?

Further consideration might be given to a 'holding' line, attempting to prevent further fire spread from occurring by directing controlled bursts of a direct straight stream into the ceiling whilst maintaining the level of tank water until a hydrant, or continuous water supply, is connected to the fire engine. The objective here is to simply take control of the fire without an all-out attempt at completely extinguishing the fire. If the fire can be completely extinguished, then do so! It's important to remember though that around half of the water we direct into the flames when using a straight sold stream plays no part in fire suppression but simply ends up on the floor or runs back out of the building. But then using a more efficient fog stream from the exterior is likely to 'push' fire deeper into the structure.

2. Reduce the heat release of the fire

We can use the above tactics to reduce the heat release rate of the fire by closing external doors to prevent air feeding in, or we can of course hit the fire from the street. However, if we turn up to find the fire is contained, all windows closed but we are able to locate the fire compartment externally, we can use water-fog (or foam) insertion tools to 'steam' the fire compartment. After we have created this 'steamed' environment to good effect, we should be safe to ventilate the compartment. This tactical approach may also be used where the fire compartment is demonstrating signs of a grossly under-ventilated fire with potential backdraft conditions. This strategy can be applied using COBRA/PYROLANCE/FOGNAIL/FOGSPIKE tools and providing you have good access to the fire compartment's outer skin, should be effective.

Many firefighters have rightly questioned the concept of 'steaming' fire compartments that may still have occupants overcome by smoke but possibly alive inside. But for the injected fog to 'steam' the compartment the temperature therein would need to be in the region where survivability is already unlikely. However, further research is needed in the area of such finely divided UHP water droplets (i.e.; not standard fog nozzles).

3. Take control of the flow-path (also see Chapter 18)

The three main ways that we may take control of the flow-path is by:

A.   Isolating the flow-path (close doors; use smoke blockers and wind control devices)

B. Ventilating the flow-path (reversing the flow-path with intent)
C. Cooling and inerting the fire gas layer existing near the ceiling (3D water-fog)

In taking these three tactical approaches, the use of positive pressure ventilation (PPV) (or Hydrovent) is most likely to have the biggest impact in controlling the flow-path to the advantage of limited staffed crews.

Search and Rescue

As already stated, in particular where staffing on the initial attendance is less than six firefighters, the primary objective should be to *take the fire first*. It has been demonstrated time and again that this action may save more lives than placing interior search ahead of firefighting. However, where there are clear reports of confirmed occupants trapped or missing then search and rescue may need to take place before fire attack. This is down to local procedural guidance and an on-scene commander's decision. During the process, tactical considerations and actions are as follows:

- Where are the occupants 'reported trapped' or missing in the building (most likely)?
- How reliable is the report of entrapment?
- Where is the fire located?
- Isolate the fire where possible prior to deployment
- Use door control tactics at the entry points to both the structure and the fire compartment to reduce the flow-path
- When firefighters are moving ahead of, or above, the fire location this should never occur until the fire compartment has been isolated (door/s closed) or a main hose-line with crew is located between the fire and their egress point (stairs?)
- Consider PPA to clear the smoke and gases via an outlet vent but there are two considerations here –
    1. The fire must be isolated if working above or beyond the fire compartment
    2. If the fire compartment is the search and rescue area then it should not be entered without fast-attack water (high-pressure hose-reel as a minimum)

Discussion

When applying the above principles of fire attack/search and rescue, even more so under limited staffing operations, the **SLICERS** concept offers a good basis for *primary tactical deployments* at all fires. In particular, residential dwellings, apartment blocks or premises with a sleeping risk may place firefighters under limited staffing arrangements with a clear moral dilemma. However, by addressing the SLICERS acronym and taking appropriate tactical actions to reduce the spread of fire, this will optimise tactical options.

It's better to have five firefighters on-scene who are trained and effective in taking positive actions to reduce fire spread and control the flow path than ten firefighters who don't make any effort to do so and deploy too quickly into the centre of a rapidly worsening situation.

Whilst undertaking a 360 of the building, take in all entrances, windows and vent status. Take note of the WIND conditions and direction onto each face of the building at this point of the size-up. Which is the best side to deploy from? Not always at the front! Is it obvious where the fire compartment is or not? From this point on the strategy for the next 60-90 seconds can be determined. Is fire coming out of the fire compartment window? Do you apply cooling and if so, with what objective? If the window is intact do you deploy internally or do you break the window to apply an exterior cooling stream? This depends on what equipment the engine carries, what staffing level you are operating from and how far away (minutes) your support engine/s is likely to be.

Make your decisions logic based and consider if it is better to contain the fire as long as possible. A recent fire in Dublin involved a large industrial shed that was signifying a fast-developing fire inside on fire brigade arrival. The IC decided to close all access points into the shed and access a constant water supply whilst awaiting the arrival of a second engine. As hoses were laid out and portable monitors were placed into position within a few minutes, the main doors were then opened. The fire belched out but was instantly knocked back by the heavy water streams and an interior attack followed that resulted in a major property 'save'. It was quick thinking by a well trained and experienced IC that ensured this company business could eventually return to trading, sooner rather than later.

1.11 EQUIPMENT TRENDS

European firefighters are becoming highly influenced by research in the USA and are trending towards an even greater use of transitional attack methods. Additionally, the delivery of higher water flow-rates onto fires is increasingly becoming a tactical priority as flow-meters become more common on fire appliances and a deeper understanding of flow over pressure. However, anti-ventilation tactics, Swedish style water-fog gas-cooling techniques, positive pressure ventilation and the use of UHP fog equipment such as COBRA and FOG-SPIKE or FOG-NAIL from exterior positions, continue to gain popularity.

1.12 CORRIDOR FIRES

In some hotels or large residential complexes or towers blocks, along with a few other buildings such as hospitals and care homes, you may come across very long corridors (20-60 metres) serving multiple door openings to variable types of accommodation. Some of this accommodation will simply be small rooms or suites whilst others will serve large apartments with several rooms off of an internal hallway. These corridors may have communal stairs at each end or just one central stair and may extend to unusual shapes in 'L', 'T', 'U' or even circular/square 'race-track' patterns. When fire extends out into these corridors many firefighters will be unfamiliar with the physical dynamics of the fire behaviour likely to be experienced as corridor fires tend to behave far differently from open-spaced room or compartment fires.

There is nearly always a ventilation flow-path pre-existing in these long corridors where the air movements may be enhanced by:

- The fire intensity within the compartment or rooms involved
- Fixed or automated smoke ventilation systems (natural or forced air-flow)
- Exterior wind conditions
- Stack effect in tall buildings
- Thermodynamic effects of the fire itself
- Pressure differentials existing between air inlets and exhaust points

This flow-path will determine flame velocity in the overhead gas layers and with air velocities at the floor of around 6 m/s (13.4 mph) heading towards the fire, there is a major cooling effect at firefighter locations. This cooling effect in the flow-path can be used to advantage by firefighters and will prevent high heat fluxes hitting the floor from burning gases in the overhead. The faster the gas layer is moving, the greater the cool air mass is drawn in below. In a room fire this would be dangerous for firefighters but in a long corridor the flaming combustion in the gas layer will stay high for a period, until an external wind reverses the flow-path or surface linings ignite, causing the thermal layer to drop down and eventually fill the space.

Chapter 2

The Building – what the Firefighter needs to know

The building regulations (fire safety) in the UK specify certain requirements for the provision of adequate access and facilities to assist firefighters during firefighting and rescue operations. It is important that these minimum levels of firefighting access and facilities are maintained during both construction and occupation phases of a new build. It is also important to review how a building is constructed and what materials are being used. At all stages of the build, firefighters should familiarise themselves with building methods and materials used, as at completion the relevant parts (to firefighters) of a fire-involved structure may be hidden from view.

As a building material, wood or perhaps more appropriately timber, has some excellent properties. It has a high strength to weight ratio; it can easily be shaped and connected; it is one of the most sustainable resources and is environmentally friendly as well as being aesthetically pleasing. Timber is increasingly becoming a more desirable construction material as international architects and designers realise that timber has significant potential benefits in sustainability and speed of construction. Traditional schemes for timber buildings as low-rise (two-stories or less) and mid-rise (three- to five-stories) are now being extended with schemes for new high-rise buildings, also referred to as tall, timber buildings (six-stories or greater). In fact, extremely tall timber framed buildings are now being constructed in many of our major cities. It is almost certain though that as these buildings age they will become more at risk of fire-spread in cavities, voids and redeveloped sections of the structure, and the future may hold some surprises for the fire service from an operational perspective.

2.1 FIREFIGHTING ACCESS AND FACILITIES (UK FIRE CODES)

Firefighting Access
It is essential that the fire service is able to effectively access the roads and approaches to any particular building or risk, so that they are wide enough and able to support the weight of fire engines. It is inappropriate to provide access routes that are blocked by parked cars or possibly cause the fire engines to enter via a very congested retail car park. On arrival, the firefighters require access within 45 metres of a residential building, measured from the fire engine to the furthest distant point, within the structure. The intention is to allow a 60-metre run of hose-reel from the fire engine to reach all points within a building. If

this isn't viable, then a dry rising fire main or perhaps sprinklers should be installed. There are also requirements to provide firefighting access to the building's perimeter (engines and height vehicles) based on a percentage of the total size of the floor-plate.

Firefighting Shaft

In buildings over 18 metres, or where deep basements are more than 10 metres below grade, or in commercial, shop, industrial or storage buildings more than 7.5 metres in height, a firefighting shaft is required. A firefighting shaft provides firefighters with protected vertical access to all floor levels from which to mount their firefighting search and rescue operations, providing two hours of protection (fire resistance) from fire, heat and smoke. Within the firefighting shaft itself certain facilities are provided to further assist firefighters:

- Firefighting Lobby (ventilated) – firefighting lobbies should have a clear floor area of not less than 5 m².
- Firefighting stair
- Rising fire main (in some cases the fire main outlets may be provided in the corridors, outside of the firefighting shaft but immediately adjacent to the shaft door. This is a non-compliant arrangement with local fire service approval/preference).
- The furthest point in any building from a fire main located in a firefighting shaft must be limited to 60 metres (three hose-lengths).

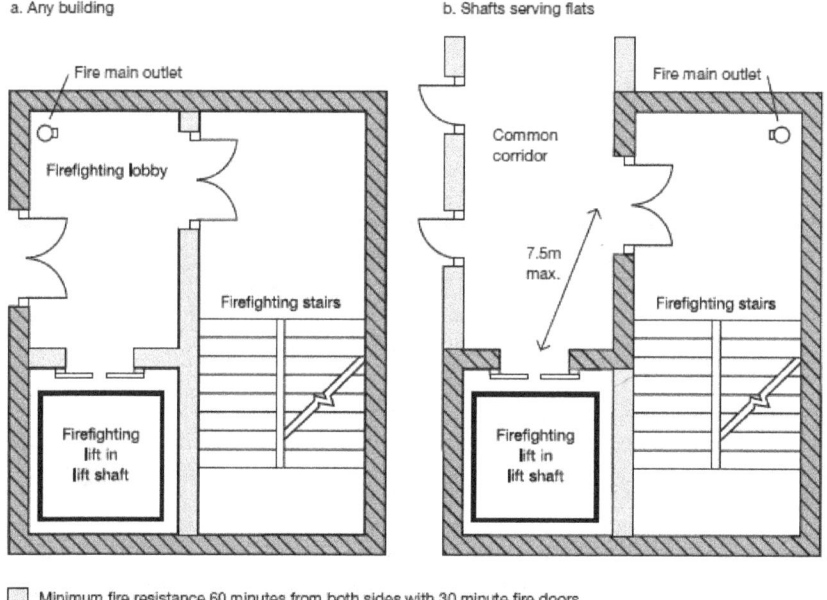

Figure 2.1: *The recommended layout for firefighting shafts in buildings other than apartments according to building regulations.* **Courtesy of DCLG (ADB)**

- May be provided with lobby protection (unlikely in apartment buildings)
- Will be ventilated either manually in multi-stair buildings, or automatically by smoke detection in firefighting shafts or single stair buildings
- May be pressurised (positive air outflow to prevent smoke inflow)
- Firefighting lift (in flats or apartment blocks the firefighting lift may not be located within the firefighting shaft, but if this is the case then the lift should be no more than 7.5 metres from the firefighting stair door.

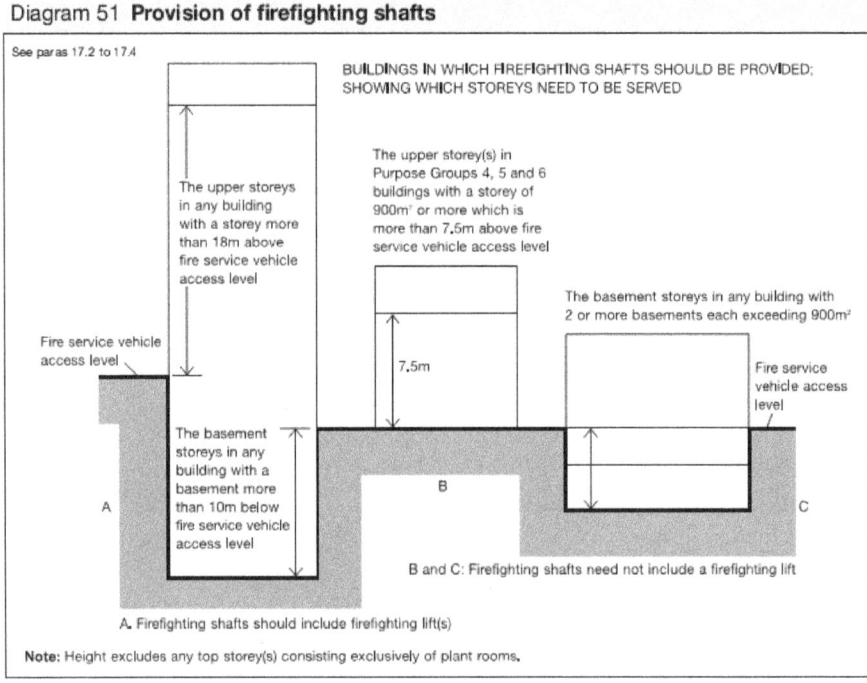

Figure 2.2: The requirements for the provision of 2-hour fire-resisting firefighting shafts according to building regulations. **Courtesy of DCLG (ADB)**

Firefighting Lift

There have been a range of design codes and standards for firefighting lifts, dating as far back as the 1970s. These were in fact ordinary passenger lifts with a firefighter recall button allocated at ground level. Currently firefighting lifts are based on a European standard BS EN 81-72:2015. A firefighting lift, unlike a normal passenger lift, is designed to operate so long as is practicable when there is a fire in parts of the building beyond the confines of the firefighting shaft, as it is used to transport firefighters and their equipment to a floor of their choice. The lift can be used in normal times as a passenger lift by the occupants of the building and during a fire, it may be used prior to the arrival of the fire and rescue services for evacuating those with disabilities and impaired mobility (BS 9999). This feature will coordinate with fire alarm activation and may not ground immediately. It should not be used for moving refuse, or for moving goods. In buildings provided with a single lift its use for the transport of goods should be avoided unless essential, lift lobbies should be kept clear, and when used for moving goods the doors should not be propped open to ensure that the lift remains at a particular level.

Typical upper floor – one escape route – storey height more than 7.5m but not more than 18m

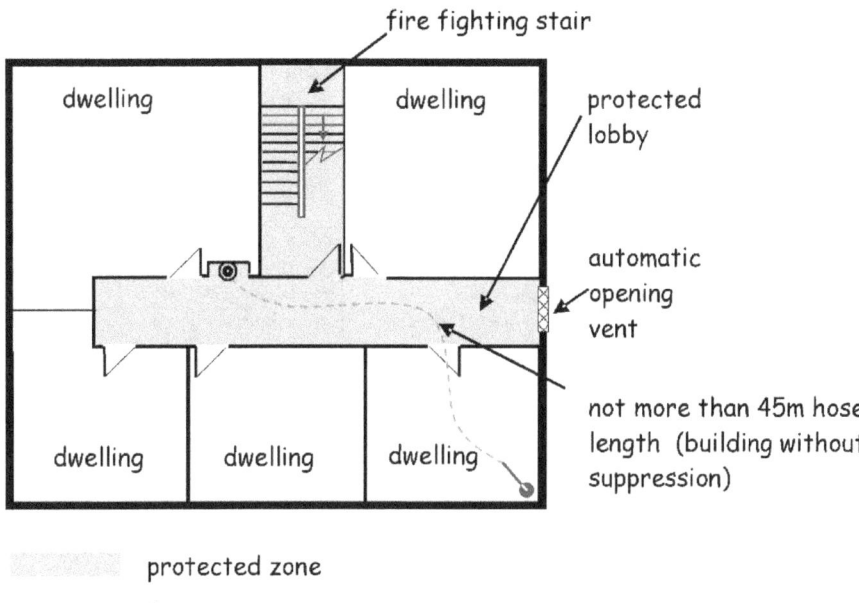

Figure 2.3: The recommended layout for firefighting shafts in apartment buildings according to building regulations. **Courtesy of DCLG (ADB)**

The BS-EN standard (2015) stipulates certain requirements that are sometimes at conflict with UK building regulations; therefore, the standard is not fully complied with, notably:

- The firefighters lift is located in a well with a 'safe area' (lobby) in front of every landing door which is used for firefighting operations. In front of every landing door a safe area, a fire shutter or a fire door shall be provided (see firefighter lift entrapment below).
- The purpose of safe areas is to protect the lift well, firefighters and those who may be waiting to be evacuated against fire, heat and smoke and in addition to allow the self-rescue of firefighters/occupants if needed. The minimum dimensions of each safe area are given by national regulations.

For detailed guidance of upgrading existing firefighting lifts to a minimum acceptable standard, reference should be made to BS 8899:2016 (section 5.4)

Summary of past provisions for lifts with operation in the event of fire			
Type of lift and relevant standard	Key lift provisions	Key building provisions	Test and maintenance
Firemen's lift BS 2655-1:1970	• 1.45 m² 550 kg min. car size • 0.8 m clear opening doors • Full height travel in 60 s • 'FIRE CONTROL' switch adjacent to lift opening at fire control floor level • Operation of switch not to override safety devices and maintenance switches; lift recalls to fire control level and parks with doors open; then lift operates from car controls only with collective control disabled	Where more than one lift, signage making clear which is the firemen's lift Electrical supply exclusive to lift and routed through negligible fire risk	No tests or checks listed
Firemen's lift BS 5655-1:1979	• Lift car at least 1.4 m², rated load at least 630 kg • 0.8 m clear opening doors • Rated speed so that complete travel time does not exceed 60 s • 'FIRE SERVICE' break glass switch on evacuation landing level • Operation of fire service switch ensures the priority recall of the car which, after its arrival at the evacuation level, operates without responding to 'landing' calls. Having arrived at the specified level, it remains with its door open until a fresh instruction is made from inside the car	Building to avoid exposing these lifts to unacceptably high temperatures and to avoid having the water used to fight the fire flowing into the lift wells Either for the 'firemen's lift' serves all the levels or several 'firemen's lifts' together allow access to all the levels of all the building divisions	No tests or checks listed
Firefighting lift BS 5588-5:1986	• 1.1 m × 1.4 m/630 kg min. car size • 0.8 m clear opening power operated doors • Fire-resistant car linings • Full height travel in 60 s • Trap door in car • One set of car doors only • Fire service communications system – two-way speed communication system between firefighting lift car, FSAL and lift machine room when firefighting switch is on. • Firefighting lift switch brings lift to FSAL and park with doors open • Firefighting control only from car – one car call only; doors open only from door open button (reclose if button released before doors fully open); new car call closes doors	Recommendations for firefighting shaft, protected lobbies at each landing, machine room location, etc. Either building measures to control water (drainage channels or ramped landings) or lift electrical equipment protected to IP03 Primary and secondary electrical supplies protected against fire	Operational tests in Annex A
Firefighting lift BS 5588-5:1991	• Based on assumption of lift machine room similar to BS 55885:1986 and • Electrical equipment within 1 m of any wall separating the lift well from a lift lobby protected to IPX3	Similar to BS 5588-5:1986 plus recommendations for either pressurization or natural ventilation	Operational tests in Annex C
Firefighting lift BS 5588-5:2004	• As BS EN 81-72:2003	As BS EN 81-72:2003	As BS EN 81-72:2003

Summary of past provisions for lifts with operation in the event of fire

Type of lift and relevant standard	Key lift provisions	Key building provisions	Test and maintenance
Firefighters lift BS EN 81-72:2003	• Allows machine room-less lifts (MRLs) • 1.4 m × 1.1 m/630 kg min. car size except 2.1 m × 1.1 m/1 000 kg if used to evacuate stretchers or dual entry firefighters lift (open on two sides at the same level) • 0.8 m clear opening power operated doors • Full height travel in 60 s • Trap door in car • Dual entry lift allowed • Fire service communications system – two-way speed communication system between firefighters lift car, FSAL and lift machine room when firefighters switch is on • Firefighters lift switch brings lift to FSAL and park with doors open • Firefighters control only from car – one car call only; doors open only from door open button (reclose if button released before doors fully open); new car call closes doors • Electronic landing controls and indicators continue to function 0 °C to 65 C • Electrical equipment within 1m of any wall separating the lift well from a lift lobby should be protected to IPX3 • Means of rescue from outside the car (under UK regulations) • Means of self-rescue from car, e.g. ladders	Building recommendations included in BS 5588-5:1991 until replaced by BS 9999:2008 and BS 9991:2011 Requirements for firefighting shaft, lobbies, machine room location, etc. Building measures to control water (drainage channels or ramped landings) Primary and secondary electrical supplies protected against fire	Installer test to meet BS EN 81-72:2003, Clause 6 BS 8486-1:2007 included tests for firefighters lifts

Continued overleaf.

Summary of past provisions for lifts with operation in the event of fire			
Type of lift and relevant standard	Key lift provisions	Key building provisions	Test and maintenance
Firefighters lift BS EN 81-72:2015	• Main technical changes from BS EN 81-72:2003: • 'Safe area' is used in place of 'lobby' to be consistent with DD CEN/TS 81-76. The two terms are interchangeable • Revision of the elements dealing with building design and the inclusion of a new informative annex on the building interface. Items to be considered in the use of pressurization of lift wells have been added including the noise level at the fire communication points • Deletion of the requirement for a firefighters lift to serve every floor of the building. The floors to be served are assumed to be determined as part of the design of the building for fire • New requirements for protection of electrical equipment against water; in the lift-well and the roof and walls of the lift car • New measures to prevent water ingress into the lift well which are strongly preferred over to measures to control the level of water in the lift pit alone. The measures considered are described in a new annex on water management • Revision of clauses dealing with the rescue of trapped firefighters with rationalized requirements for movable ladders and reduced maximum distance between consecutive landings. The use of fixed ladders and rope ladders has been removed. • New requirements for the interface between firefighters lift switches and the control system • New requirements for dual entry lift cars where not all the safe areas to be used in firefighting operations are on the same side, i.e. more than one car door could be used during firefighting operations • Revision of requirements for the control system including new requirements for when a firefighter's key switch is used in the lift car (subject to negotiation). • Revised requirements, in phase 2, for car doors to close under constant pressure from door close or car call buttons and for opening • Addition of requirements for keypad operation of the firefighters lift car in phase 2 • Inclusion of a new informative annex on maintenance requirements		

Summary of past provisions for lifts with operation in the event of fire			
Type of lift and relevant standard	Key lift provisions	Key building provisions	Test and maintenance
Evacuation lift BS 5588-8:1988	• 'Evacuation lift' switch – except for two storey. • Evacuation controls – return to final exit level, isolate landing controls, control only from car controls • Communication system except in two storey building linking each landing and lift car – controls, telephone or radio system • Controls to meet BS 5810 (disabled access)	In a protected enclosure including lift well and lobbies on each storey Primary electrical supply exclusive and independent Secondary electrical supply except on hydraulic lifts serving two storeys	
Evacuation lift BS 5588-8:1999	• 'Evacuation lift' switch – now on all evacuation lifts. • Otherwise as BS 5588-8:1988	As BS 5588-8:1988	
Evacuation lift BS 9999:2008, G.2	• As BS 55888:1999, except evacuation lift also to meet BS EN 8170	As BS 55888:1999, except: More on refuges More on secondary supplies	
Fire recall BS EN 81-73:2005	• On operation of recall device, car and landing controls inoperative, calls cancelled and lift travels to designated landing and parks with doors open. No entry sign at designated landing	Specification of manual or automatic recall device	Installer test to meet BS EN 81-73:2005, Clause 6 BS 8486-1:2007, Annex E and BS 8486-2:2007, Annex E included tests for fire recall

Continued overleaf.

Summary of past provisions for lifts with operation in the event of fire			
Type of lift and relevant standard	**Key lift provisions**	**Key building provisions**	**Test and maintenance**
Fire recall BS EN 81-73:2016	Main technical changes since BS EN 81-73:2005 are: • use of 'recall means' to denote either a manual recall device or automatic recall device e.g. fire alarm system; • changes to the assumptions including items subject to negotiations; • revision of clauses dealing with recall means, • interface requirements and designated landings. Addition of a sign to manual recall devices; • clarification that fire recall signals are not to override maintenance controls; • change in the reference of the prohibition sign; • changes to the behaviour of the lift on receipt of a recall signal including requirements for a sounder on the car if the lift is under maintenance control and a sounder in the car when doors are closing with provision to make passenger protection devices inactive if doors have not closed after a delay; • changes to the behaviour of the lift once arrived at the designated landing including requirements for audible and/or visual indication and options for the lift to park with doors open or closed; • deletion of requirements for a separate reset signal and for a 'no entry' indicator; • Inclusion of a new informative Annex B on maintenance requirements	Specification of: • type of recall means and its interface; • type and protection of switch in case of manual recall device; • number and location of designated landing(s); • suitable maintenance and verification plan is implemented; and • whether the lift parks with doors open or closed at the designated landing (see guidance in the National foreword)	Installer test to meet BS EN 81-73:2005, Clause 6
Table Courtesy of **Nick Mellor** Technical Director at the Lift and Escalator Industry Association (LEIA)			

Table 2.1: The code provisions for firefighting lifts in UK and some European buildings (1970–2017)

It is essential that firefighters familiarise themselves with the operation of lift controls when they visit buildings, the procedure for taking control of the lifts and the escape procedure from a stalled lift. Where a roof hatch for firefighter escape is installed (in some areas the installation of a firefighter escape hatch may not be approved by the local fire service) the operation of the escape hatch and the access to the car roof, both from inside the car and externally, along with electrical supply isolation is all important knowledge.

Firefighters and Lift Entrapment

There are procedures and guidance that advise firefighters to ensure the lift shaft is safe to use prior to entering a firefighting lift and not to approach closer than 2-5 floors below the reported fire floor. However, there is often initial confusion as to where the fire floor actually is. Wherever firefighters are accessing upper levels by lift elevator, it is highly recommended that ALL firefighters are equipped with breathing apparatus as a safety measure. At a recent high-rise fire in a modern residential building in the centre of London, three firefighters were trapped in a non-firefighting passenger lift that stalled on the fire floor. The corridor outside the lift was fully involved in a very intense post-flashover fire spreading out from an apartment and only two of the firefighters were equipped with breathing apparatus. With heavy smoke entering the lift car through the closed doors the three firefighters were forced to share air from the two air sets. Thankfully they were rescued just in time as their air supply was running out. On this occasion the firefighters were caught and trapped by a corridor fire impacting on a passenger lift but, could this occur in a firefighting lift? If the firefighting lift is not in a firefighting shaft, or protected in any way by lobby doors, most certainly yes! This particular lift opened directly into the residential accommodation corridor and the intensity of the post-flashover corridor fire buckled and distorted the lift doors, making them jam and stop at the fire floor. In most new residential buildings, a firefighting lift also exits directly into the corridors, so yes it could still occur. At another high-rise fire in Basingstoke (Churchill Plaza) a team of firefighters were again trapped within a stalled lift car, requiring teams of firefighters being deployed specifically to undertake their rescue.

The author has been campaigning publicly since the 1980s for ALL firefighters (including breathing apparatus control officers and crew commanders) to carry (don**) breathing apparatus with them when they use lifts to access upper levels in buildings, whenever a fire is reported. But even now (at time of writing) this is still not national operational policy.

** The term 'don' means to 'carry or prepare to wear', without being 'on air' – the term 'wear' refers to the act of starting up and breathing air from the mask[19]. In the UK, the act of 'wearing' BA under air should only occur under direct instruction from the IC, or in extreme circumstances to escape from sudden fire escalation that is transporting combustion products into occupied areas adjacent to the immediate risk zone.

Examples:

- whilst located deep within the London sub-surface underground train network or;
- if trapped in a lift where smoke is entering the lift-shaft for some reason

There have been several recorded occasions when water from a landing valve, hose lines, etc., has entered the lift well and caused malfunction of the installation when it reached electrical door interlocks, car controls, etc. It is therefore necessary to minimise both the effects of water on lift operations, and the probability of water entering the lift well in the

---

19 As defined by London's CFO Gerald Clarkson during the public enquiry into the Kings Cross Underground (LUL) fire in London 1987

first place. It is for this reason that the lift lobby is one area where sprinklers might not be installed, and special preventive measures should be installed to reduce the likelihood of lift failure (see Table 2.1).

A firefighting lift switch should be provided to enable the fire service to obtain immediate control of the firefighting lift(s) in a firefighting shaft. The car controls of the firefighting lift should become active only after it has arrived at the fire service access level and the firefighting lift switch has been operated. Once the firefighting lift has arrived at the fire service access level, its doors should open and it should then operate as follows:

- Fire personnel entering the lift car should be able to register a call to any selected landing in the building by sustained pressure on a car control until the car doors have fully closed
- If a car control is released before the doors have fully closed, the doors should immediately reopen and the call should be cancelled
- It should be possible to register a new call on the car controls which should cancel the previous call.
- The lift should travel in the shortest time to the call registered, and should stop at that floor.
- The doors should remain closed unless they are operated by continuous pressure on the 'door open' control
- It should not be possible to open the doors without sustained pressure on the control
- Release of the 'door open' control before the doors are fully open should cause the doors to automatically re-close. This allows fire service personnel to observe the situation immediately outside the lift landing doors in the firefighting lobby
- Once the doors are fully open they should remain open until a new call is registered at the car control station
- All lifts may operate slightly differently and firefighters should familiarise locally with firefighter's car key switches, where provided

Upgrading existing Firefighting Lifts

These are recommended minimum design requirements where it is entirely impractical to upgrade existing firefighting lifts to current BSEN standards:

- Primary and secondary power supplies
- Water protection measures (e.g. IP rated wiring and controls, drainage and preventive measures)
- Fire fighter recall switch at access level
- Fire fighter in-car controls
- Fire fighter communication system
- Floor indicators

For detailed guidance of upgrading existing firefighting lifts to a minimum acceptable standard, reference should be made to BS 8899:2016 (section 5.4)

Firefighting Stairs

Firefighting stairs are those that accommodate a firefighting main, where firefighting shafts are considered unnecessary. They must be at least 1100mm wide and provide a minimum of 60 minutes fire resistance. The furthest point in any building from a fire main located in a firefighting stair must be limited to 45 metres (two hose-lengths).

Rising Fire Mains (Standpipes)

The objective of having rising fire mains in buildings is to reduce the physiological stresses placed on firefighters who have to transport quantities of firefighting hose, nozzles, adaptor couplings and other equipment to the fire area within a building whilst each wearing over 50kg of protective equipment. Research has demonstrated that in doing so, inner core body temperatures are sometimes already near dangerous levels, at just about the time they are committing themselves into the extreme heat of a fire compartment. In residential buildings, firefighters should not be expected to travel further than 45 metres from a fire service appliance (fire engine) or fire main outlet, unless the fire main is within a firefighting shaft, where 60 metres is accepted as the maximum hose-lay distance into the accommodation. In commercial and industrial buildings, firefighting access limitations are addressed by perimeter access to prescribed percentages of buildings, depending on size.

In tall buildings in the UK in excess of 18 metres in height a dry rising fire main is required and in new buildings over 50 metres a wet rising fire main is now prescribed (over 60 metres pre-2006). Dry mains now have design capacity to be charged up to 12 bars (from 2015)[20] from fire service pumps but at this time, the majority of fire services in the UK are unable to charge to more than 10 bars due to pressure limitations on hose-line and pumping configurations. The ability to pump to 12 bars will add approximately a further 2 bars to outlets at the highest levels and increase flow-rate capacity from nozzles quite dramatically, in some cases doubling the flow.

A minimum 100mm bore fire main is recommended for each 900 square metres of floor plate, providing capacity for a total 1,500 Litres/minute (1.67 L/min/m^2) flow supplying two hose-lines of 750 L/min or three at 500 L/min. Importantly, the maximum hose-lay distances (45 or 60m) determine the number of fire mains needed. However, no consideration is given to occupancy type or fire load density and the same provisions are equally applicable to apartment buildings, as well as open-plan office floors. Therefore, a fire engineered strategy should be used to provide adequate water (L/min) to match the fire loading and fire growth rates for each individual project (refer to chapter 12).

2.2 SMOKE CONTROL TO ASSIST FIREFIGHTING

There are several types of smoke control systems that firefighters will encounter in their work. It is important that they gain an understanding of the working principles and the impact on fire conditions that specific system type may have. A practical knowledge of smoke system design, firefighting override control panels/switches, and fire behaviour is essential if a fire commander is to make sound tactical decisions on whether to override a system or allow it to remain in operation.

Information should be sought from the outset (360 deg. size-up) – *Where are the vent outlets located? Are they automatic? Upon which building face is the wind heading? Where is the fire in relation to the access stair? What side of the building is the fire compartment? Where will firefighters be in the flow path on entering the fire floor, or fire compartment? Will smoke infiltrate the stair if I override?* – Draw a quick plan to identify the issues.

System Categories

- Smoke clearance to justify extended travel distances in residential corridors (LS)
- Smoke clearance to justify a reduction in the number of stair-shafts needed in offices (LS)

- Smoke clearance to justify extended travel distances in large buildings with high ceilings (LS)
- Smoke clearance in shopping centres to enable zoned evacuation (LS) (PP)
- Smoke clearance in warehouses (PP)
- Smoke clearance in office and retail atriums, to allow open space floors into the atrium (LS)
- Smoke clearance from sub surface car parks to justify extended travel distances (LS)
- Smoke clearance from sub surface car parks to assist firefighters (post fire) (PP)
- Smoke clearance from basements to assist firefighters (post fire) (PP)

LS – Life Safety
PP – Property Protection

Smoke Ventilation From Basements

A system of smoke and heat ventilation should be provided from every basement storey, except for any basement storey that has:

a) a floor area of not more than 200 m^2; and
b) a floor not more than 3 m below the adjacent ground level.

Systems may be either natural, using one or more smoke outlets, or powered.

System Types

- Natural (stairs, smoke shafts or external window openings)
- Pressurisation (positive pressure to prevent smoke infiltration)
- De-pressurisation (negative pressure from within the fire compartment)
- De-pressurisation (smoke extract fans) (negative pressure to smoke shafts)
- 'Push-pull' air movements (smoke flushing system) in corridors towards smoke shafts or automatic wall vents (may be configured to work in reverse also)

System Configurations

System configuration is particularly important to firefighters where extended corridors exist. Firefighters tactically try to position their deployment approach to fires with a wind at their back. In extended corridors, many smoke control systems have been configured to extract or 'push' smoke along the corridor towards an outlet vent or smoke shaft. If the outlet vent is located adjacent to the access stair, then any smoke or heat that issues from an 'open-door' fire compartment will be directed straight at them. The SCA design guide 2015 has addressed this firefighter tenability concern (see chapter 19). In some cases, a smoke shaft or vent opening can be located at the corridor end away from the stair and a smoke flushing system as opposed to smoke extract may be able to provide equal, or better, conditions for both occupants and firefighters.

> **What does this mean to the firefighter?**
>
> Smoke extract systems are likely to place firefighters on the wrong side of a flow-path, *unless correctly configured*. It is important to know what system type is provided in any particular building, how it works and if it's manual or automatic in operation. It is equally important to know where the firefighter override controls are located and what impact on fire conditions the system may have under specific circumstances (under-ventilated fire or post-flashover fire).

Possible Impacts on Fire Behaviour

- Pressurisation systems are normally installed to protect stairs, lift-shafts and firefighting lobbies in tall buildings. These systems normally require pressure release outlets within the accommodation. This means that a fire in the accommodation could possibly receive additional air directly into the flow-path, although it is likely to function to firefighting advantage as a moderated PPV air-flow.
- Smoke extract systems are likely to place firefighters on the wrong side of a flow-path, **unless correctly configured**. A correct design configuration would place the smoke shaft or vent opening so that smoke being extracted in the corridor would head away from the firefighting access stair.
- Natural wall vents or smoke shafts may be affected by exterior wind. The opening of these vents may be automatic and should be controlled (overridden) if wind direction is influencing the firefighting approach. Some natural smoke shafts are sited in lobbies to protect the stairs from smoke infiltration but if the fire is intense, they may draw the fire, or heat and smoke, towards the smoke shaft and make access difficult for firefighters.
- Smoke flushing systems are generally designed to create a balanced or neutral pressure to atmosphere in the corridors served. This can be useful for firefighters and generally reduces the effects of harmful flow-paths. These systems may not be able to provide adequate protection to stairs and additional protection such as stair pressurisation or vented stair lobbies may be needed in support. The system design can, in some cases, configure smoke and air flows to enhance firefighting access from the stair, according to where the fire is detected.
- Basements with floor areas greater than 200m^2 at depths below grade of at least 3 metres require some form of ventilation. Such ventilation may be provided naturally using wall mounted stall board lights or floor mounted toughened glass break-out panels, or covers. It may also be provided by mechanical smoke shafts or depressurisation outlet vents. The objectives are to enable firefighters to release heat and smoke and allow cooler air to enter. Research[21] undertaken at Kingston University demonstrated that such venting configurations could lead to backdraft conditions where basement fires have become grossly under-ventilated. This is more likely in basements due to the lack of adequate openings that may enable a post-flashover fire to develop. It was noted that vertical vents were more likely to enhance backdraft and wind driven fire development than horizontal floor-mounted openings. However, in the authors experience this research throws up

---

21 Computer Modelling of Basement Fires; S. Ferraris, J X Wen, B Hume, J Fay and Kirsty Bosley, Fire and Explosion Research Centre, Kingston University, London

nothing new. It has long been known that creating openings into under-ventilated fire environments poses a risk of backdraft, or rapid fire development. In contrast though, the reduction in exposure to such high levels of heat build-up provided by such openings may greatly assist firefighters advancing upon any basement fire. The use of PPV (PPA) to control heat release via such openings, prior to firefighter entry into the basement, is worthy of consideration. However, every opportunity to isolate fire spread from the upper levels by closing doors and siting covering hose-lines should be taken.

- Roof vents or tall atriums in large warehouses, shopping centres, office buildings, retail units or industrial buildings are often used to justify extended travel distances (escape routes) beyond regulatory code requirements. The concept of smoke levels being maintained at a safe height during the evacuation phase is based on the anticipated heated smoke mass from an involved fire load leaving via roof vents of sufficient dimensions. Fire engineers and smoke system designers will use both hand calculations and possibly computer aided analysis to demonstrate this effect. This approach may also be used to assist the firefighting approach, ensuring the smoke levels remain high so that firefighters are able to advance on the fire in low heat and high visibility conditions. Unfortunately, this idealistic design approach is only valid during the very early stages of fire growth and in the average sized building the smoke layer will start to lower quite rapidly once any vertical fire load becomes heavily involved. Whilst atrium smoke control relies on low level fire loads, warehouses may have very high vertical fire loads, almost to the roof. In some buildings a sprinkler system will reduce the fire hazard but with fire service arrival and deployment times (water on fire) into non-sprinklered premises often exceeding 20–30 minutes a heavy smoke layer may be bouncing dangerously, denoting air and heat turbulence at the smoke interface as a massive flow-path develops between the point of entry and the roof vents. It is likely at this stage that the fire has grown beyond the suppressive capacity of a single firefighting hose-line.

2.3 HORIZONTAL FIRE MAINS

The provisions of horizontal (possibly sub-surface) fire mains feeding into the base of dry rising mains is often used as a fire engineering solution for compensating where there exists a practical inability of meeting the prescriptive requirement, placing the fire engine within 18 metres of a dry rising fire main inlet, or 45 metres from a fire engine to the furthest point in a residential building not fitted with a rising main. Such a strategy creates additional distance over which firefighters must transport their heavy firefighting equipment into the fire building by hand (refer to BDAG report[22] on physiological factors). Here are some considerations:

- 'The Secretary of State[23] is of the opinion that all firefighting and other rescue activities are dependent to a greater or lesser extent upon the physiological capabilities of firefighters. Thus the physiological limitations of firefighters must also be taken into consideration. Safety and efficiency are the two major operational concerns of the Fire and Rescue Service and both require judgements to be made

---

22 BDAG Physiological Assessment of Firefighting, Search and Rescue in the Built Environment; Optimal Performance Limited On behalf of the Office of the Deputy Prime Minister: London 2004
23 Building regulations determination SB-007-001-007:2011

about the workload that firefighters can undertake in different circumstances. Some of these variables include: tasks (carrying, dragging, lifting, on the level or up or down stairs); and physical load (equipment, including Respiratory Protective Equipment (RPE) and Personal Protective Equipment (PPE)). The Secretary of State is of the opinion that increased exertion for firefighters prior to carrying out operational duties would have a significant impact on their ability to carry out firefighting and search and rescue activities. This could therefore lead to firefighters not being fully fit to commit to the fire compartment'.

- Some consideration of a time delay for fire service deployment and intervention (from point of arrival) could be estimated for each individual situation, and considering physiological circumstances this could perhaps be multiplied by two for each 50 metres in excess of prescriptive access travel distances. Therefore, a primary deployment involving the 45m rule (to furthest point in a residential building) could take around 3-5 minutes, but with an additional 100 metres to negotiate prior to deployment, an intervention time of 9-15 minutes could result (depending on crew sizes and resources on scene).
- In order to ease the firefighting approach, a firefighting lift might be provided in the building (where not normally required) in addition to a rising fire main to compensate for additional distances to travel caused by a horizontal main. Additional to this, sprinklers could be used as part of a strategy but still **do not mean that excessive travel distances (>100m)** are likely to be approved. Where sprinklers are not installed then the access stair could be upgraded to a full firefighting shaft with firefighting lift.
- Would the over-provision of a dedicated firefighting lift ease the firefighting approach and reduce, or at least equal, a prescriptive intervention time? Would an over-provision of sprinklers ensure a controlled fire and enable an extended intervention time, or might the delay in intervention expose occupants to greater risks? All to be determined on a case-by-case approach.
- Where access travel distances are exceeded beyond the 45m rule, the fire service may be unable to use fast attack methods associated with high-pressure hose-reels or compressed air foam systems.
- All such deviations from prescribed horizontal travel for firefighters should be negotiated during building regulation pre-consultation with the fire service, possibly as part of a qualitative design review (BS-PD 7974 fire engineering) (QDR). However, such extended distances will generally require a mutually agreed engineered solution that is acceptable to the local fire service fire safety department in particular.
- The BDAG research concluded that the physiological load associated with climbing stairs up 28 floors in PPE both with and without BA and hose was investigated. When carrying BA and hose it took approximately 30 seconds and core temperature rose by approximately 0.02°C, per floor. When climbing unloaded it took approximately 15 seconds and core temperature rose by approximately 0.01°C, per floor.
- Climbing 28 floors with BA and hose resulted in fatigue, heat strain and physical exhaustion to the extent that committing firefighters into a fire compartment would be unwise.
- Climbing unloaded was less arduous and subsequent commitment to the fire compartment would appear to be tolerable by the majority of firefighters investigated.

- A predictive model to estimate the combination of maximum vertical and horizontal distances that firefighters could achieve, while remaining within a core temperature limit of 39°C is presented. Assuming 95% confidence in the outcomes, the model suggests that 34m is the maximum distance firefighters should penetrate into a fire compartment to rescue a casualty, where no stair climbing is required to access the point of entry. Having to climb stairs beforehand or undertake other activities reduces the maximum penetration distances proportionally.

2.4 IDENTIFYING RESIDENTIAL CONSTRUCTION TYPES

There are four main residential construction types that firefighters and inspectors should try to identify during familiarisation visits, inspection audits or emergency calls to fire, as each particular type of construction may present different risks when involved in a serious fire. Identifying modern lightweight buildings can be difficult, particularly as they begin to age, due to external cladding materials that are used to maintain a traditional look and feel to a structure that is constructed in a completely different way –

1. Traditional
2. Non-traditional
3. Modern methods of construction (MMC) (Lightweight steel or timber construction)
4. New build heavy engineered timber-frame

Traditional Construction

The majority of houses/bungalows and low-rise flats in the UK are constructed of brick, or brick block wall construction, with sloping (pitched) timber joisted roofs. Solid wall construction was common from the 1800s to 1950. Cavity wall construction was introduced as early as 1900 in some areas but in the main, it predominates from 1935 to the present day. These forms of construction are commonly seen in HMO's (Houses in multiple occupation).

Non-traditional Construction

Non-traditional dwellings built between 1918-1980 mainly consisted of either –

- Metal-framed
- Pre-cast concrete
- In-situ concrete
- Timber-framed (heavy timbers)

Between 1920 and 1944 around 8,000 timber-framed dwellings were built in the UK with a further 100,000 being constructed between 1944 and 1975. However, this form of 'heavier' timber-framed construction lost favour during the 1980's to conventional brick and block buildings.

Modern methods of Construction

Dwellings consisting of light timber-framed materials started to appear in the UK from the mid 1990s, where structural sections were often pre-engineered off-site and connected together on-site.

2.5 BEHAVIOUR OF MATERIALS AND GENERAL STRUCTURAL FORMS IN FIRE

A comprehensive knowledge of the impact an intense fire may have on the materials used to form the structural frame is essential to the firefighter and fire commander. The importance in directing safety officers, with the use of thermal image cameras, to observe any areas of possible structural failure cannot be over emphasised, where firefighters are working within the risk zone. The post-fire cooling phase is often thought of as 'safe' – it isn't! The fire resistance inherent within buildings to protect elements of structure is normally in proportion to their size, height and exposure to surrounding buildings. In very tall buildings we would require and expect far greater fire resistance to prevent building collapse whilst under attack by fire. Some smaller buildings will have much less fire resistance and might even be prone to early collapse, when under attack by fire.

Typical Structural Elements

Beams		
	A beam is an element commonly incorporated in the structure of a building to support (or transfer) an applied horizontal load to other structural elements, usually columns. Beams commonly consist of either concrete, steel or timber. Their upper surface is under compression whilst the lower surface is under tension.	Beams are designed to fail in a ductile manner at ambient temperature and should also do so at elevated temperatures that are consistent with compartment fires. Ductile materials experience extensive plastic deformation and high energy absorption before failure, whereas brittle materials do not. A building with a lateral force resisting system that is considered ductile will experience a large amount yielding during an earthquake, which would allow people to vacate the building before it fails. In recent earthquakes, non-ductile concrete buildings have proven to do the opposite. These buildings have the potential to fail without warning causing catastrophic damage and loss of lives.
		Beam failure will probably result in collapse of the load it is supporting. However, beams may fail in a non-ductile manner. Beam failure is not as critical as column failure as it will commonly be a localised failure and less likely to impact upon the entire building.
	Definitions and further guidance[25]	Recent research has shown that some buildings are susceptible to localised failure in the cooling period after fire and it should not be assumed that a building is safe during that period. Consider the use of thermal image cameras to assess temperatures of beams.

Continued overleaf.

---

24 National Operational Guidance (UK); Fires in the built environment and Fires in buildings under construction or demolition; Knowledge and information Version 1.5 - 2016; Building Research Establishment (BRE) UK

Columns	A column is an element commonly incorporated in the structure of a building to transfer loads in a vertical direction through compression. They generally transfer the load from a floor, beam or roof structure into the ground.	Columns commonly consist of either concrete, masonry, steel, cast iron or timber. By far the most common in non-domestic buildings are steel and concrete Sometimes one of these materials will be encased in another, such as steel within concrete. However, most steel columns are protected in a proprietary material such as boards or an intumescent coating.

All columns can be considered to be load bearing and will invariably play a critical role in the overall structure of the building. Failure of a column is likely to result in one or a number of other structural elements failing, particularly those listed above. Failure of one column may also result in more of the overall load of the building being transferred to other columns, which may lead to sudden or premature failure of other columns. Any loading applied other than vertically may greatly increase the stress on the column which may lead to collapse depending upon the rigidity of its supports. Such loading and collapse may occur as a result of expansion of beams or floors pushing outwards on the column, or may be a result of some other part of the building collapsing and pulling in on the column. A reduction in column cross section or material strength through fire attack can lead to serious structural failure. |
| Connections | Connections between elements may be in the form of steel joist hangers, steel nail (gang) plates connecting truss members, steel truss clips and end plates. Timber connections such as mortise and tenon connections and pin joints are likely to be in use within more historic buildings. | Modern connections do not have a defined period of fire resistance like their relevant structural members. It may be likely, therefore, that connections will behave differently in fire to the structural elements which they connect. The structural elements which they connect may also differ in material. For example steel columns connected to concrete foundations or steel beams connected to masonry or concrete walls. Resin anchors, used in connecting steel structural elements to masonry and concrete are also known to prematurely soften and weaken when exposed to fire. In the case of steel nail plates used to connect factory built timber trusses, these are known to heat up and char the timber causing them to fall out. This may lead to structural collapse of the trussed roof but it is likely there will be few or no signs and symptoms of imminent collapse. Structural members such as beams may try to move due to thermal expansion but rigid connections will not allow significant movement. This can result in the beam or connections failing prematurely during heating or cooling. Failures leading to structural collapse are more likely to occur during cooling. |

Arches	Arches can be formed of blockwork, concrete, stone or steel. Arches work by transmitting loads through the compression of the materials making up the arch.	Issues relating to arches are generally similar to beams and will be dependent upon the nature of the material of which they are constructed. However, whereas the upper surface of a beam is under compression and the underside is under tension, all of the principal components of an arch should be under compression.
Floors	Floors may be constructed from beams or joists acting as the load bearing members with a skin or boarding over the joists and a ceiling (often plaster or plasterboard) below. The specific materials used and construction (in particular the ceiling materials and beams) will impact on the fire resistance of the floor system.	Collapse of a structural floor within a building is likely to lead to partial structural collapse elsewhere in the building, particularly where the floor supports other structural elements which then support additional load above.
Load bearing walls	Load bearing walls transfer loads in a vertical direction. They generally transfer the load from a floor, beam or roof structure into the ground.	Load bearing walls transfer the load they are supporting simply through compression of the material they are made of and are similar in their basic concept to a column. Load bearing walls commonly consist of either concrete, masonry, steel, or timber. Composite products may also be used, such as structural insulating panel systems (SIPS). However, the external finish of a load bearing wall may conceal its actual load bearing structure (e.g. plaster, over-cladding or non-load bearing brickwork). The strength of a load bearing wall is reliant upon the compressive strength of its load bearing component. Any wall may be load bearing or non-load bearing. There is no easy way to identify this. The assumption should be made that all walls you see are load bearing unless it can be established otherwise.

Load bearing walls

Failure of a load bearing wall is likely to result in one or a number of other structural elements failing, particularly those listed above. Failure of part or all of a load bearing column may also result in more of the overall load of the building being transferred to other load bearing walls or parts of wall, which may lead to sudden or premature failure of the remainder of the structure. Failure of a load bearing wall tends to occur through bowing and buckling of the wall. However the stability of a load bearing wall is greatly increased if (and dependent upon) it being restrained top and bottom across the length of the wall (i.e. walls invariably fail to one side or another, not towards their end). Any loading applied across the wall will greatly increase stress which may lead to collapse depending upon the rigidity of its supports and or interaction with other structural elements. Maximum resistance to collapse from lateral loads will therefore occur at the intersection between two walls. Such loading and collapse may occur as a result of expansion of beams or floors pushing outwards on the load bearing wall, or may be a result of some other part of the building collapsing and pulling in on the wall. Thermal bowing (induced by and generally towards the fire) can also occur if the load bearing wall is fixed top and bottom and subjected to elevated temperatures. Load bearing walls may also contain cavities or combustible insulation, through which hidden fire spread can propagate and attack the load bearing materials within the wall unseen. A reduction in material strength through fire attack or collapse as a result of lateral loads can lead to serious structural failure.

Cold-form studs used within load bearing walls present the possibility of immediate failure with no warning often due to the weak nature of the connections used in their construction. Lightweight timber studs used within load bearing walls may fail suddenly if fire is able to propagate within the wall. Good workmanship can negate this problem, but any evidence of poor workmanship or fire spreading unseen (smoke issuing) should lead to further investigation by opening up the wall.

Beams built into masonry walls can promote failure outwards of a load bearing wall due to the thermal expansion of the beam. Collapsing joists which are built into the wall will provide a levering action and potential collapse of the wall. Temperature differences between the fire side of a load bearing wall and the non-fire side can reach up to 500°C. This can cause the wall to bend and potentially collapse.

Lintels

Lintels are a type of beam and generally behave similarly to other types of beam. Their upper surface is under compression and their lower surface is under tension. Lintels are usually constructed of concrete, steel or timber, although they may also be constructed of composites. Because they are found above openings, lintels are often exposed to flames exiting the building through openings. This can lead to the failure of a lintel and a collapse of the load that it supports. However, failure of a lintel is likely to cause localised structural damage rather than a complete structural collapse, and therefore lintel failure is not usually as critical as failure of a beam or column. There is potential for concrete lintels to explode under severe temperatures.

Failure of a lintel may lead to localised collapse of walls above opening or in some circumstances entire walls, e.g. where the lintel is over a large opening.

Trusses

Trusses are a collection of members and connections that work as a system under a combination of tension and compression. As a collection of members and connections that work as a system there is an increased reliance on connections. Roof trusses support the wind loads of the building and support the roof tiles. The image shown is typical of a truss in a leisure or commercial building. These are used where it is necessary to achieve long, clear spans. They are most commonly used to support roofs but are occasionally found supporting internal floors. In modern domestic buildings, trusses may consist of lightweight thin section traditional timber members connected by steel nail plate connections. These members spread the load of the roof to the walls and are connected to the wall plates by steel truss clips. Older buildings may include large timber rafter design.

Larger building trusses may include the use of laminated timber, steel girders or wrought iron. The structural economy of a truss system allows for the minimum amount of material to be used, however the performance of all materials when exposed to fire decreases with size. If one member fails within a fire this can compromise the integrity of the whole truss system. Where trusses are tied in together in order to strengthen against wind loads, the failure of one truss may lead to multiple failures at a distance from the initial failure point. Truss failure may cause non-structural collapse of load bearing walls where the trusses are tied in to the wall.

Masonry

Masonry consists of individual units laid in and bound together by mortar and can be used to construct the structural frame of the building. Masonry is unlikely to be the sole material used in a building structure as other materials are required for certain aspects, e.g. floors. Masonry units are likely to consist of brick, stone, marble, granite, limestone or concrete block. Masonry may be used in the construction of arches, beams and columns, the structural integrity and performance in fire of which may depend greatly on the quality of mortar and workmanship used. Buildings are likely to be clad in mortar due to its aesthetic appearance and may therefore in fact consist of a timber or concrete structural frame beneath. Even where masonry is used as the load bearing material in walls other materials will normally be required for the construction of floors. Older concrete (clinker) can have a high carbon content and can result in concrete 'burning' and lead to structural collapse. (Appearance of un-burnt coal). Mortar becomes friable and blocks can become loose. Later stages of fire, even after extinguished, can lead to collapse. Performance properties of masonry are highly dependent upon workmanship and quality of mortar.

Blockwork is an assembly of discrete pieces or units of natural stone, clay, concrete, red brick, calcium silicate or gypsum which are stabilised and sealed with mortar. The mortar is an inorganic setting mastic, usually containing sand and a hydraulic binder. Blockwork can be used for load bearing walls and non-load bearing walls, but it is not limited to walls. Blockwork can also be used for columns, arches and cladding systems. It is mostly inherently non-combustible and provides good fire resistance. Clay or concrete blocks have similar fire behaviour. Blockwork can undergo thermal bowing which can lead to structural collapse. Breezeblock walls may be combustible due to presence of carbon materials (e.g. coal clinker or reclaimed or recycled materials from fly ash or bottom ash). Look for appearance of unburned coal in blockwork. There is the possibility for transfer of combustion products (e.g. CO) through porous blockwork, which may be hazardous to people in other parts of the building and lead to misidentification of fire location. Mortar may become friable and weaken a wall in later stages of fire or after fire.

Steel

Structural steel may be in the form of

- Hot rolled steel
- Cold rolled steel
- Cellular steel

Each may be used in the production of steel frame structures, either separately or in combination with the other forms of steel or other structural materials. Steel frame buildings are defined by the majority of their load bearing structure comprising of steel. In typical steel frame construction, the load of the floors and cladding is carried at each level by steel beams which in turn transfer the load to the steel columns. The fire resistance of steel is a function of its mass; the thicker the gauge of steel, the longer it will take for its temperature to increase. Steel will rapidly lose strength and stiffness at high temperatures (>550°C) causing deformation and possible collapse therefore it should be assumed that after flashover, unprotected steel is weakening and is susceptible to complete collapse. Steel will also expand with increasing temperature. As an example, a 10m long steel joist will expand by 60mm under a 500°C rise in temperature. The way in which structural steelwork reacts to sustained elevated temperatures created within a fire means that the majority of steel frame construction will require fire protection. This may be solid and achieved by the encasement of concrete, the use of fire resisting boards, the application of intumescent paint or the spraying of mineral fibre etc. This may also be hollow, for example the encasement of steelwork within fire resisting boards. Each means that steelwork is often covered up, hard to see and susceptible to experiencing concealed fire spread.

Timber

Timber used in the structural framework of the building. Types of timber may include historic timber, traditional timber, engineered timber joists or laminated veneer lumber. Timber is a combustible material; however, its properties vary greatly even within a given species based upon grain direction and impurities within the timber such as knots and shakes. Solid structural timber is generally used for beams up to 100mm in width and columns of 150mm square. Laminated timber is generally adopted over greater sizes. Timber is also commonly used in roof construction and can vary greatly in size and quantity based upon roof design. It is important to note that a building may be timber framed even though it does not appear so. The materials used in conjunction with a timber frame building such as masonry, steel and cement will have an effect on the overall structural performance of the timber frame building structure.

Portal frames

A portal or rigid frame is usually constructed of steel but can be constructed of pre-cast concrete or structural timber composites. The latter are usually seen in retail buildings. Most portal frames are single storey buildings although increasingly, modern portal frame buildings will have internal mezzanine floors for offices or storage. The columns and roof members are joined with rigid connections which have the effect of passing the roof loading to the rest of the structure. The structure will consist of little or no internal bracing, whilst the roof would be supported on a series of purlins allowing for large unencumbered storage and working areas. Portal or rigid frame construction is designed for inward collapse within the UK and a basic single storey structure may be expected to collapse within 30 minutes within a fully developed fire. Such buildings are not usually fire protected unless there are other buildings nearby and fire spread must be prevented. In this situation, the walls and columns along the affected sides of the building are protected and are less likely to collapse in the early stages of a fire.

Cast Iron

Cast iron can be used in columns and arches (usually as a column to support a brick arch) and is more commonly found in historic buildings including refurbished buildings. Cast iron is not usually used in beams. Cast iron has a low tensile strength and will lose its strength quickly in fire and this can result in rapid collapse. It is considered that unprotected cast iron columns should not be exposed to temperatures above 300°C if steel beams are rigidly connected to them and should not be exposed to temperatures above 550°C if timber beams are connected to them. Failure is brittle with little or no warning. Sudden cooling from firefighters' water jets can cause cracking due to thermal shock which can result in failure of the structural member. If a building containing cast iron structures has undergone refurbishment which was notifiable under modern building regulations then there would be a requirement for exposed cast iron structures to be fire protected.

Figure 2.4: A typical steel portal frame shed-style building under construction **Photo courtesy of Graham Couchman (SCI)**

Cellular steel

Cellular steel is used as part of the structural frame and typically as beams. There a several types of cellular steel but in general they are formed by welding, together sections of steel or steel plates in certain patterns which results in a lattice or open web type structure. The open structure of cellular steel makes it easy for services to be run through the beams and to achieve long spans. Cellular steel beams are formed by the welding of cold-rolled steel bars to top and bottom plates or the cutting of a beam along a castellated line and then welding the sections back together. These beams most commonly support floors of composite metal deck. There are a number of different failure modes of cellular steel due to the construction of this type of steel, namely the aforementioned welding and/ or cutting. Predicting which failure mode will take place is possible but requires expert knowledge. Rigid connections may cause failure during heating and cooling.

Cold-rolled Steel

Steel formed by the method of cold rolling. This method is generally used to produce lightweight steel, generally no more than around 2-3mm thick. The mechanical properties of cold rolled steel are much the same as hot rolled steel and steel in general however the smaller dimensions commonly used within cold rolled steel production will provide reduced yield strengths and inherent fire resisting characteristics that come as a function of size. This is much in the same way that lightweight engineered timber will compare to large traditional timber. Cold rolled steel will generally be found within stud partitions, within external load bearing walls, as secondary beams within floor systems and cold rolled bars within reinforced concrete. If used to form load bearing members, cold rolled steel is likely to be encased within some form of passive fire protection in order to protect it from the effects of fire. The structural integrity of cold rolled steel within elements of construction is reliant on connections, which are often prone to failure (e.g. rivets or tap-in screws designed to hold at ambient temperature may snap upon expansion of steel members). The lighter nature of cold rolled steel necessarily means that once these are attacked by fire, failure will occur more quickly and potentially suddenly than hot rolled steel.

Hot-rolled steel

The mechanical properties of hot rolled steel have some similarity with cold rolled steel and steel in general. However, the larger dimensions commonly used within hot rolled steel production will provide greater inherent fire resisting characteristics that come as a function of size. This is much in the same way that large timber rafters will compare to light-weight engineered timber. Hot rolled steel is likely to be found within structural beams, columns and joists whilst cold rolled steel is likely to be found within stud partitions, external load bearing wall construction and cold rolled bars used within the reinforcement of concrete. Hot rolled structural steel is usually fire protected in multi-storey buildings using either fire-resisting boards, sprays or intumescent coatings. Steel will lose strength and stiffness at high temperatures (>550°C) causing deformation and possible collapse therefore it should be assumed that after flashover, steel is weakening.

Concrete

Concrete is a material that is formed from cement and Aggregates. Types of concrete include pre-cast, high-strength, pre-stressed and hollow core. Almost always reinforced in some way (usually steel). Concrete is a material comprised of an aggregate and a binding material, commonly cement. It possesses high compressive strength but has low tensile strength as a result of its structure. Concrete undergoes a similar degree of thermal expansion to steel when it is heated. At elevated temperatures, that are consistent with post flashover compartment fires, spalling of the concrete can occur. This effectively reduces the amount of concrete material available to support or transfer a load, which can lead to failure. Concrete possesses a high thermal mass and tends to be used with large cross sectional areas, hence tends to take some time to be significantly heated and affected by fire. Where concrete is used in a building to provide fire resistance to an element of structure this function is almost always provided by the thickness cover of concrete to the internal reinforcing material. When spalling has occurred and the reinforcing material is visible failure of concrete element of structure is considered to be imminent.

Engineered timber

Solid timber joists are being replaced by engineered products such as timber I-beams and metal web joists which are being used as floor joists. These are lightweight and rigid and can span much larger distances without the need for intermediate structural support. Engineered timber is reliant on fire protection measures. It will provide fire resistance for a defined period of time as a complete system under standard fire testing, this differs from traditional timber which has inherent fire resistance relative to the amount of excess or sacrificial timber that is present. Engineered joists usually have a smaller cross section than traditional timber joists. As these joists have a relatively open structure firefighters need to be aware that this creates a continuous floor void unlike traditional timber floor joists. Also this open structure allows for services to be easily passed through the floor voids. Holes can be created in timber I-beams to pass building services through which may weaken the beams however; these holes can be pre-drilled and part of the design. Services are easily passed through steel web beams due to their open structure.

Structural Insulated Panels (SIPS)

SIPS rely on the bond between the foam and the two layers of sheet material to form a load-bearing unit and essentially are used in the same way as timber frame panels or steel frame panels. However, these panels should not be confused with sandwich panels. They are typically found in buildings up to three-storeys high but may also be found in taller buildings. SIPS are reliant on fire protection. Maintaining the structural bond between the various layers in a SIP is a very important factor in fire resistance. Degradation of the exposed SIP face layers has a major impact on overall wall performance in a standard fire resistance test. Other SIP components, such as jointing, also impact on the fire resilience of SIP construction. Sacrificial plasterboard lining is generally installed, similar to other lightweight forms of construction, e.g. timber frame to ensure adequate fire resistance of SIPs. In addition to internal plasterboard lining, SIPS also require external cladding to complete construction and these cladding systems can include brick skin and renders. Some foams such as untreated polystyrene will melt before the panel or protection ignites and as a result the panel may lose its structural integrity. Liberation of fire gases. A breach of the fire protection and the facing panels can allow for undetected fire spread within the panels.

Stone

Stone used as a structural material can be of two main types; natural or cast stone and can be used for blockwork load bearing walls or non-load bearing walls, beams, lintels, columns or arches. Stone is generally considered to be non-combustible and will not contribute fuel to a fire but can undergo failure when affected by fire.

There are numerous types of stone used as structural material and each has individual properties. In general, stone is affected by thermal shock from firefighting activity after exposure to temperatures greater than 550°C but for most stone types this temperature is closer to 700°C. Thermal shock, either by direct cooling by water or extinguishing the fire, can result in delamination or spalling of the stone. Stone containing quartz, e.g. sandstone and granite, will be weakened by fire and may crack or become friable (crumble or reduced to powder) at temperatures greater than 573°C.

Limestone will undergo calcination of calcium carbonate at temperatures above 600°C and this will rapidly increase beyond 800°C. Calcination will reduce the strength of the stone. Cast stone is a special form of simulated stone, defined by UKCSA as any product manufactured with aggregate and cementitious binder intended to resemble and be used in a similar way to natural stone. Cast stone is either homogenous throughout or consists of a facing and backing mix. It is used as an alternative to, and manufactured to resemble, natural stone. Cast stone will behave very similarly to concrete.

Cast stone is predominately used for facings and decoration but can also be used in structural beams or lintels particularly around doors and windows. A very severe fire can result in structural collapse of stone structures.

Natural stone is affected by thermal shock and can delaminate or spall particularly when subjected to firefighting jets. This can lead to large sections of stone falling onto personnel below.

Traditional or lightweight timber-frame

Modern timber frame buildings utilise traditional or lightweight and/or engineered timber along with wood based sheathing board such as oriented strand board (OSB) or chipboard which imparts rigidity to the frame. Timber is combustible and its performance is somewhat reliant on fire protection measures (e.g. plasterboard). Lightweight timber and traditional timbers are usually covered or protected by linings either externally or internally.

Traditional or lightweight timber frames along with the sheathing board interact with air filled cavities and provide a combustible fuel should a fire enter this cavity. This is particularly the case if the cavity has not been adequately compartmented with cavity barriers and the fire is allowed to spread unhindered in the cavity. In some properties traditional or lightweight timber can also interact with composite cladding systems, which if inadequately installed can allow for concealed fire-spread. Buildings under construction present high risk of rapid fire-spread and early collapse.

Timber products, technologies, and methods of construction have evolved over time. The two most popular forms of timber framing can be categorized as:

- Light timber framing; and
- Heavy mass timber framing (Engineered wood products: EWPs).

In addition to timber, lightweight steel framing and roof trusses are now commonly seen in residential and industrial buildings, schools and public buildings. These offer different qualities but again demonstrate an ability to create fast-build schemes. But with 'fast-build' projects come 'fast-burn' fires!

2.6 LIGHTWEIGHT CONSTRUCTION

Taken from a firefighter's perspective the steady transition from heavier more solid construction involving brick or masonry walls and heavy timber or reinforced concrete floors with concrete or heavy timber columns and beams/joists, towards lightweight construction that consists of thin gauge galvanised steel framework and trusses or timber 'stick construction', presents different challenges with an increased exposure to hazard potential.

Lightweight timber Frame

Lightweight timber construction poses increased risk to firefighters because of the early collapse potential for both the roof and floors, particularly over a basement fire. Underwriters Laboratory (UL) tests in the U.S. determined some lightweight unprotected timber truss systems can collapse as early as six and half minutes after flame impingement – and without any warning. The tests conducted by UL involved exposure of components to fire in a large-scale horizontal furnace using the time-temperature curve of the American Society for Testing and Materials (ASTM) E119 test. The study included nine fire tests, two of which focused on roof and ceiling systems while the rest looked at floor joists. Methodologically, the study aimed to provide 'apples to apples' comparisons among assemblies and to show how different construction materials, including traditional timber, fared in different types of fires. In floor tests, two 300-pound (136-kilogram) mannequins simulated a pair of firefighters, in addition to a 40 psi dead load along the perimeter of two edges of the floor. The experiments documented striking differences between traditional and engineered systems. For example, a traditionally constructed floor system, without a drywall ceiling to protect its underside, withstood the test fire for 18 minutes. By comparison, a similar system using engineered wooden I-beams survived for about six minutes.

Light timber frame constructions are typically encapsulated within non-combustible gypsum plasterboard to offer sound insulation, surface finishes and protection from fire. Given the section size of typical 2'× 4' [50 mm × 100 mm] and 2'× 6' [50 mm × 150 mm] stud framing members, the inherent fire resistance of the studs alone is effectively negligible, as the members are small. Hence, unprotected, or exposed, light timber frames provide little structural fire resistance. Fire resistance is achieved by providing protection to the light timber assembly to delay the onset of heating and combustion.

Figure 2.5: A Lightweight timber-fame building under construction in the US – The New Genesis Apartments in Los Angeles, California, designed by Killefer Flammang Architects, is a mixed-use and mixed-income wood-frame building with commercial retail space.
Photo: KC Kim, GB Construction

Lightweight steel modular Construction

Lightweight steel construction is becoming popular in residential dwellings and industrial units. Compared with wood, for example, which has a somewhat predictable rate of burn or char-about 25mm in 45 minutes, thin gauge galvanised steel framework and truss members have been seen to fail without warning. When failure occurs, it appears that all the elements fail at one time. Again, comparing this to lightweight timber-frame collapses, where the loads usually shift and form a lean-to type of collapse. When a light steel structure fails, it acts more like a beer can being pressed in from both ends in a pancake-type collapse. There has always been a risk when using 'tactical time frames' to determine potential collapse. Many US fire department procedures call for a defensive attack after 20 minutes of firefighting that has yielded little or no progress. The '20minute rule' establishes that dimensional lumber used in timber-framing systems starts to become weak after 20 minutes of fire exposure. In addition to time, most collapses give some warning, such as cracking and moaning sounds and shifting loads. This is not the case with the lightweight-steel framing systems. It appears that once the compartment, the room or void space, is heated to the steel's point of failure, all the walls in the area simultaneously fail. There is little or no warning, and time is not a reliable method of determining when interior operations should be halted. The National Fire Protection

Association's *Fire Protection Handbook*[25] reports that open-web lightweight steel joists can collapse after five to 10 minutes of exposure. The predictability of lightweight truss failure is not an exact science; however, lightweight steel truss assemblies will most likely fail under the weight of a firefighter.

Modular construction uses pre-engineered volumetric units that are installed on site as fitted-out and serviced 'building blocks'. The use of modular construction is directly influenced by the client's requirements for speed of construction, quality, added benefits of economy of scale, as well as single point procurement. These benefits may be quantified in a holistic assessment of the costs and value of modular construction in relation to more traditional alternatives. Light steel framing is an integral part of modular construction as it is strong, light weight, durable, accurate, free from long-term movement, and is well proven in a wide range of applications. It is part of an established infrastructure of supply and manufacture and supported by British Standards and various design guides. Modular construction is also widely used in Japan, Scandinavia and the USA, where light steel framing is the primary structural medium, and leads to flexibility in internal planning and robust architectural solutions. There are also important opportunities for modular construction in extensions to existing buildings either by attaching serviced units to the side of buildings, or by roof-top modules. Modular or volumetric construction uses pre-engineered modular units, which are transported from the factory to the site, and are installed as fitted out and serviced 'building blocks'. In Scandinavia for example, modular construction has established a niche market in the renovation of existing concrete panel buildings, where external modular units are used to extend these buildings horizontally and vertically, and to increase their useful life and to create new habitable space economically. Modular construction is driven by the two key imperatives: to build quickly on site, and to improve quality by off-site activities.

The modular units may be room-sized or parts of larger spaces which are combined together to form complete buildings, such as residential buildings and hotels. Light steel framing is an integral part of modular construction, as it is strong, durable, light in weight and dimensionally stable. It is used as the internal framework of the units to which a variety of cladding and finishes may be attached. The framework is sufficiently stiff and robust that it protects the internal finishes against damage during transportation and lifting into place. The sizes of modular units are dictated by the economics of transportation, and units up to 4.2m wide and 12m long can be supplied. Units can also be provided with open sides to create larger internal spaces. All modular buildings are designed to be 'permanent' in terms of compliance with Building Regulations, although they are by definition relocatable and reusable.

Fire protection to light steel elements is typically provided by one or two layers of plasterboard, or similar material; three layers are provided for a fire resistance rating of more than 90 minutes. Insulation may also be provided in the wall system. Fire resistant plasterboards conforming to the requirements of Type F to BSEN 520 are designed to provide specified levels of fire protection up to 120 minutes (3 x 15mm for load-bearing wall sections).

---

25 Fire Protection Handbook 20th Edition; National Fire Protection Association USA, 2008

The Building – what the Firefighter needs to know • 65

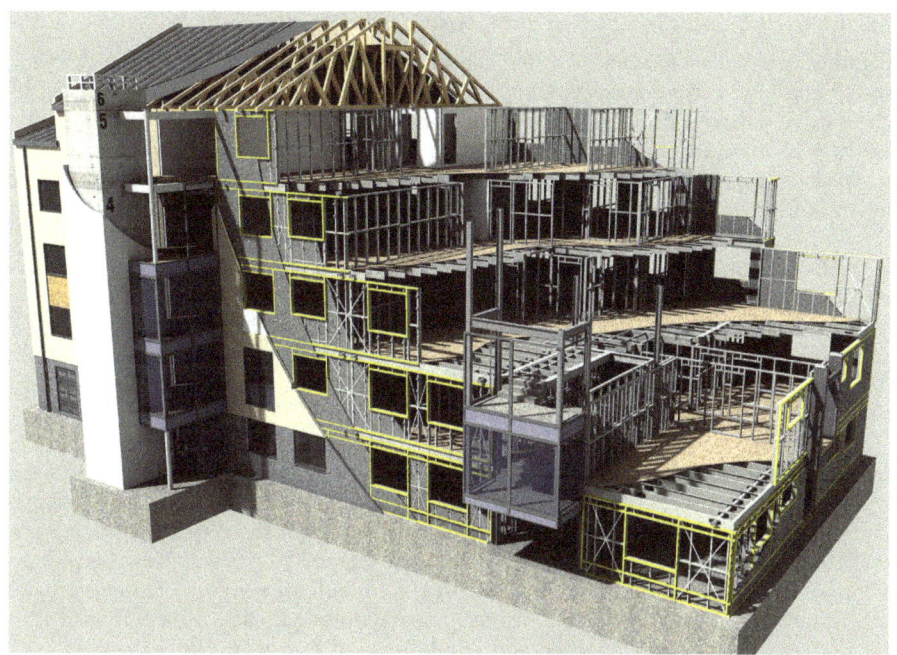

Figure 2.6: Lightweight steel construction is becoming popular in residential dwellings
Image courtesy of Graham Couchman Steel Construction Info (SCI)

Figure 2.7: Lightweight steel construction is becoming popular in residential dwellings
Image courtesy of Graham Couchman Steel Construction Info (SCI)

Firefighting Tactics in lightweight Construction

Fighting fires from the interior, being aggressive with advancing interior fire streams, search and rescue, ventilation, opening voids, and chasing down the fire until we get to clean wood – these are all sound tactics that are still applicable today. The difference for lightweight structures is in how we employ these tactics, how we access the voids, how, where and if we ventilate, the time and ability to enter and search, and the time and ability to attack from the interior. Whilst we must be *aggressive* in our actions to enter buildings to save lives and property, we need to be controlled with our aggressiveness. We need to look at every situation from an *intelligent* and calcul*ating* position before engaging in a course of action.

In his book *Fire-ground Tactics*[26] Emmanuel Fried stated in 1972: 'When the fire has substantially involved more than one floor and is out of control, after 20 minutes of inside operation back all companies out and resort to the use of exterior streams. You may lose the building, but you may save your men'.

The '20-minute rule' became a national rule of thumb guideline for firefighters in the USA for many years. It is still included in many of today's firefighting textbooks. It is the basis of structural firefighting tactics and strategy of many of today's fire departments. It matches the periodic time check firefighters receive from the dispatcher during fire operations, as recommended by the US Incident Command System. Now people are asking if the 20-minute rule is still valid. The '20-minute rule' was a valid and proven concept when it was written, in an era when most rooms and offices were equipped with furniture and furnishings of wood and cellulose-based materials, rather than with the plastics and synthetic fabrics that are in common use today. These modern furnishings contain more than twice the energy than those of wood and natural fibres, and burn faster with a greater rate of heat release. Add to this the transition to lightweight construction methods with earlier collapse potential and the 20-minute rule becomes a dangerous tactical consideration, writes US fire Chief Gregory Havel[27]. Would even a ten-minute rule be valid? Take a look at the life-loss time-lines of firefighters killed in structural collapses occurring in lightweight construction where the structural frames lack mass and have a greater surface area-to-mass ratio than traditional timber construction.

All firefighters should take every opportunity to visit new constructions in their area. Take pictures and record details of every part of the structure as it develops into a completed building. Such *Intel gathering* on buildings will enable a data set that can be recorded onto mobile data systems, providing vital on-scene information when fires occur. Without such data it may be impossible for on-scene firefighters to determine what type of construction is involved in a fire as exterior cladding may give a false impression of actual construction forms.

- Pre-planning
- Visit construction sites in your area
- Take relevant pictures during various stages of construction
- Utilize for training and familiarization sessions across all fire station shift patterns
- Enter details into the primary risk management and mobile data systems

---

26 Fried. E; *Fire-ground Tactics*; H. Marvin Ginn Corp, Chicago 1972
27 Havel. G; *Construction Concerns: The 20-Minute Rule; Fire Engineering online fire dynamics*; 11/2013

The Building – what the Firefighter needs to know • 67

Figure 2.8: Lightweight steel modular construction with bracing **Courtesy of Graham Couchman (SCI)**

2.7 HEAVYWEIGHT ENGINEERED WOOD PRODUCTS (EWPS)

Structural timber composites (STC)

There are numerous types available including glued laminated timber (glulam), laminated veneer lumber (LVL), parallel strand lumber (PSL), laminated strand lumber (LSL) and engineered timber which is explained separately. With the exception of engineered timber these STCs in a general sense are formed using adhesives to hold together smaller sections of timber with slight variations in their process.

Unlike traditional timber or historic timber, STCs are generally designed to the specific fire resistance time required and not much beyond that time. STCs are assumed to behave as historic timber but to a certain extent their behaviour is unknown. There is limited data available particularly with regard to behaviour of the adhesives used, which vary between products. There is little 'real world' experience of the behaviour of STCs in fires.

Heavy timber frame construction, also called heavy frame or heavy timber construction, is characterised by beams and columns with timber section sizes that are greater than 6' × 6' [150 mm × 150 mm]. Engineered wood products (EWPs) offer greater strength and design flexibility and have become increasingly popular as building elements in tall timber construction. Generally, engineered timber/ wood consists of derivative timber products that are manufactured to increase the strength and stiffness of the engineered wood products[28].

---

28 http://www.structuraltimber.co.uk

EWPs including glued laminated timber, finger joints, plywood, stressed skin panels, mechanically and adhesive bonded web beams and nail plated trusses, have been in existence for at least 40 years. Recently, there have been significant developments in the range of EWPs for structural applications with materials such as laminated veneer lumber (LVL), parallel strand lumber (PSL), laminated strand lumber (LSL), prefabricated I-beams, metal web joists and 'massive' or cross-laminated timber (CLT) becoming more widely available. These EWPs are typically manufactured by adhesively laminating together smaller softwood sections or laminates (e.g. glulam and CLT) or veneers or strands of timber (e.g. LVL, LSL and PSL). The varying performance of EWPs is influenced by the size of wood component used in the product. At one end of the spectrum smaller sections of timber are laminated and finger jointed to form sections of glulam, whilst at the other end, reconstituted board products such as oriented strand boards (OSBs) and medium density fibre boards (MDFs) use small wood strands or fibres bonded together.

Platform timber frame building structures comprise of loadbearing wall panels that typically have a distributed arrangement of vertical walls. The term 'platform frame' derives from the method of construction where floor structures bear onto loadbearing wall panels, thereby creating a 'platform' for construction of the next level of wall panels. Platform frame construction is particularly suited to buildings that have a cellular plan form. Internal walls may be used to contribute to this cellular layout and may be used as loadbearing elements for resistance to both vertical and horizontal loads. Vertical actions from walls, floors and roofs are supported by timber wall panels comprised of vertical studs at regular centres (typically 600mm centres or closer) that act as vertical columns. Timber frame constructions can utilise factory assembled wall panels together with floor and roof panels often referred to as 'cassettes'. Where off-site manufacturing of panels and cassettes are used, quality approval (leading to CE marking where appropriate) is required. The off-site assembled panels and cassettes may be made with joists or studs partially or fully clad, with solid panels such as cross laminated timber or composite insulation/timber structurally insulated panels. To achieve its stability, platform timber frame construction relies on the diaphragm action of floor structures to transfer horizontal forces to a distributed arrangement of loadbearing walls. The load bearing walls provide both vertical support and horizontal racking and shear resistance. Due to the presence of open-plan or asymmetric layouts or the occurrence of large openings in loadbearing walls, it may be necessary to provide other means of providing stability to the building. In these situations particular care needs to be taken by the engineer to ensure that the connections between elements are designed adequately, so that the loads can be distributed to the points of stiffness in the structure.

- **Open panels** are timber frame wall panels comprising studs, rails, sheathing on one face and breather membrane.
- **Closed panels** are timber frame wall panels comprising studs, rails and insulation with sheathings and/or linings on the faces of the panel; a vapour barrier is provided on the warm side of the insulation and a breather membrane on the outer face of the panel. Closed panels may also include fitted windows and internal service zone battens.
- **Floor cassettes** are fully assembled groups of joists, rim-boards or rim-joists with structural subdeck fitted to enable lifting as a completed assembly. Treatments to the timbers are often coloured for differentiation. Floor cassettes may also include fitted insulation and lining materials.

- **Cross laminated timber (CLT)** is a solid panel product made by laminating small lengths of timber, usually kiln-dried spruce, with adjacent layers having their grain direction at right angles to one another. These large solid panels can be used to form beams, columns, walls, roofs, floors and even lift shafts and stairs. CLT is a solid panel, capable of resisting comparatively high racking and vertical loads. The design height limit of seven storeys has, in the past, been determined by the structural robustness relating to the vertical movement and racking stiff ness and serviceability design, using working stress designs. The structural benefits of CLT over conventional softwood wall framing and joisted floor constructions, include:
    - Large axial and flexural load-bearing capacity when used as a wall or slab
    - high in-plane shear strength when used as a shear wall
    - Fire resistance characteristics for exposed applications
    - Superior acoustic properties Due to its arrangement as a panel rather than a framed construction comprising discrete loadbearing elements, CLT also distributes concentrated loads as line loads at foundation level.
- **Structural insulated panels (SIPs)** are factory-produced, prefabricated building products that can be used as load bearing or infill wall panels, floor and roof components in platform frame type construction. The benefit of the system is that the structural support and the insulation are incorporated into a single system during manufacture. This results in material efficiency but care is need for concentrated loading on the panels.
- **Thin webbed joists (I joists)** are an EWP manufactured with flanges made from softwood or LVL with glued thin webs generally made from OSB, fibreboard or plywood.
- **Metal web or open web joist** are shallow parallel-chord trusses manufactured using similar techniques to that used for trussed rafters comprising a member with flanges (or chords) usually made from softwood and with metal or timber strutting to form the webs.
- **Plywood** is a flat panel made by bonding together, under pressure, a number of thin layers of veneers (or plies). Plywood was the first type of EWP to be widely available. The structural properties and strength of plywood depend mainly on the number, thickness, species and orientation of the plies. Plywood's are typically used for roof and floor decking applications and for wall sheathing boards.
- **Laminated veneer lumber (LVL)** is a structural member manufactured by bonding together thin vertical softwood veneers with their grain parallel to the longitudinal axis of the section, under heat and pressure. In some cases cross grain veneers are incorporated to improve dimensional stability. LVL is often used for high load applications to resist either flexural or axial loads or a combination of both. It can provide both panels and beam/column elements.
- **Laminated strand lumber (LSL)** is a structural member made by cutting long thin strands approximately 300mm long and 0.8mm–1.3mm thick directly from de-barked logs. The strands are blended, coated with adhesive and oriented so that they are essentially parallel to the longitudinal axis of the section before being reformed by steam-injection pressing into a solid section. LSL is used in similar applications to LVL.
- **Parallel strand lumber (PSL)** is a structural member made by cutting long thin strands (typically 3.2mm thick, 20mm wide and up to 3.0m long) from timber veneers. The strands are oriented so that they are essentially parallel to the

longitudinal axis of the section before being coated with adhesive and fed into a continuous press and microwave-cured. PSL is used in similar applications to LVL. LSL and PSL have recently become less widely available in the UK due to a preference for LVL and supply chain considerations.
- **Glulam** is an engineered wood product, manufactured from layers of parallel timber laminations (normally spruce or pine but occasionally more durable timber species such as larch, Douglas fir or even hardwoods such as oak or sweet chestnut). Pieces of sawn timber are graded for strength, before being glued together under pressure with the grain in the laminates running parallel to the longitudinal axis of the section. Strength-reducing defects such as knots, splits and sloping grain are randomly distributed throughout the component allowing glulam to be designed to higher stresses than solid timber of the same grade. Glulam manufacturers tend to offer a range of standard sizes based on multiples of a lamination thickness for the depth and a fairly small range of widths. Depths in the range 180–630mm and widths in the range 66-200mm are common. Glulam can be manufactured in long lengths (up to 50m is possible). However, the length of a glulam member will usually be governed by transport and handling limitations and in large open spaces, glulam beams may span up to 30m. Typical structural forms include straight and shaped beams supported by glulam posts (post and beam) and single storey roof systems using rigid (portal or arch) frames. In some instances, warehouses and distribution centres with roof areas exceeding 100,000m^2 have been constructed using glulam framing.
- **Oriented strand board (OSB)** is a multilayer board made from strands of wood sliced from small diameter timber logs and bonded together with an exterior grade adhesive, under heat and pressure. OSB is manufactured in various grades with improving resistance to the effects of moisture. The minimum grade that should be used for structural applications is OSB/3. OSB is commonly used as a structural sub deck material for floors and roofs and as a sheathing material for walls. It is also used for composite constructions such as I joist webs and structural insulated panels (SIPs).

Post and beam construction tends to use columns and beams joined together with nominally pinned connections using steel plates and metal fasteners. It is possible to build structures of a number of storeys by providing braced bays for stability and steel connection shoes at each floor level, to provide ease of connectivity and to ensure durability of the timber components. Horizontal actions are resisted by the diagonal bracing members resulting in relatively stiff frames. Large open plan structures can be created by using portal frames and post and beam structures constructed from engineered wood products.

Industry guidance recommends when applying the principles of Eurocode 5 structural design standards and using high strength materials such as cross laminated timber (CLT), it is considered possible to build higher than existing prescriptive guidance of seven storeys, provided particular attention is given to connections and bearing pressures beneath wall panels. According to codes and design standards, fire resistance ratings also increase with building height and it is therefore critical to consider fire resistance of the frame at an early stage in the design approach. Local building codes may require encapsulation of the timber to achieve this.

Product	Application	Common sizes
Sawn timber	Small structural framing, studs and joists, general carcassing, door panels, joinery	Length up to 5.4m Width 25-75mm Depth up to 250mm
Finger jointed soft-wood	Floor and roof joists, ceilings, loadbearing studs, cladding support, prefabricated multi-span 'cassette floors', laminations for glulam members	Length up to 20m Width 38-75mm Depth up to 250mm
Glulam	Large structural elements, beams, columns, trusses, bridges, portal frames, post and beam structures	No theoretical limits to size, length or shape. Common size range 60-250mm wide by 180 – 1000mm deep
'Massive' or cross laminated timber (CLT)	Floor slabs, roofs, beams, columns, load bearing walls, shear walls	Length up to 20m Thickness 50-300mm Width up to 4800mm
Laminated veneer lumber (LVL)	Beams, columns, trusses, portal frames, post and beam structures, structural decking, I-joist flanges, stressed skin panels	Length up to 20m Width 19-200mm Depth 200-2500mm
Laminated strand lumber (LSL)	Beams, columns, post and beam structures	Length up to 20m Width 30-90mm Depth 90-1000mm
Parallel strand lumber (PSL)	Beams, columns, post and beam structures	Length up to 20m Width 45-200mm Depth 200-1000mm
Particleboards	Flooring, ceiling and panel infill	Board materials typically available in 1220 x 2440mm sheets and thickness ranges from 9-25mm
Oriented strand board (OSB) and Plywood	Structural sheathing and decking	Board materials typically available in 1220 x 2440mm sheets and thickness ranges from 9-25mm
Metal web or open-web joists	Floor and roof joists, ceilings, prefabricated 'cassette floors'	Length up to 20m Width 74-147mm Depth 200-400mm
I-joists	Floor and roof joists, formwork, ceilings, loadbearing studs, cladding support, prefabricated 'cassette floors'	Length up to 20m Width 38-97mm Depth 200-500mm
Box beams, thin flange beams e.g. stressed skin panels	Beams, roofs, columns	No theoretical limits to size, length or component sizes

Figure 2.9: Structural Timber Association – Structural Timber Composites (STCs)

When exposed to fire, the outer layer of wood burns and turns to char. This occurs at a temperature of approximately 300°C. This creates a protective charring layer that acts as insulation and delays the onset of heating for the unheated, or cold, layer below. This process of charring allows timber elements to achieve a level of inherent fire resistance. Primary findings from a series of experimental fires demonstrate that fire performance and charring rates of glulam, LVL and SCL are reportedly similar to that of large, solid wood sections. The engineered materials char at a constant rate when exposed to the standard test fire (curve) and form the insulating char layer that protects the unheated timber below. Therefore, in theory once the charring rate and section size are determined, the fire resistance time for exposed timber members can then be calculated. The fire resistance time for timber elements can be increased by providing gypsum board protection at exposed surfaces. Fire testing with LVL beams has shown that 30 min fire resistance can be added for a single layer of 16 mm gypsum board. Application of a double layer of gypsum board indicated at least a 60 min increase in fire resistance time. A successful ASTM E119 fire endurance test in the US of a cross laminated timber wall was highly influential in the decision to recognize CLT in the 2015 edition of the International Building Code (IBC). The wall, a 5-ply CLT specimen approximately 175mm thick, was covered on each side with a single layer of 16mm Type X gypsum wallboard. The wall was loaded to the maximum attainable by the test equipment, although it remained significantly below the full design strength of the CLT. It was then exposed to a standard fire that reached over 980 degrees centigrade in the first 90 minutes of exposure. While only seeking a 2-hour rating as required by building code provisions, the test specimen lasted 3 hours 6 minutes. Temperatures in the ASTM E119 test furnace curve follow the ISO 834 test curve but is slightly less severe, initially increasing rapidly to about 765°C after 20 minutes before beginning to level-off. After about an hour, temperatures are nearly 900°C and around 1000°C after two hours. Unlike a real fire, the temperature in the standard fire continues to increase indefinitely to ensure the specimen fails at some point during the test.

Cross laminated timber (CLT) construction has become a most popular method involving fast-build platform frame engineered products. A report[29] by Barber & Gerard (2015) describes a wide range of researcher's conclusions following experimental tests following the standard fire curve where several engineered wood products and also composite wood/concrete elements demonstrated predictable charring rates. Findings also indicated that flame spread and elevated temperatures were often restricted to the room of fire origin and that protecting the elements with non-combustible gypsum board resulted in minimal damage to CLT structures.

In October 2000, a large gymnasium fire in a glulam structure prompted a series of tests by Waseda University in Tokyo, Japan. The tests involved exposing glulam partition walls to a constant heat exposure from a propane burner to better understand the fire performance of glulam partition walls. A first exposure test resulted in charring of the wall, with no significant combustion occurring on the member. The second exposure, approximately 2.5 times more severe, resulted in full panel burnout, consistent with expected conditions within the gymnasium. Charring rates were recorded for both tests and were shown to be consistent with literature values and estimates for the case study fire.

In respect of connections, research shows that embedded connections such as screws, nails and bolts tend to demonstrate better fire performance than fasteners and plate connections. This is due to the amount of steel area that is exposed to high temperatures,

---

29 Barber and Gerard Fire Science Reviews (2015) 4:5 DOI 10.1186/s40038-015-0009-3; Summary of the [NFPA 2013] fire protection foundation report – fire safety challenges of tall wood buildings.

as steel strength rapidly decreases with an increase in temperature. The greater the steel area exposed to high temperatures, the worse the connection performance in fire. Accordingly, fasteners and plates tend to fail more quickly than nails, plates and bolts, which are generally embedded, and thus protected, by the structural timber elements.

2.8 FIRE DYNAMICS IN EWP MASS TIMBER BUILDINGS

Exposed timber has the potential to impact compartment fire dynamics throughout the entire fire duration. Further analysis is necessary to understand the effect of exposed timber elements on the early fire hazard when occupants are expected to evacuate a structure. Previous testing has shown that structural timber elements can make a contribution to the room fire behaviour and effect the structural fire resistance at later stages of the fire duration. Although charring has been shown to not have a significant effect on compartment fire dynamics, char fall-off has. This can occur where exposed CLT is used for walls and the underside of floors. This is particularly important in unprotected timber buildings where there could be a greater potential for char fall off. Char 'fall-off' occurs where sections of engineered wood products delaminate under fire exposure. When exposed CLT timber elements char and separate from the structure, they have the potential to contribute to the amount of burning material and increase the fire temperature within a compartment. While the impact of timber fall-off is assumed to occur late in the fire duration, it could have a potential impact on any structural fire resistance rating due to impact on compartment conditions and although the extent may be minor for post and beam type construction, it can be relatively significant for exposed CLT panels in panelised construction. This is largely dependent on the amount of protection provided for any timber structural elements. At present, the contribution of CLT to room fires is seen as a key issue that has limited the development of timber buildings in several countries. Previous fire testing has shown that exposed timber has the potential to contribute to the fuel load in compartment fires and can result in compartment conditions with increased rates of burn and greater temperatures. Of significance is work by McGregor, Hadjisophocleous, & Benichou in 2012 where five fire tests generated some very useful data in relation to CLT mass timber EWPs:

Primary conclusions of previous tests conclude (as reported by Barber and Gerard):

- With plasterboard concealing the CLT (encapsulation), there was no contribution of the CLT to the room fire or any influence.
- Where the CLT was unprotected, the CLT panels contributed to the fire load and increased fire growth rates and energy release rates.
- Peak compartment temperatures are not significantly impacted and compartment temperatures are directly related to the slow decay in HRR.
- After the fire was extinguished, charring continued to occur behind the plasterboard.

Results of testing suggest the following conclusions for delamination behaviour:

- When delamination occurred, the fire burned at a high intensity well after the combustible contents in the room were consumed by the fire.
- CLT increased the room energy release by about 160 %.
- The exposed CLT results in prolonged higher compartment temperatures that are higher than the ISO834 curve for the decay period.

When charring advanced to the interface between the CLT layers, the polyurethane (PUR) based adhesive failed resulting in delamination, as would be expected. The delaminated wood then falls into the compartment (mainly from the ceiling) and contributed to the fire load and the exposed uncharred timber increased the intensity of burning and duration of the fire and can produce a second flashover. The exposure of the unburnt CLT results in a faster char rate initially (as seen in all CLT fire tests) of up to 1.67 mm/min at the ceiling, before returning to normal charring rate. Char depths vary within the compartment and where measured, but average out from 0.63 mm/min to 0.66 mm/min, so very consistent with the Eurocode 5 char rate of 0.65 mm/min

The dangers of engineered wood products in building fires are not solely related to increased fire growth rates, but are also due to the smouldering effects of EWPs during decay stages. The fact that the structure itself is combustible means that fires will be more difficult to extinguish. During this process, pyrolysis continues to occur well into the decay stages and fire gases again enter their flammable ranges. This can result in supplementary flashovers or even smoke explosions, as the fire redevelops. This process will be very familiar to firefighters who learn their fire behaviour in steel containers lined with glue laden timber particle boards (now typically OSB). These training environments present a 1.5 MW gas-phase fire, allowing repeated ignitions of the flammable fire gas layers to occur as the wall and ceiling linings continue the pyrolization process. As each ignition occurs there is a flash-fire across the timber fuel linings, and this flaming immediately spreads through the gas layers near the ceiling, allowing firefighters to practice nozzle techniques and ventilation actions to knock-back and control the burning gases. In some cases, as OSB linings fall away from the wall there are momentary spikes seen in the heat release charts caused by high air currents fanning the flames if the OSB panels fail and fall off the walls and ceiling. This may also be seen as CLT panels delaminate in a real fire and this can be dangerous as air pockets are pushed into layers of CO rich fire gases, causing ignitions that may vary in their intensity.

A research project at the University of Canterbury NZ into the events surrounding both 'smoke explosion' and 'backdraft', concluded with the following recommendations:

- More research is required into the *smouldering combustion* that provides the fuel for the both smoke explosion and backdrafts. Of particular interest to is the *smouldering behaviour* of common combustibles most notably timber and polyurethane. Including experiments to quantifying the species production rates as a function of the oxygen concentration.
- The flammability limits of the enclosure gases are crucial to understanding the conditions that can result in both a backdraft and a smoke explosion. An experimental study to develop a flammability diagram similar to the diagrams developed by Zabetakis for pure mixture would greatly increase our understanding of the hazard.
- Additional smoke explosion experiments with detailed measurements of the species concentrations including unburned hydrocarbons in a number of location both high and low in enclosure and pressure to determine the level of structural damage the might be expected from a smoke explosion.

An excellent presentation[30] by Professor Luke Bisby (The University of Edinburgh) and Dr Susan Deeney (Arup) at the WTCE conference (Austria) in 2016 informed us just how

---

30 Bisby. L, Deeney. S; Needs for Total Fire Engineering of Mass Timber Buildings; World Conference of Timber Engineering (WCTE) in Austria, August 2016

wide the knowledge gaps related to the structural fire engineering of mass timber (EWPs) are at this point in time. Even so, there are many global and UK construction projects in planning for high-rise buildings based around mass timber and in fact, several mid-rise buildings are already in existence.

The WTCE presentation informed us that there are clear knowledge gaps in relation to the following:

- Furnace testing used to provide fire resistance ratings for steel and concrete structural elements were never intended for mass timber elements.
- Structural design for fire is different when 'exposed' mass timber is used.
- With exposed mass timber, the fire growth may be faster, as will the time to flashover.
- There is an increased production of volatiles and smoke.
- There is an increased severity of external flaming from windows and openings.
- There is an increased *total* heat release rate.
- The burning duration will be longer.
- Does 'fire resistance' of EWPs, as a concept, make sense?
- Whilst charring rates are typically well understood, there is still a lack knowledge concerning causes and contributory factors associated with EWP delamination.
- There remains a knowledge gap concerning the smouldering of mass timber products where the potential for continued smouldering exists once the compartment contents have burned out (a) within the fire compartment, and (b) within concealed or encapsulated adjacent spaces *(eg. by gypsum)*

These knowledge gaps provide firefighters with critical information. Current designs in construction involving mass timber EWPs may cause dramatic changes to the firefighting environment that may take firefighters by surprise, particularly if they are unaware of the risks and hazards that fire engineers such as Bisby and Deeney are working hard to reduce.

Fire testing in Canada[31] simulated room contents in an unprotected CLT building assembly to evaluate the consequences of fire in exposed CLT buildings. Results indicated that fire in unprotected rooms continued to burn at high intensity even after the combustible contents were consumed. It became necessary for the fire to be extinguished to prevent potential structural damage to the test room. It has been proposed [NFPA] that further testing is necessary to characterise and quantify this impact. This could include long duration burn-out tests to gain a better understanding. A change in fire compartment conditions could potentially impact the structural response for timber buildings exposed to high temperatures for long durations, and needs to be accounted for within the design. Improving the understanding of timber's contribution to the fuel load content and compartment fire dynamics through additional research and testing is important to achieving greater awareness of required compartment fire resistance ratings.

In the USA the National Fire Protection Association (NFPA) are undertaking research into the performance of timber frame buildings under credible fire scenarios to ensure the safety of the occupants to emissions and thermal hazards, as well as the property protection of the building and nearby structures. In 2013 their Phase One report[32] provided task based objectives of the needed research based on existing evidence. This includes a review

---

31 McGregor, C., Hadjisophocleous, G., & Benichou, N. (2012); Contribution of Cross Laminated Timber Panels to Room Fires; Fredericton, Canada
32 Robert Gerard; David Barber and Armin Wolski; Fire Safety Challenges of Tall Wood Buildings (2013) Phase One Report; National Fire Protection Association; 2013

of the safety from fire for occupants, fire fighters and emergency responders alike, as well as safety from structural failure. This research is due to conclude by report (Phase Two) in 2017.

> **What does this mean to the Firefighter?**
>
> In relation to modern construction methods utilising mass timber engineered wood products (EWPs) the firefighter and fire commander must become fully conversant with the changing fire dynamics that are likely to be experienced during serious developing compartment fires. At this point in time, fire researchers are advising that fires in existing EWP buildings are likely to burn faster and demonstrate repeated 'flashovers' (possibly smoke explosions), with hidden areas of smouldering and fire spread. In such cases, fires may become more difficult and dangerous to extinguish, requiring greater resources to handle larger and longer duration fires.
>
> - Faster growth rates
> - Earlier flashover
> - Fires of longer duration
> - Multiple flashovers more likely
> - Smoke explosions possible
> - Hidden fire spread in voids and behind encapsulation (gypsum)
> - Heavier smoke conditions
> - Greater *total* heat release rates
> - Therefore more firefighting water is required on the fire floor
> - There is an increased severity of external flaming from windows and openings
> - External fire spread and larger quantities of smoke production may require additional resources to be deployed above the fire floor than in traditional construction
>
> **The fire dynamics in this type of construction is likely to be very different to the fire development that firefighters are familiar with in traditional construction (steel and concrete buildings).**

The NFPA stated (footnote 32) that research and testing is needed to evaluate the contribution of massive timber elements to room/compartment fires with the types of structural systems that are expected to be found in tall buildings (e.g. CLT, etc.). Previous research has shown that timber elements contribute to the fuel load in buildings and can increase the initial fire growth rate. This has the potential to overwhelm fire protection systems, which may result in more severe conditions for occupants, fire fighters, property and neighbouring property. The contribution of timber elements to compartment fires needs to be quantified and compared against other buildings systems to assess the relative performance. The contribution of exposed timber to room fires should be quantified for the full fire duration using metrics such as charring rate, visibility, temperature and toxicity. This will allow a designer to quantify the contribution, validate design equations and develop a fire protection strategy to mitigate the level of risk to occupants, fire fighters, property and neighbouring property. In addition, the effect of encapsulating the timber as means of preventing or delaying involvement in the fire (e.g. gypsum, thermal barrier) needs to be characterised. In principal, the section of timber in the heated zone beyond the char layer is known as the pyrolysis zone, and corresponds to temperatures between approximately 200°C and 300°C. Within this zone, timber is assumed to undergo thermal decomposition and pyrolysis.

Figure 2.9: Typical timber pyrolysis and char zones[33]

2.9 RECENT RESEARCH INTO THE FIRE PERFORMANCE OF CLT TIMBER WALLS AND FLOORS

Experiments[34] at Empa, the Swiss Laboratories for Material Testing and Research in Duebendorf, Switzerland, were conducted to investigate the charring and fire behaviour of cross-laminated timber floor panels. The main focus of the research was to determine whether the fire behaviour of CLT is similar to that of homogeneous timber panels and how it compares to the charring calculation method stipulated in EN 1995-1-2 [14]. This involved looking at whether a charred ply layer would remain in place, similar to solid timber, or fall-off. Tests were conducted in a small-scale furnace (1.0x0.8 m) on 11 specimens consisting of 2, 3, and 5 layers. Layers were glued together using six different commonly used adhesives; five of which were one-component polyurethane (PU) and one melamine urea formaldehyde (MUF). Layers thicknesses were kept small (10, 20, and 30 mm) as opposed to traditional

---

33 Reprinted with permission of Structure Magazine, November 2013
34 A. Frangi, M. Fontana, E. Hugi and R. Jobstl, Experimental Analysis of Cross- Laminated Timber Panels in Fire, Fire Safety Journal, vol. 44, pp. 1078–1087, 2009.

layer thickness of around 38 mm, in order to reduce the amount of time that char would take to reach a bond line.

The panels were all exposed to the standard ISO 834 furnace fire and temperatures were measured between each layer at one minute intervals. Test results revealed ply fall-off occurred for all panels constructed with PU adhesives when the temperature between layers reached around 300°C. Panels constructed with MUF adhesives did not fall-off. After a layer of ply fell-off, a spike in the charring rate was observed. These increases in charring rate were not constant; the later the ply fell off during the test, the higher the increase in the charring rate. This was due to the increasing fire temperature.

Overall charring rates calculated throughout each test increased as char progressed through all panels constructed with PU adhesives. The opposite was observed in panels constructed with MUF adhesives and demonstrated the lowest overall charring rates at around 0.60 mm/min at the end of these tests. Panels with PU adhesives had higher overall charring rates, and panels with 10 mm plies had the highest overall charring rates at around 1.0 mm/min at the end of the test. This was reasoned to be due to the more frequent layer fall-off. The char depth calculation used by EN 1995-1-2 for initially protected surfaces was compared to these results and determined to provide a conservative estimate of charring in CLT. In the Eurocode, after a ply layer falls off, the standard charring rate is doubled (e.g. from 0.65 mm/min to 1.3 mm/min), up to a char depth of 25 mm before it is brought back to the standard rate. Only one test involved a ply thickness greater than 25 mm and the charring rate did reduce slightly in that test, however this reduction in the charring rate could not be completely verified since charring only continued for 5 mm.

Current interest in tall wood buildings has led to two major feasibility studies for tall buildings. Vancouver architect Michael Green (2012) has produced possible designs for 10, 20 and 30 storey timber buildings[35]. The report covers many important aspects of fire safety, but falls short of a clear strategy to meet all the Canadian Code requirements, especially for very tall buildings. The main thrust for fire safety design is to design in such a way that the timber building can be equivalent to non-combustible construction; that is to achieve 'an equal level of performance to that outlined in the acceptable solutions to the Building Code'. This is to be achieved with reliance on sprinkler systems, together with the predictable charring rate of heavy timber, and encapsulation where necessary. The report does not suggest design for complete burnout of a fire compartment. It covers the possibility of sprinkler failure by providing a 2-hour fire resistance rating to critical structural elements. In extreme events it is expected that 'fire department resources would be dispatched and able to suppress the fire condition before the 2-hour fire duration is achieved.' It does not adequately cover the case of a post-earthquake fire where the firefighting services would be unavailable, other than saying that more research is needed on built-in fire protection systems and their reliability in post-earthquake fire scenarios.

Skidmore Owings and Merrill (SOM) (2013) have produced a feasibility study for a 42 storey timber building in Chicago, *'The Timber Tower Research Project*[36]*' based on an existing reinforced concrete tower of the same size.* The goal of the Timber Tower Research Project was to develop a structural system for tall buildings that uses mass timber as the main structural material and minimizes the embodied carbon footprint of the building. The research was applied to a prototypical building based on an existing concrete benchmark for comparison. The concrete benchmark building is the Dewitt-Chestnut

---

35 'The Case for Tall Wood Buildings – How Mass Timber Offers a Safe, Economical, and Environmentally Friendly Alternative for Tall Building Structures'
36 http://www.som.com/ideas/research/timber_tower_research_project

Apartments, a 395-foot-tall, 42-story building in Chicago designed by SOM and built in 1965. SOM's solution to the tall wooden building problem is the Concrete Jointed Timber Frame. This system relies primarily on mass timber for the main structural elements, with supplementary reinforced concrete at the connecting joints. This system plays to the strengths of both materials.

SOM believes that the proposed system is technically feasible from the standpoint of structural engineering, architecture, interior layouts, and building services. Additional research and physical testing is necessary to verify the performance of the structural system. SOM has also developed the system with consideration to constructability, cost, and fire protection. Expert reviews and physical testing related to fire-safety are also required before this system can be fully implemented in the market. If they haven't already, these are the construction types coming your way.

Fire Resistance

There are four design objectives that are considered when determining the level of fire resistance, depending on the size and importance of the building:

- Time for occupants to escape from the building
- Time for firefighters to carry out rescue activities
- Time for firefighters to control the fire
- A complete 'burnout' of the fire compartment with no firefighter intervention

The most common way of designing for *burnout* is to use a time-equivalent formula to estimate the equivalent fire severity (exposure to a standard fire) for the complete process of an uncontrolled fire from ignition through fire growth, flashover, steady state burning period and decay to final extinguishment. Such time-equivalent formulae assume that the fire severity is a function of the fire load, the available ventilation, and the thermal properties of the surrounding materials of the fire compartment.

This has some problems, especially if we do not know the worst-case scenario for fuel load and ventilation. These values should be determined on a probabilistic basis, with higher safety factors for increasingly tall buildings. More research is required to assess the applicability of current time-equivalent formulae for use in multi-storey timber buildings. The fire severity, hence the time-equivalent formula, will depend on whether the wood structure has no protection, limited encapsulation or complete encapsulation.

18-Storey EWP high-rise in Canada 2016

Acton Ostry (Architects) and Fast + Epp (structural engineers) have designed an 18-storey mass timber residential high-rise, where construction was completed in Vancouver, Canada in 2016. Located on a large forested peninsula on the west side of Vancouver, the University of British Columbia is at the forefront of the global movement to revitalize mass-timber construction and be innovative in the use of engineered wood products in tall buildings. Among the large wood buildings already on campus is the Centre for Interactive Research on Sustainability, the Earth Sciences Building, and the Bioenergy Research and Demonstration Facility. The newest addition to the portfolio is the 53-m-high (18-storey) Brock Commons Phase 1 Building, featuring the first North American use of mass-timber products in a residential high-rise. Brock Commons is one of the University's five high-rise, mixed-use, residential complexes that provide housing for students while acting as academic and recreational hubs for the campus community.

The building structure is a hybrid configuration, with a dual approach to handling gravity load and lateral loads. The gravity load system for floors 2 through 18 consists of mass-timber floor plates and columns carried by point loads at the column connections. The mass-timber structure is supported by the second-floor concrete transfer slab, the first-floor concrete columns, and the concrete foundations. The lateral load system is comprised of the floor panels and the concrete cores, which transfer the loads directly to the foundation. The CLT floor panels are joined together by a plywood spline screwed and nailed to each panel, creating a single diaphragm at each floor to resist lateral forces. The lateral loads are first transferred from the floors to the concrete cores through steel drag-strap connections located at the core edges and along the building perimeter, and then from the cores to the raft slabs in the concrete foundation. The use of concrete cores enabled the code approval process to proceed quickly because concrete cores are a typical feature in standard high-rise buildings. Mass-timber cores would have been much more challenging in terms of design and in demonstrating the levels of performance required for a site-specific regulation. The use of a hybrid mass-timber structure also results in a significantly lighter building than a comparably sized concrete structure. The lower mass results in less inertia and therefore lower resistance to overturning during a seismic event. The concrete foundation and ground floor provide a counterweight to resist overturning forces.

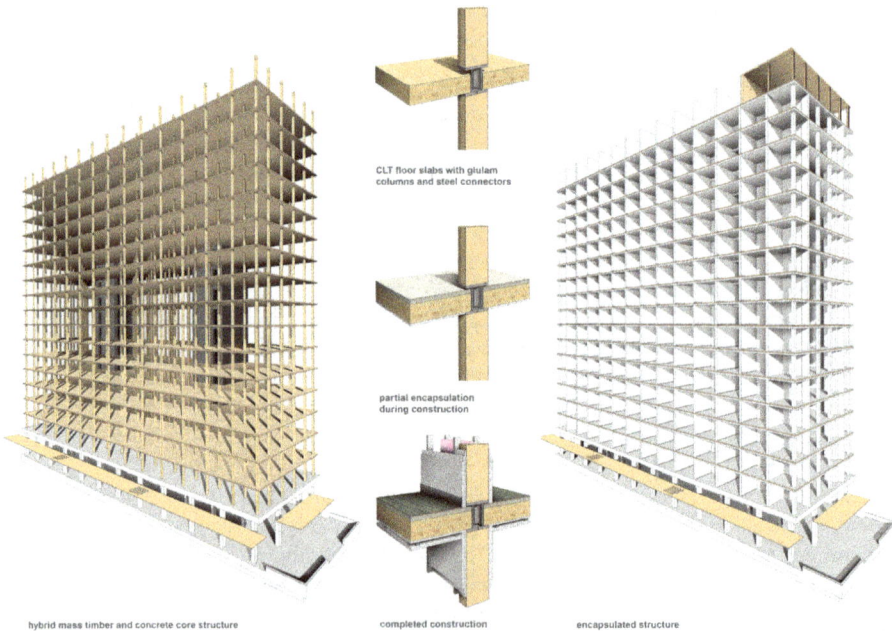

Figure 2.10

The Building – what the Firefighter needs to know • 81

Figure 2.11

Figure 2.12

Figure 2.13

Figure 2.14

Figures 2.10 – 2.14: The 53-m-high (18-storey) Brock Commons Phase 1 Building, featuring the first North American use of mass-timber (EWP) products in a residential high-rise.
Images courtesy of Russell Acton – Acton Ostry Architects Canada

The composition, sizing, and spacing of individual mass-timber components were designed to handle the anticipated loads – including gravity loads, shear loads, and lateral loads due to wind or seismic events – as well as vibrations and deflection. The approaches were validated through the performance analysis in the peer-review process.

A typical structural bay is 4x2.85m. The CLT panels come in four sizes to allow for overlap of the bays and staggering of the panel layouts between floors: 2.85x6m (spans 1.5 bays), 2.85x8m (spans 2 bays), 2.85x10m (spans 2.5 bays), and 2.85x12m (spans 3 bays). The panels are 169mm thick, with outer layers of machine-stressed timber and inner layers of SPF timber and CSA-standard wood adhesive. The characteristics of the three types of columns within the building vary depending on the location and loads: 265x215mm GLT columns on floors 10 and above, 265x265mm GLT columns on floors 2 to 9, and 265x265mm PSL columns in high-stress positions in the centre of the floor plates on floors 2 to 5.

The fire-protection strategy for Brock Commons is three-fold. First, there is no significant exposure of the wood structure. Second, the building is fully sprinklered. And third, a 30-minute, on-site, backup water supply is provided. There are two stairs and lifts serving the corridors, located in the concrete cores. The wood structural elements, as well as most of the steel connections, are encapsulated with multiple layers of Type X gypsum board to achieve a 2-hour fire resistance rating. The one exception is the exposed wood structure in the eighteenth-floor lounge, which is fully sprinklered. Additional fire-separation measures are provided by the interior wall and floor assemblies, which are designed to achieve specific levels of fire separation through layers of Type X gypsum board and concrete toppings (for floors): a 2-hour fire resistance rating for floor assemblies, suite-to-suite demising walls, and vertical shafts; and a 1-hour fire resistance rating for suite-to-corridor walls. The exit stairs are located within the concrete cores and are therefore of non-combustible construction. Similarly, the mechanical and electrical service spaces are housed on the non-combustible concrete ground floor.

The active fire-protection system includes an automatic sprinkler system (NFPA 13:2013), a standpipe system, and an external water curtain. The systems are connected to the municipal water supply and are backed by an on-site water reservoir and pump connected to emergency power. Fire extinguishers are also provided on each floor, which is standard practice. The automatic sprinkler system serves the interior of the building. Sprinkler heads in the residential units are recessed to mitigate the possibility of being accidentally hit and damaged or set off. Non-freezing sprinkler heads are used in exterior, unheated areas, including the space below the exterior CLT canopy on the ground. The sprinkler systems are electrically supervised and monitored by the fire department.

Expansion joints will be installed where the sprinkler risers exit the concrete cores, to ensure the system remains operable in case of building movement. The standpipe is a standard system within North American high-rise buildings. It is composed of pressurised pipes housed within the stair core, with special connections on each floor at which fire departments can attach their hoses. The water curtain is used on the ground-floor exterior glazed curtain wall, in areas that are in close proximity to the adjacent (six metres) parking area. A 20,000-litre tank, with a dedicated fire pump, is located on site as a backup water supply for the fire-protection systems. The capacity represents approximately 30 minutes of water supply for the entire sprinkler system, and increases the reliability of the automatic sprinkler systems to almost 100%.

2.10 ENCAPSULATION OF MASS TIMBER WITH GYPSUM – HOW EFFECTIVE?

Is encapsulation with gypsum board as effective as most fire engineers and architects believe? It is certain that most firefighters can attest to attending compartment fires where the gypsum has completely failed at an earlier stage than expected and allowed fire to spread into structural cavities. Structural failure of gypsum boards in fire is primarily due to the shrinkage of gypsum at high temperatures. Fall-off of gypsum in fire significantly affects the fire resistance of gypsum board assemblies. Gypsum fall-off is a result of structural failure and a sufficient number of discrete through-thickness cracks forming in gypsum. At present, it is still not possible to predict the time of gypsum fall-off. The standard fire curves suggested by codes, such as BS 476 (BSI, 1987), ISO 834 (ISO, 1992) or ASTM E119 (ASTM, 1988), are means which enable the comparison and categorisation of different construction systems; however, they are not reliable for predicting the real performance of gypsum board assemblies. Therefore, in line with the general tendency towards performance-based design in fire engineering practice, some researchers have addressed the performance of gypsum board systems exposed to real fires.

Dr. Ima Rahmanian reports[37] on research (2011) into how gypsum encapsulation might perform in real fires. 'Jones (2001) conducted full-scale compartment testing and confirmed that temperatures within a compartment can go far beyond those of standard curves, when subjected to typical residential fire scenarios. He employed SAFIR, a finite element programme, to predict the thermal behaviour of different gypsum board assemblies exposed to a range of non-standard fires. The results were verified by several full and pilot scale fire tests. His study showed that the temperature predictions by finite element analysis for moderate fires were in good agreement with the temperature development within the specimens; however temperature results for severe fires were much different from those of the tested assemblies.

Frangi et al. (2008) also tested a gypsum board assembly with a non-standard fire exposure more severe than the standard fire, and showed that the thermal behaviour of the gypsum board was affected considerably by the fire exposure. In order to achieve conformable results, they calibrated the thermal conductivity of gypsum used in finite element thermal analysis'.

Another effect rarely considered in experimental fire tests or test furnaces is that of exterior wind, fanning local flaming within a fire compartment. Where this occurs, as windows fail, it is just as if a blow-torch has been applied directly onto some elements of structure, causing excessive local damage near windows and other internal openings as flame velocity increases.

Fire resistance

The light-weight framing in wall assemblies is often clad by one or two layers of gypsum board on either side. There is no doubt that installing a double-layer of gypsum board on either side of a wall assembly enhances its insulation capacity in fire condition, as long as the layers stay in place. However, it is important to study the effect of an extra layer of gypsum in structural failure time of the exposed board. Having two layers of gypsum changes the temperature profile of the exposed board. According to normal insulation requirements of building construction, the cavity in a wall assembly is either left empty

---

37 Rahmanian. I; Thermal and mechanical properties of gypsum boards and their influences on fire resistance of gypsum based systems; PhD Thesis; University of Manchester 2011

or filled with insulation material. During fire, the effect of insulation is observed as a significant increase in the temperature of the unexposed face of the exposed gypsum board. The results of numerical simulation of the structural performance of the insulated board confirm an earlier failure time compared to the non-insulated gypsum. Through-thickness cracks appear after 26 minutes of fire exposure, which suggests an approximate 40% decline in the failure time of gypsum. Although higher temperatures on the back of the board reduce the thermal gradients, the sharp temperature increase reduces the mechanical strength of gypsum so rapidly that the loss of strength dominates the performance of gypsum, and hastens gypsum failure. Interestingly, an experimental study by Elewini et al. (2007) suggests that gypsum board fall-off times in floor assemblies also reduce noticeably when insulations are used in the cavity, especially when glass fibre batts were used for insulation.

There is no doubt that installing a double-layer of gypsum board on either side of a wall assembly enhances its insulation capacity under fire conditions, as long as the layers stay in place. However, having two layers of gypsum changes the temperature profile of the exposed board. The simulation of complete cracks in 'Fireline' gypsum boards coincides with the observation of a sudden increase in temperature (more than 100°C per minute) on the backside of the exposed board. As a result of the sudden increase, the temperature on the unexposed side was seen in tests to reach the temperature of the exposed side of the board. The temperature increase of the unexposed face of the board after dehydration has a similar gradient in both cases until about 40 minutes into the test. However, the temperature of the unexposed face of the exposed board in the double-layer assembly continues to rise higher than 600°C, while that of the single-layer assembly does not reach 400°C after 100 minutes of fire exposure. The higher temperature on the unexposed side in the double-layer assembly has two contradicting effects on the structural performance of the board. On the one hand, it causes a higher loss of strength in the gypsum; on the other hand, it increases shrinkage on the unexposed face, and consequently, decreases the thermal strain curvature along the board. The numerical results show that the through-thickness crack in the double-layer assembly forms after 55 minutes of fire exposure, which is a 20% improvement compared to the single-layer assembly. It appears that the favourable effect of a lower thermal gradient counteracts the negative influence of the loss in strength. An experimental study of gypsum board fall-off times in floor assemblies by Elewini et al. (2007) also yields similar results.

Type X gypsum board (Fire-resisting)

The ASTM E119 standard (USA) does not contain specific details for construction of the test furnace. Since test furnaces are subject to variation due to individual characteristics of construction and design, including ventilation, atmospheric conditions, and general thermal tendencies, test results are typically not fully repeatable or reproducible from one laboratory to another. Test results attained in an E119 test are not therefore precise predictors of future performance. Additionally, differences in assembly/system components and construction methods, the design and control features of individual furnaces, and other variables regarding the testing regimen, can cause wide fluctuations in ASTM E119 test results. A fire test, therefore, is a snapshot of a single assembly/system at a given time that includes the measurement of the performance of a specific assembly/system, composed of specific materials, constructed in a specific test furnace, on a specific day. This simply means that for a 'one-hour fire rating' of a gypsum board assembly/system, all requirements of an ASTM E119 test were successfully met in a testing laboratory furnace for at least 59 minutes and 30 seconds for that specific assembly/system and with those specific components of the assembly/system.

The ASTM E119 test method does not incorporate all dynamics essential for fire hazard analysis or fire risk assessment of the assemblies/systems under conditions in an actual fire situation. The results of an ASTM E119 test, therefore, should be regarded as one component among a variety of factors used to assess the potential of a system to perform as part of a structure.

The ASTM E119 Standard[38]

---

38 https://www.gypsum.org/wp-content/uploads/2011/11/Fire_Safety_Information.pdf

Chapter 3

Practical Fire Dynamics & extreme Fire Behaviour

'If we teach today's students as we taught yesterday's, we rob them of tomorrow'
John Dewey

Why would a firefighter need to understand *fire growth* rates? Are *fire spread* rates the same as growth rates? What about *heat release* rates, heat profiles and vent profiles, where do they fit into a firefighter's need for knowledge and understanding? The changing dynamics of fires involving higher plastic fire loads in energy efficient heavily insulated buildings is causing us to look more closely at how we deploy and engage, in order to coordinate our firefighting tactics most safely and effectively. The growth and behaviour of a compartment fire depends on several factors such as fire load, heat release rate, boundary conditions and ventilation profiles. Such characteristics may vary widely from the type of fire behaviour encountered in training facilities to that encountered at operational incidents (real fires). The ventilation profile itself may be further affected by exterior wind speed and direction as well as dimensions and locations of inlets and outlets (mostly doors and windows). It is important therefore for firefighters to grasp a wider understanding of practical fire dynamics at 'real fires'. This chapter describes these factors to enhance the existing level of awareness. It does not intend to represent a definitive text on fire dynamics but attempts to impart some basic knowledge so that key factors can be taken onto the fire-ground as well as featuring in tactical planning for local risk profiles (occupancy types and size of buildings or compartments). *It should be pointed out that key elements of fire dynamics will appear throughout this book and are not entirely reserved for this chapter alone.*

A changing environment for the fire service
- Reduced times to flashover
- Shorter time available for size up and dynamic risk assessment.
- Compartment fires burning faster and hotter.
- More rapid fire growth leading to faster fire spread rates.
- Structural stability compromised at an earlier stage in operations.
- The need for 'adequate' firefighting water.
- More intense wind-driven fires as window dimensions are increasing.
- The need for fire-ground procedures and firefighter training to be updated
- Less experience in fighting fires due to fewer fires.
- Staffing reductions in selected jurisdictions, despite increasing risks.
- New firefighter gear/tools with varying performance levels.

- Firefighter Personal Protective Equipment improvements increasing other personnel risks.
- Exposure to carcinogens from contents and construction materials.
- Reaching all the fire service with training information related to new hazards.

Does this sound familiar? This represents some of the findings of a governmental research workshop[39] in the USA where firefighters, scientists, fire engineers and 28 other organisations came together to debate some of the major factors influencing more rapid fire growth in residential fires faced by lesser experienced firefighters with, in some cases, reductions in staffing. These specific topics are now being discussed globally by fire services and researchers, with issues surrounding firefighting tactics, equipment technology, structural stability under fire attack and post-fire hazard controls being central to the debate. To create wider situational awareness amongst firefighters and fire engineers concerning the much faster fire growth rates now being experienced due to an increasing plastic content in the movable fire load and the added increase in heat release once the structure itself becomes part of the fuel load.

Figure 3.1: How we can use basic Fire Dynamics information to assist us as firefighters

3.1 FUEL-CONTROLLED FIRE

Compartment or room fires are defined as either *fuel-controlled* or *ventilation-controlled*. In the growth period or the pre-flashover stage of a compartment fire there is generally sufficient oxygen available for combustion and the fire growth is entirely dependent on the flammability and configuration of the fuel. During this stage, the fire is defined as fuel-controlled.

---

39 Changing Severity of Home Fires Workshop Report; US Fire Administration/National Fire data Center (FEMA); 2012

3.2 VENTILATION-CONTROLLED FIRE

A room heading close to, or having surpassed, flashover can either burn in a steady state phase or it may start to decay. There are two factors that determine the direction of the fire development: a lack of fuel will impede development; or the fire will become ventilation-controlled if there is enough fuel but the fire grows to a size dictated by the inflow of fresh air. Where the size of ventilation openings are not large enough to allow the required amount of air/oxygen to enter and feed the fire, the fire is now in a ventilation controlled state.

3.3 UNDER-VENTILATED FIRE

An under-ventilated fire may be referred to as an enclosed compartment fire that has ceased to continue along a growth phase as most of the air/oxygen has been used up (<10% oxygen in air). This situation may actually burn itself out, but more likely is existing in a smouldering condition waiting for a sudden inflow of air if a window fails or a door is opened. Some compartment fires can be so grossly under-ventilated that they may precede backdraft should an uncontrolled opening be made.

3.4 FIRE LOAD

The term 'fire load' refers to the amount of potential energy (heat) contained in a building's combustible contents ('movable' fire load) such as furniture and other contents, or combustible parts of the structure itself ('fixed' fire load) such as floors, walls, ceilings and other elements of structure. When this heat energy is released in a compartment fire it is measured in Mega joules (MJ), and the quantity of the 'movable' fire load potential in a building is referred to as the fire load density (MJ/m² of floor area). Typical 'movable' fire load densities for various buildings are listed in Table 3.1.

Table 3.1 – Typical fire loads in buildings of various occupancies	
Occupancy Type	**Typical Fire Load**
Container Training Facility	135–363 MJ/m²
Houses, flats and HMOs	500–800 MJ/m²
Industrial or Storage Facilities	1000–10000 MJ/m²
Town Centre Shops	1000–10000 MJ/m²
All other buildings	>800 MJ/m²

Fire loads in sports shops can be as high as 2,500 MJ/m², whilst traditional bookshops or libraries may contain a much higher fire load of around 10,000 MJ/m² which creates a fire that can be very difficult to extinguish. Such fires require the application of several jets or a portable attack monitor and high volume water (or CAFS) flow-rates within a few minutes of arrival on-scene. The ability of a fire load to release its potential energy as heat in a fire depends mainly on surface area, density in the load and the provision of an adequate amount of air/oxygen (ventilation). As an example, a large wall or ceiling area that has a combustible covering or finish is more likely to release heat quickly than a solid block of wood when involved in a fire. Also, a fire load is more likely to release its heat energy far more quickly where ventilation openings are large (windows, doors and other openings). Therefore, by making ventilation openings near to or below the fire, the added airflow will quickly increase a fire's intensity, despite any temporary and brief reduction in temperature that may occur when hot gases exit the compartment.

3.5 THE HEAT PROFILE

The relevance of a *heat profile* to firefighters will inform the likely impact on firefighter tenability in training, or in live building fire scenarios. It will also influence structural stability and safe firefighter occupation times. The point that fires may burn 'hotter' in some fire compartments than others is related to several key factors and firefighters should obtain at least a basic understanding of these:

- Fire load according to occupancy type (expected)
- Ventilation potential (number and size of openings)
- The age of the building (modern buildings may be well insulated)
- The height of the ceiling/roof from floor level
- The configuration of the fire load (high vertical stacks, or horizontal loads with low ceilings)
- The expected time to flashover (fire growth-rate) per occupancy/compartment
- The likely temperatures that may be reached at various heights from floor to ceiling
- The likely heat flux that may be radiated down, or across, to firefighter locations.
- The power, or intensity, of the fire when releasing energy at maximum *heat release* will dictate the quantity of water required to suppress, control or extinguish the fire.

3.6 HEAT, TEMPERATURE AND HEAT FLUX

Often the concepts of heat and temperature are thought to be the same, but they are not. The difference between 'heat' and 'temperature' is that;

Heat: is the **thermal energy** that is transferred from one body to another. It is measured in a metric unit termed Joules (symbol J). Heat is transferred spontaneously from objects of higher temperature to ones of lower temperature (warmer to colder bodies).

Temperature: Temperature is a **measure** of the average thermal energy (degrees Centigrade).

Heat Flux: The term 'heat flux' refers to the quantity of heat (as thermal **radiation** and measured in kW/m^2) that is received at a particular point. In this case we refer to 'firefighter locations'. It has been demonstrated that firefighters are able to work safely and effectively for specific time periods in varying temperatures and 'heat fluxes' (the amount of heat received at firefighter locations).

> **What does this mean to the Firefighter?**
>
> Achieving and maintaining tenable conditions for firefighters working inside fire buildings is an important tactical function. We must endeavour to control the fire from the outset and take further actions that ensure the working environment

> is within tenable limits for safe occupation by firefighters. The best way to monitor this is by way of thermal image cameras (ceiling temperatures) and also according to how the environment feels to each firefighter (heat flux) through PPE, compared to the 'safe' temperatures experienced in temperature controlled training facilities. If firefighters are experiencing any discomfort through their PPE that appears abnormal then immediate action is called for, such as:
> - Gas cooling the hot layer
> - Door control
> - Evacuation, or moving to a safer area whilst shutting down the flow path and closing doors as you pass them

These time periods and temperatures are shown in a tenability chart (see figure 11.7).

3.7 FIRE GROWTH-RATE (TIME TO FLASHOVER)

The speed at which a fire grows within a compartment is related to several factors but increasing ventilation and large surface area fire loads will have the greatest impact on fire development. It has been observed that fires in small compartments (less than 50m^2) can flashover almost instantly, under the right conditions. A bedroom fire in a seven room flat of 70m^2 has been seen to spread fire to the remaining six rooms in less than 60 seconds (flashover). As furnishings and contents begin to include more plastics and adhesives in the fire load, room fires are now reaching the point of flashover more quickly and placing the structure under greater stress at an earlier stage on the time-line. In addition, high amounts of energy efficient insulation contained in compartment linings are also causing building fires to retain heat and reach flashover more quickly than experienced 40 years ago. As average fire service response times are increasing and flashover times are reducing, we may begin to experience a higher ratio of post-flashover fires. In effect, we should be concerned because once a fire reaches the point of flashover, the fire spread-rate statistically increases to spread beyond the compartment and possibly even the floor of origin. Furthermore, it has been demonstrated that fires involving combustible construction, such as mass timber (engineered wood products), can lead to multiple 'flashovers' due to wood smoulder. These events are perhaps more akin to fire gas ignitions (FGIs) and possibly even smoke explosions preceding and during the decay stages of burning.

3.8 FIRE SPREAD-RATE (TIME TAKEN FOR FIRE TO TRAVEL VERTICALLY OR HORIZONTALLY)

In large compartments with floor space in excess of 300-500 m^2 the fire does not 'flashover' as it would in smaller compartments but rather it 'travels' or 'progresses' at a steady pace. A fire in a 1,400 m^2 open-plan office floor has been observed to spread across the floor-plate at a steady rate of 22m^2 a minute (taking 63 minutes to involve the entire floor area). This latter type of fire spread in large compartments is known as a 'travelling fire' and it may be observed that parts of the floor are burning at full intensity whilst other areas are in the decay stages of burning. In high-ceiling buildings >10m, with high stacked fire loads or clothing hung on racks as in sports retail shops, vertical fire spread-rates can see flames reach the ceiling in less than 30 seconds. Such vertical fire spread can be very rapid and lead to intense burning within a few short seconds. Therefore designated and well signed escape routes must immediately be accessible.

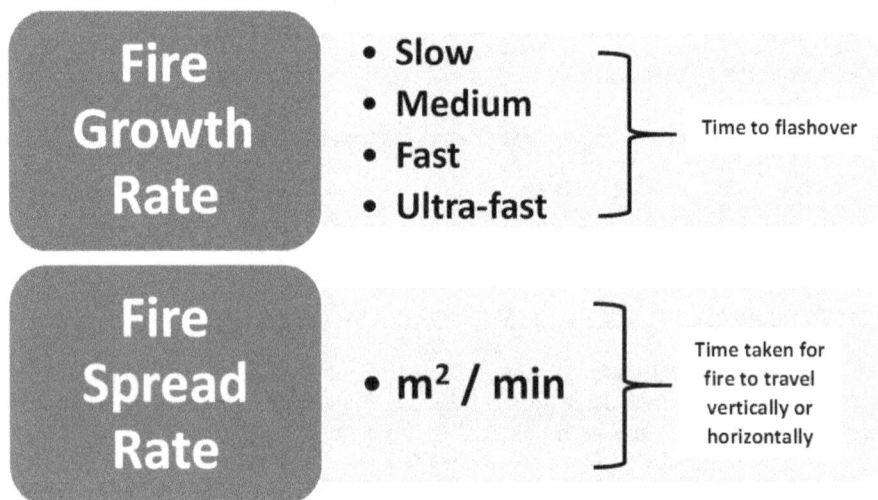

Figure 3.2: The difference between 'fire growth rates' and 'fire spread rates'.

3.9 SMOKE LAYER 'RATE OF DROP'

In a developing fire in a large industrial unit or superstore (<2,000m²) with ceilings at around 12–15 metres high, if stock or goods stored on high racking or wall hangers become well involved in fire, the average 'smoke drop rate' maybe something like **one metre/minute** *(good tactical rule of thumb but depending on floor-size)*. Therefore if the smoke layer is estimated at eight metres from the floor on deployment, those firefighters may have just eight minutes before the heavy smoke reaches the floor and reduces visibility to zero! Many firefighters have been caught out by the speed of smoke-drop rates and become lost in the heavy smoke layer.

The Hot Gas Layer and Neutral Plane[40]

As the temperature of a gas is increased it will expand, becoming less dense and more buoyant (Charles Law). If gases are confined within a compartment and heated, pressure will increase (Gay-Lussac's Law). Such pressure increases will 'push' smoke downwards towards the floor. In such cases, US firefighters will consider, or deploy, a roof ventilation team to vent the roof and raise the smoke layer again. In some buildings, automated venting is installed at roof level to create the same effect. Firefighters must familiarise with the operation of such systems at every opportunity.

Charles Law: Gases expand in direct proportion to the absolute temperature (temperature in degrees Kelvin, K° = C° + 273) applied to them. If the absolute temperature of a given quantity of gas is doubled its volume will double.

---

40 Hartin. E, Reading the Fire: Smoke and Air Track, Firehouse Magazine USA, August 2007

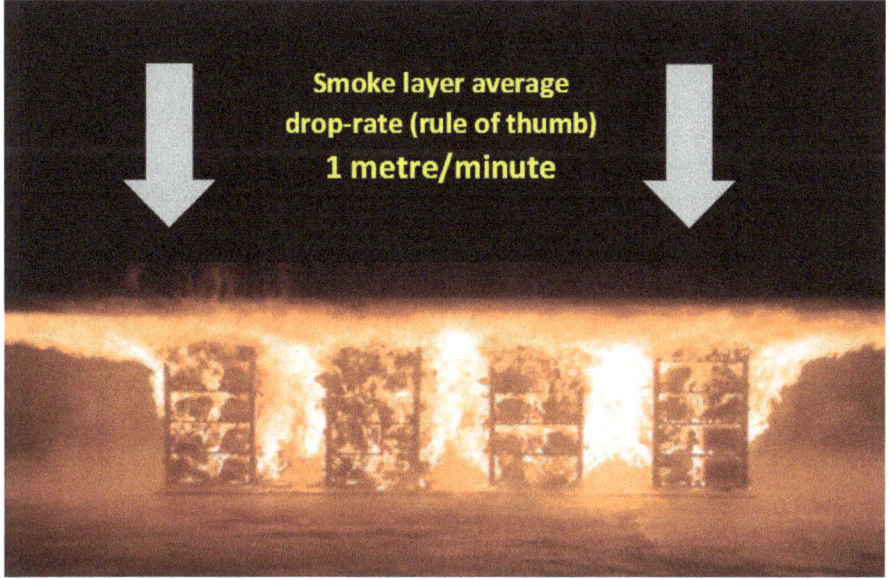

Figure 3.3: Smoke layer average drop rate (rule of thumb) under a 12–15 metre high ceiling in a non-sprinklered <2,000 m² compartment. ***Author's image***

- **Key Points for Firefighters:**
    1. Fire gases expand when heated – gases contract when cooled.
    2. Fire gases become less dense and will rise when heated.

Gay-Lussac's Law: When the volume of a gas remains the same and temperature is increased, pressure increases in proportion to the absolute temperature of the gas.

- **Key Points for Firefighters:**
    1. When gases are confined and heated, pressure increases.

What does this mean to the Firefighter?

When you next undertake a routine familiarisation visit to an open-plan office floor, take a look at how the fire load is configured. Those work-stations are they close together? How many are there across the floor? Is the ceiling higher than 3 metres? Are there plenty of windows located along the outer walls? Closely spaced work stations and a low ceiling, in areas over 300 m² of floor space, suggest that a fire will spread at a growth-rate of 22 m²/min once the fire reaches one metre in height. Think about that! This means that once the fire takes hold, it will be at 100m² within five minutes (the limit of a 500 L/min hose-line) and 200m² within 10 minutes (the limit of one 1000 L/min or two 500 L/min hose-lines).

Would you be able to deploy that amount of water in that time-frame if the fire was just beginning to take hold on the fire floor on arrival, at the start of its growth rate?

> Look further, if the floor is partially divided by cellular office partitions that may increase your time window to deploy water onto the fire by a few minutes.
>
> What if you are in a retail store or a warehouse with high vertical fire loads? Let's say products or storage to 12 or more metres? Which parts of the fire load are combustible? How quickly will a fire near the floor reach the top of the stack and then spread further? What smoke ventilation arrangements are installed and how might these affect the fire? At what time-frame will your 500 or 750 L/min hose-line be outpaced by the fire growth and can you deploy that quickly? Remember, vertical fire loads can reach 12 metres within less than a minute! Smoke layers in large volume structures can drop to the floor with a few more minutes if the fire load is high and sprinklers are not installed. If you are two hose lengths into the structure, how long would it take you to evacuate below a fast dropping smoke layer? At peak burning rate a typical warehouse smoke layer interface with clean air can drop at a rate of one metre every 45 seconds. If the layer is at six metres from the floor when you deploy, it can drop to the floor in less than five minutes!
>
> If you're in a nightclub or restaurant at Christmas and there are flammable decorations hanging from the ceiling and across walls, think of fire growth rates!

- When a fire develops in compartment, a plume of hot smoke rises to the ceiling and spreads horizontally through the compartment in the form of a ceiling jet. Increased temperature reduces gas density. Less dense gases will rise. The difference in density between hot smoke and cooler air below causes them to separate into two distinct layers. The boundary between these two layers is called the neutral plane. This is because the hot gas layer is trying to expand (Charles Law), but if it cannot, pressure will rise (Gay-Lussac's Law). Fluid pressure is exerted in all directions in an attempt to reach equilibrium. This is easy to observe when there is an opening in the compartment such as a doorway, hot smoke exits from the upper level due to higher pressure in the compartment while cooler air enters at the lower level due to lower pressure. The point at which the pressure inside and outside the compartment is the neutral plane. This is the height of the bottom of the hot gas layer at the opening. However, inside the compartment the level of the hot gas layer is dependent on the difference in density between the hot smoke and cooler air below (this may be at a considerably different level than observed at openings such as doors and windows.
- While level or thickness of the hot gas layer is categorized as a smoke indicator, the neutral plane relates to movement of smoke and air and would be categorized as an air track indicator. This illustrates the close interrelationship between smoke and air track indicators and criticality of looking at the big picture when reading fire conditions.
- Remember that smoke indicators continue to be important after making entry. In addition to the indicators discussed to this point, firefighters should consider the thickness of the hot gas layer. While often thinking about the space between the floor and hot gases (our working area), it is even more important to think about the space between the bottom of the hot gas layer and the ceiling (the volume of hot smoke over your head). Consider the difference between a compartment

with a ceiling height of 8 feet and one with a ceiling height of 24' (a considerable difference in the potential volume of smoke (think fuel) over your head).
- Early in fire development the hot gas layer is likely to be poorly defined with warm smoke defusing into the slightly cooler air in the compartment. As the fire develops, increase temperature differential between the smoke and cooler air below will sharply define the hot gas layer and it will become lower. If the fire continues to burn in a ventilation controlled state, the smoke and hot gases can lower completely to the floor.

3.10 HEAT RELEASE-RATE (HRR) (FIRE INTENSITY)

A fire's intensity (heat release rate) is measured in kilowatts kW (small fires) or megawatts MW (fully involved room fires). It is useful to get some idea how fire load and ventilation profiles can impact on the intensity of a fire. Whilst the energy release in a fire does not necessarily determine the temperatures encountered within a fire compartment, the likely burning duration and the difficulty in extinguishing a fire can be predicted and the quantities of water needed to extinguish the fire are aligned with the energy release (HRR) from a fire. High density fire loads (as in Table 3.1) can be very difficult to extinguish and the higher the fire load, the more water or CAFS will be required. In these fires the use of COBRA or Fog Spike may have little effect in extinguishing the main body of fire and fire spread to adjoining buildings along with greater impact on structural stability may occur. Typical heat release rates (MW) in a range of compartments can be seen in Table 3.2.

Table 3.2 – Typical heat release rates (MW) (Fire Intensity)	
Occupancy Type	Typical Heat Release
Container Training Facility	1.5 to 3 MW
Bedroom fire	3 to 5 MW
Fully involved 70m² flat	10 to 15 MW
240 m² Office	35 MW
500 m² Warehouse	>100 MW

It is seen in both Tables 3.1 and 3.2 that container training facilities provide a very limited fire load and this then follows on to the low heat release rate, compared to fires in real buildings. This demonstrates that demands on needed water flow-rates may be much higher in real fires than those encountered in the controlled environment of fire training facilities.

3.11 ENERGY AND POWER (FIRE INTENSITY)

The potential energy in a fire load both before or after pyrolysis, ignition or burning occurs, is measured in Joules (or Mega joules) (MJ), whereas the power (or intensity) of a fire, as this energy is released, is measured in kilowatts (or megawatts (MW).

3.12 TACTICAL ACTIONS (CONTROL MEASURES) BY FIREFIGHTERS

Parameter or event	Positive tactical actions by firefighters (control measures)
Flashover or rollover at the ceiling	• Door control (reduce air feeding into fire) • Cool the fire gases in the overhead, near the ceiling • Direct attack at the fire base • Reverse the flow-path
Backdraft	• Controlled door entry procedure • External application into fire compartment of fog insertion tools • Ventilate from outside the building, vertically or horizontally
High pressure backdraft	• Don't vent windows on the windward side of a building • External fire curtains may assist vented windows • Floor below nozzle may be used to attack fire at height
Fire Gas Ignitions	• Controlled door entry procedure • External application into fire compartment of fog insertion tools • Ventilate from outside the building, vertically or horizontally
High heat Flux	• Door control (reduce air feeding into fire) • Cool the fire gases in the overhead, near the ceiling • Direct attack at the fire base • Stay down low • Maintain a good distance from the fire • Positive Pressure Attack (PPA) ventilator • Reverse the flow-path
High temperature	• Controlled door entry procedure • External application into fire compartment of fog insertion tools
Rapid fire growth rate	• Door control (reduce air feeding into fire) • Cool the fire gases in the overhead, near the ceiling • Direct attack at the fire base • Controlled door entry procedure • External application into fire compartment of fog insertion tools
Rapid fire spread rate (large compartments)	• Gas cool and cool fuel base ahead of the fire spread if possible • Door control (reduce air feeding into the fire) • Direct attack at the base of the fire
Rapid smoke layer drop rate	• Ventilate at highest point (may already be auto-vented) • Limit time in fire building according to estimated drop-rate • Extensive cooling of high gas layers near ceiling
High heat release rate	• Direct attack at fire base
Flow-path control	• Deploy from windward side of building if possible • Firefighters to work on cool side of flow-path at all times • Firefighters not to place themselves on hot side of flow-path without isolating fire from their position (VIES) • Firefighters to anticipate and prepare for any flow-path reversal • Use flow-path reversal techniques to reduce heat flux at firefighter locations, or to change a bi-directional flow-path into a unidirectional flow (see chapter 18) • When venting, first consider what this might do to the flow-path and fire development (IMPORTANT)

3.13 'ADEQUATE' FIREFIGHTING WATER

KFRS fire engineers undertook a three year study into the firefighting water flow-rates used at over 5,000 working fires in Kent and Manchester. The data identified the ideal rates of water flow (L/min) that achieved reductions in the levels of building fire damage. In some cases, we should be aiming to apply greater amounts of water (L/min) during the early stages of firefighting than we have before. One development of this is seen in the transition from 19mm to 22mm hose-reels. In other situations we may need to reduce ventilation (closing doors feeding air to the fire) and insert water-fog from an exterior location. Following on from this research we have established an in-house flow gradient that targets the high fire loads and greater intensity of fires now being experienced in comparison to earlier fires in the 1960s-80s. These are 'target flows' aimed at achieving adequate firefighting water ahead of any likely fire development caused by additional ventilation.

This flow gradient can be met (Fig.3.4) through the use of smooth-bore nozzles; portable attack monitors and other equipment. This enables early intervention where an adequate supply of water is available.

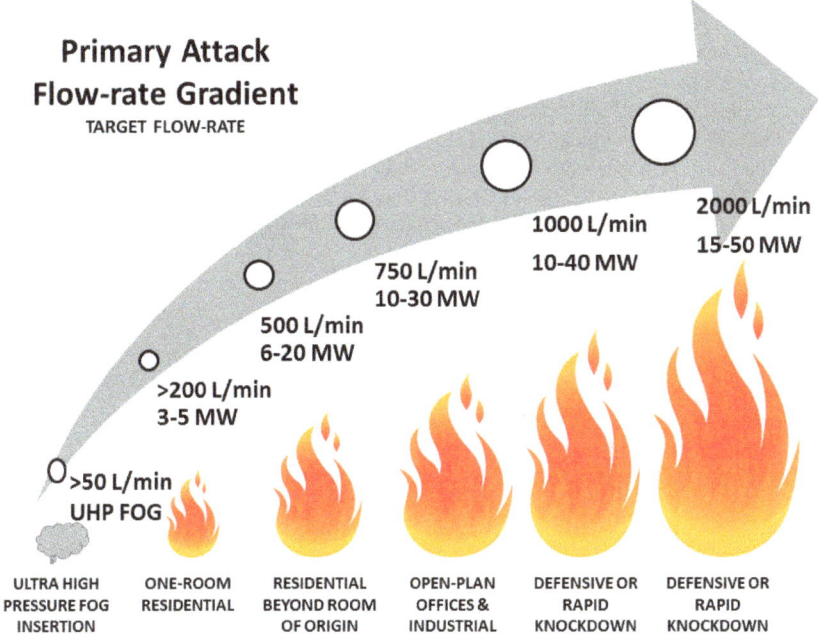

Figure 3.4: A primary attack flow gradient with target flows

The primary attack flow gradient provides a model of tactical flow objectives for various size fires in small (up to 200 LPM) to large (up to 750 LPM) compartments, housing low to high fire loads (some requiring early and rapid interventions, even up to 1000 to 2000 LPM applied within the first few minutes of KFRS arrival). An 'adequate' amount of firefighting water may be defined as the optimum quantity that extinguishes the fire without creating unnecessary water damage within areas of the building not affected by fire, but at the same time lessens any unnecessary or lengthy exposures to heat and smoke, as endured by firefighters. Whilst needed flow rates will vary between occupancy type and compartment size, an average of 5 L/min/m² of floor area fire involvement is a good target flow.

Table 3.3 below shows the practical flow rates of current firefighting equipment to allow firefighters to ensure that sufficient water flow is achieved in line with the flow-rate gradient.

Table 3.3 – Practical flow rates of current firefighting equipment	
Practical Flow Range	**Equipment**
100–140 L/min	19mm hose-reel
200–250 L/min	22mm hose-reel
350–500 L/min	Automatic nozzle
400–900 L/min	22mm smooth-bore branch
1000–2000 L/min	Portable attack monitor

Application of water flow in line with the water flow gradient supports effective firefighting during the early stages of intervention, and may be in addition to fog insertion tools and other exterior 'fire reset' options (reducing the fire's heat release from an exterior position to enable safer interior attacks). The KFRS research into adequate water capability based on fire load and fire intensity, now forms part of National Operational Guidance and the 'rule of thumb' fire-ground calculation may be used to determine approximate flow requirements for **building fires between 100–600m² of fire involvement.**

Area of fire (m²) x 5 = flow required (Litres/min)

So an area of 100m² of fire involvement would need a minimum of 500 L/min at the time water is applied, unless the fire has burned much of the fuel load down and is approaching the decay stages, when less water may be needed. In practical terms, for each 75m² of peak intensity fire involvement (floor area), one jet is required. So for 300m² of floor area fire involvement before fuel-control decay stages are reached, close to 4 jets will be needed.

As another 'rule of thumb' guide, for each MW of heat release, 25 L/min is required to counter the fire's greatest intensity. So a 4MW fire requires around 100 LPM as a minimum flow-rate for fire suppression. Other considerations are the ability to access the fire load directly and additional safety lines laid in support. A growing trend in the UK is to redevelop and construct town centre shops with greater retail floor space at ground and first levels and storage space at first or second level. The shops are extended far back from the street and now have full width 'floor to ceiling' glass at the front. Unlike shop fronts in the sixties, these new designs create major difficulties in extinguishment with higher fire load densities across larger (deep) floor space and large openings at the front that fail early, providing vast amounts of ventilation.

These modern design factors, coupled with structural redevelopment of older properties that can lead to hidden fire spread in structural voids and very intense fires at the shop front. Where occupied dwellings exist above the shops the rapid spread of fire may create untenable conditions in escape paths within minutes. In such cases, rapid and adequate water (L/min) may be required, although the depth of such shop units can soon cause the speed and intensity of fire development to outpace even 2-3 jets from the street.

3.14 VENTILATION PROFILE

A discussion with FDNY fire Captain John Ceriello concerning the term 'ventilation profile' offered an alternate view as to how and where a ventilation profile might exist. Ceriello suggested that the vent profile can exist way beyond the compartment or room of origin and this allusion is entirely correct. In EuroFirefighter-1[41] (p48) we defined an 'air track' (flow-path) as the 'point to point route that is taken by air flowing into a structure and combustion products leaving a structure'. On p43 of the same publication, the term 'ventilation profile' was defined as 'the amount of air available within a [fire] compartment'. It then went on to further define ventilation controlled burning, fuel controlled burning and under-ventilated fire compartments. In practice, the ventilation profile may relate to the quantity of air/oxygen feeding into a fire compartment and influencing fire development, just as an opening serving as an air inlet or a smoke outlet, elsewhere in the building, can create a flow-path and cause the fire to spread and travel in a variety of directions. If we create an opening at the street door into a fire building this may initiate a primary flow-path, impacting on fire development, intensity and directional fire spread. If we create a vent opening elsewhere, say an upper storey window, this again can change all of those factors and reverse the flow-path, even possibly worsening conditions and intensifying the fire further still. On the other hand, sometimes a flow-path reversal can work to our advantage.

Any enclosed fire requires an adequate amount of air/oxygen to develop and progress. In most cases, but particularly in smaller compartments, the amount of ventilation a fire receives determines the maximum intensity (Mega-Watts) a fire can reach and whether the fire will burn in a ventilation or fuel-controlled state. The greater the heat release (intensity), the more water is needed for effective intervention. Some buildings have French windows and balcony openings that enable vast amounts of air in to feed a fire should they fail, and this may cause a very intense fire, especially where high velocity wind enters to dramatically increase the burning rate. The ventilation profile of a fire compartment refers to the amount of ventilation (percentage) available through openings, as a ratio to the floor space – (area of ventilation openings/area of floor space) as A_v/A_f. So for a 6m² vent opening in a compartment with a 16m² floor space; the vent profile is 38 percent (6/16). As a general rule of thumb guide, vent profiles above 25% in a small compartment (as above) will be required to achieve fuel controlled burning. At this point the heat release from the fire is at its highest intensity. In larger compartments, a much lesser percentage in the ventilation profile will achieve fuel-controlled burning at maximum fire intensity.

As firefighters we must be mindful of the size of openings in the building and also the likely impact on fire development, should such openings fail or be force vented. In fact, the intensity of a fire can usually be estimated by the size of openings that have failed or been vented by firefighters. Some guidance on assessing the size of openings that exist, or may occur, within a residential fire compartment to determine likely fire size and safe water deployments are shown in Table 3.4.

---

41 Grimwood. P; EuroFirefighter; Jeremy Mills Publishing UK

Table 3.4 – Typical heat release rates (MW) and flow demands

Size of window and door openings feeding air to the fire	Likely Heat Release	Minimum water required (Hose-line capability) (Safety Jets additional)
1 Sq. metre	3 MW	19mm Hose-reel
2 Sq. metres	5 MW	22 mm hose-reel; or two 19mm hose-reels; or One main Jet
4 Sq. metres	7 MW	One Jet
6 Sq. metres	10 MW	One jet
10 Sq. metres	>10 MW	>One Jet

Wherever there is a need for large flow-rates >750 L/min for exterior attacks to control and reduce rapid fire spread immediately on arrival, the availability of a constant water supply is a prime consideration. The duration of an appliance water tank (prior to a constant water supply being sourced) is dependent on the equipment used (See Table 3.5).

Table 3.5 – Typical times to empty an appliance tank with a constant flow and no water supply connected

Equipment	Time to Empty a full Appliance Tank	
	1600 L tank	1800 L tank
19mm hose-reel	16 mins	18 mins
22mm hose-reel	8 mins	9 mins
Main line automatic nozzle	3 to 4.5 mins	3.5 to 5 mins
22mm smooth-bore branch	1.8 to 3 mins	2 to 3.2 mins
Portable attack monitor	0.8 to1.6 mins	1.8 to 0.9 mins

As an example, if a town centre shop is fully involved, the fire may extend to the rear of the shop and involve something like 200m^2 of high density floor space fire load.

With the entire shop front window glass failed, a flow-rate of 1000-2000 L/min may be needed to reach into the rear of the shop within the first few minutes. This can only be achieved from the street using two smooth-bore nozzles (using 4 firefighters with two on each branch) or a portable attack monitor (1-2 firefighters). In practice, without a constant supply, table five shows how quickly the tank will be emptied. In such situations it is important to have an awareness of which hydrants, or other immediate water supplies, can support these primary flow demands on the initial response. Almost all town centres will have 1-3 high-flow hydrant mains (+1200 L/min) within close reach of the shopping areas. It is useful to gain some awareness of where these high-flow mains run through local town centres.

3.15 COMPRESSED AIR FOAM (CAF)

CAF improves the efficiency of water by increasing the water surface area in contact with the fire fuel which results in an increase in cooling rate. This means that CAF provides

Practical Fire Dynamics & extreme Fire Behaviour • 101

Figure 3.5: A town centre shop fire spreading vertically, horizontally and laterally to involve surrounding buildings. **Image courtesy of Russ Talliss via Kent-online**

an excellent rapid knock down capability, and as a result its use should be considered whenever external access to the fire fuel is achievable.

While CAF is more efficient than water alone, it is the water contained within the CAF that cools the fire. This is an important consideration for larger sized fires.

When wet CAF is applied at the recommended 5 bar pressure, approximately 240 LPM of water is dispensed. When we compare this water output against the water requirements in the primary attack gradient, it suggests that only a one room fire can be tackled using CAF, when anecdotal evidence suggests that larger fires can be extinguished with CAF.

While it is clear that CAF rapidly knocks down fires (and uses less water when doing so), it is not known at what point (size of fire) CAF becomes ineffective due to the limited water content within its composition. Therefore, when considering the use of CAF for primary attack purposes at larger fires, the application of higher flow of water onto the fire may be more beneficial than a number of CAF jets.

3.16 A CHANGING ENVIRONMENT MAY REQUIRE NEW TACTICS

Over the past few decades we have seen a greater use of plastic materials in our homes and workplaces and this has led to higher fire loads in nearly all building environments. We have also experienced greater uses of natural light with larger windows in some cases and an increase in glass shop front dimensions. Coupled together, we are now seeing greater intensity in the fires we respond to as increasing ventilation profiles feeding higher fire loads become commonplace.

We must therefore address our tactics and optimise the way we apply water, or alternative means of extinguishing, suppressing or holding fires, in order to place greater control over developing fires from the outset.

The use of high-flow hand controlled branches, portable attack monitors and water-fog insertion tools are all first-strike tools that may be able to reset a fire backwards in its levels of intensity and fire growth. Sometimes a 'quick-hit' high-flow rate applied from the exterior can gain some knock-down before an interior attack is mounted. In some cases the use of a portable attack monitor fed by 90mm hose, where water is available, may be the difference in preventing a small fire becoming a large one. The use of COBRA or CAFS should also be considered at an early intervention stage as a fire development 'holding' tool.

3.17 DESIGN FIRE LOADS (FIRE ENGINEERING) AND 'DESIGN FIRES'

A 'design fire' is something a fire engineer will develop based on several variables such as fire load, compartment geometry and ventilation opening factors. This will be used as a basis for computer modelling or calculation, in the determination of fire development, size and intensity etc. 'No matter how a design fire is created, it is a *model of the real world* or an impact to be applied instead of a real world fire, and if application of it should lead to an increase of fire safety, it has to be safe compared to what can be observed and expected in the real world. A fire model is safe if it represents a fire more severe than the fires, which can be expected in the compartment and if it is based on well-documented experiments proving the basic theory and if reliable data can be found for the model. Comparing results of fire safety design for a large variety of fires and structures of concrete, steel, and wood it is a predominant characteristic that: It takes time for a fire impact to penetrate a cross-section. It means that *in general a slow fire with a low opening factor reduces the loadbearing capacity more than a rapid fire with the same fire load and a large opening factor. Only the lightest steel structures may be more susceptible to a large opening factor.* This is in contrast to design for evacuation, where rapid fires with high temperatures are usually the most dangerous[42].

A detailed analysis of variable (movable contents) and permanent (fixed structure) fire load is required when preparing or reviewing a fire strategy for a building. The 'design fire load' offers a more complex approach to evaluating actual fire load in a building' As mentioned earlier, the term 'fire load' refers to the amount of potential energy (MJ) contained in a building's combustible contents (including elements of structure). Whilst this text will not provide a complete guide to fire engineering detail, it offers some methodology and background that puts into perspective the approaches required to establish a relevant fire design strategy. It further provides basic knowledge for fire commanders in understanding how fire load can impact on fire development and structural stability.

Fire spread in buildings is a risk to life safety for which the UK Building Regulations aims to reduce to acceptable levels. For the designer, there is a responsibility to specify materials, and to provide details that:

- reduce the potential for fire ignition
- limit the spread of fire
- stop the passage of hot gases and smoke

An appropriately designed building will allow people remote from the seat of a fire to escape and provide a building from which the fire service can deal with the fire safely and effectively.

Structural fire safety is achieved either by what is called 'passive protection' e.g. fire resistant lining boards and/or 'active protection' e.g. smoke ventilation, alarm systems and sprinklers. For the structural fire engineer, the material choice within the structural solution will influence the passive and active fire protection strategy. In performance based codes, design fires are determined using engineering calculations and tools that include both computer models and experiments to demonstrate acceptable performance. In many of these calculations or tests, fires that are representative of those expected in buildings are used to evaluate building performance. These fires are known as design fires. An important input parameter that affects the design fire is the total available combustible content known as fire load (MJ), often expressed as fire load energy density (FLED) per unit floor area, (MJ/m^2). The severity of real fires (as opposed to experimental fires) is directly related to several parameters such as enclosure dimensions, ventilation conditions, building and environmental pressure differentials, fire load density and geometry, enclosure lining materials, and although experimental research attempts to replicate realistic conditions these parameters in reality vary significantly.

The design values most commonly used for fire and structural design are those termed 80% fractile values, which mean that probability of exceeding the design value is 20%, or one in five. Such high probability cannot, of course, be accepted for structural failures and this is taken care of by using safety factors that increase the capacity to such extent that the structural failure probability becomes tolerably low.

Take an open-plan office floor and consider how the fire load might perform during a growing and uncontrolled fire. A floor area of 20 x 30m (600m^2) occupied with normal office furniture and fittings with mainly work stations providing a total fire load of

42 Hertz. K; (University of Denmark) Fire Technology 48; Springer Science; 2011

17,100kg and a calorific value of 20MJ/kg (wood and plastic). Using a fire load energy density of 570 MJ/m^2 (commonly assigned to offices in British standards fire engineering documents) the total energy in the fire load is 342,000 MJ (570 x 600m^2).

Now if we allow the fire to develop to a stage where it surpasses its ventilation control limits and becomes fuel controlled, we can then see the heat release rate has peaked at 77.6MW. This requires at least 18m^2 of window ventilation (being a 0.03 opening ratio (3%) – 18/600m^2) (anything less than 18m^2 of window ventilation in this scenario and the fire is most likely ventilation controlled). A ventilation controlled fire will not be able to release the energy from the fire load as quickly and will burn at a lesser heat release rate and for a greater duration.

There are several methods that can be used to calculate a growing compartment fire but the easiest and fastest way is to use a simple zone computer model, or hand calculation spreadsheet such as the 'Firesys' programme.

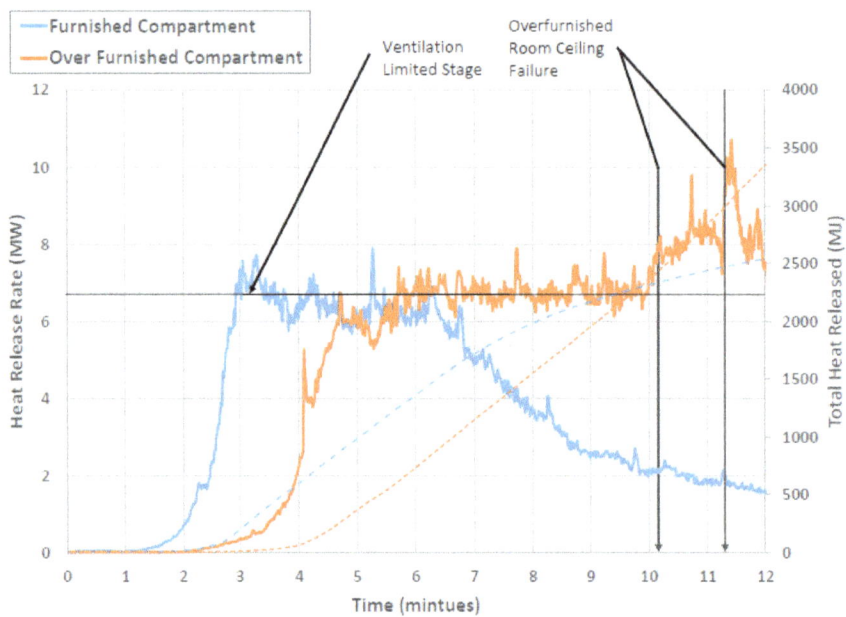

Figure 3.6: Compartment Burn Results comparing regular fire loads with high fire loads in a room fire[43] **Courtesy of Underwriters Laboratory USA**

---

43 Fire service summary report; Study of the Effectiveness of Fire Service Positive Pressure Ventilation during Fire Attack in Single Family Homes Incorporating Modern Construction Practices; Firefighter Safety Research Institute; Underwriters Laboratory, April 2016

Practical Fire Dynamics & extreme Fire Behaviour • 105

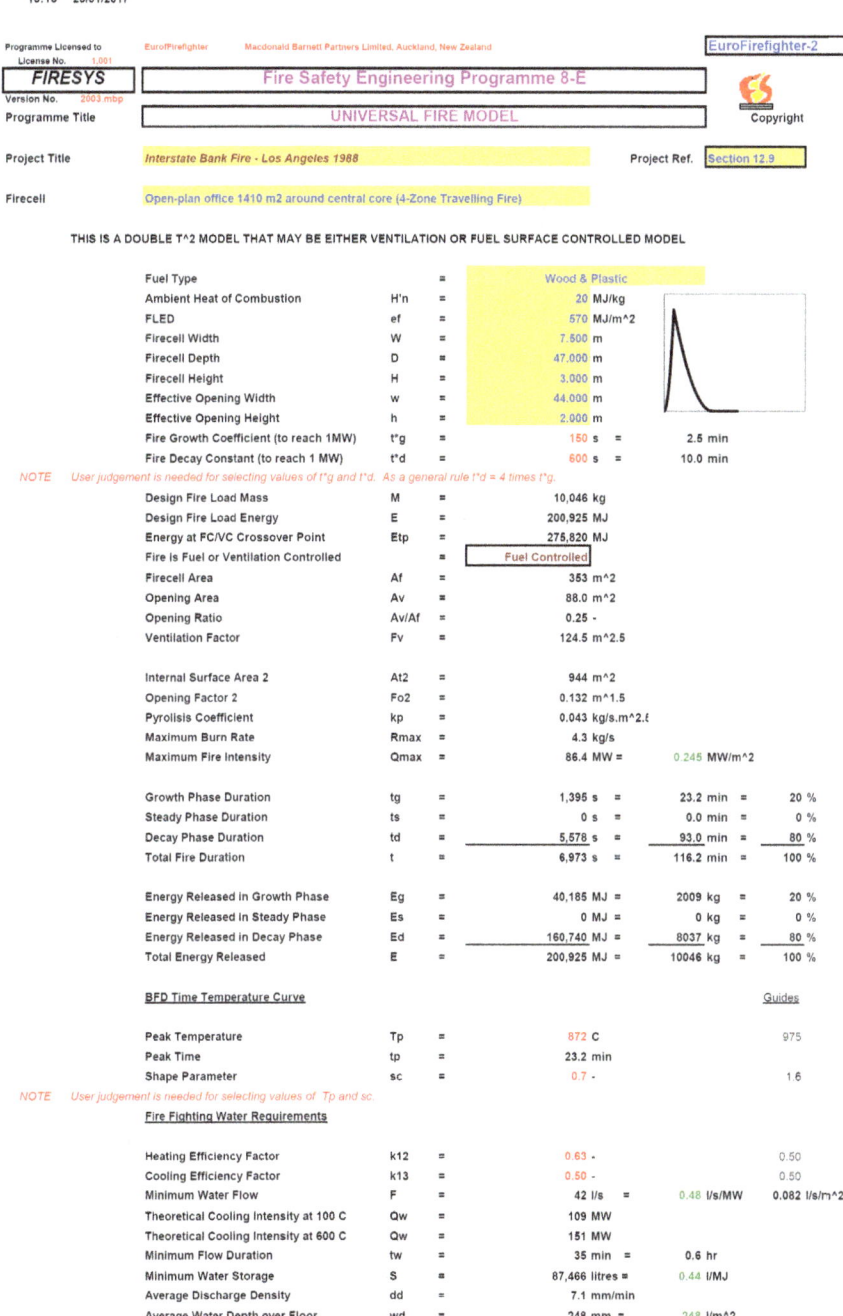

Figure 3.7: Fire compartment zone model – FireSys **Courtesy of Cliff Barnett**

The impact of including 'Permanent Fire Load' in Total Fire Load calculations

1. Light timber frame construction – studs at 400 mm centres, joists at 600 mm centres, non-combustible insulation between studs:
Variable fire load: 450 MJ/m2 × 12 m2 = **5400 MJ** Permanent fire load: 35 studs each 2.4 m × 0.045 m × 0.15 m gives a fire load of: 35 × 0.0162 m3 × 450 kg/m3 × 18 MJ/kg = **4592.7 MJ** 5 joists each 4 m × 0.045 m × 0.22 m gives a fire load of: 5 × 0.0396 m3 × 450 kg/m3 × 18 MJ/kg = **1603.8 MJ** **Total fire load = 5400 + 4592.7 + 1603.8 = 11,596.5 MJ** Percentage increase due to permanent fire load = 115%.
2. Structural Insulated Panels (SIP) construction – 150 mm deep panel with PUR insulation and 15 mm OSB skins, joists at 600 mm centres:
Variable fire load: 450 MJ/m2 × 12 m2 = **5400 MJ** Permanent fire load: Volume of OSB – (4 × 2.4 m × 4 m × 0.015 m) + (4 × 2.4 m × 3 m × 0.015 m) = 1 m3 Volume of PUR – (2 × 4 m × 2.4 m × 0.13 m) + (2 × 3 m × 2.4 m × 0.13 m) = 4.368 m3 Fire load due to SIPs = (1 m3 × 450 kg/m3 × 18 MJ/kg) + (4.368 m3 × 40 kg/m3 × 25 MJ/kg) = **12,468 MJ** 5 joists each 4 m × 0.045 m × 0.22 m gives a fire load of: 5 × 0.0396 m3 × 450 kg/m3 × 18 MJ/kg = **1603.8 MJ** **Total fire load = 5400 + 12,468 + 1603.8 = 19,472 MJ** Percentage increase due to permanent fire load = 261%. To obtain the fire load represented by all the panels (SIPS) in a building for example[44], in a flat roofed building 10 m high by 40 m long by 40 m wide entirely clad in sandwich panels, the area of the roof is 40×40 =1600 m2 and the area of the walls is 4×40×10 = 1600 m2 so the total area is 3200 m2. In this example using 50 mm thick polyurethane foam cored panels would give a fire load of: 58×3200 = **185,600 MJ**

SIPS Core material	Fire Load (MJ/m²) for core thickness of:		
	50mm	100mm	200mm
Polyurethane foam* (PUR)	58	117	234
Polystyrene foam* (EPS)	46	86	166
Stone wool	11	16	25

* The figures include a value of 6 MJ/m² for polyurethane adhesive for bonding cores to facings. The value for stone wool will be less if inorganic adhesives are used.
If the contents of the building were in the low fire load category (i.e. having a fire load density of 300 MJ/m2 of floor area) the fire load of the panels would represent over a third of the fire load of the building contents, and if the panels were 100 mm thick the factor would be two-thirds. It is clear that for thick panels, e.g. panels used in food storage rooms requiring a high level of thermal insulation, the fire load represented by combustible-cored panels is high – almost a factor of ten higher when compared with non-combustible-cored panels containing rock wool. However, insulated panels of a either combustible or non-combustible composition may increase fire growth rates and compartment fire temperatures (see chapter 3).
3. Light steel frame construction – non-combustible frame, insulated with PUR foam:
Variable fire load: 450 MJ/m2 × 12 m2 = **5400 MJ** Notional volume of PUR – (2 × 4 m × 2.4 m × 0.13 m) + (2 × 3 m × 2.4 m × 0.13 m) = 4.368 m3 Fire load due to insulation – 4.368 m3 × 40 kg/m3 × 25 MJ/kg = **4368 MJ** **Total fire load = 5400 + 4368 = 9768 MJ** Percentage increase due to permanent fire load = 81%.

Data Courtesy of NHBC Research Foundation and UK Mineral Wool Association

---

44 Cooke. G Dr, Sandwich panels for external cladding – fire safety issues and implications for the risk assessment process; Commissioned by Eurisol – UK mineral Wool Association; 2000

3.18 TRAVELLING FIRES IN LARGE OPEN-PLAN COMPARTMENTS

Where fires occur in large open-plan compartments (>300m^2) the fires rarely burn uniformly across the entire floor plate. This fact has been observed by firefighters for several decades where in such cases, fire will not generally demonstrate rapid horizontal fire spread across an entire space, akin to a flashover, but more likely spread at a steady growth rate. Although flashover may appear to occur on a more localised basis, researchers[45] have suggested that such fires will demonstrate 'near field' (at the peak fire area) and 'far field' fire conditions (behind, in front of, or surrounding the current area of burning). This type of fire spread has been termed 'travelling fires' by some researchers, or 'progressive burning' or 'fire migration' by others. The temperature distribution across zones will equally be spread over a wide range, with temperatures sometimes exceeding 1000 deg. C in the early stages at the fire (near field), reducing below 800 deg. C in the far fields. This effect has major implications for firefighting tactics, structural stability and the quantities of firefighting water required to extinguish a fire, or reduce the heat in the gas layers at any particular location.

> **What this means to the Firefighter**
>
> When fighting fire in a large open-plan office compartment (for example) a typical firefighting operation will look primarily to deploy and resource in support for (a) search for and rescue of any occupants remaining within the immediate risk-zone; (b) deploying hose-lines to apply firefighting water directly at the fire base; (c) protect surrounding risks and exposures. However, a fire commander may wish to take note of how fires travel across large floor-plates in open-plan >300 m2 where compartment fire behaviour may not support conventional post-fire 'flashover' across the entire compartment at one time. What has been observed by firefighters and fire engineers is that such fires 'travel' or progress at a steady state (producing a fire spread rate of approximately 22 square metres/minute across floor space in large offices). This means that the structure and its combustible contents are being subjected to a range of temperatures over a longer duration than the standard post-flashover fire burning at its peak (normally 20 minutes).
>
> Therefore, a 1,600 m^2 open-plan office may demonstrate 3-5 areas burning at peak heat release individually, as the fire progresses over the course of 90 minutes. In doing so, heat will be transmitted both ahead and behind (or surrounding) the main bulk of fire and it might be that hose-lines should be deployed into these areas to effectively cool the structure, either ahead of the fire or after it has burned into decay, to reduce the radiated heat transfer into the structural elements and dampen the building contents. This may actually prevent structural collapse as opposed to 'chasing' a moving fire around the compartment that is constantly transferring its energy into the structure at all surrounding locations.

The assessment of a 'reasonable worst case scenario' for fire in a building is a substantial part of the performance-based (fire engineered) design concept. Parametric fire curves

---

45 Law. A; Gillie. M; Stern-Gottfried. J; Rein. G; The Influence of Travelling Fires on a Concrete Frame, Engineering Structures 33, pp1635–1642; 2011

have been developed as a graphic tool to allow hand calculations to display how fires grow, burn steady and begin to decay once 70% of the fuel load has been consumed. They are representative of 'real' fires although their accuracy is sometimes questioned. They remain the most effective way to display fire growth predictions outside of desktop computer analysis. In using parametric curves to demonstrate likely fire development in a given compartment, the inputs include fuel load, compartment dimensions, ventilation conditions and the thermal inertia of compartment linings. Parametric fire curves therefore predict more realistic temperature-time curves than the standard fire curve, which is based on fire test data of individual items retrieved from furnaces.

However, in large compartments where fire spread presents both near and far field variances in temperature and heat release, the parametric curves are better displayed as a series of fires, where each fire is at a different stage on a growth and development curve. The Interstate Bank Fire in Los Angeles, 1988, represents a good example to demonstrate the 'travelling fire' effect, as does the Windsor (Edificio) Tower fire in Madrid 2005. The author attended the scene of the Los Angeles fire and discussed how fire spread affected firefighting operations in the building with LAFD Fire Chief Rob Ramirez. Mr. Ramirez was the fire commander during the critical latter stages of the fire, commanding operations to prevent the fire spreading beyond the 16th floor. There was also opportunity to view video tapes and observe the patterns and duration of fire spread, as well as corroborating this information when visiting the building with Task Force 3 firefighters, who advanced the initial attack hose-lines deployed into the 12th floor (floor of origin).

Time Temperature Curves

A time-temperature curve offers a means of analysis of a compartment fire rate of growth. It is used by structural and fire engineers as a simple method of calculating by hand (usually by spreadsheet or desktop) the rise in temperatures and/or heat release that may be expected in a compartment fire over a given period of time. There are several different types of curve in use, all with varying limitations in accuracy in predicting real fire behaviour. Parametric and standard fire curves were validated by test data from small fire compartments that were almost cubic and furnaces. Therefore their use in larger compartments may lead to some inaccuracy. The most widely referenced time temperature fire design curves for real fire exposure involving ventilation controlled fires are known as the Swedish fire curves.

When using Eurocode parametric fire curves to demonstrate a necessary temperature-time analysis for the design of applied fire resistance to structural elements, it has been noted by many engineers that the Eurocode parametric design fire fails to provide an accurate assessment of real fire growth in large compartments. In applying this parametric analysis to the Interstate Bank fire, some shortfalls in the actual fire development are noted.

The Interstate Bank tower has a structural steel frame with lightweight concrete slab on profiled steel deck. The external cladding system consisted of glass and aluminium. The fire started on floor 12 of the 62 storey office tower at 2225 hours and spread up to floor 16 before it was brought under control by the fire service some four hours later. An automatic sprinkler system, although installed, had been shut down awaiting installation of water flow alarms.

The open-plan office floor space was located around a central core that contained lifts, stairs and service shafts. The office space surrounding the core measured 188 metres by 7.5 metres, totalling 1,410 square metres. Although a medium growth curve is generally applied to offices in fire design specifications, the fire on the mainly open-plan fire floor (level 12) followed a 'fast' t^2 growth curve after the first phase of fire spread, wrapping

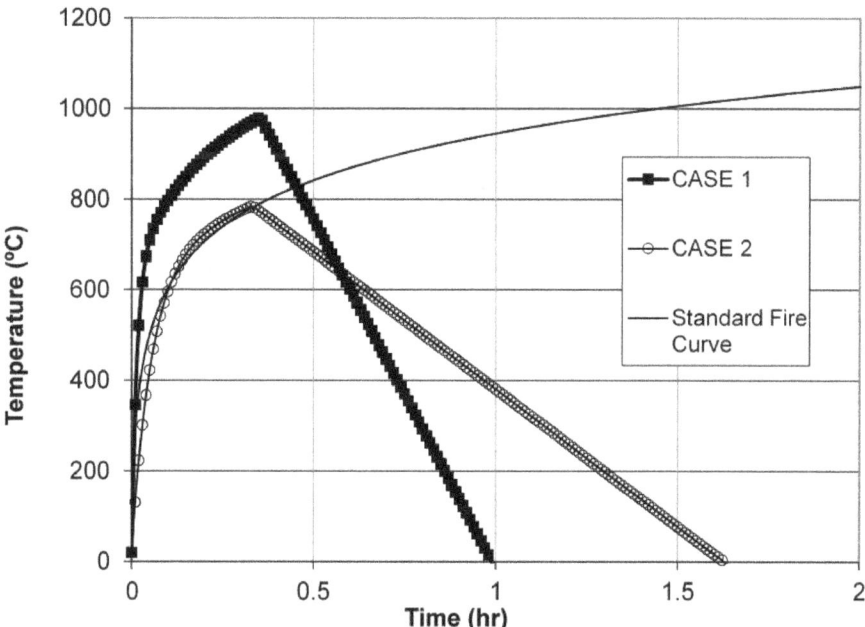

Figure 3.8 Using the Interstate Bank 12th floor fire involvement of 1,410 m² with Eurocode parametric inputs, but not accounting for a travelling fire, the fire curve (fast t-squared) demonstrates that the entire floor fire would peak at 19 minutes and burnout at 60 minutes. If the fire had been of medium t-squared growth the decay stage would end at just over one and half hours for the entire floor. Note the standard fire curve in comparison.

around the central core to eventually involve 100 percent of the floor space at an average spread rate of 22m²/min. over the duration of full fire development of the twelfth floor (This fire is also reviewed in Chapter 12).

The fire at level 12 burned for just over 120 minutes before being extinguished across four zones of 352m². The fire also spread to several upper levels but this fire spread beyond the floor of origin is not considered here. The travelling fire on level 12 is demonstrated at Fig.3.8 where it can be seen that compartment temperatures during the first hour were far higher than an ISO 834 standard fire curve suggests. However the fire duration as shown still does not account for the extended decay period (shown by the smaller dotted line) that lasted until 120 minutes from the point where fire growth first began. It was at this point when LAFD declared the fire on the twelfth level was extinguished.

The parametric curves shown in Fig.3.8 represent a fire that was being extinguished, although much of this occurred during the decay stages, after the majority of the fuel load had burned down.

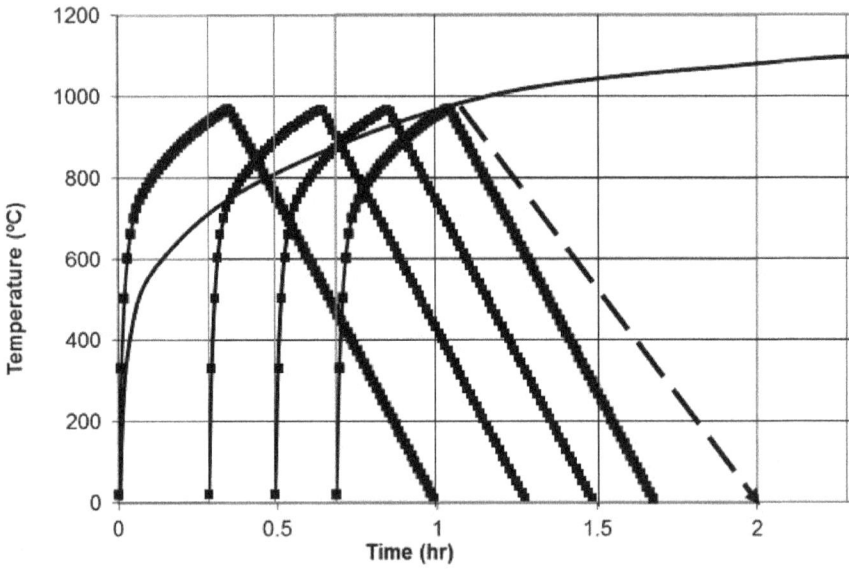

Figure 3.9: Interstate Bank fire – Los Angeles 1988 – A parametric travelling fire analysis of the 12th floor fire involvement across four zones of floor space, each of 352m². An ISO 834 standard fire curve is included for comparison where fire gas temperatures are much lower on scale than parametric curves (that take into account ventilation and boundary conditions as well as fire load). Note also the extended cooling phase that occurred during the real fire (dash line)

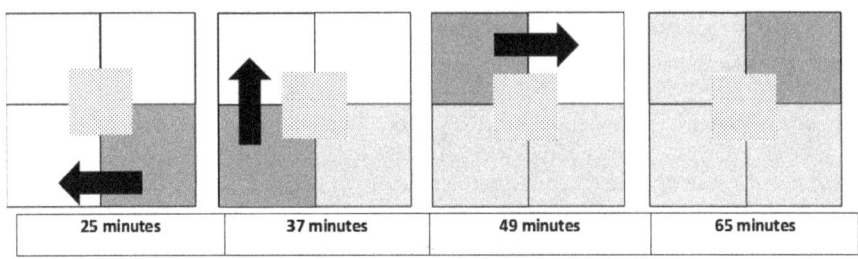

Figure 3.10: 4-zone directional fire spread around the central core of a 1,410 m² accommodation space floor-plate on the 12th floor of the Interstate Bank fire in Los Angeles (actual fire as observed)

The entire involvement of fire floor occurred in 65 minutes, followed by a 60 minute decay period. Firefighting water was being applied during the 2nd, 3rd and 4th periods, by which time the growth-rate was already outpacing the speed of deployment of a total of eight hose-lines. Due to the intense heat conditions on the fire floors, there was a limited time that firefighters could work before they required relief. This alone placed a massive demand for staffing and resources and over 400 firefighters were needed to deal with this fire. The structural frame was effectively able to support the firefighting operations throughout the fire where an unusually good application of fire resistive coating helped maintain much structural integrity and stability.

3.19 DO MODERN COMPARTMENT FIRES BURN 'FASTER' AND 'HOTTER'?

Do fires actually burn hotter now than forty years ago? Are the firefighters of today exposed to higher temperatures, experiencing a more rapid fire development where the time to flashover is reducing? Is fire spreading out of the compartment of origin becoming more likely because of this? There is clear evidence emerging that the increasing use of plastics contained in modern day fire loads and the growing use of energy efficient insulation in buildings, results in more severe compartment fires (higher temperatures and faster fire growth rates) as the compartment boundaries are more heavily insulated and thus have lower *thermal inertias* (the heat loss through wall and ceiling linings is reducing and the compartment fire temperatures are higher). Couple this with combustible or lightweight pre-engineered building materials (Modern Methods of Construction) becoming part of the structural fire load at an early stage and yes, building fires are certainly becoming more intense.

Maximum HRR as a function of available Oxygen

The development of a fire depends on the amount of oxygen available. The amount of oxygen available also affects the maximum heat release rate that could be reached. The heat release rate is therefore affected by the size of the room and the size of the openings to the room. The maximum heat release rate that can be reached in a room as a function of the amount of oxygen available can be calculated from Eq.3.1.

$m_a = 0.5 * A_o * \sqrt{H_o}$ **Eq. 3.1**

Where m_a is the mass flow rate of ambient air [kg/s], A_o is the total opening area [m²] and H_o is the weighted mean height of all the openings [m]. Multiplying 0.5, from Eq.3.1, with 13.2 MJ/kg and 23%, the expression becomes –

$Q = 1.518 * A_o * \sqrt{H_o}$ **Eq. 3.2**

13.2 MJ/kg represents the amount of energy one kilo of oxygen can produce, assuming total combustion and 23% is the mass fraction of oxygen available in the air entering the compartment.

In training scenarios, for example, firefighters often use steel shipping containers loaded with variable fuel loads to simulate a range of fire phenomena, including growth phase, ceiling jet, rollover, flashover (in part) and backdraft). In some minor cases, Fire Gas Ignitions (FGIs) occur in the form of 'ghosting flames' or 'dancing angels'. If such a container has a partition constructed at one end to enable a door sized opening, typically of 2 metres high by 1 metre width, with this door open a maximum 4.3 MW heat release rate is achievable.

$Q = 1.518 * 2 * \sqrt{2} =$ **4.293 kW**

Researchers at Underwriters Laboratories (UL) in the USA, a global independent safety science company, have concluded that homes today burn up to eight times faster than in previous decades. They report[46]: 'The changes in modern building design and materials have altered the nature of structure fires, with modern homes able to reach flashover eight times faster than homes built 50 years ago. This change is largely behind the 67% increase over the past 30 years in the rate of firefighter deaths due to traumatic injuries while operating inside structures. And although the overall fire death rate in the U.S. has decreased by 64% during the same period, it is clear that modern structure fires can be deadly to both firefighters and building occupants'.

> **What does this means to the Firefighter?**
>
> **With an increasing quantity of plastics now finding its way into the 'room and contents' of our buildings (fire load), and with energy efficient wall, floor and ceiling linings becoming more common in modern buildings we are now experiencing faster fire growth rates and much higher temperatures in fires than we did some forty years ago.**
>
> **This means we may face post-flashover fires on arrival more often, as fires begin to develop more rapidly. We may also experience fire spread beyond the compartment of origin more often as flashover is even more likely to occur before firefighters arrive on-scene. This means more firefighting water will be required and earlier structural collapse in buildings made of lightweight materials becomes more likely.**
>
> **Professor (and fire engineer) Ulf Wickstrom writes in his 2016 book 'Temperature Calculation in Fire Safety Engineering' (Springer):** *'The temperature of structures exposed to fully developed fires with gas temperatures reaching 800–1200°C will gradually increase [as lightweight energy efficient structures become more common] and eventually the structures may lose their load bearing capacity as well as their ability to keep fires within confined spaces [compartment of origin]'.*
>
> **It is worth considering the number of dwelling fires in England and Wales as a percentage of all building fires compared with the US is very similar: 71% UK versus 76% US. Where stark differences occur is in the associated financial loss. In England and Wales, these fires account for 35% of the total financial loss attributable to fire, whereas in the US it is 58% – a figure that may be relevant when considered against the 1.9% (UK) to 90% (US) lightweight timber frame domestic housing stock difference.**

The Role of Thermal inertia

Imagine a typical room fire in a building of 1950s construction. Now consider the same room but well lined with energy efficient insulation. If a fire occurs in this insulated room the question is: Will this fire behave in a different way? Heat energy (Q) produced within a room or compartment fire normally transfers (Fig. 3.11) into the hot gas layers (q_r), or into the compartment boundaries (walls and ceiling) (q_w) or escapes through openings (windows and doors) (q_L and q_r). Thermal inertia of the wall and ceiling materials (*the capacity to conduct and retain heat*) can be simply defined as the product of thermal

46 New Science Fire Safety Article, Underwriters Laboratories, USA UL.Com

conductivity, density and specific heat of a [lining or substrate] material. For compartment boundaries, the innermost layer, typically plasterboard or plaster in most homes, generally governs the thermal inertia of the compartment. However, the supporting substrate, typically masonry or timber frame and the presence of insulation, does also have relevant influence on fire development. The thermal inertia of the 'construction' is normally taken as some form of weighted average of the properties making up an element of construction and hence substrate properties carry some importance. The impact of thermal inertia (b) on peak temperature (Θ_{max}) can be shown with reference to Table 3.6 which is based upon the parametric design fire in Eurocode EN1991-1-2.

$b = \sqrt{k\rho c_p}$ (J/m2s^{½}K)	700	800	900	1000	1100	1200	1300	1400	1500
Θ_{max} (°C) temperature	1122	1083	1052	1027	1007	994	979	968	960

Courtesy of NHBC Research Foundation
Table 3.6: The impact of thermal inertia (b) on peak temperature (Θ_{max})

It is demonstrated in Table 3.6 how compartment fire temperatures are likely to increase as energy efficient linings become heavily insulated, as in most modern residential buildings and even more so in large warehouse cold stores for example. Quite simply, the amount of heat lost through compartment walls, floors and ceilings is reduced by heavy insulation or low thermal inertia materials and this fact causes an increased build-up of heat within the compartment to radiate back into the fire compartment and raise temperature. We can see that calculated temperatures for a lower thermal inertia of 700 will result in compartment temperatures of 1122 deg. C and this aligns with a mixture of linings using gypsum board and lightweight concrete. However, where lightweight materials are used in a timber frame building using gypsum or OSB board and heavy insulation in the wall and floor linings, the thermal inertia (b) will be closer to 400 J/m2s^{½}K (or less), resulting in fire compartment temperatures approaching 1400 deg. C, where ventilation opening factors exist between 0.04 and 0.08.

The UK guidance NAD[47] on thermal insulation factors used in the Eurocodes refers to a k_b conversion factor[48] depending on the thermal properties of each particular enclosure. The highest k_b factor recommended in the UK NAD is 0.09 where the thermal inertia is less than 720 J/m2s^{½}K. This is higher than that required by the Eurocode (0.07) although there is an alternate view that k_b factors greater than 0.10 would more accurately address the thermal inertias and true compartment temperatures now being experienced in modern lightweight construction. Not only may structural elements be affected by higher compartment fire temperatures, but so too are firefighters who are advancing into untenable areas without realising. Therefore, accurate assessments of heat losses (or heat retention) in modern buildings are also needed from an operational firefighting perspective.

The 0.09 k_b factor in the UK NAD resulted from research reported on by British Steel plc where a series of nine live fire tests were conducted in a compartment built inside the BRE large experimental building facility at Cardington in Bedfordshire. Overall, the compartment measured 23m long x 6m wide x 3m high and was designed to represent a 'slice' through a much larger compartment 46m deep, of infinite width and having an effective (internal) depth to height ratio of 16:1. The prime objective was to determine the

---

47 UK National Annex Directive to BS EN 991-1-2:2002 refers to BS 6688-1-2:2007
48 K_b is a lining materials factor ranging from 0.04 to 0.10, being a function of the thermal inertia of the compartment boundaries

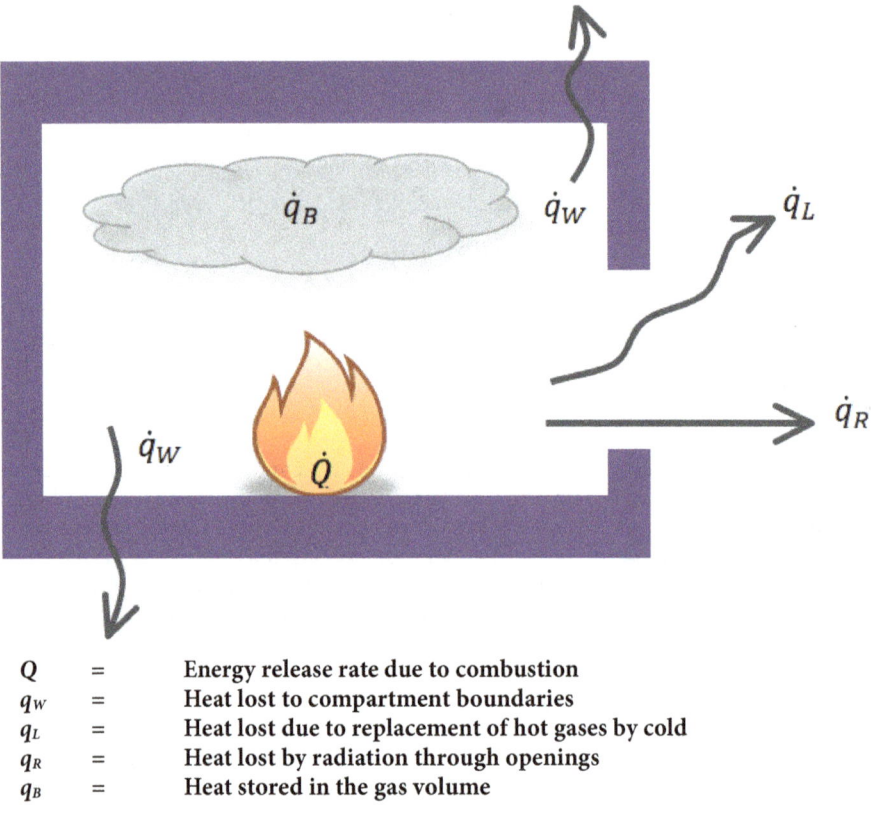

Q	=	Energy release rate due to combustion
q_W	=	Heat lost to compartment boundaries
q_L	=	Heat lost due to replacement of hot gases by cold
q_R	=	Heat lost by radiation through openings
q_B	=	Heat stored in the gas volume

Figure 3.11: Mechanisms of heat transfer in a room or compartment fire

impacts of *time equivalent fire exposure* on structural steel elements whilst taking a more accurate assessment of thermal properties of the boundary lining materials into account, along with their impact on compartment fire temperatures, in comparison to those used in the Eurocodes.

The concept of thermal inertia in building fires is clearly relevant where fires occur in converted cold stores or highly insulated refrigerated compartments. The author has experienced several such fires where heat retention in the fire compartment causes extreme temperature build-up and multiple flashovers as fire gas cycles repeatedly ignite. Where leakage paths are few, such fires may smoulder under extremely warm conditions that can generate backdrafts when opened up or vented.

One important observation of the tests was that the aspect ratio between the ventilation height and compartment depth is much greater in smaller compartments than in large compartments. This will give rise to a higher rate of burning and higher temperatures in the fire compartment but result in a greater amount of heat lost through openings, as opposed to remaining inside the compartment and heating the structure. However, if vent openings occur in adjacent rooms or surrounding compartments, the increase in temperature caused through the effects of reduced thermal inertia is still maintained locally.

Material	Thermal Inertia ($\sqrt{kpc}$ J/$m^2s^{0.5}$K)
Steel	12,650
Stone	2,423
Marble	2,273
Ordinary Concrete	1,650–2,192
Fireclay Brick	1,432
Glass	1,312
Clay Brick	961
Lightweight Concrete	931
Gypsum Plaster	761
Fireline Plasterboard	520
Lightweight Concrete Blocks	660
Vermiculite Plaster	650–742
Wood	736
Mineral wool	86–426
Wool (pine)	426
Aereated Concrete	386
Polyurethane Foam	30
Polystyrene Foam	26

Table 3.7 Typical thermal inertias for building materials and insulators

The Six Main Parameters that have the Greatest Impact on compartment Fire Intensity are:
- Fire load density
- Calorific value of the fire load
- Vent size (opening factor), height and configuration
- Geometry and size of the fire compartment floor area and surface area
- Thermal properties (thermal inertia) of the compartment boundary linings and substrate
- Combustibility of the structural frame

An engineering research project[49] by Anna Back (Lund University, Sweden) reported on how increased thermal insulation can affect the fire development in a compartment. The objective was to compare the fire development in an insulated compartment to a non-insulated compartment. It should be mentioned that the research was undertaken using

---

49 Back. A; Fire development in insulated compartments: Effects from improved thermal insulation; Report 5387, Lund 2012

an insulated steel-lined shipping container. However, consider the relevance of this when new homes and hotels have been constructed in London (and elsewhere) using these very materials. The possibility to carry out and compare hand calculation methodologies and simulations were also to be evaluated. The following four questions were therefore answered:

Does increased thermal insulation lead to significantly higher gas temperature in the fire compartment? In each of the experiments with different fire sources, wood crib and heptane pool, the gas temperature reached a higher level in the insulated compartment than in the non-insulated compartment. In the insulated heptane pool fire experiment the maximum average gas temperature was 26% higher than in the non-insulated experiment. For the wood crib fire experiment the maximum average gas temperature was 18% higher in the insulated compartment. Even in the hand calculations and simulations, higher gas temperatures were obtained in the insulated compartments. Therefore, increased thermal insulation does lead to significantly higher gas temperatures in the fire compartment.

Will increased thermal insulation lead to a significantly larger and quicker heat release rate of a fire? This was only possible to investigate in the experiment as the heat release rate is used as an input in both hand calculations and simulations. The results from the heptane pool fire experiments show that a larger heat release rate is reached in the insulated compartment compared to the non-insulated compartment. The maximum average heat release rate reached in the insulated compartment was 13% higher than in the non-insulated compartment. A higher fire growth rate is also experienced in the experiment with the heptane pool fire. The fire growth rate in the insulated compartment is approximately twice as large as the fire growth rate in the non-insulated compartment.

In the insulated wood crib fire experiment the maximum heat release rate was only 2.5% higher than in the non-insulated compartment and the fire growth rate did not change between the insulated and non-insulated experiment. Larger and quicker heat release rates are reached in compartments with increased thermal insulation where the fire source is sensitive to incident radiation.

Is it plausible that the condition flashover is reached earlier in an insulated compartment than in a non-insulated compartment? This question stands in connection with the question above. As the heat release rate increased faster in the insulated compartment with the heptane pool fire, flashover would also be reached earlier in this compartment compared to the non-insulated compartment. In the experiment with the wood crib fire on the other hand, the heat release rate curve has the same shape for both the insulated and non-insulated compartment. It then comes down to if the heat release rate is high enough for flashover to be reached. In the experiments carried out, flashover was reached in the insulated compartment but not in the non-insulated compartment. Hence, the provided insulation resulted in an increased heat release rate, sufficient for flashover to occur. This implies that improved thermal insulation could make the difference of whether flashover occurs or not.

Do hand calculations and simulations give similar results to full scale experiments when comparing the fire behaviour in an insulated compartment to a non-insulated compartment? Both hand calculations and simulations are very sensitive to the chosen inputs and therefore the results did not correspond to the ones from the experiments. Calculations using the MQH method [McCaffrey, Quintiere and Harkleroad] give as good results as the FDS simulation, and are far less time consuming, if there is a high uncertainty to the heat release rate and properties of the boundary materials. The $\sqrt{k\rho c_p}$

value needs to be within the range of 100 J/m²s^½K and 2,200 J/m²s^½K according to the Eurocode which means that calculating for such insulation $\sqrt{k\rho c_p}$ values is likely to result in over estimations in compartment temperature and heat release.

Further research by Arthur Ting Kung Kii[50] explored how the parametric curves used in Eurocodes for temperature predictions in compartment fires appear as if they are back to front (the decay stage should be far longer where thermal losses through the walls, floors and ceiling are less – Fig 3.12). An observation he reports that was also made by Professor Andy Buchanan in previous work. The researcher informs us: 'the law of physics would tell us that for a fixed fuel load and ventilation factor, the gas temperature in a well-insulated fire compartment (ie; with low thermal inertia value) would be higher compared to that in a poorly insulated compartment. The reason being that, the heat generated by the fire is lost through conduction into the walls at a much slower rate than in a poorly insulated compartment. The rate of fire decay demonstrated in the Eurocodes on the contrary, suggest a faster decay rate for fire curve in a well-insulated compartment than that in a poorly insulated one'. This is demonstrated in Figure 3.12 where a Eurocode parametric curve with a ventilation opening factor of 0.07 displays a reduced rate of compartmental fire decay for the well-insulated gypsum linings with low thermal inertia compared to the reduced insulation of concrete.

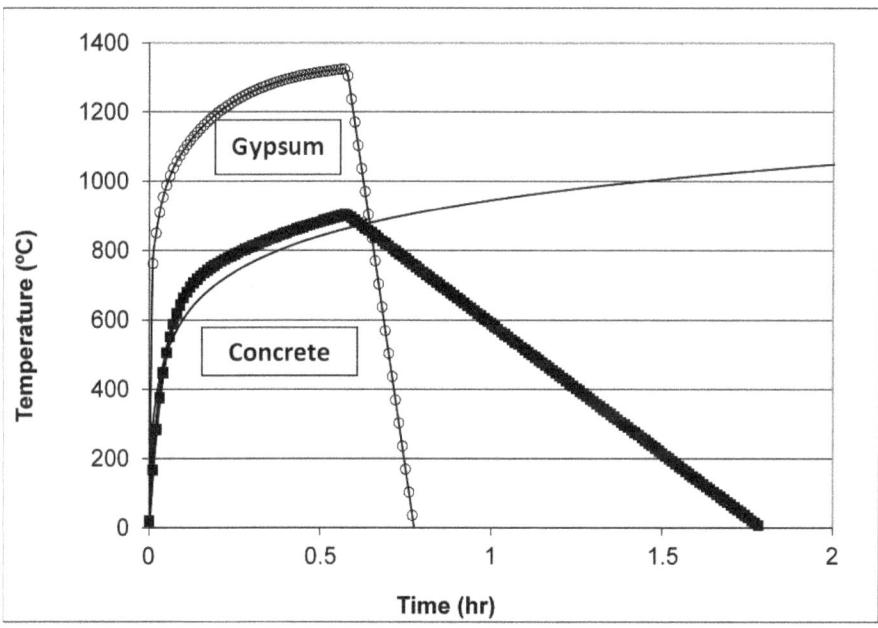

Figure 3.12: A Eurocode parametric curve with a ventilation opening factor of 0.07 displays a reduced rate of compartmental fire decay for the well-insulated gypsum linings with low thermal inertia compared to the reduced insulation of concrete.

---

50 Ting Kung Kii. A; Time Equivalent for Protected Steel and Reinforced Concrete Structures; University of Canterbury MSc, New Zealand, 1999

Research by Anna Back (Lund)[51] comparing both wood crib and heptane fires in a steel firefighter training shipping container, with insulated walls (Rockwool) and non-insulated walls, concluded that neither hand calculations nor FDS simulations are able to reflect the incident radiation towards the fire source and therefore neither the increase in heat release rate. As seen in graphs over the heat release rate, it does reach a higher level in the insulated compartment compared to the non-insulated compartment, which likely occurs due to the incident radiation from the hot gas layer and the hot compartment surfaces. It is not possible to recreate this effect in FDS simulations as the heat release rate is defined in the input file. The same applies to the hand calculations, where the heat release rate is one of the inputs needed to be able to calculate the gas temperature. Both hand calculations and simulations are very sensitive to the chosen inputs. If there is a high uncertainty to the heat release rate and properties of the boundary materials, hand calculations using the MQH method[52] give as good results as the FDS simulation, and are far less time consuming.

Thermal runaway

Thermal runaway is a condition of rapid fire spread directly linked to thermal feedback from the hot gas layers and compartment boundaries (walls, floor and ceiling). It is a situation where an increase in temperature changes the conditions in a way that causes a further increase in temperature, often leading to a destructive result. In other words, 'thermal runaway' describes a process which is accelerated by increased temperature, in turn releasing energy that further increases temperature. This is one good reason to consider water cooling applications directed at both the fire gas layers and compartment linings during a pre-flashover fire. In a ventilation controlled fire, there may come a turning point at which the heat release rate exceeds the energy lost through room openings and surfaces, resulting in thermal runaway with increased compartment temperatures that may lead to an earlier point of flashover, than might normally be expected.

Two series of full scale room fire tests[53] comprising 16 experiments were used for a study of the onset of flashover and thermal runaway. The fire loads were varied and represented seven different commercial applications and two non-combustible linings with significantly different thermal inertia. The test results showed that by lowering the thermal inertia and thereby lowering the heat loss from the room and at the same time increasing the thermal feedback, a thermal runaway occurred involving modern compartment fire loads that included plastics. In these cases the onset of thermal runaway was found to occur at room temperatures in the range 300°C to 420°C, supporting that the room temperature at the *onset of thermal runaway is strongly dependent on the thermal inertia*. It also shows that the onset of thermal runaway cannot in all cases implicitly be predicted by the traditional flashover temperature criterion of 500°C to 600°C. For fire loads composed of pure wood/celluloses the onset of flashover in the tests occurred at significantly higher room temperatures (725°C). This can be explained by flammability parameters making wood/celluloses less sensitive to thermal feedback.

---

51 Back. A, Fire development in insulated compartments – Effects from improved thermal insulation; Report 5387, Department of Fire Safety Engineering and Systems Safety, Lund University, Sweden 2012
52 McCaffrey, Quintiere and Harkleroad method of hand calculation to estimate gas temperatures in a ventilated compartment
53 Poulsen, A; Bwalya. A; Jomaas. G; Evaluation of the Onset of Flashover in Room Fire Experiments; Fire Technology (2013) 49: 891. doi:10.1007/s10694-012-0296-3

3.20 PHASE CHANGE MATERIALS (PCMS)

It is well known that the use of adequate thermal energy storage (TES) systems in the building and industrial sector presents high potential in energy conservation. Now that all new builds have to be highly insulated to meet the code for sustainable homes, the issue of overheating of housing during summer months is of great concern. Air-conditioning and automated ventilation can be expensive options to solve this problem but phase-change-materials (PCMs) are now being seen as the most innovative way to deal with summer heat build-up. Microscopically small polymer spheres contain in their core a wax storage medium. When there is a rise in temperature over a defined temperature threshold (21°C, 23°C or 26°C), the wax absorbs the excessive heat energy and stores it in a phase change. When the temperature falls under the temperature threshold, the capsule releases this stored heat energy again.

Recently, gypsum plasterboards with incorporated paraffin-based (wax) PCM blends have become commercially available in the UK and globally. In the high temperature environment developed during a room fire, the paraffins, which exhibit relatively low boiling points, may evaporate and, escaping through the gypsum plasterboard's porous structure, emerge into the fire region where they may ignite, thus adversely affecting the fire resistance characteristics of the building. Aiming to assess the fire safety behaviour of such building materials, an extensive experimental and computational analysis has been performed by researchers[54]. The fire behaviour and the main thermo-physical physical properties of PCM-enhanced gypsum plasterboards are investigated, using a variety of standard tests and devices (Scanning Electron Microscopy, Thermo Gravimetric Analysis, Cone Calorimeter).

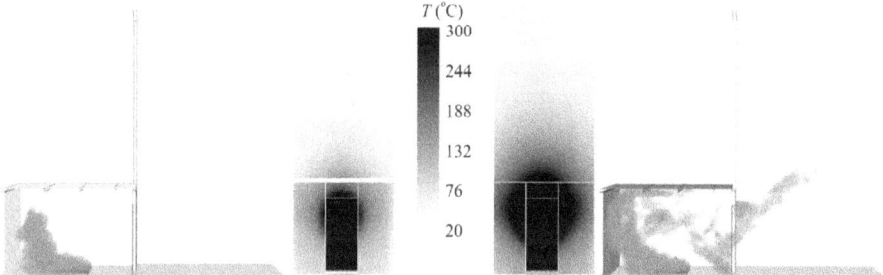

figure 3.13: Compartment fire simulation slices with wall temperatures of gypsum board lining (left) and gypsum board parafin based PCM (right). **Courtesy of School of Mechanical Engineering, National Technical University of Athens**

---

54 G Dionysios I. Kolaitis*, Eleni K. Asimakopoulou and Maria A. Founti; Gypsum Plasterboards Enhanced with Phase Change Materials: A Fire Safety Assessment using Experimental and Computational Techniques; School of Mechanical Engineering, National Technical University of Athens

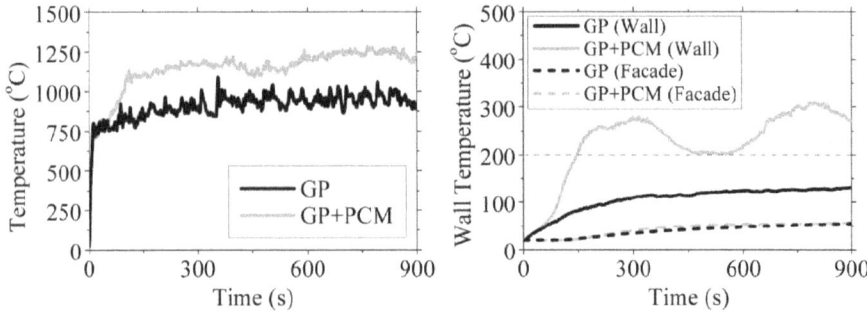

Figure 3.14: Compartment fire simulation slices with room fire and wall temperatures of gypsum board lining (GP) and gypsum board parafin based PCM (GP+PCM). Courtesy of School of Mechanical Engineering, National Technical University of Athens

It was evident in the research that the produced amount of 'combustible' paraffin vapours enhances the fire power, thus resulting in higher wall temperatures. The authors wrote: 'CFD [computer modelling] tools allow the estimation of the fire resistance characteristics of wall assemblies. In this context, the performed simulations are used to investigate the fire resistance of the utilized GP [Gypsum Plasterboard] wall assemblies. GP exposed to fire are considered to exhibit mechanical failure when cracks or openings are observed through the wall; however, since cracking phenomena cannot be accurately simulated in the FDS [computer modelling] code, alternative failure criteria are used in this study. According to the Eurocode standards, fire safety regulations regarding the 'integrity' of a compartment wall assembly specify that the maximum temperature rise at the unexposed side (ambient facing side) should not exceed 180 °C. In the current simulations, the ambient temperature was considered to be 20 °C; therefore, the aforementioned 'failure' criterion for a GP wall assembly corresponds to a temperature of 200 °C on its unexposed side. Predictions of the temporal evolution of the unexposed side temperatures for the compartment front wall and the façade are depicted in Figure 20 and 21; the illustrated numerical results are obtained at a height of 1.2 m and 2.6 m, respectively. It is evident that the façade does not exceed the Eurocode fire resistance 'failure' criterion (temperature at the unexposed side higher than 200 °C) in both cases. However, in the PCM-enriched GP case (GP+PCM), predicted temperatures at the unexposed side of the compartment front wall, which lies close to the opening, exceed the critical failure limit of 200 oC approximately 150 s after fire initiation. In the GP case, no 'failure' event is observed at the front wall, thus demonstrating the potential adverse effects of PCM enrichment on the fire safety characteristics of GP clad compartments.'

3.21 EXTREME FIRE BEHAVIOUR

What is 'extreme' fire behaviour and can you even use the term 'extreme'?

The definition of 'extreme' is reaching the highest degree of something; very severe or serious; furthest from the centre or a given point. It is clear then that even though it may be predictable, some form of rapid fire phenomena can and should be termed extreme, or severe.

So how does the existence or creation of a flow-path impact on the **three classes** of extreme fire behaviour?

Event	Definition	Flow-path effects
Flashover	Flashover represents a **heat** or **ventilation** induced rapid development of a compartment fire leading to sustained combustion and a fully developed fire. It will take a high amount of heat flux at the floor to achieve flashover, >20 kW/m^2. In rooms or spaces greater than 300m^2 in floor space, a progressive travelling type of fire is more likely to occur than a full compartment flashover. This means that the fire will spread across the space at a steady rate, depending on fuel and air availability.	We can see that flashover, by definition, may be induced by an increase in heat or an increase in air feeding into the fire. An increase in heat may occur due to the involvement of additional fuel load or because the room is lined with insulating material, as in timber-framed buildings or cold stores. An increase in air may occur because doors to the building or fire compartment are left open and unattended, or because a window is opened or vented by the fire, or by firefighters. Where a vent opening is made, or occurs above the fire floor in particular, this may create negative pressures that 'pulls' fire out from the room of origin, into the stairs and upwards, possibly catching firefighters or occupants on the wrong side of the flow-path.
Backdraft	A negative pressure **ventilation** induced ignition of fire gases following air transport into an area containing fuel-rich gases and an ignition source. This can also occur in a situation where a wind is forcing air into a closed compartment (with one opening facing the wind). This positive pressure backdraft may enable pressure to build-up within the fire compartment and as the ignition occurs, the sudden unleashing of this very high pressure may cause a very intense period of combustion (**high-pressure backdraft**). This may be very intense, or short-lived. A period of normal post-flashover burning may or may not follow a backdraft.	The difference between a ventilation induced flashover and a backdraft may exist along a very fine line. Usually, backdrafts cause a blow-torching fire-ball to emit from an opening, either internally or externally to the fire building. On occasions the pressure expansion caused by a backdraft can cause structural failure. A flashover involves less pressure release and is observed as flames moving down from the upper gas layer and filling a space floor to ceiling. At this point, the fire may transition into a ventilation controlled fire where the majority of flaming combustion occurs at the vent opening until the fuel load burns down somewhat. Where signs of an under-ventilated fire exist, precautionary control measures, such as safe door-entry procedure, water-fog or foam injection from an external position, or vertical ventilation of the highest **fire floor** or roof might be considered.

Continued overleaf.

| Fire Gas Ignitions (FGIs) | The flammability of smoke is still not fully understood. What is smoke and why do some fires seem to have more smoke than others? Smoke is a collection of tiny solid, liquid and gas particles. Visible smoke is mostly carbon (soot), tar, oils and ash.

Smoke occurs when there is incomplete combustion (not enough oxygen to burn the fuel completely). In complete combustion, everything is burned, producing just water and carbon dioxide. When incomplete combustion occurs, not everything is burned. Smoke is a collection of these tiny unburned particles, along with a mixture of flammable gases including carbon monoxide. Each particle is too small to see with your eyes, but when they come together, you see them as smoke.

Types of FGI –
• Smoke explosion
• Auto-ignition
• Flash-fire
• Rollover
• Ghosting flames
• Forced draught (draft) fire

Following his fire behaviour research in 2002, Grimwood **introduced and defined** the FGI terminology working with Dr Martin Thomas of BRE. The author grouped the various range of events associated with ignitions of the fire gas layers, or transporting fire gases, and titled this group under the heading of **Fire Gas Ignitions (FGIs)**[*] This term then became a welcomed addition to the UK Fire Service guidance document[**] relating to extreme fire behaviour events. Even so, the definition provided therein failed to conform to the author's original definition of 2002 – | • The high pressure from a room fire may cause smoke to move out of the room if a door is opened, travelling into other rooms or spaces such as shafts, voids or attics.
• The high pressure of an external wind can drive smoke further into the structure if a window fails or is vented.
• The flammability of the smoke produced from a building fire is widely variable but fires that have smouldered for long periods may produce the most flammable mixtures.
• A situation can occur where hot smoke has cooled to ambient temperatures and transported into areas remote from the fire. This smoke, although cooled, can retain its flammability level. These areas can include: high ceilings, adjacent rooms (on the same floor level or above or below), corridors, stair-shafts and hidden voids (especially behind false ceilings).
• This transporting of smoke into other parts of the building is driven by density (hot air rises) and pressure from the fire compartment forcing smoke into, or through, low-pressure areas or openings. It may also be driven by external wind currents or building stack effect.
• This smoke from a fire can seep into areas adjacent to, or quite someway from, the fire compartment.
• An ignition can occur in several ways, for example:
A **flashover** in the fire compartment could force flaming combustion into adjacent areas containing flammable smoke. At several fires, the fire has burned through a floor to the upper level where smoke that has travelled or pyrolysed and on ignition, the burning gas layers then travel across and down the stairs, involving all levels within 3-4 seconds.
Smoke from a smouldering fire could accumulate at high level, near the ceiling. If a burning brand develops and rises on the thermal into the gas layer, an ignition could occur that may accompany an explosive pressure wave.
In a mattress store fire in 1974 three Kent FRS firefighters were killed as they vented windows and uncovered some latex foam mattresses to extinguish a fire that had been smouldering underneath. The area before the explosion was very tenable with low temperatures. The resulting ignition of accumulated fire gases was termed as a typical **smoke explosion**. |

Fire Gas Ignitions (FGIs) *(Continued)*	**Fire Gas Ignition:** *'an ignition of accumulated fire gases and combustion products, existing in, or transported into, a flammable state. Any such ignition is usually caused by the introduction of an ignition source into a pre-mixed state of flammable gases; or the transport of such gases towards a source of ignition; or the transport of a fuel-rich mixture of gases into an area containing oxygen and an ignition source, or an accumulation of heated fire gases transporting through an opening and mixing with air to auto-ignite'.* It was stated in GRA 5.8:2009 that *'Fire gas ignitions occur when gases from a compartment fire are 'leaked' into an adjacent compartment and mixed with the air within this additional area. This mixture may then fall within the appropriate flammable limits that if ignited, will create an increase in pressure either with or without explosive force. Where this process occurs it is not necessary for an opening to be opened for such ignition to take place. If an explosive force is experienced, this is commonly termed a 'smoke explosion'. Where an ignition occurs with much less pressure, the term 'flash fire' is more appropriate'.* In fact, the above event may also occur within the fire compartment itself and does not always require fire gases to leak into an adjacent compartment. **In principle, both flashovers and backdrafts are also a form of FGI, although the two events are more clearly defined individually and are generally better understood by firefighters as separate phenomena.**	• On occasions smoke will not need the introduction of any ignition source where the smoke exiting an opening is so hot, it may **'auto-ignite'** as it reaches air containing adequate oxygen. • In some cases the mix of a smoke layer with air will cause the fire gases near the ceiling to ignite in a **'flash-fire'** that lasts just a few seconds, before disappearing. • Where sustained burning exists in a smoke layer existing near the ceiling, this is termed 'rollover' and this generally precedes a full compartment flashover. • The event associated with **'ghosting flames'** refers to tiny pockets of fire gases that may exist within a flammable range in the fire compartment that can auto-ignite in short pops or bursts of flame (often blue flames) that may last 3-4 seconds. These flames are often termed dancing angels by firefighters. The smoke surrounding these brief ignitions is generally too rich a mix to ignite. However, ghosting flames serve as a warning sign for an impending ignition of the entire fire gas layer. A **forced draught fire**[***] is one resulting from a high-velocity wind, (wind driven) or a PPV air-flow, forcing flaming combustion several metres out of an opening, which is igniting on meeting outside air and also through mixing from a forced internal air-flow. The event can be observed to appear as a 'flame-thrower' type flame emitting from an opening. It is literally a rapid burn-off of fire gases as they meet air.

Table 3.8: Extreme fire behaviour events

---

[*] Grimwood. P; Desmet. K; Tactical Firefighting; A comprehensive guide to compartment firefighting and live fire training (CFBT) Firetactics.com – CEMAC; 2003 - http://www.olerdola.org/documentos/cemac-kd-pg-2003.pdf
[**] Generic Risk Assessment 5.8; Flashover, backdraught and fire gas ignitions, Department for Communities and Local Government, August 2009
[***] Grimwood. P, 'Forced draft fires', EuroFirefighter, p288-9, Jeremy Mills Publications, 2008

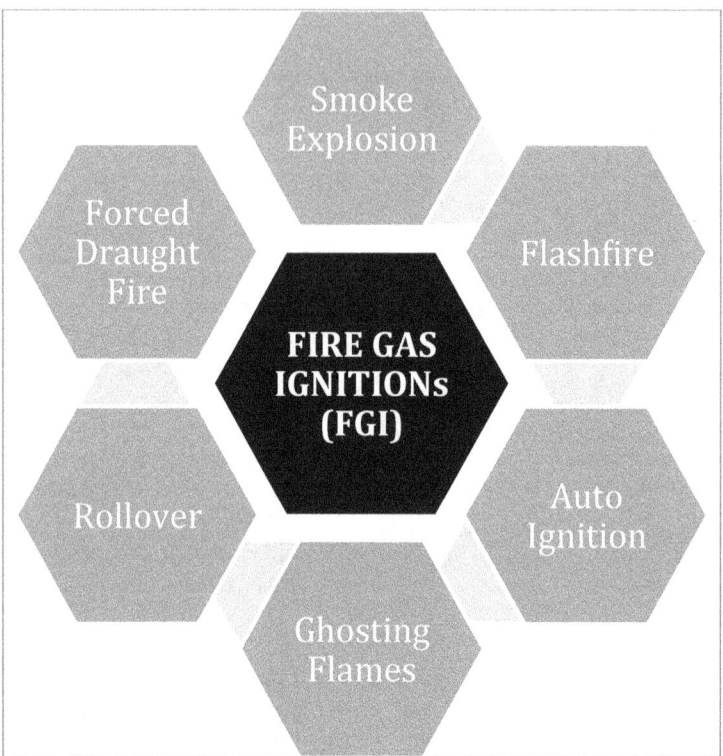

*Figure 3.15: The various forms of **Fire Gas Ignition (FGI)** that may occur at a building fire, each bringing their own warning signs, variations of intensity and threat in the combustion process.*

The differences between a Backdraft and a Smoke Explosion

This is particularly an area that causes much confusion amongst firefighters. There is definitely some grey areas on the fire development curve where either event crosses over to the other.

There remains a great deal of misunderstanding about backdraft which is often used synonymously with the term smoke explosion. A research project[55] undertaken at University of Canterbury, Christchurch in New Zealand focused on improving an understanding of these poorly understood events and the conditions that precede them. A series of small scale experiments were conducted burning a timber crib inside an enclosure with tightly controlled ventilation. Under certain fire conditions, the compartment would suddenly erupt, ejecting smoke and flames from the small openings in the compartment. The research describes the experimental results from the smoke explosion research and compared the smoke explosion to the more familiar phenomena known as backdraft. The research paper concluded as follows:

---

55 Fleischmann. CM; and Chen. Z; Defining the difference between backdraft and smoke explosions; The 9th Asia-Oceania Symposium on Fire Science and Technology; Procedia Engineering 62 (2013) 324–330; Science Direct, Elsevier

'The results presented in the research shows that it is possible for a fire in a closed compartment to undergo a sudden unexpected explosion <u>with no change in the available ventilation</u>. This phenomenon is known as a smoke explosion. A smoke explosion can occur when a fire starts in a closed compartment where the only ventilation is from the leakage through the compartment boundaries. The fire develops into a smouldering state and produces large quantities of excess pyrolyzates and carbon monoxide. Overtime, oxygen rich air leaks into the compartment and the mixture of compartment gases reaches the flammable range, suddenly ignition occurs on the burning item sending a rapidly developing flame front through the enclosure.

The *smoke explosion* is a separate phenomenon to *backdraft* which requires a change in the ventilation. <u>For a backdraft to occur there must be a change in the ventilation</u> such as a window breaking or a firefighter opening a door as they enter the compartment. The new opening allows cold oxygen rich air to form a gravity current that will propagate across the floor mixing with the compartment gases and igniting when the current reaches an ignition source. Once ignited, the flame propagate through the flammable mixture pushing the flammable compartment gases in front of the flame and out of the compartment which culminates in a large fireball as the flammable gases are forced from the enclosure. Understanding the differences between these two phenomena, will help firefighters to better understand the hazardous environment they are entering and be able to take action to mitigate the potential hazard'.

A smoke explosion is one category of Fire Gas Ignitions (FGIs), as reported above.

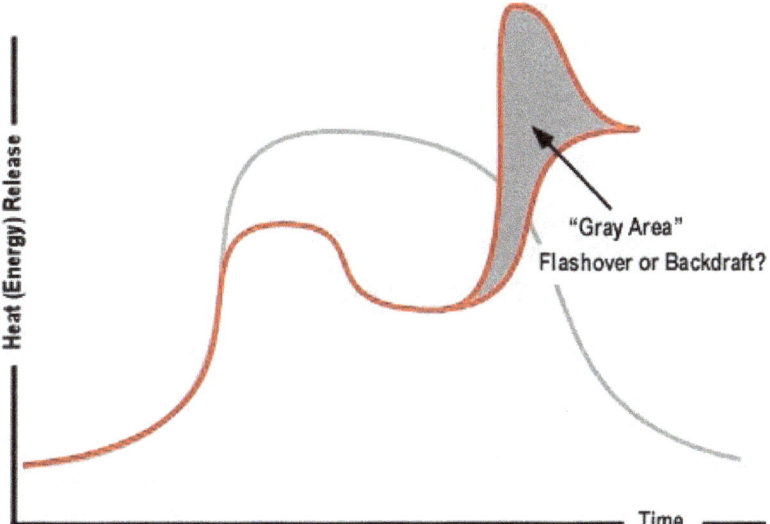

Courtesy of Ed Hartin CFBT-US.com
Figure 3.16: A time/energy release fire growth curve showing the 'grey area' where definitions between events become vague.

The paper provided some recommendations for future research:

- More research is required into the *smouldering combustion* that provides the fuel for the both smoke explosion and backdrafts. Of particular interest to is

Figure 3.17: Flashover, backdraft or Fire Gas Ignition (FGI)? **Photo courtesy of Glenn Ellman**

the *smouldering behaviour* of common combustibles most notably timber and polyurethane. Including experiments to quantifying the species production rates as a function of the oxygen concentration.
- The flammability limits of the enclosure gases are crucial to understanding the conditions that can result in both a backdraft and a smoke explosion. An experimental study to develop a flammability diagram similar to the diagrams developed by Zabetakis for pure mixture would greatly increase our understanding of the hazard.
- Additional smoke explosion experiments with detailed measurements of the species concentrations including unburned hydrocarbons in a number of location both high and low in enclosure and pressure to determine the level of structural damage the might be expected from a smoke explosion.

The photograph at Fig. 3.17 was used in one of the author's CFBT training manuals (2003)[56] and this was linked to an internet poll of over 300 firefighters who had been asked the question – *Flashover – Backdraft or Fire Gas Ignition?* Although unscientific, the results were interesting and demonstrated just how hard it can be to determine precisely what type of event occurred! What would you say?!

Flashover	29%	91 votes
Backdraft	35%	108 votes
Fire Gas Ignition (FGI)	34%	105 votes

Table 3.9: *Online poll of over 300 firefighters ref: Fig. 3.17*

3.22 FLASHOVER OR BACKDRAFT – CAN ONE FOLLOW THE OTHER?

I had a most enlightening discussion the other day with a group of firefighters, fire engineers and fire behaviour instructors. The question of debate was; can a backdraft follow a flashover in the same compartment, and vice versa. The general belief is that flashover may follow a backdraft but not the other way around. Then we went on to discuss multiple compartment flashovers that is, what can cause several flashovers to occur within the same fire compartment. Firefighters generally refer to the 'total involvement of a compartment in fire' as a way of defining the event relating to flashover. How the compartment initially became totally involved in fire is often not referenced or discussed on a tactical level. This is a thought-provoking question that will generate good debate!

---

56 Grimwood. P; Desmet. K; Tactical Firefighting; A comprehensive guide to compartment firefighting and live fire training (CFBT) Firetactics.com – CEMAC; 2003 - http://www.olerdola.org/documentos/cemac-kd-pg-2003.pdf

Flashover	A flashover is caused by thermal heat transfer from a fire plume and ceiling jet, radiating heat (kW/m2) down and across onto the contents fuel load and compartment linings. This causes pyrolysis where solid fuels break down to phase change mainly into combustion products and fire gases. As they are hot they normally rise to the ceiling where they form a hot gas layer, which further radiates heat downwards. In some situations, heat cannot escape from the boundaries because of high insulation levels and will be radiated back into the compartment. Where sufficient air is mixed with these gases they will ignite, usually along the smoke interface with clear air. This flaming will transfer even more heat down and the process speeds up dramatically. Eventually all surfaces reach their auto ignition temperature and start to burn. The entire process depends on room size, ceiling height and the make-up of the fuel load, but normally takes less than 60 seconds in an average sized residential room, once the fire plume reaches the ceiling.	Can a backdraft follow a flashover in the same compartment? It's highly unlikely but yes, I would imagine it's possible. If we can imagine a basement compartment with small quarter windows at ceiling level to the outside. If a flashover occurred in the basement and vented out the quarter windows, the small openings may not be enough to allow adequate air to enter and enable steady state burning, and the fire may become under-ventilated. A similar 'pulsing cycle' could develop as the fire searches for air. The opening of a door into the basement could initiate a backdraft if conditions are right. Alternatively, a ventilation induced second flashover could occur, or even a smoke explosion may occur if a burning brand is uncovered. Whichever event occurs, it would probably be too rapid to determine on-scene and would be referred to as a generic 'flashover'. It would take some computer modelling to demonstrate the most likely outcome.
Backdraft	Where the fire is being starved of air/oxygen and the fire is becoming grossly under-ventilated, a situation may arise where a negative pressure arises in the fire compartment. This can cause air to be drawn in through small openings and under doors to feed the fire. As it starts to burn freely again it emits smoke through these openings but is stifled again by lack of air and a 'pulsing' effect of air in/smoke out cycles that last a second or two. The classic warning sign of backdraft. However, when an opening suddenly occurs, or is created, the negative pressure in the room causes a large mass of air to head in and feed the fire. The resulting backdraft of air can result in a fireball or even an explosion of fire gases causing roofs to lift and other structural damage to occur.	Can a flashover follow a backdraft? Almost certainly yes, this occurs in most (not all) situations.

Continued overleaf.

| Fire Gas Ignition (FGI) | Fire gas accumulations can occur in voids or reservoirs formed under the ceiling/roof that are just waiting for an ignition source. In some instances, these accumulations of gases will be dispersed at lower level to entirely fill a compartment or space. If they are pre-mixed with air within the limits of flammability, then they can ignite or explode as soon as an ignition source is revealed and/or makes contact with the fire gases. Similarly, very hot gases can auto-ignite as soon as they meet with sufficient air. In this case the gases move towards the air supply. If the air moved towards the gases caused by a negative pressure in the room this is a backdraft. If the air is forced into the gases by an external positive pressure (a wind) the resulting blow-torch fire is called a forced draft. | A Fire Gas Ignition (FGI) may occur at any stage of a fire without any warning signs. This can even occur sometime after the main fire has been extinguished.

Commonly, the use of modern methods of construction (MMC) such as cross laminated timber (CLT) panels have been linked to the potential for multiple 'flashovers' to occur in a fire compartment. As each laminated panel chars and peels away in a fire a new surface is exposed. In a long duration fire these repeated pyrolysis and ignition cycles might be defined as either 'flashovers' or FGIs.

The dangers and potential associated with smoke explosion (FGI) in such a fire can only be increased. |

Table 3.10: Can a backdraft follow a flashover?

Multiple flashovers can also occur in fire compartments, particularly where fires occur in converted cold stores or highly insulated refrigerated compartments. As mentioned elsewhere in this book, the author has experienced several such fires where heat retention in the fire compartment causes extreme temperature build-up and multiple flashovers occurring, as fire gas layers repeatedly ignite. Where leakage paths are few, such fires may smoulder under extremely warm conditions leading to a grossly under-ventilated state. This can generate backdrafts when opened up or vented. Both of these events might equally be defined as auto-ignitions (FGIs) or smoke explosions.

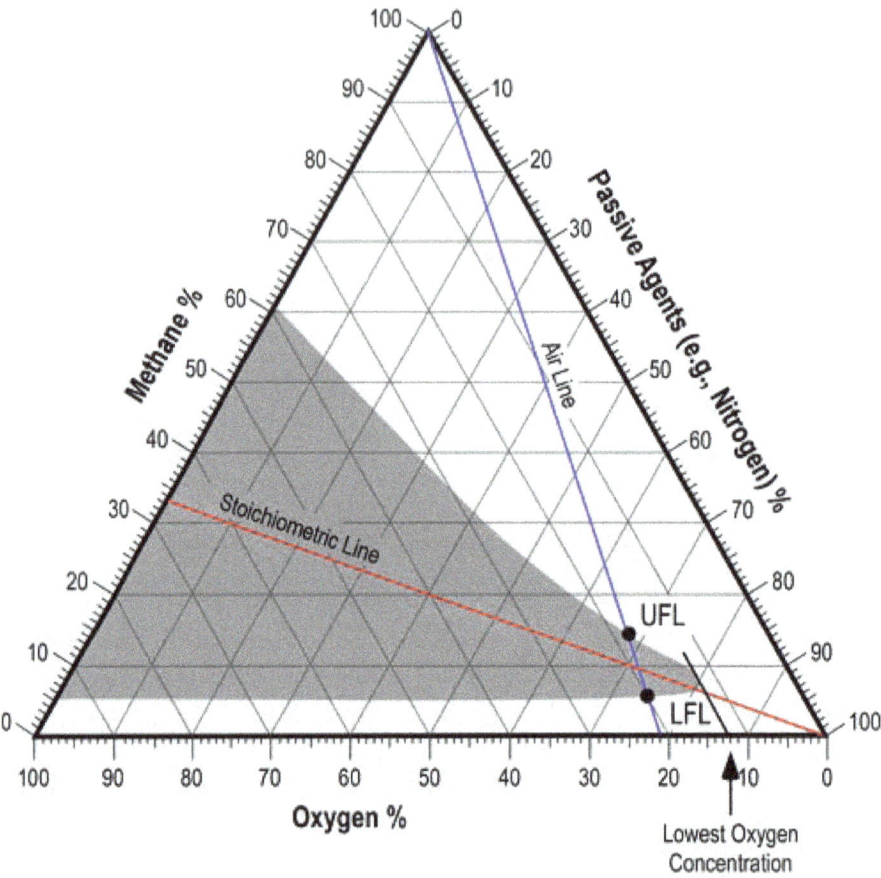

Figure 3.18: Explosive limits (Methane)

3.23 THE KINGS CROSS UNDERGROUND RAILWAY FIRE – LONDON 1987

On the 18th of November 1987, a fire at the King's Cross Underground Station in London exhibited behaviour that was both unexpected and disastrous, ultimately resulting in the loss of 31 lives and the grievous injury of many people including commuters, London Underground staff and firefighters. The fire had ignited within a wooden escalator trench and burnt for approximately 15 minutes, during which firefighters assessed its size and behaviour and concluded that although it was a significant fire, there was nothing to suggest that it would rapidly develop. Indeed, the first responders actually described the fire as like a 'small campfire' or a 'cardboard box on fire'. However, within an unexpectedly short amount of time the fire spread with extreme ferocity up the escalator trench and into the ticket hall and surrounding areas with tragic consequences. Investigations into the King's Cross fire revealed that the extreme fire growth was primarily caused by a distinct phenomenon, subsequently termed the 'trench effect', which caused the flames and combustion products to be confined and concentrated within the escalator trench below the balustrades of the escalator.

Flame attachment driven by the Coanda effect on the buoyant plume in enclosed trenches (at least with side walls) with a pitch angle of at least 30° or greater is at the source of this unusual fire spread. Such extreme fire behaviour of this nature was not known or understood by firefighters at the time of the fire and is demonstrated in an excellent online video produced by the University of Edinburgh[57] where the importance of trench angle and side walls become clearly apparent. The reader is advised to view this video link in order to appreciate where and how such extreme fire behaviour might occur.

Investigator's report

The author was a fire investigator with the London Fire Brigade at the time of the fire and formed part of the two-person team that located the fire's cause and the six-person team that further investigated the fire behaviour over the following six months. Then, working with a team of forensic scientists the unusual fire behaviour (trench effect), although previously known to fire scientists, was established and defined. Some years later, the author's online report[58] gave further information:

> Kings Cross underground station is one of the busiest on London's 'tube' railway network serving over 100,000 passengers during peak hours. At approximately 7.32 pm on the evening of the fire, smoke was seen coming from one of the wooden escalators that was transporting passengers up from the platform levels to the ticketing hall.
>
> The London Fire Brigade dispatched 4 engines and an aerial ladder as the call was received at 7.36 pm and the first of these arrived on scene at 7.42pm. A team of firefighters went down from street level into the ticketing hall from where they could see a fire burning about 20 feet down the escalator shaft with four feet high flames emerging from the escalator stairs. At this stage, there were still passengers exiting from the platforms below in an orderly manner.
>
> As firefighters returned to street level to collect hose and breathing apparatus three officers remained in the ticketing hall to supervise the evacuation of passengers. Whilst two of them began a descent down towards the platforms to prevent further use of this escalator as an exit route, the senior officer – Station Officer Colin Townsley – remained in the ticketing hall at the top of the escalator shaft.
>
> At 7.45pm the fire suddenly erupted up into the ticketing hall and created severe conditions likened to that of a flashover. At street level, thick volumes of black smoke began to emerge from the station entrances and a large number of screaming passengers exited into the street. The fire burned for several hours killing 31 people including Soho's Station Officer Colin Townsley, who died trying to rescue a woman from the blazing ticketing hall as the fire suddenly erupted. Firefighters attempting to re-enter the ticketing hall to fight the fire likened the conditions as similar to climbing down into a volcano.
>
> Various theories were put forward as to what caused the 'flashover like' conditions as the fire suddenly erupted into the ticketing hall. In terms of scientific definition, the event did not conform to the universal acceptance of what constitutes a 'flashover'. Consideration was given to the possibility of a 'backdraft' or a Fire Gas Ignition (FGI) in the ticketing hall's false ceiling. Further thought was also directed at the likelihood of the escalator fire being pushed upwards in the shaft by a 'piston effect' (forced draught fire)

---

57 https://www.youtube.com/watch?v=BJ6VSOkpDYs
58 http://londonfirejournal.blogspot.co.uk/2005/07/kings-cross-fire-1987.html

as trains arriving at platforms forced a major airflow out of the tunnels and up into the ticketing hall. However, mathematical modelling and computer simulations promoted a new theory of rapid fire development within inclined shafts with combustible surfaces termed trench effect. indicating that the existence of the **trench effect** *was strongly dependent on the trench geometry. Indeed, the trench effect was found to only occur in trenches that were sufficiently steep and more readily in trenches that were more enclosed.*

It was established that once the trench effect became established, the 'piston effect' from trains would not have played an important role in the rapid spread of the fire up the escalator and into the ticketing hall. This trench effect was seen to cause hot gases in the buoyant plume to lay along the escalator surface and create a rapid airflow which caused these gases to curl over and over towards the next steps above. The airflows in the trench increased in proportion to the size of the fire, eventually creating a flamethrower type effect up and into the ticketing hall.

However, it is certain that a large quantity of unburned combustion products (pyrolyzates) existed in the smoke and fire gases forming at ceiling level in the booking hall at the head of the escalator as the fire reached this stage of rapid fire progress. It is also apparent, from firefighter's accounts, that the piston effect did play a major part at various stages of the fire's development in creating a more intense fire as trains passed through the station tunnels some levels below the escalator itself.

The cause of the fire was discovered by London Fire Brigade investigators to have resulted from the most likely discarding of a lighted match (smoker's material) by a passenger on exit from the system and strong evidence existed that demonstrated several smaller fires of this type had occurred in the past on this escalator but never progressed beyond a small self-extinguishing fire that probably went unnoticed. Subsequent tests confirmed the likelihood of such a fire being able to develop beyond a stage of self-extinguishment when located in the grease track that existed under the escalator. There had also been several small fires on other underground stations in London that involved wooden escalators.

Although the evidence was heavily weighted towards the careless discarding of a lighted smoker's match as a cause of the fire there had been reports of an earlier incident in the evening on an adjacent escalator in a nearby shaft that raised the question as to whether the Kings Cross fire could have been caused maliciously. There had been several accounts by witnesses who described a small fire existing at the base of an escalator in the Victoria line shaft just minutes before the reported fire in the Piccadilly line shaft that progressed into a major conflagration.

This smaller fire involved an item of burning paper that appeared to have been rolled up and thrown down the shaft from or near the top of the escalator.

The Fennell Investigation into the fire prompted the introduction of the Fire Precautions (Sub-surface Railway Stations) Regulations 1989 (usually referred to as the Section 12 Regulations because they were introduced under section 12 of the 1971 Fire Precautions Act). These led to: the replacement of all wooden escalators on the Underground; the mandatory installation of automatic sprinklers and heat detectors in escalators; mandatory fire safety training for all station staff twice a year; and improvements in emergency services liaison.

Chapter 4

The 'House Fire'

The vast majority of life losses through fire occur in our homes every year where young children and elderly persons are most at risk. The 'house' fire is predominantly the most common and basic building fire a firefighter can encounter. It is generally small by nature compared to commercial and industrial premises, requiring less resources, but often the challenges can be extreme as this is where we are most likely to undertake search and rescue (S&R) operations. Therefore a firefighter's potential for exposure to high risk is increased and since the early 1990s it has been demonstrated in the US, where lightweight construction has become commonplace, that the ratio of firefighters (to the number of fires) killed by 'structural collapse' in residential buildings has actually increased.

Working in thick smoke, even in small compartments, a firefighter can easily become disoriented and even lost where the ability to work safely in such conditions relies on regular and realistic training. It is also dependent on crews working and remaining in close touch with each other. Where smoke levels are within reason it is sufficient for firefighters to remain in visual contact with each other. This old maxim should guide us:

'Go in together, stay together, protect your escape routes at all times, take enough water, and come out together.'

Now there alone is a one hour evening classroom session for firefighters, based on the above statement! Take that maxim and debate amongst each other exactly what it means to you.

4.1 'GO IN TOGETHER – COME OUT TOGETHER'

When analysing why firefighters get killed or injured at house fires we can see a common pattern of command and tactical errors occurring over and again. It's not so much about the quantity of water at the nozzle, any particular extinguishing technique, or anything more than basic tactical actions. We so often fail to learn from past fire experience and implement the lessons of tragedy into our own operational approaches. Where crews are deployed to the interior for S&R:

- The commander's 360° view of the fire building is a critical and primary strategic action giving vital information
- View the sides and rear, as well as the frontage of the building, noting relevant features
- Try and locate the fire compartment/s during the 360 and form a deployment plan

- Important – smoke issuing from an opening does not always denote a fire's location!
- Don't deploy into the building from the front if there is a safer access point elsewhere, depending on fire location and <u>wind direction/speed</u>- this is CRITICAL!
- Are sufficient resources on-scene to deploy safely and effectively?
- Has an <u>adequate</u> firefighting water supply been located and connected?
- Prior to deployment, communicate and brief the crew/s (face to face where possible) exactly what their role is – be very clear and ensure the brief is understood – this is also critical
- Is the flow-rate being deployed to the interior adequate to deal with the involved fire load?
- If the fire rapidly increases in size due to fire related ventilation or further fuel load involvement, is the deployed flow-rate adequate to meet this increasing heat release?
- Monitor all interior crews location and air-supplies, and establish effective communication with them at the earliest opportunity
- Where crews are to work beyond or above the fire for search and rescue, assign a protection hose-line with responsibility to protect their escape route at all times
- Never ventilate or make openings without anticipating the potential impact on fire spread and flow-path direction and ensure the interior crews are aware and support such an action
- As soon as possible have relief crews ready to deploy, to relieve interior crews in good time
- Anticipate bad outcomes, rapid fire spread, lost or trapped firefighters and prepare actions and assignments ahead of such events

4.2 WORKING IN SMOKE

The ability for firefighters to work and search safely and effectively in smoke and heat is something that demands regular and thorough training, learning to work blindly using 'touch and feel' techniques. When doing so, firefighters should remain in close contact with a wall, where possible, as well as each other. Searching a house for unconscious victims must be undertaken with controlled haste, ensuring that no area is left untouched before moving on to the next room in a structured search pattern. Should firefighters take a hose-line with them when searching? Many instructors or SOPs recommend that they should.

- A hose-line is cumbersome and slows the search beyond and above the fire compartment
- For searching or entering the fire compartment itself, a hose-line in support is highly recommended
- In some cases where the fire location is unknown, it is important to protect escape routes by siting additional crews with hose-lines at the stairs or at critical junction points in the house, to protect other firefighters and possible victims
- The greatest use of fire containment actions should be made whilst search is underway. At every opportunity, close all doors that may feed air into the fire compartment.
- If passing the fire compartment to search beyond or above the fire, close all doors leading to into it where possible
- If crews are working in the fire compartment, assign a door control firefighter/crew to reduce the fire's heat release rate progressing towards flashover – partial

closing, inserting a tool to prevent full closure. This support team may be equipped with a secondary safety hose-line.

4.3 THE RISK PROFILE OF A HOUSE FIRE

Based on your country or area, the average house has a lot in common but within a range of variable possibilities in both construction and size. These could typically be brick built, or timber framed with exterior cladding, demonstrating room sizes between 10-100m². They may have multiple or split interior levels with some having basements. In some cases the basement at the front of the building is the ground level to the rear. They may also be detached from surrounding properties or attached with common roof voids between all properties. Each particular structural design presents its own challenges and risks in terms of fire spread.

Typical compartment size	Typical fire load (80% fractile)	Typical HRRPUA*	Design fire growth rate
10-100m2	870 MJ/m2	250 kW/m2	Medium t2

Table 4.1: A typical *HRRPUA – Heat release rate per unit area (m2) – for ventilation controlled post flashover fire

4.4 CHANGING LEVELS

Whenever a team of firefighters working internally are planning to change levels, that is go up or down stairs from the access level, there are some critical rules of engagement they must first consider:

- Has the fire been **located**?
- Has the fire been **isolated** from their position and confined behind a closed door/s?
- Is there a **wind** impact assessment?
- Is there **water** being applied to the fire?
- Is firefighting water-flow off the fire appliance tank or has it been augmented?
- Has a **hose-line** been located between them and the fire?
- What is the **reason** to change levels?

If the fire is a basement fire then the protocols for following such a situation should be pre-planned and documented through SOP, ensuring extreme caution and coordination between fire attack and venting actions. If the fire is on an upper floor the strict protocols should exist for such an operation. However, if firefighters are moving with intention to go above a confirmed fire floor for search and rescue purposes, then again strict protocols should be followed. Failing to follow these protocols can expose firefighters to increased hazards.

Chapter 5

The 'Apartment (Flat) Fire'

Fires in modern mid-rise apartment blocks, residential high-rise tower blocks, low-level flats, or houses in multiple occupation represent a broad range of occupancies that create a range of challenges for the fire and rescue service. There are often access related problems surrounding the buildings, or within the structures themselves. There are also water transport issues in getting adequate amounts of firefighting water to fires at great height. However, as the design of modern apartment blocks evolves we begin to experience even greater challenges in lightweight combustible frames, more open-plan compartments incorporating increased fire loads and larger windows that encourage faster and more intense fire development than ever experienced before. The potential for multi-floor fires in such buildings would seem to be eradicated or at least lessened. However, recent experience shows this not to be true.

5.1 RICE (RESCUE – INTERVENTION – CONTAINMENT – EVACUATION)

One key feature of the design codes stipulates that a fire attack hose-line should never be longer than 45 metres from the fire engine to the furthest apartment, or where a rising fire main is installed, never more than 18 metres from the engine to the pipes inlet and a maximum of 60 metres to the furthest point on a fire floor from a two-hour fire protected [stair] shaft. These are critical distances and UK firefighters have established national tactical approaches and equipment set-ups to meet these code requirements. As an example, the first-aid attack hose-reel used to control over 80 percent of building fires is 54 metres long, effectively meeting the 45 metre rule laid down in the codes.

Dealing with fires in apartment blocks of low to mid to high-rise demands a generic approach strategy. In newer buildings it is more common to see large floor plates with long corridors served by one or two stairwells. It is certainly not uncommon to see smoke control systems protecting long open runs of corridor, sometimes exceeding 60 metres in length (two stair) or 30 metres (one stair). A centrally located stair could, in fact, provide sole egress to 12–15 apartments in a sixty metre long corridor, with smoke control provided. In such cases, the tactical approach needs to consider the four basic elements of **RICE** which are *Rescue – Intervention – Containment and Evacuation*. These tactical options should be evaluated on the situation presenting on arrival and must be prioritised according to a primary risk assessment.

The most likely tactical considerations to be made are:

- Is an immediate rescue or search operation in the fire apartment required?
- Is there good information or belief that persons might be involved?

- Is an immediate firefighting intervention needed?
- Is there a rising fire main (standpipe) or will hose need to be laid up stairs from the exterior?
- Should assisted evacuation take place on the fire floor as a priority?
- If a defend in place (stay put) strategy for the building is in place, be aware of uncontrolled free movement of occupants at untimely points when opening the fire apartment
- Is the building set-up for simultaneous evacuation or defend-in-place operations?
- Is the situation and are fire conditions supportive of either of the above?
- Are adequate resources available to consider evacuating the fire floor prior to opening the fire compartment into the corridor?
- What will be the time-delay in accessing the fire?
- Is the smoke control system designed to maintain a tenable escape route for fire floor occupants from a post-flashover apartment fire during firefighting operations?
- Or, is the smoke control system purely to protect the stair only?
- How is the smoke control system configured to direct smoke to an extraction point?
- What smoke containment options might be used, or deployed?
- Have you considered a smoke-stopper door control option?

5.2 SMOKE BLOCKER – DOOR CONTROL CURTAIN

The 'smoke-blocker' door control device[59] has been a tactical option used by German firefighters for twenty years and this clever device is now gaining popularity across Europe and also in some parts of the USA. The uncontrolled movements of occupants who decide to self-evacuate at the exact moment firefighters are opening the fire apartment, will cause major concerns on the fire floor. Placing the fire resistant portable curtain over an apartment doorway takes a few short seconds and protects the corridor or stair from heavy smoke or even flaming combustion. With the curtain in place the fire apartment can be entered for search and rescue and firefighting with the corridor/stair remaining relatively smoke free and tenable for escaping occupants. Without the curtain in place and even with a mechanical smoke removal system protecting the corridor, this would not normally be possible during the firefighting phase of operations. Past experience and computer aided analysis both confirm that corridors served by smoke shafts and mechanical smoke clearance systems are soon over-powered by heavy smoke emitting from an open fire compartment.

5.3 NHBC FOUNDATION FIRE RESEARCH

The NHBC Foundation[60] was established in 2006 by the NHBC in partnership with the BRE Trust. Its purpose is to deliver high-quality research and practical guidance to help the industry meet its considerable challenges.

---

59 http://www.smokeblockingdevice.eu/
60 www.nhbcfoundation.org.

Figures 5.1: The smoke blocker is a tool for controlling smoke emissions from the fire compartment or main entry door in an apartment block. It serves to protect the corridor and stairs from smoke during firefighting intervention taking place. **Images courtesy of Dr. Michael Reick http://www.rauchverschluss.de/index_e.htm**

Since its inception, the NHBC Foundation's work has focused primarily on the sustainability agenda and the challenges of the government's 2016 zero carbon homes target. Research has included a review of microgeneration and renewable energy techniques and the ground breaking research on zero carbon and what it means to homeowners and housebuilders.

The NHBC Foundation, in association with the Building Research Establishment (BRE) and a panel of industry experts, including London Fire Brigade, is also involved in a programme of positive engagement with government, development agencies, academics and other key stakeholders, focusing on current and pressing issues relevant to the building industry. Its research into issues surrounding fire safe building design and construction is highly commendable and well worth reading.

In 2009 the foundation produced a guidance document[61] relating to open-plan apartment designs. They inform, 'to ensure a more coherent approach and highlight potentially unacceptable risks, this research report has been published for the benefit of designers, housebuilders and building control practitioners. It is hoped that it will contribute to more consistent design and a more straightforward approval process and so

---

61 Open plan flat layouts – Assessing life safety in the event of fire; NHBC, 2009

benefit the industry and its customers'. It continues, '[the report] addresses layout, size, travel distances, enhanced detection options and sprinkler use. In addition it addresses the human implications, including the various reactions, wake up and response times from people occupying the building'.

A number of arguments had been put forward by designers and fire engineers to justify why departures from AD B (UK fire safety prescriptive guidance) should be allowed, enabling open plan flats to be built.

The arguments include the following points:

- Travel distance could be limited to that allowed for a bedsit (9 m), although with inner room(s) (which AD B does not allow).
- Enhanced detection and alarm systems could be provided, beyond what is specified in AD B.
- Sprinklers or other suppression systems could be provided.
- 'Upside-down' (two level) flats, with bedrooms on the lower level and lounge, kitchen, etc. on the upper level, would reduce the consequences of fires while people are asleep.
- AD B no longer includes the requirement for fire-resisting doors to have self-closers. Therefore, it is assumed that these doors will be open. The question is how this is different from an open plan layout.
- Fire modelling may be employed to demonstrate acceptable safety.

Each of these arguments attracts a counter-argument to justify why open plan flats should not be allowed. Fire and smoke movement modelling, where it is carried out correctly, can provide a quantitative, deterministic analysis of alternative approaches. These analyses can be used in comparative studies to demonstrate equivalency. However, the assumptions made in practice by the users of these models can be highly subjective and do not typically address human factors, with the result that the arguments may not be conclusive one way or the other.

The main conclusions of the study were:

These results indicate that open plan flats with a sprinkler system (in accordance with BS 92514 or BS EN 12845,13 as appropriate) and an enhanced detection system (LD1 system in accordance with BS 5839-63) can provide a level of safety that is at least as good as that of a similar AD B compliant design.

- Flat size/travel distance has been shown not to be a significant factor (up to the largest size considered here: 12m x 16m). However, this result should not be extrapolated to larger designs without further analysis.
- It is not possible to state with sufficient confidence that enhanced detection alone could satisfy the requirements of the Building Regulations – Open plan flat layouts. Assessing life safety in the event of fire
- A fire engineered solution should consider all aspects of the whole fire system, including fire growth, smoke movement, detection, suppression, human behaviour and interactions between them.
- Without consideration of human behaviour, depending on the scenario, fire models might not give an adequate measure of the risk.

This study did not investigate the effect of:

- apartments larger than 12 m × 16 m (192 m^2)
- multi-level apartments

- smoke control systems
- water mist and other suppression systems
- an open plan kitchen close to the front door
- self-closing doors on the inner rooms

Case	Description	ADB compliant or open-plan	Size of footprint and ceiling height	Active protection systems
1a	Studio (9m travel distance) using a range of window sizes	ADB compliant	8 x 4 x 2.4m high	LD3 system (smoke alarm in circulation space)
1b	One inner bedroom* using a range of window sizes	Open-plan	8 x 4 x 2.4m high	LD1 system (smoke alarm in each room)
1c	One inner bedroom* using a range of window sizes	Open-plan	8 x 4 x 2.4m high	LD1 system (smoke alarm in each room) plus sprinklers
2a	Two bedrooms using a range of window sizes	ADB compliant	10 x 8 x 2.4m high	LD3 system (smoke alarm in circulation space)
2b	Two inner bedrooms using a range of window sizes	Open-plan	10 x 8 x 2.4m high	LD1 system (smoke alarm in each room)
2c	Two inner bedrooms using a range of window sizes	Open-plan	10 x 8 x 2.4m high	LD1 system (smoke alarm in each room) plus sprinklers
3a	Three bedrooms using a range of window sizes	ADB compliant	16 x 12 x 2.4m high	LD3 system (smoke alarm in circulation space)
3b	Three inner rooms using a range of window sizes	Open-plan	16 x 12 x 2.4m high	LD1 system (smoke alarm in each room)
3c	Three inner bedrooms using a range of window sizes	Open-plan	16 x 12 x 2.4m high	LD1 system (smoke alarm in each room) plus sprinklers

*9 metre travel distance measured to back of inner room

Table 5.1: Summary of the 9 modelling scenarios analysed during the NHBC research
Courtesy of NHBC Research Foundation

It was demonstrated in the research that the computer model CRISP or an equivalent tool that considers all aspects of the whole system, including human behaviour, can be utilised to compare risks in open plan and conventional AD B compliant flats.

The detailed conclusions that apply to the specific layouts and the simulations carried out are:

- Flat size/travel distance does not seem to be a significant factor.
- The open plan designs with enhanced detection have risks of death and injury similar to those for ADB compliant flats. At present it is not possible to reach a robust conclusion that enhanced detection can offer equivalent or better levels of safety compared with AD B compliant flats.

- The open plan flats with a sprinkler system (in accordance with BS 9251 or BS EN 12845, as appropriate) and an enhanced detection system (LD1 system in accordance with BS 5839) can provide a level of safety that is at least as good as that of a similar AD B compliant design.
- More work is needed if conclusions about comparisons of risks of death and injury can be drawn for generic cases rather than the specific cases generated.
- Whether sprinklers would be a cost-effective option has not been addressed by this study. The greater value of an open plan flat compared with one of the same floor area that includes a hallway could compensate (or exceed) the cost of installing a sprinkler system.

As an engineering document offering benchmark guidance on open-plan apartment layouts, the NHBC research is extremely detailed in its coverage and the reader is recommended to view this document for a more detailed analysis.

The Fire Performance of new lightweight steel and timber framed residential Buildings

A further programme of detailed research was reported in another NHBC Foundation document[62] in 2011 where both *pre and post* construction issues concerning the fire safe design of medium-rise lightweight timber and light gauge steel framed residential buildings were addressed.

It was a stated objective of the research that; 'In terms of internal fire spread the importance of specifying and installing the correct passive fire protection has been highlighted. Compartmentation may be prematurely breached in the event of a fire either by incorrect specification of linings, poor workmanship or inadequate supervision. It is essential that cavity barriers are installed and located correctly. Any discontinuities will provide a route for fire spread through cavities bypassing all other passive fire protection systems'.

This important point has proven time and again where fire has spread dramatically through such construction, following occupation.

The research itself takes a close look at fires in roof voids, structural insulated panels, exterior cladding, lightweight steel and timber engineered floor joists amongst other structural features. It also reviewed the levels of fire resistance requirements. It then goes on to offer guidance on post-flashover fire development, contents and combustible structure fire loading and the influence of compartment boundary thermal inertia on fire spread in dwellings.

For example, the report continues; 'A number of studies have been performed in the USA and Canada to investigate the fire performance of light engineered floor systems. These studies, however, are of limited use as they predominantly investigated unprotected floors as these are allowed under the US and Canadian regulatory frameworks. Indications from such research suggest that the failure of engineered timber can occur in as little as 6 minutes when left unprotected, as is the case in many US and Canadian basements. Comparatively solid timber (traditional) floors have been shown to fail in 19 minutes for approximately the same fire load conditions. Similar conclusions were found in research undertaken in Canada'.

Data from the FPA suggested that problems with fire service access and, to a lesser extent, inadequate water supply, are some of the chief impedances to firefighting in dwellings.

62 Fire Performance of New Residential Buildings, NHBC Foundation, 2009

5.4 STAIR-SHAFT FIRES

The likelihood of fire in a residential stairwell is rare but in houses of multiple occupation they do occasionally occur. In some situations, with open flow-paths from fire floor to roof, smoke and hot gases, and sometimes fire, can be drawn towards and into the stairwell. A stair-shaft fire often presents as illegal storage within the stairs or through the failure of self-closing devices leading into flats off the stair. Such fires are of course extremely likely to create a major hazard to occupants in single stair buildings but are generally fairly straightforward for firefighters to deal with as the stairwell generally acts as a 'chimney' if open at the top. Sometimes this vent opening will be natural or may even be part of the ventilation strategy to the stair. In other cases the fire service may create the vent themselves where wired glass roof-lights may be opened to advance the approach of interior firefighting crews. As such vents are opened, the smoke interface with clean air will rise in the stair but the gases may also ignite at the roof opening if hot enough. As air feeds in to the base of the smoke interface, there may be an ignition of the gases if they exist in an under-ventilated state (rich-mix) however this burning will only occur at the base of the smoke layer, until the layer itself rises further up the stairwell and out of the vent, eventually becoming well vented and alight as the ratio of *smoke stack height* to *air supply* balances out.

The simplest way to deal with such fires is for the firefighter to apply 2-3 short bursts (2-3 seconds each) of a narrow fog cone up into the smoke stack and follow the advance up the stairwell as the smoke interface continues to rise. This approach maintains a smoke layer that is always above the firefighter's heads and the water droplets are carried up on the thermals created by the fire vented from below and above. This tactical approach will be extremely rapid as the hose-line advances upwards. However, a second hose-line should follow immediately behind to deal with any heavy storage on the stairs, or directly into apartments located off the stair where fire may have spread to, or from. It is important never to go above these points with the first line for fear of the fire redeveloping behind your advance compromising your egress route.

Un-vented Stair-shaft Fire

Sometimes the head of the stairs will not be vented, or the vent if it is open is too small to release the smoke and hot gases effectively. In such cases the smoke stack will drop further down the stairwell and may completely fill it. This makes for a more difficult and slow approach through a hot gas layer that will hang in the stair and it is here that the benefits of vertical ventilation are soon realised. An effective venting action in such cases will raise the smoke stack far more quickly, assist the firefighting approach and also protect occupants located on lower floors.

5.5 CORRIDOR FIRES

What can burn in a sterile corridor? Is a question I am frequently asked by fire design engineers and architects? Quite simply, it's the flammable fire gases transported there when a door's self-closing device fails. There is a trail of destruction that has recently taken the lives of multiples of firefighters in Paris (5 firefighters), New York City (3 firefighters) and places such as St. Petersburg (7 firefighters), where corridor fires have dramatically extended out beyond the room of origin due to pressure differentials created between the fire and other openings. Corridor fires can be very difficult for firefighters and like stair-shaft fires, they are rarely encountered. However, the creation of a flow-path between the stair, a smoke shaft or a service shaft and the fire compartment, if enhanced by an exterior wind in particular, can wreak havoc on the fire floor. As building design moves to longer corridors created by the removal of stairwells in residential buildings, firefighters can be exposed to very intense and difficult to handle fires. It is not uncommon to see >30 metre corridors in single stair buildings and 60 metre corridors in multi-stair buildings. These are generally protected by various smoke control arrangements, some natural and some mechanical with many being automated by smoke detection in the corridors. It is critical for firefighters that these smoke control systems are correctly configured (see chapter 19) by design and that firefighters fully understand the various operating principles, so that they may be overridden by firefighter control panels if they are not supporting the firefighting tactical advance. As stairs are removed and corridors become longer, alternate vantage points from which to approach fires are reducing in number. If an exterior wind is forcing smoke and hot gases from the fire-involved apartment, reaching temperatures of >400°C floor to ceiling along a corridor towards an opening, the option of approaching from an alternate stairwell located at the other end of the building may not now be available.

As well as exterior winds, fire gases can be forced into and along corridors by flow-paths created by the pressure differentials associated with natural stack effects in tall buildings, mechanical smoke control systems or even natural smoke buoyancy effects. It is important for the fire commander to establish control over such smoke movements by taking account of openings into the stair, into any corridor vent including smoke shafts and also at the head and base of the stairs. A good understanding of smoke movements in tall buildings is therefore required.

Taking a look at fire videos of simulated, or training, corridor fires one might ask; what makes a fire turn right along a corridor when leaving a fire apartment instead of left (or vice versa)? Have you ever considered how heat flux bearing down on a firefighter in a corridor is lessened by cool air flowing in towards the fire-involved apartment at low level? That is not to say that intense corridor fire conditions are ever at all bearable! However, unless there is a wind driven scenario pushing *floor to ceiling* heat in your direction, there is always a fast moving and cooler air current at low level. Fire, smoke and heat will always head towards the lowest pressure differential, which in a corridor will be a smoke shaft, a window AOV or an open stair door. In this way it may be possible to control the direction of smoke and heat movements by closing off such openings if such tactical advantage is needed. Close the stair door; close any openings in the stair; close an incorrectly *design configured* smoke shaft using a control panel; open a vent at the opposite end of a corridor in a two stair building. These actions can sometimes redirect combustion products away from the hose-line advance where the stair with the rising main is the tactical vantage point. (Some smoke shafts are inappropriately located adjacent to and alongside a firefighting stair (rising main) in a long corridor and will cause combustion products to flow towards this vantage point). An example of this is given in the author's first book (reference author's first book – *Fog Attack* – Empire State building fire p263-4) where an exterior wind was creating difficult conditions for firefighters when approaching a fast-developing fire high up in a city centre office tower. This approach was eventually changed to an alternate stair that offered some protection where firefighters were able to make good headway and extinguish the fire, due to the wind driven flow-path's preferred direction towards the original stair.

The principle of designing effective fire doors, where fire resistance actually extends to the floor slab above, should be considered in extended length corridors (beyond codes) for these may serve to save firefighters from injury or even loss of life. If the stair is central to the corridor then fire doors should ideally be sited either side of the stair to protect the wings. If the corridor is longer than 12 metres then fire breaks (even on hold-open devices) should be located. Whilst these doors may not be critical features during the escape phase, they will surely assist the firefighting and evacuation processes if needed.

What does this mean to the Firefighter?

Fires in apartment blocks should be considered high-risk, even though life losses and injury have reduced over the past three decades. However, the buildings are generally designed to remain occupied whilst firefighters deal with any outbreak of fire and occupant safety relies on the building meeting its design standard and firefighters being effective in preventing the fire spreading.

Tactical options should follow the RICE strategy –

- Rescue
- Intervention (fire)
- Containment
- Evacuation

Which of these tactical options comes first depends on fire conditions, floor layout, the building evacuation plan, viable reports of missing or trapped occupants and environmental factors such as wind velocity and direction. An effective pre-plan should also cover water requirements, flow-testing the nearest three hydrants and anticipating the potential for aerial access used for rescue or as a water tower.

Importantly, the functioning and status of any smoke control system should be determined and considerations around its likely impact on the firefighting operation should be made. As a benchmark, no such system should be deactivated or overridden unless it is having a negative impact on firefighting.

It is worth noting that residential sprinkler systems may only have 10 or 30-minutes water supply, depending on the age and type of building. Therefore, compartment access should be within these time-scales where possible.

Another important consideration should be for occupants adjacent to the fire and particularly in the apartment/s immediately above or on the fire floor itself. Human behaviour has shown that occupants may self-evacuate at any point during the firefighting operation, if not pre-evacuated before firefighting occurs.

Chapter 6

The 'Office Fire'

The standard 'office' fire is not seen in most circles as presenting a serious risk. Working for the UK Home Office, Wright and Archer63 analysed 1990s UK fire data and derived relationships between the "age" of the fire (time from ignition until fire brigade attendance at the fire) and the level of fire loss incurred (average area of damage in m2) for a range of different building occupancy types. A linear regression line was fitted to the data for each occupancy type, with the slope of the line representing the rate of damage (m2/min) produced by a delay in attendance. Occupancies were then categorised in accordance with this (linear) fire growth rate –

- **High**: public buildings, factories and universities
- **Medium**: retail, hotels and schools
- **Low**: hospitals, licensed premises and offices
- **Very low**: care homes.

The researchers also examined the level of fire loss incurred in terms of average area of damage in m2 and financial loss (using insurance claim data) for a range of different building occupancy types. Using this information, they were able to derive a relationship between fire brigade response time and financial fire loss for application to fire cover risk assessment.

A sample of incidents containing suitable fire spread data was extracted from the London Fire Brigade's real fire library (RFL) for the five-year period between 1996 and 2000[64].

Fire Damage	<1 m² (%)	1–10 m² (%)	10–100 m² (%)	>100 m² (%)	Total Number of Fires
Percentage	30	44	22	3	63

Table 6.1: 63 office fires in London – data taken from the real fire library

Of 63 fires in office buildings, just 19 were analysed for fire growth rates where a medium *t-squared* growth rate was observed on two occasions and faster growth rates were not recorded at all. However, it has been seen that there are clear differences between fire growth rates in cellular office space and open-plan layouts. The GCU research described in Chapter 12 points out *t-squared* fire growth rates are often 'medium' in cellular floor layouts but may approach or even exceed 'fast' growth rates in open-plan office space. When designing and planning for fire resistance ratings and firefighting operations in

---

63 Wright MS, Archer KHL. Further Development of Risk Assessment Toolkits for the UK Fire Service: technical note –financial loss model. UK Home Office Technical Note C8155/D99052, March 1999.
64 Holborn. P.G, Nolan. P.F, Golt. J; An analysis of fire sizes, fire growth rates and times between events using data from fire investigations; Fire Safety Journal 39 (2004) 481–524; Elsevier

office buildings, a *medium* fire growth rate is always applied (BS 9999 for example) and no consideration is given to cellular versus open-plan floor layouts.

6.1 OFFICE FIRE (CCAB) CHICAGO 2004

On October 17, 2003, in the Cook County Administration [open-plan office] Building (CCAB), 69 West Washington, Chicago, Illinois, a fire resulted in six fatalities and several injuries. In response to a request from the Governor of Illinois, the National Institute of Standards and Technology (NIST) agreed to provide technical assistance to the Governor's review team. NIST's focus was the simulation of the fire using the Fire Dynamic Simulator (FDS) and graphic visualizations to provide insight into the fire growth and smoke movement[65]. A multiple workstation enclosure experiment was designed to simulate the ignition and heat release rate of multiple workstations exposed to a high thermal insult in a partially enclosed area. The fire spread throughout a 240m² open-plan office space in less than 12 minutes. There were no sprinklers in the building.

Figure 6.1: The high-rise fire at 69 West Washington (CCAB) Chicago in 2003. **Image courtesy of National Institute of Standards and Technology (NIST)**

---

65 D. Madrzykowski W.D. Walton NIST Special Publication SP-1021 Cook County Administration Building Fire, 69 West Washington, Chicago, Illinois, October 17, 2003: Heat Release Rate Experiments and FDS Simulations

The 'Office Fire' • 147

Figure 6.2: Multiple workstation enclosure fire experiment at CCAB at 20 secs into the fire growth. **Image courtesy of National Institute of Standards and Technology (NIST)**

Figure 6.3: Multiple workstation enclosure fire experiment at CCAB at 300 secs demonstrating a heat release rate of 16 MW with four workstations fully involved just five minutes into the fire growth. **Image courtesy of National Institute of Standards and Technology (NIST)**

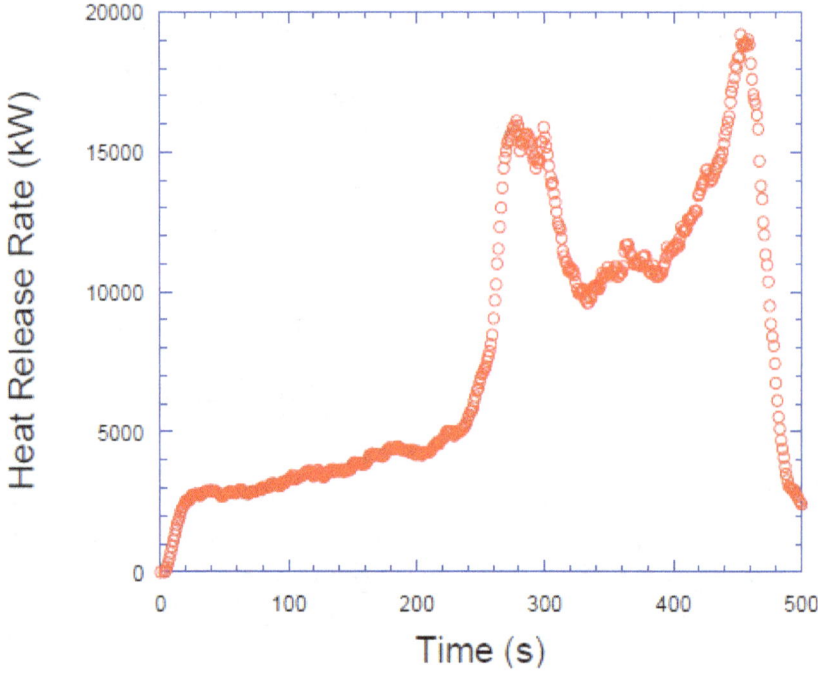

Figure 6.4: The measured heat release rate of the Multiple workstation enclosure fire experiment at CCAB demonstrated a 19 MW fire within 8 minutes of the start of the fire growth period. **Image courtesy of National Institute of Standards and Technology (NIST)**

Based on the estimated timeline of fire development, the fire spread throughout the open plan office space took approximately 10 min to 12 min. The fire department was unable to suppress the fire from the southeast stairway. These two facts are indicative of a fire that developed rapidly and generated high temperatures in and around the 240m² fire compartment. This raised the question, was there something special or unusual about the furnishings and/or the interior finish, or other factors, which led to the fire conditions encountered by the fire department? Or were the conditions consistent with normal office building fires?

At this stage, all items in the fuel load had been closely analysed for their contribution in such a fast-developing fire. In previous experiments, the peak heat release rate of workstations with a floor area of 3.8 m2 (40.9 ft2) ranged from 2.8 MW to 6.9 MW range. The range was dependent on the number of sides enclosed. The workstation with the lower heat release rate had two sides enclosed while the workstation with the higher heat release rate had 4 sides enclosed with an opening for an entry way. The side panels of the workstations tested were in similar construction to those from the Cook County Administration Building. The single workstation from the Cook County Administration Building covered a floor area of 3.15 m2 (33.9 ft2) and had a similar fuel load per unit area of 83 kg/m2 (17 lb/ft2). The Cook County workstation had two sides enclosed and two partial sides enclosed. The peak heat release, 3.5 MW, falls within the lower quartile of the data range from the previous experiments.

Based on the heat release rate comparisons, the furnishings in Suite 1240 of the Cook County Administration Building would be considered 'typical' or representative of office furnishing or interior finish materials that have been tested previously, with a total fuel load per unit floor area of approximately 46 kg/m2. In fact, several other serious office fires with open-plan designs were discussed in the NIST report where fire spread throughout the floor space had been extremely rapid and far greater than would normally be expected from a medium t-squared fire. Based on the heat release rate data and the historical information from previous high rise, office building fires, the fire load and the fire development follows the trend of being a typical fire in this type of occupancy.

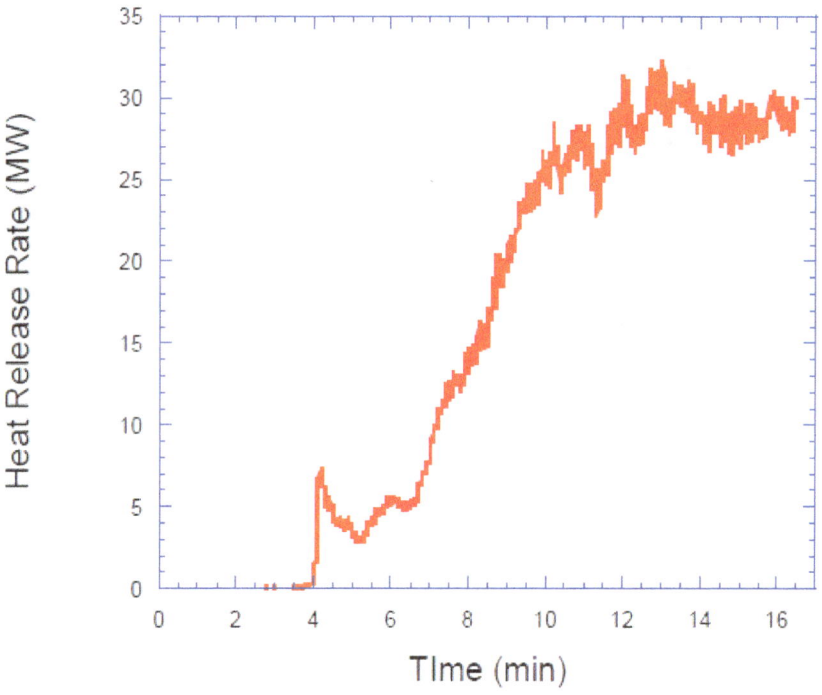

Figure 6.5: Estimated heat release rate based on an FDS computer model, with fire fully involving the 240 m² 12th floor office floor plate at the CCAB fire – reaching over 30 MW within 12 minutes from the start of the fire growth period. **Image courtesy of National Institute of Standards and Technology (NIST)**

What does this mean to the Firefighter?

- When pre-planning, visiting or firefighting in offices take note of the floor layouts in terms of the following:
- Is the building cellular or open-plan in floor-layout (or even mixed occupancy/layout)?

- Where are the access stairs and firefighting mains in-line with the floor entry points?
- Are firefighting lobbies separating the stair from the accommodation?
- Could a fire be fought from that location, applying a high-flow stream onto the floor-plate?
- Where are the firefighting lifts?
- Where are the riser inlet points?
- Imagine how firefighting hose will be laid into/onto the floors and what are the hindering factors, if any?
- What flows are available from nearby hydrants (1, 2 and 3)?
- Pre-plan for aerial access and flow capability
- A fire entering a development curve may spread beyond the capability of a single firefighting hose-line within 10-12 minutes (500 L/min) or 12-14 minutes (700 L/min)
- Office buildings of just two levels may only have 15 minutes of fire resistance to the structural elements

The key to success in such fires is pre-planning! Familiarisation visits, in-house training where possible and a tactical water-plan for the first three nearby hydrants, taking into account a possible need for an aerial water tower in the early stages of a tall building fire.

Chapter 7

The Town Centre 'Shop Fire'

Over the past few decades we have seen a greater use of plastic materials in our homes and workplaces and this has led to higher fire loads in nearly all building environments. We have also experienced greater uses of natural light with larger windows in some cases, and an increase in glass shop front dimensions. With these factors coupled together, we are now seeing greater intensity in the fires we respond to as increasing ventilation profiles feeding higher fire loads become commonplace. In fact our town centre shops have gone through a phase of redevelopment during the past twenty years that has resulted in greater floor space, far deeper shops going back as much as thirty metres and floor to ceiling window glass at the front of the units. We are now seeing a sudden trend in serious town centre shop fires that escalate far beyond the original shop unit to involve upper floor levels (sometimes residential) and surrounding buildings.

We must therefore address our tactics and optimise the way we apply water, or alternative means of extinguishing media, when suppressing or holding fires, in order to place greater control over fast developing fires from the outset.

7.1 THE TOWN CENTRE SHOP PROBLEM

The problem with town centre shop fires lies in the fact that they are now so heavily loaded with combustibles, ranging from shoes, to sports equipment and clothing, to paper goods and card shops that demonstrate an elongated floor space with a large glass opening at the front that generally fails early in a developing fire. However, the actual opening to floor space factor (A_v / A_f) demonstrates a very low ventilation factor for a high fire load. What this means is that when a fire occurs, the air supply is initially ventilation limited and the fire load will burn for a very long period (at around 15 MW heat release rate) in an area that is difficult to reach from the street using exterior streams. However, the fire intensity in the shop creates a funnel of heat heading towards the front and entry is often difficult for firefighters due to untenable conditions. In effect, such fires burn for a very long time and may well breach fire resisting partitions and spread both horizontally and vertically into adjacent areas. Unfortunately, it is rare to find sprinkler protected shop units in town centre developments outside of covered shopping centres.

7.2 FIREFIGHTING TACTICS AT TOWN CENTRE SHOP FIRES

In some cases, particularly where the fire is located in the front section of the shop, a high-flow stream of water or CAFS is required to gain rapid knockdown at an early stage of fire development as fire and heavy hot smoke breaches the front windows. At this point the firefighter is faced with applying a quantity of water that will extinguish a major part of the fire whilst maintaining a constant supply of water as the water tank heads towards dry. If there is an 1800 litre tank on-board this will run dry within 3.5 minutes unless an immediate hydrant or water supply is located. A second fire engine may be used to augment the supply at this point, if immediately on-scene.

Tactical options at fast developing town centre shop fires:

1	One or two high pressure hose-reels from the street in an attempt to 'hold' fire spread	A hydrant should be available before the water tank runs dry
2	A main 500 L/min jet applied from the street using tank water and then hydrant	3.5 minutes before the water tank runs dry
3	A portable monitor (1500 L/min) applied from the street using tank water (BLITZ attack)	Just over one minute before the water tank runs dry

Table 7.1: Tactical options for fast developing town centre shop fires

One will ponder over these three options but they are generally the three tactics used where a shop front is spewing large amounts of heavy smoke and fire on arrival. It is however clear that very few firefighters or fire commanders will choose option 3 (BLITZ) although this is quite often the best and only way to deal with such fires. There are countless videos on social media of firefighters attempting to control intense shop-front fires immediately following arrival with one and even two jets, to no avail. The decision to use option 3 will of course depend on (a) equipment availability and stowage; (b) time to support engines arriving on-scene; (c) location of and access to nearest hydrants or water supplies, and; (d) fire conditions as they present. However, this scenario is certainly one to be planned for and prepared for. Is your crew actually prepared to deliver 1000–1500 L/min in one quick hit? It isn't that long ago that we had roof-mounted monitors on our fire engines to deal with exactly this kind of situation.

What does this mean to the Firefighter?

There is a national trend in the UK where town-centre shops have become deeper (longer) following redevelopment over the past two decades and the fire loads have increased dramatically in-line with plastic content. Add to this the larger windows than these shops used to have, sometimes with full-width glass, and we have a recipe for very intense long duration fires that are difficult to reach with firefighting streams. What generally happens is that 2–3 firefighting jets are unable to achieve knockdown from the street as the fire reaches structural boundaries to spread vertically and horizontally back and sideways, involving all the surrounding buildings. Again, common roof voids add to the problems and in some cases, firefighting may be delayed due to the typical 'shop and dwellings' construction that places a life risk directly above. Tactical options may need to address these fires in ways that are not commonplace to UK firefighters these days. In fact, going back to WW2 firefighting tactics using high-flow >1500 L/min external streams immediately on arrival may be required.

This will require some elements of pre-planning in relation to the location of adequate hydrant grids and the provision of portable monitors fed by a single 90mm hose-line.

Chapter 8

The 'Warehouse or Industrial Unit Fire'

Warehouses are properties that are used for the storage of commodities. Despite their common purpose, warehouses vary on the basis of size, types of materials stored, design, storage configurations, construction and other factors. As warehouse designs evolve towards lightweight construction with increases in both volume and floor space, their contents and layouts are conducive to fire spread and present obstacles to manual fire suppression efforts. Large buildings such as retail or storage warehouses present one of the greatest challenges to firefighters and fire engineers. These mainly consist of older traditional style construction with brick facings and trussed roofs or lightweight steel 'shed' structures with exterior cladding. The hazard exposure to the fire service in these occupancy types is extreme and multiples of firefighters have tragically lost their lives when fighting such fires. Four firefighters killed in the Warwickshire warehouse fire in 2014; six firefighters died in a Glasgow warehouse fire in 1972; and in 2016 eight firefighters die in a Moscow warehouse fire with three more firefighters killed in a Peru warehouse/factory fire a few weeks later. The risks to firefighters in such large buildings are immense, where UK high bay storage commodities may be stacked almost to the high ceiling (>10m), or high fire loads may exist closely packed in areas with just 3-metre-high ceilings. Heights of 5 pallets is quite common and is the maximum that can be reached by a turret truck. This is equivalent to just under 11m in storage height. Storage heights of 8 pallets to 18 m, and occasionally as high as 30m may be seen. Most new builds are 12m to the eves, older buildings are 6 m, 9m, 10m. 12m eve height is to some extent the limit for a *standard* warehouse as this is the limit of the mechanical handling equipment. Floor areas are typical around 25,000 m² on average (application of the 45m escape travel distance, to some extent, limits the floor area) and up to 450,000³ volume spaces have been seen.

The economic losses through fires in warehouses are enormous. Once a fire develops into a growth curve the fire spread is usually vertical during the initial stages of growth and such a fire will normally be beyond the control of an internally deployed primary attack hose-line within 15-20 minutes. Unless firefighters can access the fire from a safe distance (possibly from the exterior) using high-flow hand-lines or monitors, the fire will spread horizontally and break through the roof very early on and interior stacked storage will begin to collapse. At this point, structural stability is severely compromised and the building will inevitably be a total loss. Warehouses pose substantial challenges for fire protection due to their building layouts, storage configurations and technologies, ceiling heights, and types of commodities stored, with the specific challenges influenced by the characteristics of a given warehouse. Properly designed sprinkler systems are an essential element of warehouse fire protection.

One in five warehouses in England, approximately 621 premises, will have a fire requiring the attendance of fire fighters over the course of its lifetime[66]. The total annual cost to the UK economy of fires in English warehouses without fire sprinklers is £232 million. The main finding from a three-year study conducted independently by BRE Global[67] and commissioned by the Business Sprinkler Alliance (BSA) has shown that sprinklers are, on average, a cost effective investment for warehouses with a floor area above 2,000 m², with the greatest benefit arising from the reduction in direct fire losses. The current guidance on fire sprinklers in warehouses in the UK only applies to warehouses over 20,000m². Across Europe and competitor economies the regulations apply to much smaller sizes which mean they are far better prepared and able to recover from fires that threaten their businesses. The BRE study looked at the whole-life cost benefit analysis for fire sprinkler installation in three ranges of warehouse sizes. It is noted in the codes that currently, warehouses in the UK have to cover more than five football pitches before they require sprinklers (>20,000m²). In contrast, throughout Europe the majority of fire safety codes will insist on sprinklers being installed into far smaller warehouse floor areas.

Key findings from the BRE study include:

- The whole life costs for warehouse buildings larger than 2,000m² (around half a football pitch in size) with fire sprinklers are on average 3.7 times lower than ones without them
- Fire sprinklers were, on average, not cost-effective in warehouses with an area below 2,000m² Environmental benefits from sprinklers include a reduction in CO_2 emissions from fire, reduced size of fire and reduced quantities of water used to fight fire
- Only 20% of warehouses between 2,000 and 10,000m² are fitted with fire sprinklers. For warehouses above 10,000m², the estimated fraction with fire sprinklers is 67%
- If all warehouses above 2,000m² were fitted with sprinklers, the annual saving to businesses in England could be up to £210m
- Dr. Debbie Smith, Director of Fire Science and Building Products at the BRE, said: 'Despite a year-on-year decrease in the number of commercial fires, the estimated annual cost of these fires is rising along with related societal and environmental impacts. This project has broken new ground in terms of evaluating these broader sustainability impacts of fire in warehouses and demonstrating that, on average, sprinklers can be shown to deliver a net benefit.'
- Iain Cox, BSA Chairman and former Chief Fire Officer of Royal Berkshire Fire and Rescue Service, said: 'The findings of the BRE study scratch the surface in terms of the return fire sprinklers bring to business. What is clear from the current research is that insurance alone is not enough to fully protect companies from the long-term impacts of fire.'

8.1 SPRINKLER EXPERIENCE IN US WAREHOUSES

Approximately one-quarter of fires in US warehouses were identified as confined or contained incidents (23%), while 14% of these were confined to the object of origin.

---

[66] http://www.business-sprinkler-alliance.org/news/study-shows-that-on-average-fire-sprinklers-are-sound-investment-for-larger-warehouses/

[67] An Environmental Impact and Cost Benefit Analysis for Fire Sprinklers in Warehouse Buildings; Building Research Establishment (BRE) 2013

Another one-fifth of the incidents (21%) were confined to the room of origin, with 6% to the floor of origin. Fires that extended beyond the room of origin represented 41% of the total, but caused 91% of direct property damage, as well as 49% of civilian injuries[68]. In the most recent NFPA report on the U.S. experience with sprinklers[69], John Hall calculated that sprinklers were present in 32% of warehouse structure fires from 2007 to 2011, and that sprinklers operated 86% of the time when properties were protected by wet pipe sprinklers and fires were large enough to activate the equipment. Wet pipe sprinklers were effective in 84% of the fires in which they were present and contributed to a 61% reduction in dollar loss in those fires. Other protective measures generally applicable to warehouse properties include reducing compartment sizes by fire-resisting construction, smoke control and fire detection systems.

8.2 THE LIFE SAFETY ISSUE

The purpose of the UK building regulations is to provide a uniform standard for the purpose of securing the health, safety, welfare and convenience of people in or about buildings. This only applies to firefighters when undertaking the hazardous work of extinguishing building fires where Part B5 of the fire safety (building) regulations is aimed at assisting and protecting firefighters to undertake this work in with some elements of safety and efficiency. However, in terms of safety firefighters are not considered as 'relevant persons' according to the Regulatory Reform (Fire Safety) Order 2004, as in legal terms it may seem unreasonable to place a duty of care on the building's owner or 'responsible person' to protect firefighters whilst in their building for the purposes of fighting a fire[70].

In 2012 the UK government legally enacted the removals of a vast range of local building acts, restricting the powers of local authorities to enforce certain fire protection features that were seen as non-life safety measures. One of the main intentions of the UK government in removing the powers of local building regulations (acts) was to provide a uniform standard of regulatory building control throughout the country, with the primary function for the health and safety of people. The provisions in local Acts were seen as over and above the requirements relating to national building regulations i.e. over and above the health and safety of people. Business/property protection was considered a matter of consideration for the business sector and not local authority enforcement[71]. This ruling had a major impact on existing and future provisions of sprinklers and smoke control systems in warehouses where they had been installed for property protection and to assist/protect firefighters.

8.3 REMOVAL OF SPRINKLERS (FIRE ENGINEER'S TABLE-TOP EXERCISE)

We are now seeing the removal of sprinklers, and in some cases smoke control systems, in existing warehouses where they are not considered necessary for life safety purposes according to current building regulations. In some circumstances, insurers will insist they remain or are installed. However, at a time when research conducted independently

---

68 Campbell. R, Structure Fires in Warehouse Properties, NFPA 2016
69 John R. Hall, Jr., *U.S. Experience with Sprinklers, Division of Fire Analysis and Research*, June 2013.
70 http://www.publications.parliament.uk/pa/cm200304/cmselect/cmdereg/684/4061504.htm
71 Removing inconsistency in local fire protection standards – Final impact assessment, Communities and Local Government UK (DCLG) 2012

by BRE Global has shown that sprinklers are, on average, a cost-effective investment for warehouses with a floor area above 2,000 m², this appears nonsensical.

An example of a fire engineering strategy[72] undertaken as a proposal to remove sprinklers from a warehouse, used both hand calculations and computer aided field analysis to demonstrate that the existing smoke removal system would be sufficient to maintain tenable conditions for firefighters to enter the structure at a point between 20-30 minutes to extinguish a 'worst case scenario' fire and that the sprinklers were an unnecessary over provision. This is a common fire engineering approach and one that can be used as a *table top exercise*, particularly for student fire engineers, architects or building regulators. It is a most useful exercise to view this fire strategy from both perspectives, firstly the fire engineers who developed the strategy for their client and then the regulatory fire service perspective who would consider approval or rejection. (See link to online engineer's report in footnote below).

A Fire Engineering Brief (FEB) following the guidance set out in the International Fire Engineering Guidelines[73] was adopted to justify the removal of fire sprinklers from a distribution and storage warehouse in the UK. The warehouse is a single storey double height space with a maximum floor to ceiling height of 13 metres at the apex, and a floor area of 7,670 m². The warehouse is used for the receipt, storage, and distribution of marine and vehicle plant equipment. The warehouse contained a variety of sections consisting of major unit storage areas (A), variable beam racking (B), with vehicle loading areas (C) and delivery bays (D) (both internal and external), as shown in Fig.8.1 below.

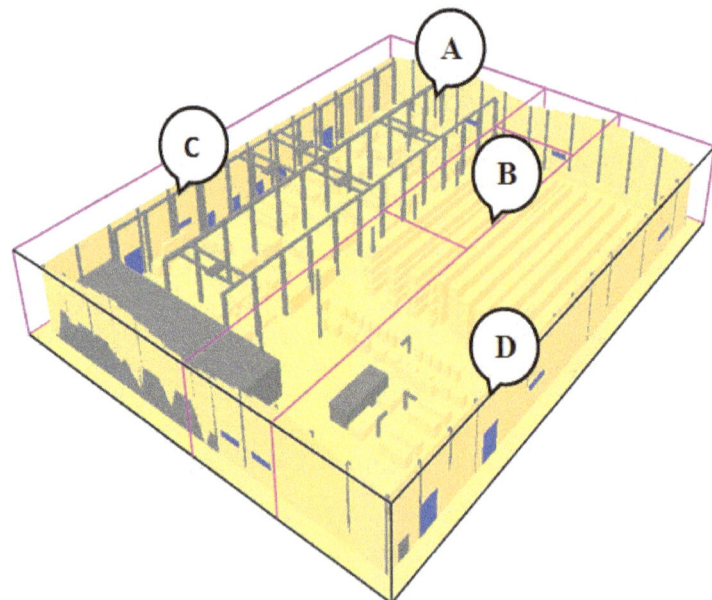

Figure 8.1: 3D model and plan view of the warehouse where the removal of sprinklers was proposed

---

72 http://www.fireseat.eng.ed.ac.uk/sites/fireseat.eng.ed.ac.uk/files/images/FIRESEAT2010pp077Chennell.pdf
73 International Fire Engineering Guidelines, Edition 2005. Australian Building Codes Board (Canberra, A.C.T)

The objective of the brief was to demonstrate compliance with *local codes* by demonstrating that the omission of sprinklers would not impact on the operational procedures of attending fire crews. To achieve this, a Quantitative Assessment was conducted of the projected fire load within the warehouse and compared with typical fire loads for a number of other occupancies which did not fall within the jurisdiction of local acts or codes. A deterministic analysis of smoke movement was conducted with the use of Computational Fluid Dynamic (CFD) modelling in support of the proposal. The FEB model was developed to heighten communication between fire engineering practitioners and building regulators. It was recognised that although fire engineering analysis are often complex and require extensive use of engineering judgements, guidance was required to improve the standard of the fire engineering applications. In turn this would increase the understanding of the fire engineering process, and importantly what constitutes an acceptable analysis, by those required to assess the output, i.e. the building regulators.

As agreed between the fire engineers and the building regulators the intention of the local code in question is the protection of attending fire crews in the event of a fire. Life safety of the building occupants was not considered within this scope of works as the building was in compliance with the functional requirements of Part B (Fire Safety) The Building Regulations 2000. Therefore, it was discussed and agreed as to what hazard the heat and smoke from a likely fire scenario in the warehouse would pose to fire fighters with the omission of sprinklers and how this risk to their safety could be mitigated. The following acceptance criteria were set out for the analysis.

- The existing smoke control system will maintain a reasonable clear layer height above the heads of attending fire crews.
- The smoke layer temperature will be such that it will not subject attending fire crews to a radiant heat density such that operational procedures are impeded.
- Air flow rates that prevent movement through the building access routes should not be generated.

A smoke control system was installed to maintain a clear smoke layer at 2.5 m above floor level, i.e. above head height of attending fire crews, at a visibility of 10m beneath the smoke layer. This ensured that the location of the fire can be determined quickly by the attending fire crews. The seat of the fire will be visible from the periphery of the warehouse when fire crews are committed, assumed to be of the region of 20–30 minutes post fire ignition. Smoke layer temperature was controlled such that it will not subject fire crews to a radiant heat density of 2.5 kW/m^2 or greater, translating to no greater than 200 °C. By the adoption of this limit a conservative tenability criteria has been chosen, as, any firefighter committed to the warehouse will be wearing full Personal Protective Equipment (PPE) and thus afforded a greater level of protection. It is typical to expect that fire fighters can withstand greater temperatures whilst in PPE. The Australian Fire Brigade Intervention Model (FBIM)[74] manual states that received radiation up to 4.5kW/m2 can be withstood by fire fighters donned in full PPE, translating to an upper smoke layer temperature of 280 °C. At no time will an air velocity greater than 5 m/s[75] exist at the fire service entry point, to ensure that access for the attending fire crews is not impeded. By the adoption of this limit a conservative tenability criteria has been chosen, as, any fire fighter committed to the warehouse is assumed to be of a given physical fitness. The 5 m/s parameter is commonly used in buildings for egressing occupants, of all ages and mobility, with an upper limit of 11 m/s adopted within passenger tunnel escape scenarios.

74 Fire Brigade Intervention Model (FBIM); Metropolitan Fire & Emergency Services Board Australia; 2003
75 CIBSE Guide E (2003), Second Edition. The Chartered Institute of Building Services Engineers. London

Although the predominant load within the warehouse was large metal machine parts for vehicles and marine engines, a large quantity of wooden crates and cardboard packing material formed the bulk of the potential fire load as follows:

Variable Beam Racking (section B on plan)

The engineering report stated that on average cellulose material (wood and paper) such as these have a calorific value between 13–18 MJ/kg. Palletised units are stored in the variable beam racking using wooden pallets on racks <u>up to 10 metres above floor level</u> (4 levels).

- 2 small pallets per shelf, 14 shelves per rack, 4 levels of racks per high bay racking, 7 rows of racks;
- 2 small pallets x 14 shelves x 4 levels x 7 rows = 784 pallets.
- 784 pallets x 300 MJ per pallet = 235,200 MJ (Total fire load)
- Total fire load / Floor area; 235,200 MJ / 710 m2 = **331 MJ/m2.**

The report went on to estimate the average fire load across the entire warehouse (69 MJ/m^2) and then then compared this data with the fire load densities of occupancies taken from BS 7974 Part 1. The total fire load density for the warehouse at 69 MJ/m2 was calculated as a sixth of the typical fire load found in an office, and 5% of the fire load found in a typical manufacturing and storage occupancy. The report suggested that by comparison this demonstrated that the warehouse did not present the same risk as a typical storage occupancy would:

- Office space = 420 MJ/m2
- Retail space = 600 MJ/m2
- Manufacturing and Storage = 1180 MJ/m2

Fire Scenarios

The Comparative Assessment concluded where the worst case fire scenarios would be within the warehouse, based on fire load density and orientation of combustibles. Therefore, two fire scenarios where proposed and agreed upon as the worst cases. Fire area (47 m^2), perimeter (24 m), heat release rate density (255kW/m2) and total heat release rate (12 MW) used for the scenarios were those provided in BRE 368, Table 3.3 for an unsprinklered fuel bed controlled fire as this had the closest fire load density (420 MJ/m2) to that of the two fire scenarios;

Scenario 1, Variable Beam Racking

The variable beam racking is used to store palletised units that have been delivered to the site. The major fire load on the racking system will be the pallets. Each shelving level and projected number of palletised units has been assessed and maximum of four shelf levels assumed. The area of 47 m^2 will be applied to the racks and involve two racks back to back and twenty shelf units in total. Over the given floor area that this racking system will be located a fire load density of 331 MJ/m^2 has been calculated. To represent the height of the fire load all four levels were assumed involved and up to twenty shelf units in two racks back to back where included at each level.

Scenario 2, Major Units 1

Crated units were stored prior to shipping adjacent to the vehicle loading areas. This scenario consisted of wooden packing crates stacked in pairs in close proximity with other crates, producing a fire load density of 235 MJ/m^2. The 47 m^2 fire area would be situated in a corner to replicate reduced entrainment into the smoke plume and thus increase flame height and smoke temperature.

Stage 1 – Scoping Calculations

Prior to the CFD simulation a number of scoping calculations where conducted using the CIBSE Guide E calculation method. This empirical calculation method assesses conservation of mass and energy, predicting how the smoke layer would descend. The design fire parameters were discussed and agreed with the regulators based on the two scenarios listed above assuming a <u>fast growth rate fire</u> within the geometry of the warehouse. The model was run over <u>30 minutes</u> to include roof vent operation <u>after 180 seconds</u> to simulate time to smoke detection and define a critical smoke layer height of 2.5m, and a reduced critical smoke layer temperature of 185 °C.

The results from the two scenarios both produced clear layer heights above 2.5 m, with Scenario 1 having a greater smoke volume due to radial air entrainment into the smoke plume. Scenario 2 produced a smoke layer temperature in excess of 185 °C, whereas Scenario 1 did not, but as the smoke layer height was approximately 11 m above floor level the radiant heat flux at 2.5 m was concluded to be within the acceptable criterion. As the volume of smoke produced was greater in Scenario 1, Scenario 2 was discounted and Scenario 1 was used within Stage 2 of the deterministic analysis. This decision was taken as it was agreed that the Stage 1 results showed that visibility through the smoke layer to the seat of the fire was of greater concern than the heat flux from the smoke layer.

Note: The fire is located in the variable beam racks. The upper-most surface of the fire is located 1.5m above ground floor level. The 2.5m of shelf space directly above the fire has been removed to facilitate air flow to the fire.

Stage 2 – CFD Analysis

After discussion with the regulators the parameters for the CFD analysis were agreed upon. Scenario 1 was agreed to be the worst-case fire, and the fire load, density, orientation, building geometry and ventilation provisions were all discussed and agreed prior to running the computer models, as listed below;

- Building plan area = 105.0m x 80.5m x 15m = 126787.5m^3 (8450m^2)
- Building height = 12.0m. (Taken to lowest point of roof structure – conservative.)
- Maximum fire total heat release rate = 12MW (48m^2 at 250kW/m^2).
- Radiative fraction = 0.33.
- Maximum convective heat release rate = 8MW.
- 'Fast' t^2 growth rate (= 0.04689kW/s^2).
- Fuel – predominantly timber – with limited plastic content.

Typical properties for timber are:

- Heat of combustion = 13,000 kJ/kg;
- Soot yield = 0.01 – 0.025kg/kg.

To allow for the presence of a limited plastic content, the following properties were adopted.

- Heat of combustion = 15,000 kJ/kg;
- Soot yield = 0.045kg/kg.

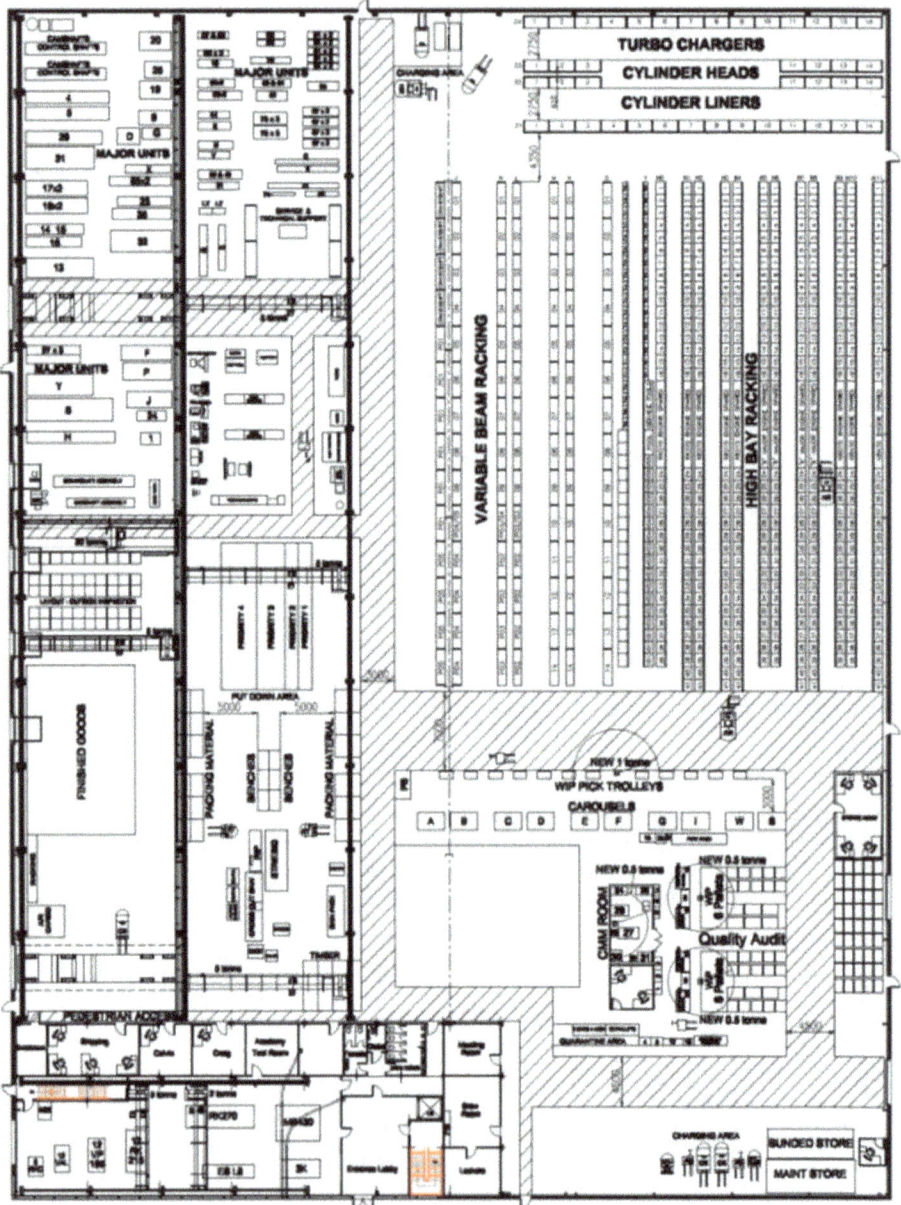

Figure 8.2: The warehouse floor plan with the variable beam racking and high-bay racking shown in detail

The key parameters, for this study, are the height and temperature of the smoke layer. Assessment of the results took place with respect to the anticipated time at which the Fire Service commences firefighting operations: taken by the engineer to be 20–30 minutes after fire ignition. (In this model, ignition is considered to represent the post-smouldering phase of fire development.). A sensitivity study into varying natural ventilation inlet and outlet arrangements was conducted resultant in the design ventilation provisions required to meet the acceptance criteria. Also, a study into variations in the model geometry, vent orientation, fire growth rates and fire time lines was conducted to ensure the analysis results had considered a number of eventualities and the results were sufficiently robust to be considered representative of a real fire scenario. The results from the CFD simulations, defined that the agreed acceptance criteria for the models had been met; the smoke temperature did not exceed 185°C; the visibility at 2.5 m was greater than 10 m.

When the results from the Quantitative Assessment comparing fire loads, and the Deterministic Analysis, substantiating the deviation from local fire codes with CFD simulation, were presented to the local fire and rescue service and regulatory building control it was agreed that the acceptance criteria set out through the Fire Engineering Brief had been met.

- Firefighting operations would not be impeded with the omission of sprinklers from the warehouse.
- A significant cost saving had been made for the client with a design which still meet the intention of The Great Manchester Act.
- These both had been achieved through the successful application of the Fire Engineering Brief model.

The Firefighter's Perspective

There are regular fire strategies being developed by well-meaning fire engineers working for a client who wishes to remove or downgrade various features of their fire protection installations. To maintain these annually in effective working order can be expensive. However, care should be taken to ensure that the origin and primary purpose of these installations such as sprinklers, automated roof vents and fire detection systems are investigated as sometimes they may have originally intended to allow extended escape travel distances, or to assist firefighters in (a) firefighting operations; (b) post-fire smoke clearance.

The following acceptance criteria were set out for the analysis.	
• The existing smoke control system will maintain a reasonable clear layer height above the heads of attending fire crews.	• A critical smoke layer height of 2.5m was demonstrated in the CFD analysis with a 10m clear visibility path below.

Continued overleaf.

The following acceptance criteria were set out for the analysis.	
• The smoke layer temperature will be such that it will not subject attending fire crews to a radiant heat density such that operational procedures are impeded.	• A point was made in the FEB that firefighters can work safely against a heat flux of 4.5 kW/m2 and the engineer referred to an Australian (FBIM) document in support. However, see Figure 11.7 where it is demonstrated that firefighters subjected to such a heat flux are located within a hazardous environment and may have less than 60 seconds before burn injury occurs. • The Australian FBIM document referred to by the fire engineering proposal also makes this point and if any firefighting actions taken are not immediately effective in reducing this temperature, the firefighting team/s safety is greatly compromised. • However, the FEB acceptance strategy targeted a maximum 200°C in the smoke layer with a maximum heat flux to firefighter locations of 2.5kW/m². This enables a maximum 10 minute working duration for firefighters but allowance may then be given for firefighting actions that will reduce the received heat flux and extend the working duration for firefighting operations. • A critical smoke layer temperature of 185°C was demonstrated in the CFD analysis at maximum fire service deployment times (30 minutes). • A sprinkler controlled fire will reduce this heat exposure to firefighters down to far safer and normal routine levels <1 kW/m².
• Air flow rates that prevent movement through the building access routes should not be generated.	• Achieved in the CFD analysis

What wasn't included in the FEB was any assessment of the fire service capability in attacking and extinguishing the fire with a single hose-line, as deployed at 20-30 minutes into the simulation. We can also see that a fast growth fire was used for the initial input to the CFD and that at time of firefighter deployment, the fire would, according to the computer analysis, have reached an estimated 12MW (48m² at 250kW/m²). However, BS PD 7974-1-2003 (Table 3) recommends an *ultrafast growth* input (0.188kW/s³) be used when calculating or modelling rack storage fires and this is seen in the literature to commonly be the case where rack storage exceeds 6 metres in height. This would demonstrate a much faster spread of fire and at 20-30 minutes into the incident, the area of fire involvement would therefore be greater in volume. The 47m² floor area fire proposed by the fire engineer is slightly inappropriate. It is based on a 40-pallet fire, each pallet being 1.2m² in area (with stacks of two pallets per shelf). The engineer's fire strategy was based on a typical 'office fire' taken from BR 368[76] (table 3.3) producing a 255 kW/m² heat release rate per unit area.

There is a useful amount of experimental research data available that should have been

---

76 BR 368; Design methodologies for smoke and heat exhaust ventilation; Building Research Establishment; 1999

used in this case that will produce greater accuracy. Results from full-scale fire tests[77] carried out at the Building Research Establishment had shown that the vertical spread of fire through materials stored in racking can be very rapid, reaching the top of a 12m tall rack in less than 2 minutes. Much field work of rack storage fires has been undertaken by researchers in the USA and Sweden in particular and a host of Factory Mutual (FM) videos are available online that graphically demonstrate the speed and extent of both sprinkler controlled rack fires and free-burn rack fires. A more accurate assessment for the rack storage of cardboard carton commodities with steel content would be at least 550 kW/m² to a height of say 8 metres (actually 10 metres to top of stack). A more accurate calorific value for cardboard cartons in rack storage seen in the literature is 22 MJ/kg. At around 1,964 kg of involved fire load, this would result in a more representative figure of 4,400 kW/m² and with the engineer's estimated fire involvement of 20 shelves (40 pallets), the 6 or 8 m² fire area would result in a heat release rate of at least 26–35 MW, at the time of fire service intervention. BS PD 7974-1-2003 provides a method of calculating heat release (equation 5 in 7974) as follows:

$Q = Q'' A_{fire}$ (per one metre height)   Eq. 8.1

Where:

Q = The total heat release rate from the fire (kW)

Q'' = The heat release per unit area (kW/m²)

A_{fire} = Area of fire (m²)

So: for fire spread of 6 and 8 square metres –

$Q = 550*6 = 3,300$ kW per metre height

3,300 x 8 metres (actual height of storage) = 26,400 kW

$Q = 550*8 = 4400$ kW per metre height

4,400 x 8 metres (actual height of storage) = 35,200 kW

= 35.2 MW

The engineer's model also failed to account for the effects of any amount of non-vented ceiling jet radiating heat downwards, as well as flying brands being carried into other areas of the storage area on thermal currents (clearly apparent in the FM online videos). The heat release and heat flux would therefore be even higher at 30 minutes into this fire and certainly beyond any viable fire service intervention. In effect, without the sprinklers and just relying on roof vents, this entire facility would likely end in a total loss.

Therefore, the prime objectives of this fire engineering brief to maintain tenable limits firefighters during firefighting operations have not been effectively demonstrated in this case. The sprinklers in this case should therefore be retained, unless alternative passive measures, such as effectively reducing fire resisting compartment sizes to at least below the recommended BRE maximum of 2,000m², are implemented.

Research undertaken in Sweden's SP National Testing and Research laboratories[78] using

---

77 Hume. B; Eady. M; The Use of CFD Computer Models for Fire Safety Design in Buildings Large Warehouse Case Study; ODPM Fire Research Division; 2002
78 Lönnermark. A,Ingason H,; Fire Spread in Large Industrial Premises and Warehouses; BRANDFORSK project 630-021; 2005

cardboard boxes loaded with polystyrene cups demonstrated the hazards of rack storage fires. The report concluded: 'The results show that the distance between the top of the goods and the ceiling has a decisive importance in determining the rate of spread of fire between the racks. When the flames reach the ceiling, they are deflected to the side and increase radiation to the top of adjacent racks. As a result, these racks catch fire at the top, with the fire spreading downwards from them. The tests show that the fire spreads most rapidly with the distance above the top of the racks equivalent to 3m in full scale. It took approximately the same time for the fire to spread when the heights above the goods were 1m or 6m: the longest time occurred when there was effectively no ceiling above the racks. The fire spread with a clear height equivalent to 1m above the racks was retarded since the combustion is less complete. When the impinged flames reach the layer of hot fire gases formed below the ceiling, they start to be affected by the lower oxygen levels. It is clear from the tests performed here that the only way to protect the goods against this type of galloping [fire spread] process is to use sprinkler systems'.

8.4 FIREFIGHTING OPERATIONS IN LARGE INDUSTRIAL BUILDINGS AND WAREHOUSES

The most hazardous Environment

Fires such as these in large industrial buildings, warehouses or retail superstores may present one of the most hazardous environments in which the firefighter may be tasked to enter. The difficulty in locating the seat of the fire and hauling hose across the already hot environment can place firefighters at great risk of heat exhaustion within a few short minutes. Add to this the fact that fire in the overhead area near the ceiling may be hidden behind a smoke layer and will be placing the roof under great heat stress without firefighters below realising it. The early collapse of the roof, or the high stacked shelves, or rapid fire spread across the floor plate can cause a sudden drop of the smoke layer, causing firefighters to lose their bearings and become lost. As fires in warehouses may not be encountered that regularly by firefighters, their awareness of such risks are generally confined to the classroom.

Structural Stability and interior Fire Service Operations

The design and construction of steel warehouse buildings requires structural members to achieve various levels of fire resistance depending on the size of the building and the function of the member. The fire resistance of each member is dependent on purpose group, occupant level, height and other fire safety precautions taken. Fire resistance is determined by subjecting the member to standard fire test conditions and measuring the time to failure. The building codes of practice detail the fire resistance ratings of any member by three criteria: stability, integrity and insulation. Each criterion is described in more detail below:

- The **stability** criterion applies to primary members and is concerned with the ability of the member to support applied loads without collapsing during a fire.
- The **integrity** criterion applies to secondary members. These members are required to prevent hot gases and flames from transmitting through the member by means of cracks, fissures and the like.
- The **insulation** criterion applies to both primary and secondary members. This criterion is concerned with the ability of a building member to provide an adequate barrier between a fire compartment and an adjacent compartment

Note: Primary members are defined as building elements providing the basic load bearing capacity to the structure (i.e. columns, beams). Secondary members are defined as building elements not providing load bearing capacity to the structure (i.e. non-load-bearing fire separation walls).

Where provided, sprinklers are expected to extinguish or to at least control the fire to the growth area at the time of activation. In a sprinkler controlled fire the compartment temperature is not expected to exceed 300°C. It is stated[79] that if sprinklers are present in an industrial building, a fire resistance rating of 60 minutes for boundary walls may be sufficient. However, sprinkler systems have been overcome by some manufacturing and warehouse fires due to the following reasons:

- Incorrect installation according to hazard classification resulting in inadequate protection
- Incorrect storage for the sprinkler hazard classification
- Failure of sprinkler system such as blockage in pipelines
- Explosion being the cause of fire, either resulting in an excessive number of heads to operate or causing critical damage to the sprinkler system

It is important to point out that although such buildings will normally have fire resistance of 60-90 minutes, some UK codes of practice may allow structural fire resistance of just 30 minutes in non-sprinklered industrial warehouses, depending on maximum height and ventilation conditions. It may therefore be difficult for any immediate assessment on structural stability expectations to be undertaken by fire commanders arriving on-scene and these buildings are likely to collapse early in a fire, even as soon as 20-30 minutes following fire service arrival. Any interior operations, other than a command decision to deploy for an urgent search and rescue following a <u>confirmed</u> report of persons reported as missing, should be carefully risk-assessed and avoided where necessary. Such buildings made of lightweight materials are not designed or constructed to support interior fire service operations and should be identified and flagged up on premises risk systems, immediately informing responding firefighters, prior to their arrival on-scene and any interior deployment occurring.

Firefighting Access

Adequate access for firefighters is always important but never more so than where large volume buildings are concerned. An adequate amount of access points should be provided along a large percentage of the building's perimeter. Battalion Chief Jerry Tracy (FDNY) emphasises that the following tactical benefits arise from such vantage points:

- Firefighters will be able to reach the fire more easily
- Less hose and staffing are needed to penetrate in towards the fire
- The fire can be located more quickly and the time for 'water on the fire' is lessened
- Relief cycles for SCBA wearers are reduced
- There will be less likelihood of firefighters suffering from heat exhaustion
- The distance to an exit, in case of firefighter emergency, is reduced
- The extended reach of fire streams should be taken advantage of by maximising pressure, yet not losing flow
- Chief Tracy recommends a 10-minute relief cycle for crew turnaround

---

79 Cosgrove, B.W.; Fire design of single storey industrial buildings, Fire Engineering Research Report No.96/3, University of Canterbury, Christchurch, New Zealand. 1996

An important strategic element at large volume structures is ensuring you have located, and are using, the nearest access point to the fire. If the fire location is still to be determined then reconnaissance is essential using overhead drones and internal crew deployments to transmit thermal heat images, in order to get an overview of the plan layout according to optimum points of access. This has repeatedly been noted as a strategic failing at such fires.

Firefighting Water

The quantity of water, when matched to typical warehouse fire loads, suggests a cautious approach should be made with a minimum hose size of 51mm flowing to 700 L/min. A secondary support hose-line should be following close behind to observe the smoke layer and protect the primary attack team. This initial approach to the interior demands at least 8 firefighters to advance the hose-lines.

Firefighting Deployments

Where a fire has entered a clear growth-phase and is developing at a steady rate, any interior approach beyond one hose-length (25 metres) (if any interior approach is possible) should be carefully considered by the incident commander and a high-flow attack at long range may be called for.

Firefighter Tenability

With high heat at 12 metres near the ceiling, the heat flux at ground level may not be felt for several minutes. Therefore, firefighters may feel they are working in a safe and comfortable environment. Regular temperature checks should be made, directing fog droplets into the overhead and looking or listening for water 'drop-back'. If there isn't any, the situation may be approaching a call for evacuation of all firefighters from the interior.

Firefighter Physiology

Any consideration to commit a crew after they have already experienced heat within the last hour should be monitored by medics, who should keep a record of firefighters inner-core temperatures.

Crew relief Cycles

In solid construction fire buildings, it is essential to be able to establish an ongoing and uninterrupted interior offensive operation where safe to do so. It is certainly something we aspire to do on the fire-ground but working durations of breathing apparatus are variable between individual wearers and it is sometimes difficult to accurately determine when any particular crew is likely to be exiting. However, it is critical that we maintain a constant flow on the fire where possible, to prevent the fire escalating between reliefs. If this does occur, crews may be deploying into worsening conditions without realising. To achieve this the IC will require sufficient resources from the outset to maintain constant water on the fire, but at the same time recognise when the fire attack is having little effect over a long time-scale and it is time to pull crews out of the structure. As time passes, the structure may begin to fail from high heat damage, particularly at the roof, and if the interior fire attack is having no real effect it is pointless (and potentially catastrophic) to maintain a lengthy offensive approach. Just as important is the need to deploy additional resources to work just behind the attack team/s, assisting in hose advancement or providing support from a safety line, observing and cooling fire gases in the overhead beyond the attack team's position.

Where crews are unable to maintain constant and adequate amounts of water on the fire, due to ineffective or interrupted flow-rates, or late reliefs, the fire will; have time to regenerate on a growth curve. In large volume buildings with high-ceilings the temperatures at low level may be tenable and even comfortable for firefighters but at high level they may exceed 600°C where the structural roof members are being weakened by the minute. Therefore, a relief cycle that sees firefighters relieving crews at the nozzle is essential. To deploy crews for 10 or 20 minute cycles and relieve at the nozzle it is a resource heavy commitment. To deploy and maintain a single hose-line operating for 60 minutes can take at least 10 to 20 firefighters.

	Time in	Time arriving at nozzle	Time leaving the nozzle	Time exiting the structure
1st Crew	10.00	10.05	10.15	10.20
2nd Crew	10.10	10.15	10.25	10.30
3rd Crew	10.20	10.25	10.35	10.40
4th Crew	10.30	10.35	10.45	10.50
5th crew	10.40	10.45	10.55	11.00

Table 8.1: A crew relief cycle for a maximum of 20 minute deployments demands more than five crew entries per hose-line exceeding one hour

	Time in	Time arriving at nozzle	Time leaving the nozzle	Time exiting the structure
1st Crew	10.00	10.05	10.10	10.15
2nd Crew	10.05	10.10	10.15	10.20
3rd Crew	10.10	10.15	10.20	10.25
4th Crew	10.15	10.20	10.25	10.30
5th crew	10.20	10.25	10.30	10.35
6th crew	10.25	10.30	10.35	10.40
7th crew	10.30	10.35	10.40	10.45
8th crew	10.35	10.40	10.45	10.50
9th crew	10.40	10.45	10.50	10.55
10th crew	10.45	10.50	10.55	11.00

Table 8.2: A crew relief cycle for a maximum of 15 minute deployments demands more than ten crew entries per hose-line exceeding one hour

It can also be seen that where crews are working in conditions that are likely to increase inner core body temperatures to dangerous levels and expose them to the risks of heat exposure in a very short time, the number of firefighters needed for an ongoing deployment using a *'no nozzle down-time'* relief cycle (per hose-line) can be far greater than anticipated and requires careful pre-planning and resource allocation. At the Warwickshire warehouse fire in 2007 where four firefighters lost their lives due to heat exhaustion and/or becoming lost in smoke, four firefighters were deployed with a flow-rate of <100 L/min, for three periods of entry not exceeding 12 minutes each in BA, during the first hour on-scene. They reported 'warm to hot' conditions within the building as they exited and that the fire had not yet been located. The next four firefighters deployed into the burning warehouse never made it out as the fire suddenly escalated, causing intense heat.

Crew briefing and debriefing

It is absolutely critical that crews leaving the fire building should immediately be debriefed by a fire commander and any relevant information recorded. A plan should be used to work from so that firefighters are able to point out where they have been to in the building. Equally, time should be taken to brief crews before they begin to start their BA donning and starting process. The communications in briefings should be two way – confirmed – relevant – precise and not too long. The level of heat exposure should be discussed and observed for crews exiting (signs of sweating etc).

National Guidance[80] on the Deployment of Firefighters Wearing BA now States (2014):

- Relief BA teams should be planned to take account of the physiological load imposed on the BA teams and their need for rest and recovery and the Incident Commanders should be aware of the hazards associated with wearers being re-committed.
- In order to optimise recovery between BA deployments and maximise performance and safety prior to being redeployed as a BA wearer, personnel should be afforded a period of time in which to rest and rehydrate. The actual time afforded to rest and recuperation should be based on the task undertaken; time committed; and conditions encountered by the wearers.

Ideal rest and recovery periods prior to redeployment of BA wearers		
Initial deployment conditions	Rest period	Water to be consumed
Ambient	30 minutes	500 ml
Hot and humid	60 minutes	1000 ml

Table 8.3: Ideal rest and recovery between redeployments into a fire building based on DCLG sponsored research 2008

2014 Guidance provided by the Fire Brigades Union for extended relief firefighting operations[81]

A detailed strategy document was published in response to the national guidance document by the Fire Brigades Union in 2014 that included proposals in relation to extended relief operations.

---

80 Operational Guidance -Breathing Apparatus; DCLG CFRA 2014
81 Fire Brigades Union (FBU) Guidance Document; Respiratory Protective Equipment; Breathing Apparatus; Safe Operational Procedures 2014

- Exhaustion from wearing BA is a common occurrence and may not become immediately apparent. Consequently, BA teams re-entering an incident for a second-wear should be an exceptional occurrence and should only take place to save an identifiable saveable life.
- The BA Entry Control Officer (ECO) should inform the IC of any information on adverse/worsening conditions relayed from BA teams.
- The Fire and Rescue Service (FRS) should pre-plan for such incidents and be sufficiently well-resourced to ensure that adequate relief crews are mobilised to ensure that no SCBA wearer should be required to be deployed for a second-wear.
- The FRS shall have procedures which require Incident Commanders (ICs) to manage an incident such that 'second wears' are not required
- The IC should ensure that sufficient crews are mobilised to the incident to ensure that no BA wearer should be required to be deployed for a second-wear.
- Second wears should be treated as a near miss event and subject to full and investigation and report procedures
- Liaise with the IC/Sector Commander to ensure reliefs are available and that briefing of relief team takes place at least five minutes before they are due to be committed;

Author's note: Where firefighters are required to redeploy into a burning building, following an initial BA wear, a safer solution would be to monitor body core temperatures using non-evasive methods, such as ear thermometers. Although the accuracy is never as good as evasive methods, it is a procedure well used by medical professionals and with limited training, can also be used by firefighters to assess when it is safe to redeploy firefighters into fire buildings. Where body core temperatures are above normally accepted averages (36°C–37.5°C) firefighters should be assessed before re-entry is allowed. There are also methods to ensure effective cooling takes place prior to redeployment such as cooling chairs and rehydration stations.

Research into Firefighter Body Core Temperatures

A DCLG sponsored research project in 2008[82] looked at both thermal and cardiovascular responses of 21 firefighters exposed to moderate fire temperatures whilst undertaking search and rescue operations in a training building at the UK Fire Service College (FSC). This research provided the national guidance on redeployments, as detailed in Table 8.3 above. In only 4 percent of cases under non-fire ambient test conditions did firefighters exceed a maximum recommended upper-core temperature of 39°C (this is for trainers who are normally able to exit training fire facilities far more quickly than a real fire building). In subsequent tests involving live fire, where firefighters were exposed to moderate amounts of heat for short durations, 9 percent of firefighters exceeded the recommended maximum upper-core temperature of 39°C.

Earlier research (2004)[83] of a similar nature, again sponsored by government (ODPM), suggested that 32 minutes was a maximum time-frame within which a deployment should start and finish before the 39°C core temperature was reached. Mean ambient temperatures throughout the live fire scenarios, by floor, as measured by the body-borne probes, were between 27°C and 53°C, while mean peak temperatures ranged from 65°C to 103°C. In these particular trials, 39 percent of firefighters exceeded the maximum core temperature

---

82 Core Temperature, Recovery and Re-deployment during a Firefighting, Search and Rescue Scenario; DCLG 2008
83 Physiological Assessment of Firefighting, Search and Rescue in the Built Environment, Office of the Deputy Prime Minister (ODPM) 2004

of 39.5°C agreed as an upper level. A further 40% were stopped for safety reasons, either by the safety officers or by the firefighters themselves, most of which were heat-related. Rates of rise of core temperature averaged 0.054°C/min and 0.045°C/min for firefighting and search and rescue teams, respectively, which is a statistically significant difference. Although both teams started the scenario at the same core temperature (approximately 37.5°C), the FF team ended hotter averaging 39.1°C compared to 38.9°C for SR. The greater proximity of the FF team to the fire may have accounted for the higher rise and rate of rise in core temperature. No differences were found in core temperature response between floors, even though temperature data from both the instrumented compartment and the body-borne external sensors showed differences, with the basement being the hottest and the fourth floor being the coolest.

Further work in this research examined the physiological load associated with climbing stairs up 28 floors to explore further the vertical component of firefighting and rescue operations. This assessment did not cover the physiological component of returning to fire service access level. Climbing stairs may be required where either no firefighting lifts have been provided or in the case of their failure. Two separate assessments were Physiological Assessment of Firefighting, Search and Rescue in the Built Environment conducted in PPE both with and without carrying EDBA and hose. When carrying EDBA and hose it took approximately 30 seconds and core temperature rose by approximately 0.02°C, per floor. When climbing unloaded it took approximately 15 seconds and core temperature rose by approximately 0.01°C, per floor. Climbing stairs in PPE while carrying EDBA and hose is very physically demanding. Operational planning assumptions, including levels of resources, should take account of the physiological demands of reaching the upper floors of tall buildings with RPE and PPE including any equipment carried.

Heat strain among the firefighters was the greatest single source of performance limitation in the scenarios investigated, causing the premature termination of approximately 65% of firefighters involvements in the tests.

> **What does this mean to the Firefighter?**
>
> When deploying into 'big box' high volume warehouses here is a list of key points that firefighters should take into account, for their own safety:
>
> - <u>No building is worth your life!</u> If there are no building occupants reported as missing, then locating and extinguishing the fire becomes the primary objective. This can become a lengthy and time consuming operation in many cases.
> - Teams should always deploy with adequate water, matched against the potential fire load involvement at time of deployment, realising that the fire is still increasing in size.
> - A safety team with adequate water should deploy in support to protect the attack team in observing the velocity and movement in any high-level smoke layer, smoke gas cooling where necessary.
> - Any sudden changes in the velocity of the flow-path, or at the base of the smoke interface such as a 'bouncing' action, with smoke moving up and down, is warning of a fast-developing fire and possible rapid fire behaviour in the overhead.

- Hose penetration distances are dependent on staffing and local SOPs. Safe interior working durations (time) and hose penetration (distance) are dependent on fire conditions, temperature profiles, state of the smoke layer, size of the building and how arduous the working environment is becoming.
- Any internal deployment beyond 25 metres should demand additional control measures in staffing, safety lines and command assignments. Any deployment beyond 50 metres should again bring an additional level of resources in support.
- The deeper the hose-line penetration the less time crews can spend firefighting and the greater number of relief crews are required to ensure crews are relieved at the nozzle. (See crew relief cycles).
- As each crew leaves the building – debriefing is essential and commanders should ensure key information is received and passed on to crews about to deploy, particularly in relation to tenability. Any signs of heat stroke or exhaustion should be a major cause for concern.
- Interior crews must always be wary of the smoke in the overhead that can hide flame travelling over and behind them, cutting off escape routes. They should also listen for noises, such as creaking or rivets popping out in the steel structure near the ceiling that might signal an impending roof collapse.

Chapter 9
The 'Car Park Fire'

Fires in car parks are rare, and, although there have been few deaths or injuries recorded to date in the UK, there are concerns regarding new and emerging risks from modern cars and alternative fuels. There was a need to gather up-to-date information on fires involving cars in car parks in order that the current fire safety guidance could be reviewed and, if necessary, updated. A series of full-scale fire tests carried out by the Building Research Establishment (BRE) in 2006 demonstrated the ease with which a car fire in a car park might spread to nearby cars. Tests included side by side parking arrangements and automated car stacker systems, where cars can be parked in a space saving environment, stacked three or more high. Once a very severe fire has developed, ignition will occur on cars separated by an un-filled parking bay. In this situation, where a number of cars are burning simultaneously, the fire is exacerbated by heat-feedback and heat release rates in excess of 16 MW might be achieved from two or three cars.

Monica Wills House served as a care home in Bristol and was opened in October 2006. Two months later it was subject to a tragic fire which spread through the building from a lower ground floor car park, causing one fatality and an estimated £5 million worth of damage. Had it not been for the sprinkler system installed in the care home development above the car park, together with the other fire safety measures in the building along with the prompt attendance of the fire and rescue service, many more lives would have been lost. As firefighting began, breathing apparatus (BA) entry control was established, with 22 firefighters wearing BA. In total, 70 firefighters were engaged in search, rescue and offensive operations. Three main jets and four hose reels were used to extinguish the fire, and two positive pressure ventilation fans were put into action to disperse smoke logging. Following the initial attendance, the eventual number of engines was increased to 12 over the next 40 minutes. Senior Fire Engineer Simon Lay commented on this fire as follows[84]:

'In this particular case, it is noted that the car park soffit was sprayed with insulating foam, and that this was a Class 0 product. It is suggested that this was in accordance with standard guidance. This is not the case. Clause 11.3e of ADB 2006 (12.3 of the ADB 2002) states that materials used for the construction of car parks should be non-combustible, while the surface spread of flame classification recommended is Class 3 (or Class 1 if you apply the 'garage' definition). One would hope that a combustible product has not been justified as being acceptable on the grounds that it has a higher surface spread of flame characteristic than the standard recommendation? A Class 0 material can readily burn if exposed to sustained heating. Most insulating foams used in this application are precisely the type of materials which have caused concern for the fire service when used in sandwich panels, and yet the potential influence of this insulating foam on the fire spread in the car park does not appear to be fully considered in the public report of the fire. The application of insulating foam to the soffit of car parks is a common practice for open-sided car parks under buildings and it may be that the ADB guidance is not being correctly applied. Insulating foams are much less common in basement car parks (as they are enclosed spaces). If the insulation was a significant factor in this case, then it suggests that current interpretation of standard guidance may make open-sided car parks potentially more dangerous than enclosed, subterranean car parks. This incident may also raise questions about what constitutes a naturally ventilated car park. The images provided with the report suggest that many parts of the car park were remote from natural openings. Some mechanical vent ducts are also visible in the images, suggesting parts of the car park may have exhibited restricted ventilation. It is plausible that the ventilation satisfied the standard ADB guidance, but that distribution of vents was poor. This could increase the fire severity in locations, perhaps to a level where even sprinklers would not have stopped the fire spread (noting that the insulation would have probably been installed above any sprinkler installation).

The Bristol care home incident demonstrates how a fire engineered approach that properly addresses the risks in a scheme can lead to enhanced safety levels where it really counts (leading to the sprinklers in the care home units), while the absence of a rigorous risk assessment to elements of the design which were perceived to be compliant with ADB perhaps left the car park areas vulnerable.

9.1 CAR PARK FIRE STATISTICS (UK)

In the 12-year period 1994-2005 there were 3096 fires reported in car parks that were subject to the Building Regulations. Of these, 1592 started in a vehicle (i.e. 1504 did not start in a vehicle, but a few of these did spread to a vehicle). The average number of fires in car parks in buildings over the period was therefore 258 per year. However, the number of car park fires per year shows an overall (but not consistent) decline, with 401 in 1994 but only 142 in 2005. Most car park fires (68%) occurred in buildings reported as 'car park buildings', as identified in the FDR1 forms completed by Fire and Rescue Services following incidents. 6% of fires in car parks occurred in 'flats'. Most injuries from fires in car parks occur in 'car park buildings' (45%), followed by 'flats' (26%). When the ratio of injuries to number of fires is examined, it is found that the 6% of fires in 'flats' cause 26% of the total injuries resulting from fires in car parks, while the 68% of the fires in 'car park buildings' cause only 45% of the injuries. In other words, fires in car parks that are part of flats are 6.5 times more likely to cause injuries, compared with fires in dedicated car park buildings.

To summarise the statistics:

- The number of fires in car parks reported by UK fire and rescue services represents a very small percentage of all fires in the UK (i.e. 426,200 in 2006 – hence less than 0.1%).
- Of these fires in car parks, about 50% did not start in a car.
- Most fires in car parks do not spread (to a car or another car).
- Most fires are in buildings classified as a 'car park building'. Only 6% occur in car parks within 'flats'.
- About 7 people are injured in car park fires each year. (There are very few fatalities; on average, less than one per year.) Fires in car parks for which the building is classified as 'flats' show an injury rate which is quite high compared with other

types of premises. However, fires in car parks for which the building is classified as 'car park' show an injury rate which is low compared with other types of premises.
- The fire statistics have shown that there are very few reported injuries in car park fires, and very few fatalities. However, fires in car parks associated with flats are causing 6.5 times more injuries (per thousand fires) than fires in purpose-built car parks.
- The ease with which a car fire in a car park might spread to nearby cars has been demonstrated. Once a very severe fire has developed, ignition will occur even to cars separated by an un-filled parking bay.

9.2 VENTILATION OF UNDERGROUND CAR PARKS

Over the last 3 to 4 years, impulse ventilation using Jet-fans, has been replacing the more traditional method of ducted extract in providing for the ventilation of enclosed and underground car parks.

Impulse fan ventilation uses a number of small Jet-fans suspended from the car park ceiling, replacing the traditional ductwork. The system sometimes retains the main exhaust fans to provide the air change rate, whilst the Jet-fans create and control the air movement within the space.

When using the ducted extraction method of smoke removal, the mass of smoke produced by a car on fire is calculated using the simplified expression:

$m = 0.19 * P * Y^{1.5}$ **Equation 9.1**

Where:

m = mass of smoke produced Kg/sec.

P = perimeter of vehicle metres.

Y = height of the smoke layer above ground.

Having calculated the mass flow, the volume extraction rate can be determined. A ventilation system designed to provide 10-air changes/hour will only provide the correct level of smoke extraction at a particular car park size. Table 1 illustrates these points:

Having calculated the mass flow, the volume extraction rate can be determined. A ventilation system designed to provide 10-air changes/hour will only provide the correct level of smoke extraction at a particular car park size. Table 9.1 illustrates these points:

TABLE 1 – SMOKE EXTRACT RATES CAR PARK SIZE m²	5 AC/HOUR m³/SEC	10 AC/HOUR m³/SEC	SMOKE PRODUCTION m³/SEC
1000	4.17	8.34	12.6
2000	8.34	16.68	12.6
3000	12.5	25	12.6
4000	16.68	33.36	12.6
8000	33.36	66.72	12.6

Table 9.1: Smoke extract rates from car parks at five and ten air changes per hour per cubic metre per second **Courtesy of Flakt Woods Group**

Typical considerations for Car Park Fire inputs

- Car park height 3.0m
- Smoke layer height 2.25 m
- Fire perimeter 12.0 m
- Fire size 3.0mw
- Radiation losses 25%

Only in car parks with a floor area greater than 3000m² will there be sufficient ventilation to remove the smoke produced. Car parks below this size would smoke log. Above 3000m² the system would increasingly be oversized and expensive. Replacing the extract ductwork with Jet-fans has an immediate advantage of making the extract system 100% efficient. With the extract points at high level, all car parks down to about 1500m² in area will be provided with sufficient ventilation to prevent smoke logging.

Design for Ventilation by Code – BS 7346

The primary design code for use in relation to **ventilation** in underground car parks is BS 7346-7:2006, where from a fire service perspective;

Clause 9 Systems: impulse Ventilation to achieve Smoke Clearance (Post-fire)

The objective of the clause 9 smoke <u>clearance</u> system design is to:

α) *assist firefighters by providing ventilation to allow speedier clearance of the smoke once the fire has been extinguished;*
b) *help reduce the smoke density and temperature during the course of a fire.*

This system is not intended to maintain any area of a car park clear of smoke, to limit smoke density or temperature to within any specific limits or to assist means of escape. It is possible that some smoke clearance systems, if set in operation too early, might actually worsen conditions for means of escape by encouraging smoke circulation and descent of the smoke layer. For this reason, it could be preferable to delay operation after automatic detection of fire.

Clause 10 Systems: impulse Ventilation to assist Firefighting Access (during Firefighting)

*The objective of the clause 10 smoke <u>control</u> design is to **aid access** by the fire service to more quickly locate and tackle a fire and carry out search and rescue as necessary.*

*The extract rate should be calculated for the removal of the mass of mixed air and smoke impelled towards the exhaust intakes. Calculations should be based on a **steady state design fire** from Table 9.2, or another design fire acceptable to the approving authorities. All supporting calculations and justifications should be fully documented.*

Fire Parameters	Indoor car park without sprinkler system	Indoor car park with sprinkler system
Dimensions	5 x 5m	2 x 5m
Perimeter	20 m	14 m
Heat Release Rate	8 MW	4 MW

Table 9.2: Recommended steady state fire design inputs for a clause 10 system

Note: The BRE research (below) has since provided more up-to-date data on likely fire spread between closely parked cars, on a time-line, and should be considered for use in any design fire aligned with fire service response and intervention data.

Time-dependent design fires should be based on an experimental test fire, which should be described and justified in the documentation specified in Clause 18 (of the code). Where the experimental data has been placed in the public domain a reference to the publication may be used as justification.

Where the car park employs a system that is designed to assist firefighting it is imperative that sufficient information is provided to enable attending firefighters to understand the system and operate any override controls as necessary. In the case of smoke clearance systems, a simple plan with a description of the system, override controls and their location in the building will normally suffice. For systems designed to assist firefighting or to protect means of escape, suitable plans showing the extract points and fans should be provided for each level of the car park, together with a brief description of the system's function. Additionally, where jet fans are employed, their location should be indicated on the plan and information should be provided to identify the preferred firefighting access point and direction of approach for a car fire in any particular fire alarm/smoke control zone that is activated.

This information can be provided in the form of plans for fire service use, held at a suitable location accessible to the fire service 24 h per day. Alternatively, for more complex systems an electronic graphical representation could be provided adjacent to the fire alarm panel showing the zone involved and the preferred access stair core/direction etc.

Clause 11 Systems: impulse Ventilation to protect means of escape

The objective of the clause 11 smoke and heat control system is to provide for the protection of escape routes for occupants within the same storey as the car on fire, to preserve a smoke-free path to either the exterior of the building, or to a protected stairwell which leads to a final exit to a place of safety.

9.3 SPRINKLERS IN UNDERGROUND CAR PARKS

Car stackers are becoming extremely popular in the crowded cities of Europe, termed *automatic car parks* as they are generally computer controlled, where typically park cars in multiple levels. Consequently, the fire will not only spread horizontally but also vertically, and even more rapidly. In Germany, *automatic car parks* with more than 20 car spaces are required to be protected by an approved sprinkler system. The installation of a sprinkler system customised to the car park geometry can limit a fire to a single car whilst simultaneously giving an alarm. This gives the Fire Brigade the necessary time to locate the fire and effect extinguishment.

Sprinklers are not considered a requirement under UK building regulations although design best practice suggests they should be installed where car stackers are used or where hose distances exceed 60 metres to the furthest point in the car park as laid around and through the configurations of parked cars.

The findings of the BRE Research included:

'The ease with which a car fire in a car park might spread to nearby cars has been demonstrated. Once a very severe fire has developed, fire will spread to other cars, even where separated by an un-filled parking bay'. 'In this situation, where a number of cars are burning simultaneously, the fire is exacerbated by heat feedback and heat release *rates in excess of 16 MW might be achieved from two or three cars. In Test 1 the initial car fire, Car 1, burned at around 2 MW for about 20 minutes and it was only then that Car 2 became involved (although Car 3 then ignited very soon after). However, in Test 3, all three cars were burning after around 10 minutes. In Test 4 (Buxton – LPG), Car 2 was alight after 21 minutes and all four cars were burning after around 23 minutes. In Test 8, an engine fire test with a nearby car 'nose to nose' the fire spread to the second car within 5 minutes.' 'The ventilation limitations on such a fire in an enclosed car park result in a very hot ceiling jet, which spreads the fire to nearby cars with the dominant mechanism of heat transfer being radiation from the flames and hot gas layer, but with some direct flame contact.*

There were only a limited number of cars in each of the tests (a maximum of four); however, escalation to many cars within a specific proximity in an actual car park must be expected under these conditions.' Gas temperatures in the enclosed rig (beneath parts of the ceiling) reached 1100°C in all the enclosure tests, exceeding 1200°C briefly in Test 4.

Such temperatures and fire spread rates mean that basement car-park fires can be very intense as firefighters are arriving on-scene and may already be beyond the control of a single hose-line. Although such fires are rare, they may place extreme demands on fire service resources. Therefore, high levels of incident command, additional safety hose-lines and <u>reduced duration</u> crew relief cycles in very hot environments will be key to a successful operation.

Chapter 10

The 'High-rise Fire'

In 2008 the author opened the chapter on High-rise fires in his book *EuroFirefighter* with the following paragraphs: 'When we are faced with a serious fire at ground level, our firefighters often encounter great difficulties and exposure to some element of risk. When they are faced with that same fire, thirty stories above ground, the physiological [and logistical] demands are much greater and the difficulties and risks greatly magnified. There may be long time delays between a fire commander's chosen strategies becoming tactical operations on the fire floors. There may be changing circumstances during this delay that requires the strategy to be altered. There will be a great demand for effective staffing to accomplish even the most basic operation and then, where firefighters are working hard, the need to support them in a sustained attack on the fire will treble the resources operating on the fire floors.

To be effective you must have a pre-plan that is based on the experience of those who have fought these types of fires and learned many lessons. The pre-plan must be well understood by everybody and to achieve this requires frequent practice in such buildings. The communication process at a high-rise fire will inevitably break down and the pre-plan must ensure that critical tasks, such as searching stair-shafts, elevators and roofs, are documented as written assignments into the pre-plan. The objective is to enable firefighting teams to adapt and function in small teams with pre-assigned tasks and, on occasions, without [overall] fire command supervision.

Above all, avoid complacency! This is inevitably the firefighter's worst enemy. Approach every situation with care and professionalism and always try to be at least one step ahead of the fire's next move'.

Since then, nothing has changed and high-rise firefighting is every bit as demanding and complex as stated above. As with all calls to fire, they must be approached consistently, with a non-complacent demeanor. It is so common for things to go wrong when we approach fires in a routine manner and we have probably all experienced this. However, a review of firefighter line of duty death reports at fires quite often throws up some elements of complacency that led to a chain of events, resulting in tragedy. Nowhere is this truer than at a high-rise fire where there can be time-lags set in motion due to the command and operations structure that needs to be established in order for a safe approach to be made to the fire floor from the strategic bridgehead set up, 1-3 floors below the fire floor. Sometimes things aren't all that they might first appear. My colleague Karel lambert (Brussels) takes up the story.

10.1 KAREL LAMBERT (FIRE OFFICER WITH THE BRUSSELS FIRE SERVICE – BELGIUM)

'One day, I was dispatched as Incident Commander to a high-rise fire in a 20-storey residential building. More often than not, these calls turn out to be nothing at all. When we turned up, the first arriving station was already going up the stairs. The incident was on the 6th floor. Upon arrival, it seemed to be 'food on a stove'. Nothing was suggesting this was a real fire. It was probably one of those typical calls which would turn out to be nothing at all. I sent our guys in to have a look but the radio traffic gave me an awkward feeling and I decided to go up myself to assess the situation.

When I climbed the staircase, everything confirmed my initial assessment:

- There was no smoke in the staircase.
- People were descending but in a calmly manner.
- When I passed the floor below the incident floor, there was no high-rise pack. In a real fire situation, the first arriving station would have connected the high-rise pack to the water riser at the floor below the fire. Here, there was no hose in the staircase.
- No smoke was to be seen on the landing of the fire floor, even not after I opened the first fire rated door. (In Belgium, two fire rated doors separate the hallway from the stair-case).

This all changed when I opened the second fire rated door to the hallway connecting the six apartments on that level. A thick smoke layer had descended to 1 metre above the floor. I could hear the guys working in the fire apartment. Things were not going well and they were yelling for more water. Two firefighters ended up in the hospital. One had to be rescued by his colleagues. Finally, a high-rise pack was connected and the fire was brought under control.

Afterwards, it turned out that the initial crew took the building hose-reel that was present on the fire floor (it flows 60 L/min) to attack the fire. At first, I was very angry with the crew commander who led the first crew in. 'How could he do such a thing?'

After a while I realized that he went through the same line of thinking as I did. He climbed to the fire floor thinking it was nothing at all. And he too was surprised that there was so much smoke when opening the second fire rated door. Probably, the smoke layer was higher when he entered the hallway. Continuing his line of thinking (a very small fire), he chose the hose-reel of the building, which was right in front of his crew. Both of us were biased by our size up, by routine. This was an important lesson for all of us. Sometimes things start not the way they are supposed to be. The lesson is to stop and rethink the selected approach when confronted with new elements.

A second lesson for me is that it is so easy for a battalion chief to be hard on the crews. When we arrive, the situation has changed. Often we have more information than they had when they took their decisions.

And a third lesson was that high-rise incidents should always be handled with sufficient flow-rate.'

10.2 RISING FIRE MAINS (WET AND DRY PIPES) – DESIGN SPECIFICATIONS

One of my biggest concerns is to see firefighters repeatedly pump inadequate firefighting water to height in tall buildings. In the UK and many parts of Europe, the wet (*permanently charged with water*) rising main standpipes are only provided in buildings in excess of 50 metres in height (UK). Prior to 2015 the rising main stand-pipes were all dry to a height of

60 m. This height was reduced in 2015 in recognition of the higher pressures required by US nozzles commonly being imported into the UK since 1987. During a period of at least 15 years, the UK fire service became devoid of any awareness that their water deployments at height had reduced dramatically, where the rising main pressures failed to match the transition to new branch nozzles. The original design specifications for dry rising mains were based on 1960/70s tactics where firefighters in London were specifically using a 19mm smooth-bore nozzle in high-rise buildings. It had been determined by the Fire Research Station (FRS) that a flow of 100 gallons/min was required to control the fire size of the time. At a maximum height of 60 metres the 19mm smooth-bores were flowing 455 L/min (100 gallons/min). No other demands other than a 100 gallons/min flow for one hose-line were required. However, a wet rising main was initially expected to provide three hose-lines, each of 100 gallons/min at 4 bar outlet pressure. As the wet riser was designed to flow three firefighting hose-lines at that flow-rate, the design flow was actually 1,365 L/min (or 300 gallons/min (imperial). However, when metrication arrived in the UK in 1969 this figure of 1,365 L/min was rounded up to 1,500 L/min and this flow-rate is currently referred to in design codes[85]. The original design specifications (converted to L/min) are below:

Outlet height (m)	Floor Level	Inlet pressures (Bar)					
		15 bar	15 bar	10 bar	10 bar	7 bar	7 bar
		Outlet pressures (P) – (and flow-rate (Q) from a 19mm smooth-bore nozzle)					
		P (bar)	Q (L/min)	P (bar)	Q (L/min)	P (bar)	Q (L/min)
63	22	7.78	642	3.36	422	0.71	194
59.5	21	8.1	655	3.66	440	1.02	232
56	19	8.41	66	3.97	458	1.32	264
52.5	18	8.72	679	4.28	476	1.62	293
49	17	9.04	692	4.59	493	1.93	320
45.5	16	9.35	703	4.90	509	2.24	344
42	15	9.67	715	5.21	525	2.54	367
38.5	14	9.98	727	5.52	540	2.85	388
35	13	10.3	738	5.83	555	3.16	409
31.5	11	10.62	750	6.15	570	3.47	28
28	10	10.94	761	6.49	586	3.78	447
24.5	9	11.26	772	6.77	598	4.09	465
21	8	11.57	782	7.09	612	4.40	482
17.5	7	11.90	793	7.40	626	4.71	499
14	6	12.21	804	7.72	639	5.03	516
10.5	4	12.54	814	8.04	652	5.34	531
7	2	12.86	825	8.35	665	5.65	547
3.5	1	13.18	835	8.67	677	5.97	562

100mm dry-pipe rising main diameter
10 bar charging pressure into base of main from fire engine pump
Twin parallel 70mm x 25m hose-lines from pump to inlets at base
1 x 25m 70mm hose-line supplying ⅜" (19mm) smooth-bore branch nozzle from ea. outlet

Table 10.1: The original design specification for dry-rising mains in the UK (BS 5306)

85 BS 9990:2015

The pressure available at the riser outlet in a dry rising main is largely determined by the performance of the fire appliance pump supplying the main. In the UK charging pressures used have typically ranged from 7 to 10 bar (pre-2015 dry risers are based on a design specification of a 10-bar maximum whereas, post-2015 risers can be charged safely to 12 bar). As the change-over to a wet rising main is at 60m (50m from 2015) the pressure available for firefighting at higher elevations will vary significantly due to static head loss. Pressure loss through static head at 60m equates to approximately 6 bars and this is reflected above. By charging to 10 bars enables just under 4 bars at the highest outlet (60m) or if charging to 12 bars (50m risers) just over 6 bars will be available from the highest outlet.

$$P_{loss} = \frac{Height\ in\ metres}{10} = \frac{60}{10} = 6\ bar\ loss \quad \textbf{Equation 10.1}$$

Current codes require that where fire mains are installed and there are no floors higher than 50 m above fire service access level, wet or dry fire mains may be installed. Where there are floors higher than 50 m above fire service access level, wet fire mains should be installed owing to the pressures required to provide adequate firefighting water supplies at the landing valves at upper floors, and also to ensure that water is immediately available at all floor levels. Other changes in the 2015 BS 9990 code for rising mains stipulated that the wet-riser should be capable of providing a flow of water of at least 1,500 L/min in the fire main, i.e. sufficient to supply hose-lines from two separate landing valves simultaneously (pre-2015 codes asked for three hose-lines). A running pressure of (8 ±0.5) bar should be maintained at each landing valve when fully opened.

NOTE 1 *The reduction to 50m height is based upon the fire service using 51 mm hose and a firefighting branch having hydraulic characteristics of K-value = 230*

NOTE 2 *Guidance on when a fire main should be installed is given in BS 9991 and BS 9999.*

Any proposed use of horizontal fire mains should be discussed and agreed with the local fire and rescue service.

Other changes in the 2015 code required dry risers to a design specification meeting a 12-bar charging pressure, as opposed to the original 10 bar pre-2015. The problem will now lay in operational firefighters identifying the post-2015 dry risers from the pre-2015 versions! The charging pressures are different but how can firefighters know which system is installed on the fire-ground.

It is worthy of note that when using 15/16th smooth-bore nozzles on 65mm hose-lines, firefighters in New York City will pump their dry rising stand-pipes to the following pressures –

Floors	Pump Pressure	Pump Pressure
1-10	10 bar	150 psi
11-20	13.5 bar	200 psi

Table 10.2: FDNY pumping pressure guidance into dry-risers

However, if using combination fog-nozzles the pressures will be proportionally higher.

- Increase in Pressure per metre below grade – 0.1 bar
- LOP per metre height – 0.1 bar
- LOP per floor height – 0.3 bar
- LOP per 5 floors height – 1.5 bar
- LOP per 10 floors height – 3 bars
- LOP per 15 floors height – 4.5 bars
- LOP per 20 floors height – 6 bars

When using 22mm smooth-bore nozzles below the 10th floor level from dry risers	Recommended dry-riser charging pressures
Ground to 2nd floor	7 bar
3 – 5 floor	8 bar
6 – 9 floor	9 bar
10 or higher	10 bar
15 or higher	12 bar if system allows (post 2015 risers)

Table 10.3: Kent Fire and Rescue Service pumping pressure guidance into dry-risers using 22mm smooth-bore nozzles

10.3 NOZZLE REACTION FORCE

1990 I completed a research project[86] (Fire Magazine UK – November 1992) that evaluated the operational capability of firefighting hand-line streams as used by London Fire Brigade. At that time we had main-line options of 45mm (1 3/4') hose-lines with 12.5mm (1/2') nozzles and 70mm (2 3/4') hose-lines with either 19mm (3/4') or 25mm (1') nozzle options. One of the basic laws of physics –Newton's third law – states that for every action there is an equal and opposite reaction. Quite simply, to the firefighter this means that as water is projected from a nozzle to form a 'jet' or firefighting stream, the nozzle tends to recoil in the opposite direction. This effect, termed nozzle or jet reaction (or kick-back) requires the firefighters at the nozzle to exert sufficient effort into overcoming this reaction force. The entire force of this reaction takes place as the water leaves the nozzle and whether or not the fire stream strikes a nearby object has no effect on the reaction. Thus, whether or not a hose-line's stream is allowed to strike a wall whilst a firefighter is working it from the top of a ladder is immaterial to his stability on the ladder, which is governed solely by the reaction at the nozzle.

By evaluating maximum flow capability for a hose-line that could be effectively directed and safely handled whilst *advancing and working* inside a fire-involved structure It was observed that there was a maximum nozzle reaction force that could be handled by one, two and three firefighters as follows

- **One firefighter – 266N (60 psi)**
- **Two firefighters – 333N (75 psi)**
- **Three firefighters – 422N (95 psi)**

---

86 Grimwood. P, Fire Magazine, UK, November 1992

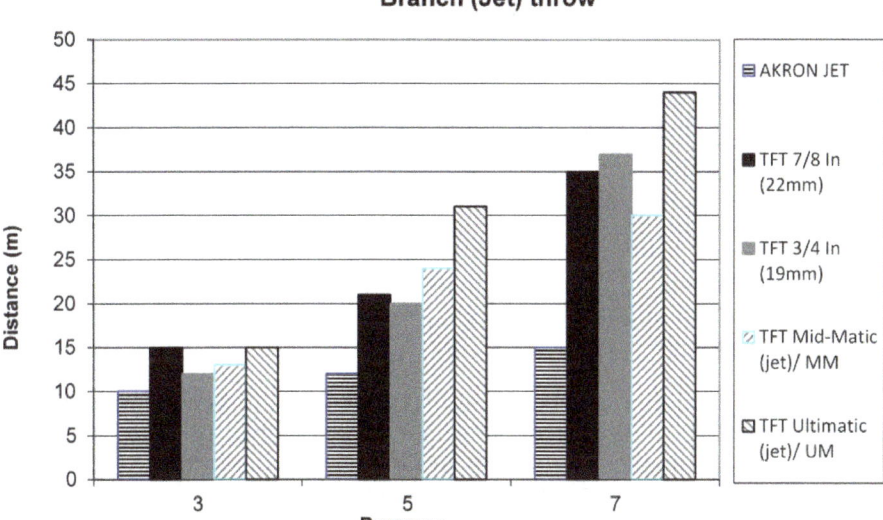

Figure 10.1: The automatic nozzles (TFT Ultimate and TFT Midmatic) *demonstrate greater stream reach than the TFT 22mm smooth-bore at nozzle pressures in excess of 3 bar, demonstrating the effectiveness of the automatic facility (However, flow-rates at 3 bar were five times higher from the 22mm smooth-bore compared to the automatics).*

These were interesting findings and from these I was able to establish baseline flows for interior firefighting operations. To achieve this it became necessary to take acceptable pumping practice into consideration without contravening the limitations placed upon European pumps, hoses and equipment available at that time. This resulted in baseline flows of 277 L/min (73 gpm) on 45mm hose-lines with 12.5mm nozzles, 650 L/min (172 gpm) on 70mm hose-lines with 20mm nozzles and 750 L/min(200 gpm) on 70mm hose-lines with 25mm nozzles, as *advanced* by two-man crews. However, these 'base-line' flows were rarely, if ever, achieved in practice as tradition had established a base-line pumping pressure of 3-4 bars (45-60 lbs psi) to which a small adjustment may have sometimes been made for frictional loss and pressure head. Actual firefighting flow-rates were in fact far lower than had been previously thought – ref: SRDB codes of the period.

Interestingly, similar research has been carried out by other fire departments, notably San Francisco, Los Angeles and Chicago, who proposed that a safe and practical baseline flow for a workable firefighting hand-line would be around 550 L/min (150 gpm). More recently (1996), the City of St. Petersburg in Florida USA have established that, for their purposes, the ideal baseline flow is around 600 L/min (160 gpm) using a 7/8' (22mm) nozzle with a 50 lbs psi nozzle pressure on a 45mm (1 3/4') hose-line. This set-up will create an acceptable reaction force of 266N (60 psi) and offers a hose-line that is easily advanced for interior position.

The firefighter is able to calculate the amount of nozzle reaction (NR) by resorting to various formulae:

$NR\ (Newtons) = 1.57 \times NP \times d^2/10$ (European Smooth-bore), or;

$NR\ (Newtons) = 0.22563 \times L/min \times \sqrt{NP}$ (European Combination fog/jet or automatic Nozzles)

These are metric formulae where NP = Nozzle Pressure; d = Nozzle Diameter; L/min = Flow in Litres Per Minute and NR is in Newtons.

In the USA, different formulae are used as follows -

$NR\ (psi) = 1.57 \times d^2 \times NP$ (US Smooth-bore), or;

$NR\ (psi) = 0.0505 \times gpm \times \sqrt{NP}$ (US Combination fog/straight or automatic Nozzles)

10.4 BDAG RESEARCH 2004 – MINIMUM K-FACTOR NOZZLE REQUIREMENTS

In 2004 there was some realisation that modern (US design) branch nozzles in the UK were not effective when operating at the low pressures available at height, particularly if using dry-pipe rising main standpipes. At the highest floor levels, just 3.66 bar pressure was available at the outlets and by the time friction loss was accounted for in the 45-60m length interior attack hose-lines, the flow-rate at the nozzle was often below 100 L/min. In fact, many UK fire services are still operating (2017) with mismatches existing between risers and nozzles and are failing to take account of design flow-rates. A national UK BDAG research project[87] addressed the problem in part but the full impacts have not been fully addressed. This research raised some key points but did not offer comprehensive solutions –

'There is concern that internal firefighting from a firefighting shaft may not be adequate to control or extinguish a fire due to the extent of fire development at the time of fire and rescue service intervention. In particular, the use of generic criteria in defining the number of firefighting shafts does not account for:

- the fire load and rate of fire growth for the particular occupancy, and
- the time of intervention against the time of ignition of the fire (and thus its potential size and heat release)'.

The research report continued –

'Firefighting shafts containing rising mains are provided to assist the fire and rescue service in accessing and fighting fires in tall buildings. The rising mains may be either a dry main, which is supplied from a fire appliance pump during an incident, or a wet main, which is permanently charged with water. Where dry mains are provided the pressure available at the firefighting branch reduces with increasing elevation due to the static head of the water in the rising main and frictional losses. In very tall buildings these losses will ultimately exceed the pressure supplied from the fire appliance pump supplying the main. For this reason, in the UK, in buildings over 60m, the mains are permanently charged to provide a pressure-regulated flow. However, there is little data currently available to establish whether the current provisions and corresponding fire and rescue service procedures are appropriate. There is also a pressing short-term need to ensure that firefighting techniques in tall buildings align

---

[87] Hunt. S, Roberts. G, Effect of reduced pressures on performance of firefighting branches in tall buildings – Aspects of high-rise firefighting, Building Disaster Assessment Group (BDAG), Office of the Deputy Prime Minister London, UK 2004

to the equipment used and facilities provided. Where techniques do not align with equipment then the changes needed to support firefighting in tall buildings should be identified'.

The main findings of this research stated that:
- *The subjective performance of the firefighting branches assessed decreases with decreasing pressures.*
- *There is significant variation in the pressure beneath which the subjective (from a panel of professionals) performance of firefighting branches is considered inadequate to undertake the techniques taught for compartment firefighting. The majority of branches tested required a minimum operating pressure of 4 bars at the branch to perform satisfactorily in the panel's view.*
- *The techniques that are taught for compartment firefighting may not be appropriate at low pressures with some of the branches assessed. This will depend upon the size and length of hose line supplying the firefighting branch.*
- *When firefighting in tall buildings fitted with dry rising mains there will be an elevation beyond which there is inadequate pressure to undertake adequate compartment firefighting techniques with some firefighting branches. This elevation will depend upon the size and length of hose used for the attack line, the flow and the specific performance of the firefighting branch used. For the same firefighting branch, where 45mm hose is used, this elevation will be significantly less than that where 70mm hose is used. If 51mm hose was used a firefighting attack could be mounted at higher elevations than could be achieved with 45mm hose currently used by most fire and rescue services.*
- *When firefighting in tall buildings fitted with wet rising mains, the pressure at the riser outlet is regulated between 4 and 5 bars. Depending upon the size of the hose and the specific performance of the firefighting branch, there may be insufficient pressure available at the firefighting branch to undertake techniques that are taught for compartment firefighting. This situation will be exacerbated where smaller diameter hose is used for the attack line.*
- *There appears to be limited correlation between the running pressure and flowrates specified for wet rising mains indicating that the performance criteria specified is not empirically based and should be reviewed.*
- *The results highlight the fact that fire and rescue services may need to evaluate the performance of the branch types that they use during high-rise firefighting operations to comply with their obligations under Section 4 of The Provision and Use of Work Equipment Regulations. This will include other influencing factors such as the pressures available from dry/wet riser systems and the diameter and lengths of hose used.*
- *Further research should be conducted into the performance standards required of both dry and wet rising mains in tall buildings to develop standards which will support the use of compartment firefighting techniques required to support the safety of firefighters. This work should also include contingency arrangements for possible failure of facilities designed to support firefighting in tall buildings.*
- *The generic risk assessment for high-rise firefighting and search and rescue procedures produced by HM Fire Service Inspectorate should be revised in light of the results of this work and the future research identified. Effect of reduced pressures on performance of firefighting branches in tall buildings*

- *Finally, the report recommends that an agreed national high-rise firefighting and search and rescue procedure should be developed, which reflects:*

  - *the type, performance and limitations of firefighting facilities provided in tall buildings,*
  - *the physiological limitations of firefighting and search and rescue procedures in tall buildings,*
  - *the performance and limitations of fire and rescue service equipment designed to support firefighting in tall buildings and*
  - *contingency arrangements for possible failure of facilities designed to support firefighting in tall buildings.*

The basis of the research had been to test 27-firefighting branches/nozzles in current use (2004) in the UK, including manual flow select 'combination' and 'automatic' spray/jet branches. In establishing the subjective criteria which would be used to appraise the performance of the firefighting branches at different pressures, the panel decided firefighting branches were required to produce:

- An effective jet, as this would be required to undertake a direct fire attack.
- An effective spray on full cone, as this would be required to provide a defensive spray to protect firefighters should it be necessary to withdraw from a compartment if the conditions deteriorated.
- An effective spray pattern at a 70° cone angle, which would be required to undertake 3D gas cooling. This angle was chosen as being within the 60° to 75° optimum range that is recommended by other research work that has been undertaken into compartment firefighting (Grimwood 2000).
- The research highlighted the wide variation in hydraulic performance of the branches tested. For the majority of the branches tested to attain the performance criteria set by the panel a minimum operating pressure of 4 bar at the branch would be required. **This did not however account for branches which are designed to operate between 5-7 bar pressure at the branch.**

Based on this research, it was stated that the use of **automatic branch nozzles** may not be appropriate for use in high-rise building fires. This research report also stated that it could find no correlation to the code requirements (BS 9990) for a 500 L/min flow-rate at 4–5 bars pressure – however the correlation has been explained by this author at 10.2 above.

51mm Hose – Optimised for High-rise Firefighting

The researchers had interviewed this author prior to compiling their report and based on his personal experience gained on a study assignment in Los Angeles in 1990[88], the use of 51mm diameter hose-lines was further researched by BDAG and included as a recommendation for all UK fire services and then from 2015, rising main design specifications were configured (according to BS 9990) based on 51mm attack hose-lines being used.

However, the use of 'K-Factors' in the 2015 version of BS 9990[89] to determine the performance factor of any particular firefighting nozzle may be inappropriate. The minimum K-factor that is acceptable for a UK high-rise firefighting nozzle is given in the code as 230. The original riser design specification from the 1960s was for a 19mm nozzle

---

88 Grimwood. P, Fog Attack, DMG Publications, Redhill Surrey UK, 1992
89 BS 9990 – Non-automatic firefighting systems in buildings – Code of practice 2015

operating at 3.45 bar (50 psi) at the nozzle. This provides a K-Factor value of 245 and that was based on residential apartment fires. The requirement for an effective high-rise nozzle for use in open-plan office or commercial fires suggests that minimum K-Factor values of 350-440 may be far more appropriate, in accordance with 2015 wet-riser design specifications (two firefighting jets from a 1,500 L/min supply).

The K-Factor performance rating of firefighting nozzles demands a *flow-rate* divided by *nozzle pressure* calculation, so in high-rise terms this calculation should not occur at ground level but should be based on a nozzle operating at the riser's <u>highest</u> outlet.

Table 10.4: K-Factor values for high-rise branch nozzles

K-factor is a branch design requirement for dry rising mains (BS 9990:2015) and is defined as any particular nozzle's firefighting capability at a given volume flow rate and is generally calculated using a formula as follows:

	Nozzle	Nozzle	NP bar	L/min	'K' Value
$K = Q / \sqrt{p}$ Where: K = nozzle constant Q = volume flow rate (L/min) p = pressure at the nozzle (bar) Flow (LPM) is found by multiplying $\sqrt{p}$ by the K factor (K) $\sqrt{p} = Q/K$ $Q = K \times \sqrt{p}$	19mm	3/4'	3	423	244
	22mm	7/8'	4	665	332
	24mm	15/16'	3.5	700	374
	25mm	1'	3	756	436
	Automatic	Fog/Jet	4	115	58
	Automatic	Fog/Jet	3	100	58

- The benchmark requirement for a UK high-rise firefighting nozzle meets a minimum K-value of **230** (2015 rising main design specification)
- A 19mm (3/4') smooth-bore nozzle at 3.45 bar (50 psi) nozzle pressure meets a K-value of **245** (1960s rising main design specification at 100 gall/min; or 455 L/min)

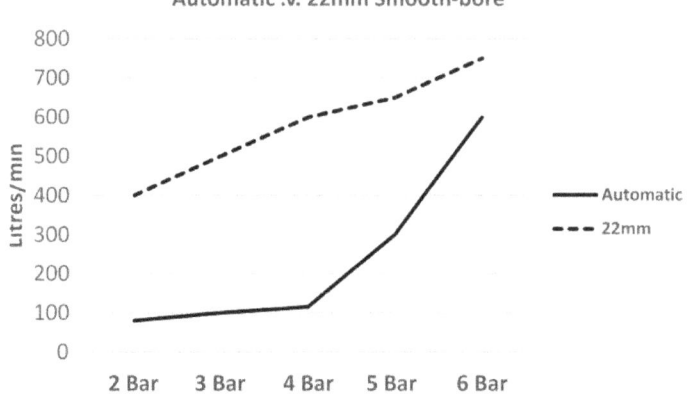

Figure 10.2: The 22mm smooth-bore nozzle will flow far greater amounts of water at lower pressures than an automatic nozzle – which optimises the 22mm for high-rise flow-rates above the tenth floor

22mm Smooth-bore 2 x 45mm Hose	Pump or Fire Main Outlet Pressure	Nozzle Pressure	Flow-rate
2 FF on Branch	4 bars	2.2 bars	420 LPM
2 FF on Branch	5.5 bars	3 bars	520 LPM
2 FF on Branch	8 bars	4 bars	600 LPM
3 FF on Branch	10 bars	5.5 bars	700 LPM
22mm Smooth-bore 2 x 51mm Hose	Pump or Fire Main Outlet Pressure	Nozzle Pressure	Flow-rate
2 FF on Branch	4 bars	2.7 bars	480 LPM
2 FF on Branch	5.5 bars	3.4 bars	550 LPM
3 FF on Branch	8 bars	5.5 bars	700 LPM
NOT RECOMMENDED	10 bars	6.8 bars	770 LPM

Table 10.5: The 22mm smooth-bore nozzle will operate effectively at a range of nozzle pressures. In general, two firefighters (2 FF) can handle the nozzle reaction when advancing at high flows, but any flows above 650 L/min may require 3 FF to advance the hose-line safely. (Based on practical tests by Kent Fire and Rescue)

10.5 TELSTAR HOUSE OFFICE FIRE – LONDON 2003

The fire at London's Telstar House tore through five floors of a thirteen-storey high-rise office building after firefighters were forced off the fire floor due to inadequate flow-rate[90]. The ongoing firefighting operations suffered further as the sole 100mm rising main serving each 1,500m² floor-plate at all levels were unable to cope with the requirements for copious amounts of firefighting water. Deputy Assistant Commissioner Terry Adams stated that the lack of water available from the rising main stand-pipe created a need for additional staffing and resources to lay hose-lines up the stairs, as high as the eleventh floor. He stated that greater flow capacity at an earlier stage would almost certainly have enabled crews to deal more effectively with this fire. Would the 2015 K-Factor values (BS 9990) and the existing rising main provisions of ADB and BS 9999 provide this additional capacity?

10.6 HOSE-LINE AND NOZZLE CONFIGURATION FOR HIGH-RISE

Having determined that 51mm hose is optimised for high-rise firefighting[91,92], the author coordinated a research project in 2008 in Kent Fire and Rescue Service, to compare their automatic firefighting nozzles with a range of smooth-bore nozzles for use in high-rise building fires to 60 metres in height (dry rising standpipes). After a lengthy period of

---

90 Grimwood. P; EuroFirefighter, Jeremy Mills Publishing, UK, 2008
91 Hunt. S, Roberts. G, Effect of reduced pressures on performance of firefighting branches in tall buildings – Aspects of high-rise firefighting, Building Disaster Assessment Group (BDAG), Office of the Deputy Prime Minister London, UK 2004
92 Grimwood. P, Fog Attack, DMG Publications, Redhill Surrey UK, 1992

Figures 10.3 and 10.4: Telstar House fire in London **Photos courtesy London Evening News**

testing it was finally determined that 22mm smooth-bores would out-perform the automatic nozzles above the tenth level, or in any situation where pressure was inadequate. This changed their approach to high-rise firefighting and compartment firefighting tactics accounted for this necessary change where K-Factor values in automatic nozzles above the tenth floor were recorded as <60. If using anything smaller than a 51mm hose-line, serious consideration should be given to running larger hose into the final length of compartment hose-line (say 65 or 70mm hose feeding 45 or 38mm hose) to reduce friction loss and increase flows at height.

Serious consideration was given, to the viability of 3D compartment firefighting techniques at height where natural air-flow dynamics, stack effect, forced draft fires and automated mechanical smoke ventilation systems had the potential to worsen fire conditions for firefighters, particularly in lengthy corridors. In fact, two fires within a month in our southern response area had created very intense fire conditions, as wind driven fires were forced deep into the accommodation causing fire doors to burn through within a few minutes and pressurised smoke to push through expansion joints in the concrete firefighting shaft. In one case smoke was 'flow-reversed' **down** 12 floors in the stair-shaft to ground level, exiting into the street at extreme velocity.

These type of fire conditions cannot be approached with low-flow streams or bursts of water-fog. What is needed is the high-flow reach of a solid stream. In environments such as this, solid streams can be bounced around off walls and ceilings to create some effective gas cooling as well as reaching shielded fires. This is nothing new to firefighters who are used to working in high-rise buildings and when an advance need to be made along a long corridor to reach the fire compartment, high-flow solid streams should be the first choice every time. Having had this conversation with fire behaviour instructors from several countries it is clearly not a strategy they would appear to prefer. However, I would ask them this question and it is usually one they cannot answer positively – 'Have you ever experienced such fire conditions as these (described above)?' – and the answer has been 98% 'No I haven't'! I think if ever they do, they will change their opinion very quickly in support of high-flow solid stream attack.

Chapter 11

Compartment Fire Behaviour Training (CFBT)

'The Swedish fire service taught us a great deal about fire behaviour. However, when it comes to fast developing building fires – and I've used both options – if I had to choose between low-flow spray patterns with finely divided water droplets or high-flow solid streams – I'd go with the solid stream every time.'

Paul Grimwood, 2014

There were 549 fires in London 1984–1994. *Image courtesy of TA Hydronics*

11.1 ANDY FREDERICKS FDNY SQUAD 18 NEW YORK CITY

Andy Fredericks was a great firefighter and an inspiration to generations of firefighters worldwide. He worked and tragically died, alongside 342 other FDNY firefighters, as a member of Squad 18 attending the scene of terrorist attacks on the World Trade Center in New York City, in September 2001. Andy and I had a series of endless, but enjoyable, exchange of views not long before his death. We used to talk for what seemed like hours over water-fog tactics versus high-flow solid stream fire attack. We also took our hardened beliefs and opinions into the pages of *Fire Engineering magazine* (US) where he would write about the benefits of solid streams over water-fog and I would counter with an alternative viewpoint.

Andy wrote, quoting my previous publications, in Fire Engineering[93] 2/2000 –

'As a result of two Swedish firefighters being killed in a flashover in the early 1980s, fog nozzle techniques were devised to counter the effects of fire gas ignition and prevent injuries from flashover and backdraft. Termed 'offensive' or 'three-dimensional' water fog application, these techniques have been explained in great detail in the writings of Paul Grimwood, a retired veteran firefighter from the London Fire Brigade.

Grimwood was kind enough to address my questions and concerns about '3-D' fog techniques. Although I agree with his assessment of the modern fire environment and its attendant hazards-particularly the volatile nature of fire gases and the increasing hazards of flashover and backdraft, I disagree with several of the specific tactics he advocates. The brief examination of 3-D water for techniques contained here is taken from a pamphlet entitled 'Flashover & Nozzle Techniques' prepared by Grimwood. Offensive fog application requires that small (around 400 micron) droplets produced by special fog nozzles be directed into the overhead gas layers in short bursts or 'pulses.' The objective is to suspend the droplets in the gases to cool them and retard their ignition (in other words, putting water on smoke as a preventive measure). While ideally 3-D fog application will prevent ignition of the fire gases, Grimwood states that the technique is suitable for both pre- and post-flashover fires. As the water fog turns to steam and expands in volume, it is accompanied by a corresponding decrease or contraction in volume of the fire gases, reportedly avoiding the debilitating effects associated with steam production caused by fog streams during interior firefighting efforts. In addition, by avoiding contact between the water and the heated walls and ceiling (opposite of what the combination method of attack requires), unwanted steam production is further reduced, thereby maintaining tenable conditions for the nozzle team.

Offensive fog techniques require rather precise execution for success. Grimwood states that firefighters employing 3-D fog techniques should be 'extremely well practiced in nozzle handling and 'pulsing' actions'. Given the wide spectrum of distractions faced by the modern fire service (EMS, hazmat, technical rescue, and so on) and the youthful look of many fire departments, hand-line and nozzle techniques must be kept as simple and straightforward as possible. Regardless of its reported effectiveness, offensive fog application does not fit this description. I believe a more traditional approach is in order'.

---

[93] Fire Engineering Magazine, Penwell Publications, Penwell Fire Group USA

Author's response to Andy Fredericks in Fire Engineering 10/2000 –

'Most recently, I read with great interest Andrew A. Fredericks' two-part discussion of water fog applications and the direct form of attack. I noted that generally the U.S approach to compartmental (interior) firefighting currently seems to favour the low-pressure, high-flow, solid stream attack. The results of a recent Internet survey suggested that 54 percent of US firefighters preferred smooth bore over the 42 percent who support fog nozzles. However, these figures are mainly influenced by the direct attack vs. the indirect attack experience. I have been fortunate enough to have served on both sides of the Atlantic [as a firefighter] and am fully aware of the varying approaches made and the differences in construction and cultures that exist between our nations to appreciate why preferences may evolve. Having advanced hose-lines alongside Fire Department of New York firefighters into burning South Bronx tenements during the 1970s and having experienced the effects of smooth bore streams directed into heavy fire fronts [in both London and New York], I can testify to the reasons why such an approach is effective and remains popular. I have enjoyed conversing with Andrew Fredericks in the past and fully respect his views and opinions in relation to compartment firefighting, which are based on many years of sound experience. I certainly do not oppose his viewpoint that the flow from a straight stream attack, at the base of a fire, is the most effective (and safest) application a firefighter can use to control a room-and-contents fire.

The new wave use of water fog is not a reinvention of the indirect water fog tactics that became popular in the 1950s, where the intention was to evaporate water droplets on the heated surfaces (walls, ceilings, and so on) within a fire-involved compartment (room). This effect of producing large amounts of steam to smother the fire to extinction most certainly had its drawbacks. Although the concept generally worked, it definitely wasn't popular with the majority of firefighters. The reports of nozzle operators and building occupants' receiving steam burns and the 'pushing' of smoke and fire farther into the structure are well-documented; I won't discuss these negative aspects further here. The aim was to achieve a 10 to 35-percent concentration of water vapour within the fire compartment; the application of water was measured in GPM/per square foot of surface contact.

The new wave concepts of water fog applications [also termed three-dimensional (3-D) water fog] are aimed essentially at flashover control and may be applied both defensively (pre-flashover gaseous cooling) and offensively (post-flashover suppression of gaseous combustion). In simple terms, the water droplets are placed directly into the fire gas layers that form throughout a fire-involved structure, with the applications being measured three-dimensionally in GPM/cubic foot of compartmental volume. Such gases will exist both in the compartment where the fire is burning and adjacent compartments (rooms, corridors, hallways), although their flammability limits will be dynamic and somewhat unpredictable as the firefighting operation progresses.

The changing ventilation parameters as firefighters 'open up' and the development of the fire itself alter the flammability limits of these gases as they form and transport throughout the structure. To counter the hazards of fire gas ignitions, water droplets are applied in short bursts using rapid on/off motions at the nozzle-termed pulsing. This effect ensures that the droplets suspend in the gases for several seconds, not on the walls and ceiling, avoiding the formation of excessive amounts of 'wet' steam. Placing fine water droplets into the gas layers cools the gases, taking them below their ignition temperature and outside their limits of flammability. Although there is massive expansion of the water as it turns to steam, this is immediately countered by the effect on the gases, which actually contract as they cool. Where the application is precise, water vapour is seen to be dry and humid (as in a sauna) as opposed to wet and cloudy (as in a Turkish steam bath). The nozzle operator is easily able to recognize when too much water has been applied and adjusts pulses to compensate. The results are a safe and comfortable working environment for firefighters and an effectively maintained thermal balance.

These applications may be applied offensively to deal with burning gas layers (post-flashover), but it is important that this approach not be viewed as a replacement for the straight stream direct style of attack. The true qualities of pulsing water fog tactics are realized in their defensive role when approaching a fire through heavy smoke conditions. Placing fine water droplets into the smoke (gases) on the approach route will create a much safer corridor for firefighters to traverse, allowing them to revert to a straight stream pattern as and when flaming combustion is encountered.

I have heard so many ill-informed arguments from firefighters who have 'read articles and seen videos' concerning these techniques but who have never actually been trained to apply them or have not experienced their practical effects. Having personally suffered the effects of misapplied indirect water fog applications is no reason to refute the new wave approach. It is essential to keep an open mind about such an approach to compartment firefighting. The techniques described above are complementary to the more traditional and accepted methods of fire suppression and are not intended as replacements. What is important is that readers stand back from their previous experience and knowledge of water fog applications to re-evaluate this innovative new wave approach with a view of not passing professional judgment until such techniques have been seen, tried, and tested first hand under strict and qualified supervision.

I do agree with Fredericks though, that although our tactics may differ, 'our goal remains the same – to keep firefighters alive'.

I wish Andy was still here … inspiring others and always promoting sound firefighting guidance.

11.2 LITTLE DROPS OF WATER – LLOYD LAYMAN – ROYER/NELSON – UK FRS

The early work into fog tactics (certainly not the earliest) was undertaken by Lloyd Layman as well as Keith Royer/Floyd W. 'Bill' Nelson (at Iowa State University) and the UK Fire Research Station (FRS) during the 1940s–60s, is widely documented. The late Chief Lloyd Layman of Parkersburg, West Virginia, presented a paper entitled 'Little Drops of Water' at the Fire Department Instructors Conference (FDIC) in Memphis, Tennessee, and in the process stood the fire service on its collective head. In his paper, Layman introduced what he termed the indirect method of attack to suppress interior building fires using the tremendous heat absorbing properties of expanding and condensing steam, produced in great quantities by fog (spray) streams. Most of the theory and methodology of indirect fire attack was based on the Coast Guard experiments (Layman was in charge of the Coast Guard's wartime firefighting school at Fort McHenry), as well as additional testing conducted jointly by the U.S. Navy and other agencies in San Francisco under the project name 'Operation Phobos'.

Royer and Nelson developed a formula for estimating, with a high degree of accuracy, the amount of water required to control an interior fire based on the following: a) the amount of heat liberated by common fuel materials burning in ordinary air within a compartment, b) the extinguishing (heat absorbing) capacity of water, and c) the cubic foot volume of the fire compartment. In the 'critical rate of flow' formula, as it came to be known, Royer and Nelson determined that the amount of water (expressed in gallons per minute) needed to control (not completely extinguish) a fire in the largest open space within a structure can be determined by dividing the cubic foot volume of the space by 100. Royer and Nelson explained the formula and its scientific basis in Engineering Extension Service Pamphlet #18 'Water for Fire Fighting-Rate of- Flow Formula' (1959, Iowa State University). They also introduced the fire service to a fire extinguishment technique they called the 'combination method of attack'.

Royer & Nelson[94] based their approach around early European research into water-fog tactics where in general, an application of 0.5 L/min/m^2 of floor area would be considered an optimum fire stream. It was then concluded on the basis of both theory and practical research that to apply water into a post-flashover fire compartment, so that 80 percent of that water is turned to steam, it was assumed that one gallon (US) (3.78 Litres) of water would absorb the heat presented in 200 cubic feet (6m^3) of space in a model room fire.

This resulted in the IOWA formula for estimating needed flow-rate when attempting to create high volumes of steam to smother the fire –

$$\frac{cubic\ feet\ of\ fire}{100} = GPM\ flow\text{-}rate \qquad \text{Eq.11.1}$$

Or in metric format –

F (L/min) = Room volume (m^3) / 0.75

or: 1.33 L/min/m^3

or: 0.55 L/min/m^2 floor space based on a 2.4m ceiling height

1.3 x m^3 L/min flow-rate   Eq. 11.2

Royer/Nelson (Iowa) methods of fire attack –

- Direct attack method; applying firefighting water directly onto the materials involved in fire.
- Indirect attack method: applying firefighting water into the heated fire gases, avoiding contact with surfaces as far as possible.
- Combination attack method: striking the walls and surfaces as well as the fire gases.

So, an un-ventilated room of 4 x 4 x 2.4 m^2 would require a flow of just 28 L/min for around 20-30 seconds (Iowa guidance) by creating enough steam to smother any fire.

During the 1950s a series of fire tests were undertaken by UK fire research scientists to simulate some of the earlier work by Lloyd Layman in the USA and to further evaluate spray and straight stream (jet) patterns in extinguishing one room fires. In 1954 Thomas & Smart reported in UK FRN 86/1954 a series of post flashover room fire tests in a 58m^3 (24m^2) room loaded with furniture with 5m^2 of venting. An application of 15 litres of water/30m^3 was required for fire control and a further two thirds this amount was needed

---

[94] Royer. K, Nelson. F; Iowa State University Engineering Extension, Bulletin No. 18; 1959

for full suppression. It was noted that in doubling the flow the extinction time was reduced by 15% although overall water consumption was increased by 70%. It was also noted that approximately twice the room volume in steam volume was needed to control the fires. This correlated closely with the 32 litres of water/30m³ used by Lloyd Layman to extinguish room fires ranging from 100 to 950m³. Further tests were reported by Thomas & Smart in FRN 121/1954 in larger domestic rooms that confirmed the above application rates of water needed to control and fully suppress fires using the Lloyd Layman techniques of filling rooms with water vapour (water fogging) until the fires were unable to progress. Even to this day the Layman methods of indirect fire attack and water application rates are referred to in UK Fire Service training manuals. In 1959 Thomas researched a large number of building fires in the UK where fire damage occurred in excess of 200m² areas, and based on the actual number of jets (J) in use, reported that the number of main attack hose-lines needed (J) could be calculated from a formula:

$$J = 0.33 * \sqrt{A_{fire}}$$ Eq.11.3

Where:

J Number of jets needed (firefighting streams) flowing 600 l/min
A_{fire} area of fire involvement (m²)

In order to provide such an estimation, each jet was assumed to be flowing 10 l/s. Current research (chapter 12) would suggest that firefighting jets will begin to apply less per m² of floor area fire involvement as the fire area increases and water demands are outpaced by fire growth (see Figs.12.16 and 12.17).

In a series of test burns using water sprays in 1962 Rasbash noted that it was never really clear whether the most effective form of suppression arose from flame cooling and subsequent smothering by water vapour, or by direct cooling of the fuel base. In summary, he stated that based on the test data, direct cooling of the fire's base was probably the best way to extinguish a compartment fire.

In FRN 492/1963 Fry & Lustig report on water used in firefighting by eleven UK fire brigades over a period of 12 months (1960-61) where it is stated that over 80 percent of fires attended were extinguished using 455 litres or less. It was further noted that 1,800 litres water tanks on fire engines dealt with 95 percent of all fires in dwellings and 81 percent of fires in other building types (overall 86 percent of all fires attended).

A series of test burns in laboratory settings (full-scale rooms) that were later taken into real buildings produced what appeared the most accurate data of the period concerning *critical, minimal* and *optimum* firefighting flow-rates in 1970. These test burns by Salzberg, Vodvarka and Maatman consisted of one and two room fires joined by a one metre wide corridor. The rooms each measured 3.66 x 3.66 x 2.44m high (floor area of each room being 13.4m²). The fire load consisted of furniture, books and clothing to 22kg/m². Each room had two window openings and a standard door. All fires were post-flashover and subsequent tests also took place in small commercial buildings with realistic fire loads.

Using 38mm hose-lines and 25mm low-pressure hose-reels effective 'fire suppression' was considered achieved at the earliest point firefighters could enter and remove smouldering furniture to the exterior for damping down to complete final extinguishment. A firefighting water flow-rate of 1.9 L/min/m² was found to be most efficient in achieving control with minimal water damage. However, the control times were longer and firefighters were exposed to physiologically untenable conditions at this flow-rate. Therefore, the *optimum flow-rate* used by firefighters during the live fire tests was determined at 5 L/min/m².

The conclusions of the research determined that **2 L/min/m²** was the critical flow-rate in order to achieve control of the room fires (at a cost of severe physical punishment to the firefighters); **3.7 L/min/m²** was considered a minimal effective flow-rate; but **5 L/min/m²** was considered to be the optimum flow-rate. In real fires in residential settings, flow-rates of **4 L/min/m²** (one room) and **6 L/min/min/m²** (two rooms totalling 32m²) were considered optimum. In the case of commercial fire tests (50-100m² buildings) an optimum flow-rate of **6.5 L/min/m²** of the fire involved area were determined.

11.3 FIRE BRIGADE INTERVENTION MODEL (FBIM) AUSTRALIA-NEW ZEALAND

The Fire Brigade Intervention Model (FBIM) is commonly used and accepted as a building design tool in Australia and New Zealand, as a means of evaluating the effectiveness of fire service intervention, including receipt of alarms, weight and speed of response and the effectiveness of preparation and command functions. The FBIM model is utilised through a computer programme and the actual firefighting capability is determined through a firefighting water flow calculation that references the work of Sardqvist.

The general approach is that a 300 L/min attack hose-line is the minimum that should be deployed internally and that this line has a maximum extinguishing capacity of 8 MW. In terms of external firefighting, FBIM estimates that a 600 L/min hose-line has an extinguishing capacity of 5.25 MW.

It was noted in research by Sardqvist that the theoretical extinguishing properties of water should not be used in isolation for modelling purposes, but that efficiency factors should also be applied. The Sardqvist research further suggested that, of the water used, only between 5% and 30% was likely to be efficient when applied to the fire. The reasons for this vary but include, nozzle size, water droplet size, firefighting skill, fire size and room geometry. It is for this reason that the efficiency factors of 15% for firefighting within an enclosure and 5% for firefighting external to the enclosure are recommended in the FBIM.

The calculation process used in the FBIM is based solely on the suppression of flaming combustion, mostly in the reaction zone at 550°C (3.5 x 3 refers to 30-35% of heat being removed from diffusion flames).

For internal fire (15% efficiency); or exterior fire stream (5% efficiency) applications -

- 3.5MW/L/s x 3 = 10.5MW x 5L/s = 52.5MW x **15% (efficiency)** = 7.87MW (8 MW)
- 3.5MW/L/s x 3 = 10.5MW x 10L/s = 105MW x **5% (efficiency)** = 5.25MW (5.25 MW)

In order to relate this to time, it is assumed that there are 3 possible outcomes of water application:

a. The heat release rate will decay at an appropriate rate (given the heat release rate at time of water application) and the fire be extinguished over time if, at the time of water application, the cooling capacity is equal to or greater than 110% of the heat release rate.

b. The heat release rate will be controlled and remain constant if, at the time of water application, the cooling capacity is within a ±10% range of the heat release rate.

c. The heat release rate will remain unaffected and the fire continues to grow if, at the time of water application, the cooling capacity is equal to or less than 90% of the heat release rate.

Note: *±10% is a range for uncertainty*

*Table 11.1: Effect of **internal** fire attack according to FBIM where examples are provided for various heat release rates (HRR).*

Applied water L/s	HRR (MW)	110% HRR	90% HRR	Maximum Extinguishing Capacity MW	Result
5	5	5.5	4.5	8	Decay
10	10	11	9	16	Decay
20	30	33	27	32	Constant
20	40	44	36	32	No effect
30	40	44	36	48	Decay
30	50	55	45	48	Constant

Note: Where the extinguishing capacity is not met effectively then the result will likely be B or C above.

Table 11.2: Effect of external fire attack (not exposure protection) according to FBIM where examples are provided for various heat release rates (HRR).

Applied water L/s	HRR (MW)	110% HRR	90% HRR	Maximum Extinguishing Capacity MW	Result
10	5	5.5	4.5	5.25	Constant
20	10	11	9	10.5	Constant
30	10	11	9	15.75	Decay
40	15	16.5	13.5	21	Decay
40	30	33	27	21	No effect

Note: Where the extinguishing capacity is not met effectively then the result will likely be B or C above.

Although an appropriate efficiency factor is applied to the interior fire attack (15% in flaming combustion) the FBIM approach fails to account for any additional cooling that may take place at the fuel base. It is suggested this may result in an over estimate in the actual amount of water needed for suppression. Furthermore, the FBIM application of 0.625 L/s per MW actually equates to a 31% overall flow efficiency factor if compartment combustion efficiency is accepted as 50%.

Extinguishing efficiency taken as 0.5 (k_f) / 2.6 MJ/Kg x 0.625 L/s/MW

FBIM = 31% efficiency

11.4 LONDON, OCTOBER 1984

The summer of 1984 had been a busy one for us in London's west-end. There were still plenty of fires to fight as my fire station at that time, located just off Baker Street W1, frequently attended incidents around the Oxford Street area, into Euston and Paddington to the north and Soho and Knightsbridge to the south. The structural stock was referred to as 'A' risk buildings, all densely packed within a highly-populated area full of high-class retail, shops, restaurants, hotels theatres and night clubs, along with a broad range of residences and apartment blocks. It was October and I was spending much of my time piecing together the tactical ventilation strategy I was proposing following my detachments to the FDNY. I would spend many hours in the London Fire Brigade library, based at our HQ building on the Albert Embankment, filling in time between the busy night shifts. It was a nice quiet place and it wasn't unusual for me to gain a quick 'power nap' for half an hour between reading unending articles and fire journals. In fact, everyone who knew me would know how and where to find me between shifts, searching the shelves for the next 'golden nugget' in firefighting tactics. I would read every word from the books of ex-London chief officers and some other legends like Bill Clark and Vinny Dunn (FDNY). The shelves were stacked out with knowledge and I wanted to go through it all!

I was a very inquisitive firefighter at work, wanting to learn something about everything if I believed was relevant. I spent much of my time searching out roof-top access to buildings and learned how our west-end structures intertwined, providing access to almost all the building's roof-tops from adjacent areas. It soon dawned on me that I could reach rooftops that were beyond the reach of ladders, simply by entering the building next door and searching out their own roof access points, be it by fire escapes or roof hatches. In the coming years, this information would prove vital as I would lead teams of firefighters up into key positions from where we were able to mount early and coordinated venting operations of the fire building, based on the incident commanders needs. The strategy (learned in New York City) was highly successful, particularly in some of the older hotel buildings, offices and houses in multiple occupation (HMOs).

It was October 1984 in the LFB library when I first saw the article[95] by Mats Rosander and Krister Giselsson that described new methods of extinguishing room fires being developed by these two Swedish fire engineers. Their academic paper detailed the scientific principles associated with *indirect extinguishing* (evaporating water against hot surfaces); *offensive firefighting* (flame cooling); *inerting the fire gases* (preventing a flammable fire gas layer from igniting) and *cooling off fire gases in the smoke existing long the approach route* (fire gas cooling). They discussed the Royer/Nelson/Iowa indirect methods but made it clear that their objectives were different in gas and flame cooling. They proposed using a series of short 2-4 second bursts of a fog pattern from an interior position around one metre into the room, advising that the applied water flow-rate should be variable, depending on the intensity of the fire (the *Fogfighter* nozzle they proposed flowed between 100-300 L/min although a larger version flowed up to 475 L/min). They also recommended that no planned venting of the fire room should take place whilst firefighters occupied the room, until after the fire had been extinguished. A series of firefighter training booklets were published in Swedish in Stockholm, describing these techniques from 1978 and one of these first became available in English in 1990[96].

---

95 Giselsson. K, Rosander. M; Making the Best Possible Use of Water for Extinguishing Purposes; Fire Magazine (UK) October 1984
96 Giselsson. K, Rosander. M; The Fundamentals of Fire; Giro-rand ab; 1990

Figure 11.1: 3D fog bursts into the overhead gas layers to cool and contract hot gases
Image courtesy of Bart Noyens

The concepts appealed to my inner curiosity and they actually made sense from a scientific perspective. But putting water into smoke? This was something we had all been trained not to do for fear of wasting water and causing unnecessary damage within the lesser affected areas of the building. However, we discussed the viability of this approach at station level that very night shift (October 1984) and it was agreed at all levels on shift, that this was perhaps something worth trying at building fires.

It was six weeks later on a night shift when we were presented with the 'ideal' scenario where these Swedish methods might actually work.

11.5 THE FIRST USE OF SWEDISH WATER-FOG TACTICS IN THE UK

It was a six-storey hotel fire in Baker Street, less than 500 yards from the fire station. As soon as the fire station's bay doors opened we could smell the fire and drove out into a haze of smoke filling the street. The fire was at roof level and lit the night sky with a familiar orange glow. However, flames were also issuing from a fourth storey window at the front. It was clear this was a stair-shaft fire and normally we would have pulled a main hose-line off for this one. However, I wanted to get a quick hit on the fire and a rapid ascent up the stair so we laid the 110 L/min Ultra High Pressure (UHP) 19mm hose-reel into the stairs to be followed by the larger line. We had two engines and a third on the way with additional engines being called in to assist, along with an aerial ladder.

As we advanced up the stair through the first and second levels there was nothing other than a strong airflow up the stair-shaft, feeding air into this furnace. There was no door control at this fire although with hindsight this served to our advantage. As we arrived on the third storey landing the smoke layer was hanging lazily in the stair and at the base of it was a sea of flame. It was a classic stair-shaft fire that was vented at the top by the small roof hatch, allowing a steady flow of air in at the base to maintain the burning smoke layer that was slowly lowering itself. This was it, I gave three brief one second bursts of a 30-40-degree fog-cone into the smoke and the flames subsided instantly. We moved further up the stair and as we advanced I maintained the 'burst and pause' cycling of the fog pattern, almost chasing the fire up the stair as fast as we could advance. As we arrived on the fourth floor I trimmed the stream down to a solid pattern and directed it into the fire room. It was at this point the main hose-line arrived up the stair behind us and instead of passing us, they took over in the fire compartment which was now under control. This allowed us to advance further up the stair and extinguish the fire in the stair-shaft, again using the Swedish burst and pause fog pattern.

We had extinguished three floors of stair-shaft fire using a very small amount of water and when the salvage team arrived on-scene they were astonished at the amount of fire damage compared to the lack of water damage in the building. The water-fog tactics had been very successful indeed and a new ear was being born. However, over the next few years our experiences were not always this good using such low flows. It was the ideal scenario but there were fires where the techniques might not be appropriate. So, we went on to use the methods at 549 working fires in London from 1984–94 to further test the Swedish tactical approach. At this point in time, there were no steel shipping containers in which we could practice and develop nozzle techniques. Our learning ground was the actual fire-ground!

11.6 A DECADE OF FIRE-GROUND EXPERIENCE AND THE '549 FIRES'

Over the course of that ten-year period that the Swedish firefighting water applications were trialled and used in central London, for the rapid knockdown of burning fire gases (flame cooling) and as a preventive measure to reduce compartment fire temperatures as a means of avoiding flashovers (gas cooling). A trial that began with just one fire station spread to involve an additional five surrounding stations, where we established some operational limitations through experience in using the tactics at building fires:

- The flame-cooling tactics were ideally suited to small to mid-sized one-room fires, or stair-shaft fires
- The methods of attack were only to be used in conjunction with 120 L/min High Pressure (HP) 19mm hose-reels – increasing the flow-rate may have resulted in wider successes
- Control (limit) the ventilation and allow the 120 L/min High-Pressure (HP) flow of water to achieve the maximum impact
- Ventilate the compartment immediately the fire was extinguished or under control
- Gas cooling tactics were most effective and could be used wherever a motionless hot gas layer existed (for example; pre-flashover room fires and at the head of stairs with upper level room fires)
- Wherever there was a fast-moving fire gas layer, or overhead flame-front, the low flow HP fog patterns appeared ineffective and direct attack stream applications were utilised
- Shielded or hidden fires were generally approached using a combination of fog, and solid stream bounced of walls and ceilings
- That direct fire attack (solid stream) was still the preferred method in penetrating, suppressing and controlling the base of the involved fire load

It was important to recognise that placing water into smoke was unnatural to us back then and was something we had been taught NOT to do, for fear of creating increasing amounts of water damage. However, as each experience provided us with greater confidence and knowledge, the tactics became extremely popular as surrounding fire stations began to take an interest and experiment themselves with the new firefighting techniques. It wasn't something where any detailed training was provided, it was simply learning the application technique in open-air with a quick five-minute demonstration. After that, it was down to each individual nozzle operator to decide the best stream application at any particular time. It was almost certain that some flashovers were prevented by using short bursts of water-fog into the gas layers near the ceiling of many room fires. It was also clear that water damage in buildings was dramatically reducing, rather than increasing, when using these methods.

The data associated with the use of Swedish water-fog tactics during the 549 fire trial in London's west-end was very subjective (personal opinions based on visual and sensory observations) as building fires don't come fitted with scientific recording equipment, thermo-couples and smoke collection hoods. Each nozzle operator was questioned for their personal experience and views following each fire and this information was recorded by the author. It was noted that firefighters most commonly utilised a direct attack at the base of the fire as the primary tactic on 39 percent of occasions.

In 1989 a team of three London firefighters (including the author) visited Sweden to train in the newly developed fire container training simulators but it was ten years later, after the 549 building fire trial, before the tactics were rolled out across the entire London Fire Brigade at the UK Fire Service College in Moreton-on-Marsh.

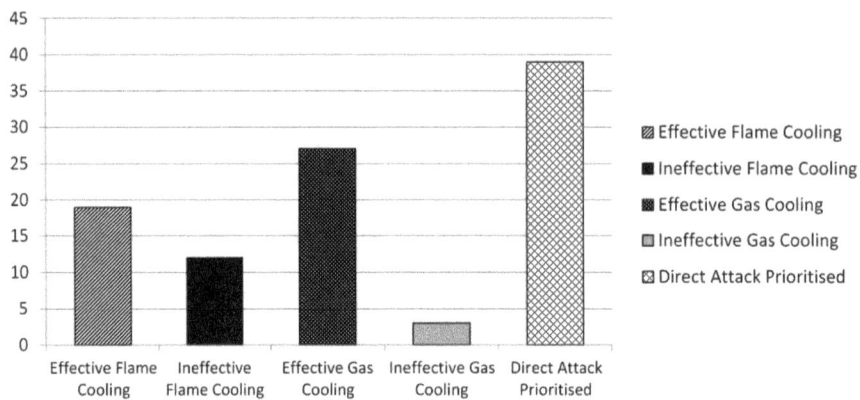

Figure 11.2: The subjective data percentages recorded by the author that demonstrated the impact of Swedish firefighting methods during a trial period by firefighters in London's west-end from 1984–94

11.7 '100 FIRES' IN LONDON 1989

In 1989, as a London Fire Brigade fire investigator, Grimwood undertook a research project with the primary objective of producing a simple fire-ground flow-rate formula that could be used by on-scene fire officers in order to estimate an effective firefighting water flow-rate, in situations of escalating building fires 50–500m². This resulted in 'rule of thumb' fire-ground calculation for fires >120m² in area:

$F_{FIRE} = A_{fire} * 5$ **Eq.11.4**

Where

F_{Fire} = L/min (Quick fire-ground approximation)

A_{fire} = floor area fire involvement in m²

5 = flow coefficient

The research involved attendance at 100 working building fires in London where firefighters wearing breathing apparatus used variable quantities of water to deal with compartment and building fires in a broad range of purpose groups and occupancies. These included fires in residential dwellings, offices, warehouses and industrial units. The fires spread to involve and damage floor space from 16 to 1,400 m². It was noted that 34% of these fires were dealt with using 19mm high-pressure hose-reels with the remaining 66% of fires requiring higher flows from mainline firefighting jets.

The water flow-rates used to control and extinguish the fires were calculated at the scene and a subjective approach was used to estimate the extent of fire damage in various fire compartments. It was noted that 62% of fires demonstrated extensive fire damage to building contents with most of these showing fire damage beyond the fire resisting boundaries and/or within the structural frame. It was observed that these fires were finally suppressed during the decay stages of burning.

A further 21% of fires were suppressed effectively but it appeared with great difficulty, possibly due to inadequate water deployments at an early stage of operations. There were 17% of fires where water was deployed effectively, during the fire's growth stages and these fires were termed 'good stops', as much of the building's contents remained only partially damaged by fire and the fire had not penetrated the structural fire resisting boundaries.

This research data suggests a critical flow-rate at real fires exists at or below 2 l/min/m² and that by increasing the applied water flow-rate to 3.75 l/min/m² or more, effective control and extinguishment may be achieved during the fire's growth or early steady state phases.

Table 11.3: Research data of 100 large building fires in London in 1989-90 demonstrated that 62% of fires were extinguished during their decay stages where a critical flow-rate of >2 l/min/m² had not been met. (reported in 'Fog Attack' (Grimwood) 1992)

Adequate Suppression	17%	<3.75 L/min/m²	200 kW/m²
Difficult Suppression	21%	<2.75 L/min/m²	150 kW/m²
Critical	**Critical**	**2.0 L/min/m²**	**Critical**
Decay stage suppression	21%	<1.87 L/min/m²	100 kW/m²
Decay stage suppression	25%	1.25 L/min/m²	70 kW/m²
Decay stage suppression	16%	>0.625 L/min/m²	35 kW/m²

Note: Subjective assumptions were made on the Heat Release Rate per Unit Area (HRRPUA) data based on an applied flow-rate of 3.6 l/m² at 200kW/m² being a reasonable estimate of developing 100m² fires existing in a ventilation limited state (Table 11.5).

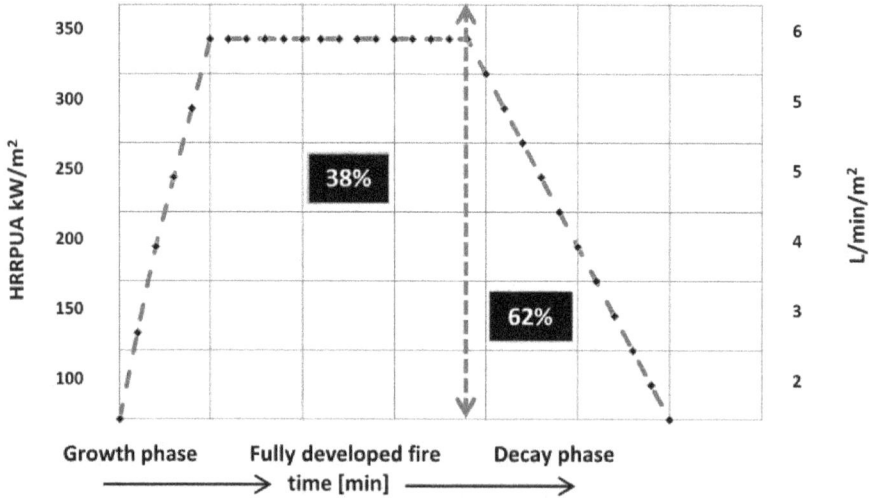

Figure 11.3: 100 Fire study London *(1989): A design fire growth curve demonstrating Heat Release Rate per Unit Area (HRRPUA (kW/m²)) and flow-rate (l/min/m²) during the growth and decay phases.*

Note: *During the 100-fire study in London, it was apparent that 62% of the fires were finally extinguished during the cooling (decay) phase. These building fires were categorised as those requiring resources above that of the pre-determined attendance.*

It was noted that firefighting water flow-rates between 0.6 and 3.75 L/min/m² were used at 63% extinguishing efficiency, in what were commonly fire involved buildings where the fire service attempted to 'contain' fire spread through a controlled *anti-ventilation* strategy. Where venting of the fire did occur, this was usually a non-fire service event where the fire self-vented. On a few occasions firefighters did resort to tactical venting actions to assist their interior approaches.

With this in mind a subjective approach was considered whereby optimal flow-rates could be linked to Heat Release Rate per Unit Area (HRRPUA) as follows:

It is important to remember that cooling efficiency alone does not offer a means of evaluating how effective a firefighting operation might be, as the effective cooling and final suppression may take place on the decay side of the natural fire development curve as was often the case (62% of 100 fires) in the London research. It is also worth noting that tactical objectives in London, and in general throughout the UK, are to limit fire spread by controlling and reducing ventilation and air feeding the fire wherever possible. This may lead to lower HRRPUA in some cases than general fire design guidance may suggest.

Based on a 63% extinguishing efficiency at 200kW/m² in an average 100m² office compartment fire, the 100-building fire study in London 1989-90 suggests a minimum deployed flow-rate of >4-6 L/min/m² to be effective, for medium office type fire loads (may be required on either side of fire development curve, growth or decay).

Table 11.4: BS PD 7974-1:2003 (table 4) offers some suggested HRRPUA for fuel controlled fires.

Occupancy	HRRPUA kW/m²
Shops	550
Offices	290
Hotel rooms	250
Industrial	90-620

Table 11.5: Assumptions made on the flow demands according to the 100 Fire research in London 1989

HRRPUA	Rounded Flow-rate L/min/m²	Flow-rate estimates based on an estimated 63% extinguishing efficiency of deployed flow-rates (l/min) against an average 100m² enclosed fire (100 large building fires in London 1989-90)	Actual Flow-rate L/min/m²
100 kW/m²	2	0.30 x 10MW x 60 = 180 L/min (63% flow efficiency) as deployed	1.8
125 kW/m²	2	0.30 x 12.5MW x 60 = 225 L/min (63% flow efficiency) as deployed	2.25
150 kW/m²	3	0.30 x 15MW x 60 = 270 L/min (63% flow efficiency) as deployed	2.7
175 kW/m²	3	0.30 x 17.5MW x 60 = 315 L/min (63% flow efficiency) as deployed	3.15
200 kW/m²	4	0.30 x 20MW x 60 = 360 L/min (63% flow efficiency) as deployed	3.6
225 kW/m²	4	0.30 x 22.5MW x 60 = 405 L/min (63% flow efficiency) as deployed	4
250 kW/m²	5	0.30 x 25MW x 60 = 450 L/min (63% flow efficiency) as deployed	4.5
275 kW/m²	5	0.30 x 27.5MW x 60 = 495 L/min (63% flow efficiency) as deployed	4.95
300 kW/m²	5	0.30 x 30MW x 60 = 540 L/min (63% flow efficiency) as deployed	5.4
325 kW/m²	6	0.30 x 32.5MW x 60 = 585 L/min (63% flow efficiency) as deployed	5.85
350 kW/m²	6	0.30 x 35MW x 60 = 630 L/min (63% flow efficiency) as deployed	6.3

11.8 THE CATALYST FOR TACTICAL CHANGES IN THE UK – FEBRUARY 1996

In February of 1996, two fires occurred in the UK within two weeks of each other where three firefighters tragically lost their lives to rapid fire development. The two events were likened at the time to 'backdrafts' and it was suggested at governmental level (Home Office), following fire investigations, that UK firefighters may be failing to understand basic fire behaviour indicators and could be exposing themselves to unnecessary risk levels because of this tactical omission. All UK fire brigades were instructed to address their training of firefighters in live fire behaviour and it was from this point on that the Swedish firefighting methods, the author had been promoting vigorously via the national fire journals through the 1980s–90s, started to be seen at various training locations. The Essex Fire and Rescue Service were the first UK fire service to initiate the training brigade wide and then they assisted other brigades in setting up their own training programmes.

The overriding objectives of the nationwide training were to increase the UK firefighter's awareness and understanding of fire behaviour and tactical approaches made into fire compartments, with a primary intention to reduce the life losses associated with rapid fire development, flashover, backdraft and fire gas ignitions (smoke explosions, rollovers, auto-ignitions and flash fires). However, there have been a number of similar incidents in the UK and Ireland since those two 1997 fires where more than a dozen firefighters have lost their lives to rapid fire behaviour events across the two decades 1997-2017

Year	Occupancy	Firefighter Life Losses	Firefighting tactics and causal factors***
1996	House	2	• Entry from front in a terraced property • Access to rear difficult and time consuming • VEIS not considered (rear bedroom) • Prioritised rescue of children over firefighting • Did not isolate ground floor fire before ascending stairs to upper floor level • Did not isolate street entry door • Limited staffing prevented hose-line protection of egress route • Fire behaviour was indicating rapid fire development • Uncontrolled flow-path • Eventual rapid fire development trapping firefighters
1996	Large retail	1	• High ceiling with fast developing low-level fire towards rear of main entry point • High level smoke layer initially • Deployment into commercial building fire with low flow-rate (100 L/min hose-reel) • Fire self-venting at plastic roof vents • Smoke layer descending • Fire behaviour was indicating rapid fire development • Uncontrolled flow-path • Eventual rapid fire development trapping firefighters
2004	Basement retail	2	• Basement storage area with racking to ceiling • Deployment into commercial building basement fire with low flow-rate (100 L/min hose-reel) • Low-flow hose-line in place for long period of time without increasing flow to gain control • Tactical venting actions above the fire in adjacent stairwell created flow-path down to basement • Fire behaviour was indicating rapid fire development • Uncontrolled flow-path • Eventual rapid fire development trapping firefighters
2005	High rise flats	2	• Prioritised rescue over firefighting • Use of rising main initially restricted • Did not control door opening to apartment • Delay in support team arriving at fire floor due miscommunication between commanders • Falling of ceiling cables trapped firefighter • Fire behaviour was indicating rapid fire development • Uncontrolled wind driven flow-path • Eventual rapid fire development trapping firefighters

2010	High rise Flats	2	<ul><li>Lack of information gained from occupants in street</li><li>Unable to locate small fire on lower level of duplex maisonette</li><li>Right hand wall search took firefighters with hose-line straight up to second level without searching lower level first</li><li>Fire self-vented as window failed (candle fire behind curtains)</li><li>Did not control door to apartment</li><li>Second team delayed with second hose-line that laid short when deployed from two floors below fire</li><li>Falling of ceiling cables trapped firefighter</li><li>Tactical ventilation placed firefighters the wrong side of the flow-path</li><li>Uncontrolled wind driven flow-path</li><li>Eventual rapid fire development trapping firefighters</li></ul>
2007	Warehouse	4	<ul><li>Large warehouse with various commodities stored on pallets</li><li>Hose-line distance to interior pallet fire > 100 metres</li><li>Deployment into industrial building fire with low flow-rate (100 L/min hose-reel) extended in length to reach fire</li><li>Low ceiling</li><li>Unable to reach seat of fire due length of penetration in dense smoke</li><li>Fairly warm conditions increasing with several crews deployed over time</li><li>Debriefing of crews failed to determine level of firefighter exposures to heat</li><li>Unable to access adequate water supplies</li><li>Unable to determine alternative entry points</li><li>Unable to locate fire before firefighter BA emergency</li><li>Heat exhaustion and increasing smoke density caused 4 firefighters to become incapable of self-rescue</li><li>Firefighter suffering from apparent heat exhaustion located by firefighters but directed out and not taken out – became lost and collapsed inside</li><li>Fire behaviour was indicating rapid fire development</li><li>Uncontrolled flow-path</li><li>Eventual rapid fire development trapping firefighters</li></ul>
2007	Derelict building	2	<ul><li>Derelict single-storey factory with surrounding exposures</li><li>Limited staffing</li><li>Exterior venting actions – defensive mode</li><li>Team of two firefighters advanced their hose-line into the structure against ICs instructions</li><li>Fire behaviour was indicating rapid fire development</li><li>Uncontrolled flow-path</li><li>Eventual rapid fire development trapping firefighters</li></ul>

| 2013 | Large retail | 1 | • Unable to reach seat of fire due to heat exposures to firefighters
• Extended duration firefighting operation
• Repeated transitions from offensive to defensive operations
• Effective tactical plan restricting interior firefighting a maximum of 20 minutes per entry
• Transition period from defensive to offensive firefighting occurred at time when changes in staffing were being made across the fire-ground
• Defensive from 1918 hours to 1935 hours
• Offensive from 1935 hours to 1952 hours
• Defensive from 1952 hours to 2004 hours
• Offensive and Defensive from 2004 hours (high-flow exterior stream from front with firefighters entering from rear) at staffing changeover
• Poor communication of tactical plan to key command staff and BA control officer at staff changeover on fire-ground
• First crew of (two) firefighters deployed into severe internal fire conditions (2004 hours) following a 12-minute defensive action (exterior streams) and Tactical venting actions
• Firefighter suffering from apparent heat exhaustion located by firefighters but directed out and not taken out – became lost and collapsed inside
• Fire behaviour was indicating rapid fire development
• Uncontrolled flow-path
• Eventual rapid fire development trapping firefighters |
|---|---|---|---|

Table 11.6: UK and Ireland Firefighter life losses where fire behaviour was a major causal factor (shaded area were the two fires that preceded the national fire behaviour training programmes being triggered. ***Causal factors are not necessarily seen as 'errors' but rather as tactical options (shaded area just prior to the introduction of national fire behaviour training programmes).

11.9 IS THE KNOWLEDGE GAP MET BY FIRE BEHAVIOUR TRAINING (CFBT) BEING EFFECTIVELY TRANSFERRED ONTO THE FIRE-GROUND IN ITS CURRENT FORMAT?

Taking a close look at these fires and there are, in fact many more similar incidents that are classified as 'near misses', we must ask ourselves if the way we deliver fire behaviour training has had any real impact in reducing the numbers of firefighters killed by such tragic events? We have, in fact, trained an entire generation of firefighters this way and would therefore expect to see some return. A close analysis of both the coroner's and fire service reports for the above incidents suggests that there are more causal factors than simply a lack of awareness, understanding or appreciation of how fire behaviour may impact on firefighting deployments and interior firefighting operations. At the very core of these incidents one can see some commonality in the following strategic and tactical areas of concern:

- The quality of communication between fire commanders and firefighters
- A knowledge gap in water flow-rate suppressive capacity versus involved fire load
- A knowledge gap in what is adequate firefighting water in different occupancies
- A knowledge gap in relation to tactical ventilation and flow-path formation
- A knowledge gap in relation to flow-path control measures

- A knowledge gap in relation to strategic hose-line placements to protect interior crews
- A knowledge gap in the applications of Vent-Enter-Search (VES) or (VEIS) tactics
- A knowledge gap in relation to what is adequate firefighting water when deploying into working fires involving a variable range of occupancy types
- A knowledge gap in dealing with under-ventilated fires and flow-path management
- A knowledge gap in how fire behaves in high-ceiling buildings as opposed to low ceilings

> *'The Swedish fire service taught us a great deal about fire behaviour. However, when it comes to fast developing building fires – and I've used both options – if I had to choose between low-flow spray patterns with finely divided water droplets or high-flow solid streams – I'd go with the solid stream every time.'*
>
> Paul Grimwood

Now let me qualify that statement! It has been the case that an imbalance has occurred, certainly in the UK, over two decades of fire behaviour training in placing an unqualified bias on using low-flow nozzles to extinguish fires. This is the result of two generations of [many] fire behaviour instructors over-simplifying and teaching the *fire attack process* by using the steel shipping container as a means of replicating the actual fire-ground. This has created an environment whereby many firefighters believe they can handle just about any compartment fire with finely divided spray patterns, flowing as low as 45 L/min. Some UK fire services have actually worked with manufacturers on several occasions to develop such hand-line nozzles and are taking these to building fires and deploying firefighters with such low flow-rates.

The steel shipping container is not a real fire it is a simulator of how compartment fires enter the growth phase of fire development, preceding flashover. The actual fire growth is limited for safety reasons to a maximum of 1.5 MW in most cases. The involved fire load is mostly burning in the gas-phase and the 45 L/min nozzles will deal effectively with such a fire. However, when we get to the fire-ground where building fires are affected by new energy efficient construction materials, higher fire loads and increasing ventilation profiles than we had in the 1980s, the potential for more rapid fire growth than the training simulators provide us with is becoming more prevalent.

Add to this the difficulties in achieving adequate water flow and pressure in UK high-rise buildings above 30 metres (new buildings above 50 metres are now generally adequate with permanently wet rising fire mains), the solid stream from a smooth-bore nozzle or *low-pressure* fog nozzle becomes the favoured choice. If the pressure is too low the fog pattern is not finely divided and the solid stream can be bounced off of surfaces, walls and ceiling to provide some gas cooling effects.

Let me say, gas cooling is undoubtedly something we should all teach our firefighters to do for ignoring the dangers of heat and transporting fire load in the smoke and fire gas layers above our heads is foolhardy, and may lead to firefighters becoming trapped by rapid fire behaviour. However, recognising and defining the limited opportunities to utilise small amounts of water for 'flame-cooling' is equally important. Yes! If I'm in a fast moving fire the greatest thing you can do for me is provide adequate flow-rate at the nozzle!

11.10 ANALYSIS OF 3D WATER-FOG'S GAS AND FLAME COOLING CAPABILITIES

It had become apparent that the Swedish method of attack, when translated into English, resulted in the term 'offensive firefighting'[97]. This created some tactical conflicts with existing UK terminology. In fact, there were several conflicts of translation and some incorrect scientific definitions when the Swedish fire service began to write about fire behaviour and water-fog tactics in English. Terms such as 'delayed backdraught' and 'hot-rich flashover' had no meaning in English and the term 'brand gas explosion' in English translated to the historic use of smoke explosion. There was a meeting during the 1990s between UK and Swedish fire scientists to provide some clarification on the incorrect scientific terms and definitions being translated to English. This resulted in mutual agreements and Bengtsson's excellent book[98] in 2001 confirmed the agreed terminology to be used. However, the references to 'offensive firefighting' were not clarified at this meeting and the author worked with the UK fire scientists at the Fire Research Station, introducing and developing further terms for clarification. In 1999 the author introduced the term used to replace 'offensive firefighting' on his website as *3D Firefighting*[99] and also in an article[100] responding to FDNYs Andy Frederick's previous article[101] in Fire Engineering magazine, USA (2000). The National Research Council (NRC) of Canada went on to produce a technical paper – a review on 3D water-fog tactics, in 2002[102] with the author's involvement, and in 2005 the firefighting manual – 3D Firefighting[103] broadened the three-dimensional concepts further with guidance on taking control of the flow-path (air-track) using water-fog and tactical ventilation combined. The wider scope associated with fire gas ignitions (FGIs) was also introduced.

The effectiveness of 3D firefighting water-fog was assessed during a live fire test in a real structure by researchers Alarifi, Dave, Phylaktou, Aljumaiah and Andrews in 2014[104]. Their research (Leeds University) is discussed in some depth here due to its relevance and importance to the firefighter. Jim Dave, one of the researchers (serving with the States of Jersey FRS), is a long term colleague and we have discussed this research personally. Whilst demonstrating the benefits of gas-cooling on the approach route, this research clearly confirmed some limitations in flame cooling, if used in a post-flashover fire compartment. It was shown that not only is adequate water flow-rate (L/min) needed but the method of applying water into the fire compartment is also relevant. The research informed that; *'A typical room enclosure was used with ventilation through a corridor to the front access door. The fire load was wooden pallets. Flashover was reached and the fire became fully developed before the involvement of the firefighting team. The progression of the firefighters through the corridor and the main room suppression attack – in particular the*

---

97 Giselsson & Rosander
98 Bengtsson. LG; Enclosure Fires, Raddningsverket, Swedish Rescue Services Agency, 2001
99 www.firetactics.com (now offline)
100 Grimwood, P., 'New Wave 3-D Water Fog Tactics: A Response to Direct Attack Advocates,' Fire Engineering, October 2000
101 Fredericks, A. A., 'Little Drops of Water: 50 Years Later, Part 2,' Fire Engineering March 2000
102 Z.Liu, A.Kashef, GD Loougheed and N.Benichou . National Research Council (Canada) . Research Report RR124 . 3D Water-fog Review
103 Grimwood. P; Hartin. E; McDonough. J; Raffel. S; 3D Firefighting; IFSTA Oaklahoma University Publications, 2005
104 A. A. Alarifi, J. Dave, H. N. Phylaktou, O. A. Aljumaiah and G. E. Andrews (2014) Effects of firefighting on a fully developed compartment fire: temperatures & emissions, Fire Safety Journal (2014;68: 71-80), http://dx.doi.org/10.1016/j.firesaf.2014.05.014

effect of short, medium and long water pulses on either the hot gas layer or the fire seat – was charted against the compartment temperatures, heat release rates, oxygen levels and toxic species concentrations. The firefighting team was exposed to extreme conditions, heat fluxes in excess of 35 kW/m² and temperatures of the order of 250°C even at crouching level, from the 3 MW Q_{max} fire. The fire equivalence ratio showed rich burning with high toxic emissions in particular of CO and unburnt hydrocarbons very early in the fire and a stabilisation of the equivalence ratio at about 1.8. The firefighting operations made the combustion temporarily richer and the emissions even higher'.

The most severe fire conditions normally encountered in residential and commercial fires, not involving explosions, are conditions where a room or rooms are in flashover. The heat release rate required for flashover of the standard international room (ISO 9705) is about 1700 kW and the flow of hot gases from the doorway or window may exceed 400°C (752°F). NIST experiments show that in the flashover environment, gas temperatures in the room may reach 1000°C (1832°F) and heat flux at the floor can measure 170 kW/m².

Significant temperatures and fluxes can occur even for rooms not at flashover and Quintiere investigated heat fluxes several metres from the doorway of a room containing a fire not at flashover where fluxes as high as 4.5 kW/m² were measured. In tests conducted on UK firefighters by Foster and Roberts, it was found that a flux level of 10 kW/m² could only be tolerated by fully equipped firefighters for one minute. They also found that there was damage to their PPE clothing and masks at this flux level. Experiments by Gross and Fang showed that severe conditions may exist even for low intensity fires, where flame temperatures are about 700 °C (1292 °F). At the edge of a burning wastebasket they measured heat fluxes of 10 kW/m² to 40 kW/m² and air temperatures of 100 °C (212 °F) to 400 °C (752 °F) at the ceiling above the wastebasket fire.

The thermal performance of fire fighters' protective clothing has been a point of interest and discussion for several decades. However, limited scientific information is available on the technical issues. Much of these discussions are based on fire service field experience and many of these studies are therefore difficult to reproduce. There exist a range of fire environments normally faced where firefighters can be burned by radiant heat energy that is produced by a fire, or by a combination of radiant energy and localized flame contact exposure as replicated by the Thermal Protective Performance (TPP) test[105] for US firefighter clothing. Some injuries also occur as a result of compressing the protective garment against the skin, either by touching a hot object or by placing tension on the garment fabric until it becomes compressed against the skin. In addition to these mechanisms, moisture in protective clothing can significantly change the garment's protective performance. Firefighter protective clothing that is wet may exhibit significantly higher heat transfer rates than garments that are dry. Burn injuries that result from the heating and evaporation of moisture trapped within one's protective clothing is also significant. These injuries are generally referred to as scald or steam burns. Moisture may also help to store heat energy in protective clothing[106]. Protective clothing standards are essentially the highest level of safety measures but perhaps what is equally as important is the methodology used by firefighters to either work at close quarters with a fire to achieve cooling effects in the overhead fire gases, or to remain at safe distances from the fire using the effective reach of adequately flowed firefighting streams and the protective cover of solid barriers such as at room doorways. The former tactic clearly provides greater

---

105 American Society for Testing and Materials, (1987) 'ASTM D 4108-87 Standard Test Method for Thermal Protective Performance of Materials for Clothing By Open-Flame Method,' West Conshohocken, PA
106 Mell. WE; Lawson. JR; A Heat Transfer Model for Fire Fighter's Protective Clothing; NISTIR 6299:1999

efficiency in the use of limited water whereas the latter approach creates greater water run-off but reduces heat exposure to nozzle operators.

11.11 LEEDS UNIVERSITY (UK) RESEARCH INTO 3D WATER-FOG TACTICS

The Leeds University live fire tests by Alirifi et.al. were carried out in abandoned bungalows about to be demolished. The bungalows were constructed in the 1960's and were of traditional build, 100 mm brick wall outside and 100 mm concrete block work inside with 50mm cavity between the two layers. The bungalow consisted of a small hallway with kitchen and bathroom off of this, two small cupboards and a single main living room (17.6m^2), as shown in Fig.11.4. The ceilings in the burn room (living room) were double lined with 12.5 mm plaster board. The back wall to the living room was also double lined. This effectively gave the room one hour fire protection and also ensured that any air for the fire was only coming from the door. The thermal inertia is estimated at around 700 J/m2s$^{0.5}$K which would create post flashover ceiling temperatures in excess of 1100°C for an established fire load according to occupancy type and an adequate opening factor, ensuring maximum heat release is achieved. The fire load for the tests consisted of wood pallets which were stacked on top of one another (9 in each of two stacks) with a total weight of 143kg per stack; the pallets measured 1.22 m x 1.22 m x 0.140 m. The primary stack was ignited using a small metal tray (200 mm square) with 400 ml of methanol to the centre of the fuel mass. A secondary wooden pallet stack (intended to be identical to the first one) with a total weight of 144 kg was positioned on the opposite corner, to assess the pyrolysis effect between the two stacks, did not ignite in the fire. The British Standards guidance (BS PD 7974) suggests that the average fuel load in dwellings is 780 MJ/m2, which for this compartment is 786 kg of wood. The fire compartment was therefore very lightly loaded and the fuel involvement low. Nevertheless, severe fire conditions were reached during the test. The front door was the main ventilation path and it was shown that the fire was ventilation controlled so that the relatively low fire loading was not a major factor in the fire development. Water application in the firefighting phase was performed by the attending FRS personnel using 3D firefighting tactics; hot layer gas cooling was carried out to make safe entry into the compartment then when the team reached the ideal position direct attack on the fire was conducted, using a cone approximately 30° alternating to 60° as needed with an estimated droplet size of 30μm (0.3mm). Key fire compartment conditions (ceiling and lower compartment temperatures, oxygen levels and toxicity levels) were continuously monitored and communicated to the fire ground incident commander, firefighting and support crews. Dura-line lay-flat 38mm low pressure hose was used with internal diameter 38 mm, 15 m length x 2 (30 metres in total), giving a flow-rate of 340 L/min at 7 bar. Tests on the flow rate meter gave 1 L/s with a short pulse 30/60° and 2 L/s with a long pulse.

Approximately 50 kg of wood was consumed in the duration of the test, 60% of which was lost before the start of the firefighting operations. Firefighting was initiated when it was deemed that the fire had reached steady burning rate, which was at 320 s. The progress into the access corridor and the fire compartment, of a group of 3 firefighters (with one charged water line), was tracked from video recordings and the length of hose fed into the enclosure.

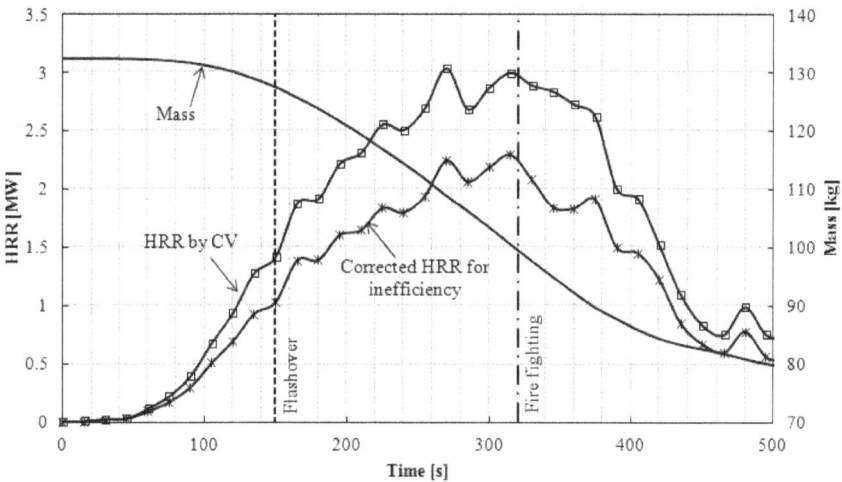

Figure 11.3: Heat release rate recorded during the live burn test undertaken by Leeds University

On entering the corridor, the firefighting team adopted a crouching or kneeling position, trying to keep below the outflowing smoke layer, whilst directing a series of short water pulses towards the corridor ceiling and then the compartment ceiling ahead of them. The spray had an immediate effect in reducing the smoke layer temperature at the ceiling. It can be seen that the water pulses were more effective in dropping the temperature in the corridor by about 100 degrees, but the temperature drop achieved by using longer pulses in the fire compartment were much smaller. On entering the fire compartment the firefighters attempted to manoeuvre and position themselves in the near right hand corner of the room close to the door. This would have allowed all three firefighters to be inside the room during fire extinguishment. However, for the few seconds that it took the leader to adjust his position he stopped pulsing water and this, in combination with the prevailing conditions resulted in the team experiencing unbearable heat levels and an immediate retreat was ordered, accompanied by a long water pulse directly to the seat of fire. From the fire room entry to room exit there was only a 20s interval.

The team retreated all the way to the outside regrouped and re-entered the corridor immediately starting with a direct pulse towards the fire and then 3 short pulses as they positioned themselves in the entrance just inside the room. It should be noted that the heat release at this point was demonstrating the fire decay phase had begun and compartment temperatures were reducing. The average lower layer temperature of the gases surrounding the crouching firefighters was in the range of 242 to 267 °C. This is above the accepted 235 °C tenability limit and therefore within the critical range, as defined by DCLG in their report[107].

To define the focus of the thermal conditions experienced, the researchers also determined the likely heat flux at the firefighter level within the fire compartment, both from the hot layer and the flames, using view factors and flame and hot layer temperatures.

---

107 DCLG. Measurements of the Firefighting Environment. In. Measurements of the Firefighting Environment. Department for Communities and Local Government, 1994.

This resulted in estimates of heat fluxes ranging from 15 to 36 kW/m², for vertical and horizontal body parts at varying heights from the floor, as depicted in Fig. 2. This heat flux is well above the 10 kW/m² limit delineating the 'extreme' from the 'critical' firefighter tenability conditions.

Research in the UK by the Fire Research and Development Group (FRDG) in 1994[108] reported the findings from a series of tests by the Fire Experimental Unit (FEU) in which they arranged for a firefighter to carry specially designed instrumentation whilst taking part in fire training exercises. With regard to tolerated conditions they reported that in tests at ambient temperature, 10 kW/m² was tolerated for 1 minute but damage was sustained to equipment and these conditions would not be acceptable operationally (note: 2 kW/m² corresponds to 'strong sunlight'). The report identifies critical conditions where fire gas temperatures at firefighter locations exceed 235 °C and/or thermal flux >10 kW/m2. This environment is recognised as life threatening and it was suggested that a firefighter would not be expected to operate in these conditions. However, in a rapidly changing environment fire fighters may encounter conditions which are much more severe than the above and this research demonstrates that under these conditions exit timing is extremely critical for survival and it is therefore important for firefighters to appreciate this. Work by Krasny et al. at NIST suggests that firefighters will likely receive serious burn injuries in less than 10 seconds when exposed to a heat flux of 84 kW/m². It should be noted that the temperature and heat flux conditions shown in Fig.11.5 refer to those measured on the body of the firefighter and NOT to the compartment conditions.

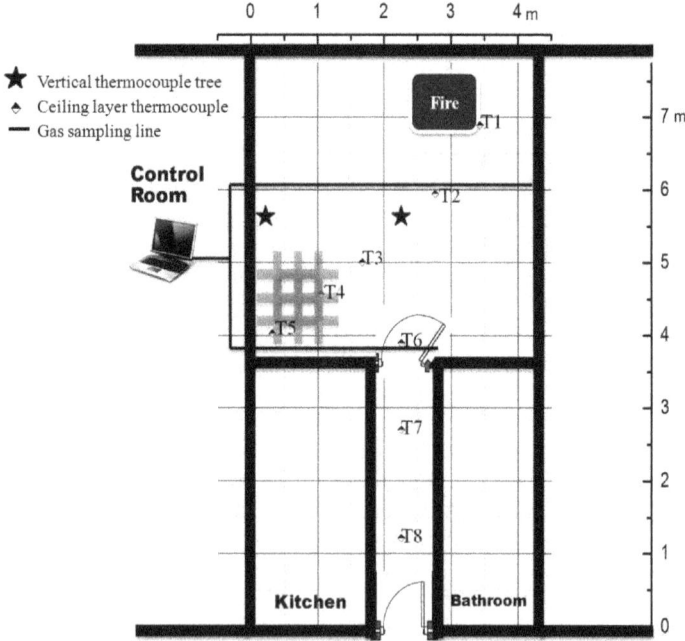

Figure 11.4: Floor-plan of the bungalow fire compartment used in the Leeds University live fire test

108 Foster. A, and Roberts. G.V; Fire Research and Development Group (FRDG) Research report 61; 1994

The level of thermal radiation required to produce a given level of damage is commonly defined in thermal dose units[109]:

$Thermal\ Dose\ (TD) = I^{4/3} * t$ **Eq. 11.5**

Where, I is the incident thermal flux (kW/m²), and t is time (s).

1 Thermal Dose Unit (TDU) = 1 (kW/m²)$^{4/3}$·s

The limit of **3500 TDU** coincides with the calculated values from *Chang et al.*[110] for significant damage to firefighters PPE, and consequently a large coverage of 3rd degree burns. *Chang et al.* tested different types/makes of firefighter clothing under engulfment conditions. He states that the incident heat flux was 84 kW/m² (2 cal/cm².s) but he does not list the exposure time. He refers to the standard test requirements provided by ISO DIS 13506. The standard provides for exposures for engulfment times of 2 to 10 seconds. Assuming that Chang used the longest time this would correspond to a maximum thermal dose of 3679 TDU.

$Thermal\ Dose\ (TD) = 84^{4/3} * 10 = \mathbf{3679\ TDU}$ Using **Eq. 11.5**

In terms of the thermal dose received by the fire fighters in the Leeds University live fire test it was estimated that during the first 15 seconds in the compartment they received 1800 TDUs which built up to around 2400 TDUs during the next 5 seconds of retreat time. This is marked on Fig. 1. The calculation shows that they would have exceeded the threshold limit of damage to their protective equipment (PPE) (3500 TDU) if they delayed their exit by 10 seconds more. This is congruent with the very fast build-up of physical discomfort that the firefighters reported on debrief. They also reported experiencing hot temperatures on their knees where their clothing was compressed against the skin. This again agrees with the high ambient temperatures measured at low level. The very short time to unbearable conditions experienced by the team and the researcher's estimate of 30 s to PPE thermal damage levels, demonstrates and quantifies the very short time available for fully protected firefighters to move to a safer location in an escalating or fully developed fire (*note that some damage to PPE was recorded at 1293 TD in the FRDG tests*).

Figure 11.5 shows the calculated thermal doses for the range of heat fluxes likely to be encountered in compartment fire for exposure times of 1, 3, 5 and 10 s. These are compared with the 100% fatality limit for offshore workers, which also approximately coincides with the thermal dose limit shown to result in significant heat damage of firefighting PPE, as discussed above. It is clear that in post-flashover fires with incident heat fluxes of the order of 150 kW/m2 are likely to result in severe injury even for fully protected firefighters for short exposures of the order of even a few seconds.

---

109 O'Sullivan. S, Jagger. S; Human Vulnerability to Thermal Radiation Offshore; Health & Safety Laboratory UK; HSL/2004/04
110 Chang YC, Lin YW, Lin GH, Jou GT, The Study of Flame Engulfment Protection of Firefighter's Clothing, Hawa Kang Journal of textile, 2007.

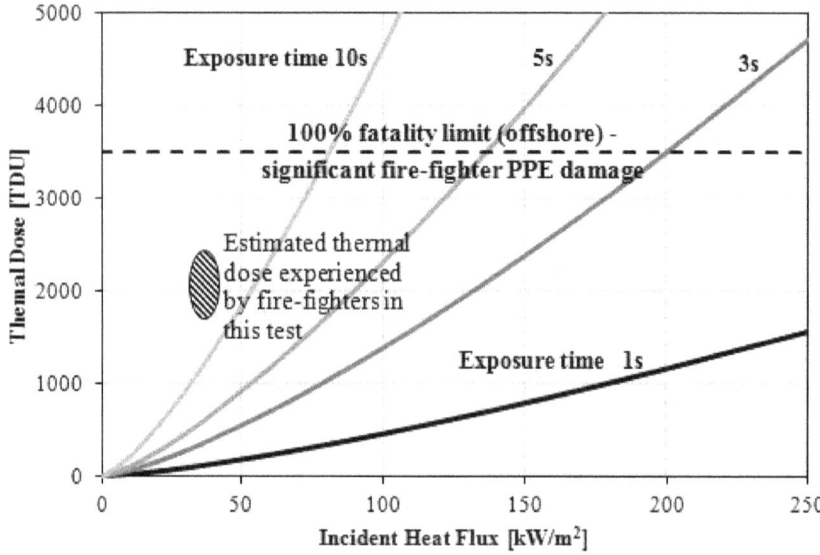

Figure 11.5: Incident heat flux and exposure times (TDUs) (Leeds University live fire test)

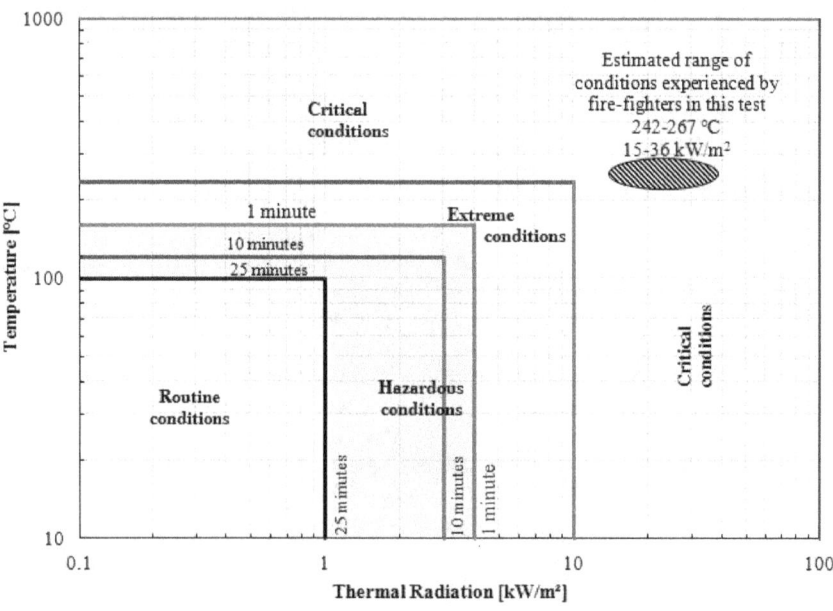

Figure 11.6: Firefighter tenability and exposure times (Leeds University live fire test)

11.12 THERMAL CLASSES

The concept of Thermal Classes was first introduced to us in 1980 by Fred J. Abeles[111] through the NASA[112] contracted 'Project Fires' research venture. It is a most valuable indicator as it provides temperature/time targets for some equipment manufacturers as well as firefighters. The issue that needs to be resolved is to define the classes in a manner that is realistic for conditions experienced by firefighters and is also not so broad as to lose its usefulness for equipment manufacturers. The USFA[113] and FEMA[114] defined a three-class system while a number of other researchers proposed a four-class system. There appears to be an advantage to defining a four-class system in that life safety conditions are better defined and breaking the ordinary category defined by USFA and FEMA into two categories provides better definition for the operational capabilities of different electronic equipment.

The proposed four *Thermal Class model*[115] is presented in Table 11.8.

The rationale for **Thermal Class I** is based on the analysis of Foster and Roberts (FRDG)[116] for firefighters. It is also appropriate for an equipment standard as it was noted in tests using the FEE that the display for the oxygen detector tended to fail at temperatures just over 100 °C (212 °F) making devices using these types of displays Thermal Class I devices.

The UK Fire and Research Development Group (FRDG) recommendation was also used to define **Thermal Class II** conditions. From an equipment standpoint, this may be the Thermal Class for unprotected emergency radios but additional testing is required in order to understand the failure modes for this type of equipment.

The recommendation for **Thermal Class III** was based on a number of issues. Because Thermal Class III is a region where firefighters work for a short time (5 min) and the heat flux must be less than 20 kW/m², which is considered the onset of flashover, the 10 kW/m2 value for heat flux was chosen. Work done at NIST has shown that over short exposure times, exterior surface temperatures of garments above 100°C resulting from 250°C air temperatures caused interior garment temperatures of 55°C (131°F) to 60°C (140°F) when human skin inside the garment may feel pain and/or burn. The 260°C (500°F) criteria was selected based on recommendations by FEMA, USFA, and IAFF117 suggesting that firefighters could work for 5 min at this temperature. This temperature also fits nicely with the NFPA 1982 standard for PASS devices and the NFPA 1971 standard for protective ensembles.

Class	Air Temperature °C / °F	Radiant Heat Flux kW/m°	Exposure Time	Thermal Dose Units
1	40 / 104	0.5	30 mins	714
2	95 / 203	1.0	15 mins	900
3	250 / 482	1.75	5 mins	632
4	815 / 1500	4.2	10 secs	1459

Table 11.7: Thermal classes according to Abeles[118] Project Fire 1980

111 Abeles. F.J; PROJECT FIRES, VOLUME 4: PROTOTYPE PROTECTIVE ENSEMBLE QUALIFICATION TEST REPORT, PHASE 13; Grumman Aerospace Corporation 1980
112 National Aeronautics and Space Administration
113 United States Fire Administration
114 Federal Emergency Management Administration
115 M. K. Donnelly; W. D. Davis; J. R. Lawson; M. J. Selepak; Thermal Environment for Electronic Equipment Used by First Responders; NIST Technical Note 1474; 2006
116 Foster. A, and Roberts. G.V; Fire Research and Development Group (FRDG) Research report 61; 1994
117 International Association of Firefighters
118 Abeles. F.J; Project Fires, Volume 4: Prototype Protective Ensemble Qualification Test Report, Phase 13; Grumman Aerospace Corporation 1980

Class	Air Temperature °C / °F	Radiant Heat Flux kW/m²	Exposure Time	Thermal Dose Units
1	100 / 212	1	25 mins	1500
2	160/ 320	2	15 mins	2268
3	260 / 500	10	5 mins	**6463**
4	>260 / 500	>10	<1 min	***

Table 11.8: Thermal classes according to Donnelly et.al. NIST 2006[119]

*** Thermal Dose Units for an exposure of a pre-flashover heat flux of 20 kW/m² at the floor for 60 seconds' results in 3257 TDUs – close to the maximum recommended by Chang of 3500 TDUs before injury occurs

Class	Air Temperature oC / oF	Radiant Heat Flux kW/m2	Exposure Time	Thermal Dose Units
Routine	100 / 212	1	25 mins	1500
Hazardous	120 / 248	3	10 mins	2596
Hazardous	160 / 320	4	1 min	381
Extreme	235 / 455	10	<1 min	1293
Critical	>235 / 455	>10	Avoid	***

Table 11.9 Thermal classes according to Foster and Roberts, FRDG UK Fire Service College tests, 1994[120]

*** Thermal Dose Units for an exposure of a pre-flashover heat flux of 20 kW/m² at the floor for 60 seconds' results in 3257 TDUs – close to the maximum recommended by Chang of 3500 TDUs before injury occurs

The accuracy of 'Thermal Classes' and Temperature Monitoring against Firefighter Thermal Exposures in Live-Fire Training Scenarios

In a most recent research series[121] (2016) data from typical firefighter thermal exposures were collected from 25 live-fire training exposures during seven different types of scenarios. Based on the collected data, mild training environments generally exposed firefighters to temperatures around 50°C and heat fluxes around 1 kW/m², while severe training conditions generally resulted in temperatures between 150°C and 200°C with heat fluxes between 3 kW/m² and 6 kW/m². For every scenario investigated, the *heat flux data portrayed a more severe environment than the temperature data* when interpreted using established thermal classes developed by the National Institute for Standards and Technology for electronic equipment used by first responders. Local temperatures from a firefighter worn portable measurement system were compared with temperatures measured by stationary thermocouples installed in the training structure for 14 different exposures. It was determined the stationary temperatures represented only a rough approximate bound of the actual temperature of the immediate training environment due to the typically coarse distribution of these sensors throughout the structure and their relative (fixed) distance from the fire sets. The portable thermal measurement system

---

119 M. K. Donnelly; W. D. Davis; J. R. Lawson; M. J. Selepak; Thermal Environment for Electronic Equipment Used by First Responders; NIST Technical Note 1474; 2006
120 Foster. A, and Roberts. G.V; Fire Research and Development Group (FRDG) Research report 61; 1994
121 Willi. J, Horn. GP, Madrzykowski. D; Characterizing a Firefighter's Immediate Thermal Environment in Live-Fire Training Scenarios; University of Illinois Fire Service Institute and National Institute of Standards and Technology (NIST); Fire Technology, 52, 1667–1696, 2016

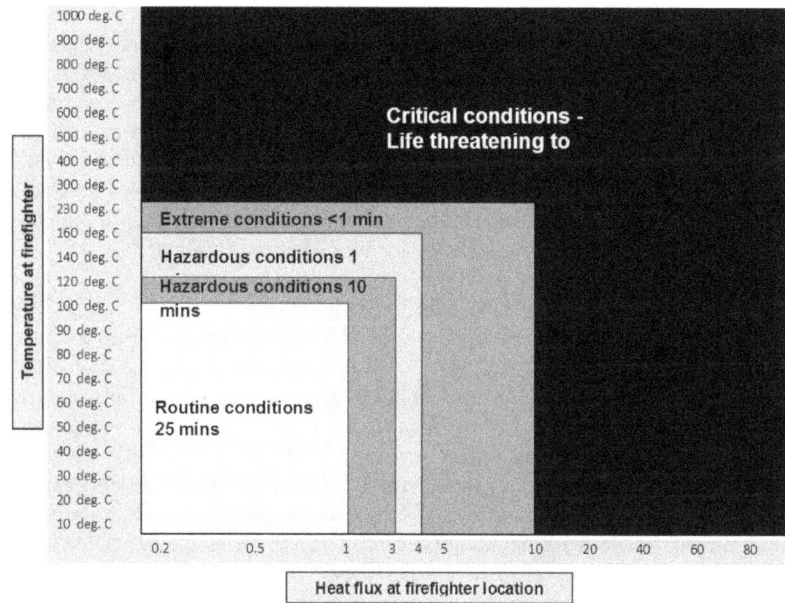

Figure 11.7: Firefighter tenability based on the FRDG research at the UK Fire Service College 1994*

has provided new insights into the integration of electronic sensors with firefighter PPE and the conditions experienced by firefighters in live-fire training scenarios, which has promise to improve the safety and health of the fire service. In this research, to characterise a firefighter's continuously changing local thermal exposure during live-fire training, the heat-flux data acquisition system was designed to be used with standard firefighter PPE to allow for portability with minimal impact on mobility. Neither the thermal protection of the PPE nor the user's range of motion were significantly compromised by the utilization of the system.

The fuel loads for the test consisted mainly of wooden pallets and straw. Fire sizes varied widely within and between each scenario but based upon previous experiments conducted by NIST to characterise the heat release rate of various wooden pallet and excelsior fuel load configurations, it was estimated that the fire size never exceeded 3 MW for any of the studied scenarios.

Comparing the ambient temperature data to define the thermal environment of each exposure produces drastically different results than using the incident heat flux data. Based on ambient *temperature data*, the environment of a steel container 'flashover simulator' training exercise was seen to be quite mild. However, the incident *heat flux data* suggest the actual environment exposure of the 'flashover simulator' training exercise

---

*   Foster. A, and Roberts. G.V; Fire Research and Development Group (FRDG) Research report 61; 1994

was severe, containing average heat fluxes of 3.0 kW/m² and 2.3 kW/m² (NIST Class III in both cases) and maximum heat fluxes of 11.3 kW/m² and 8.3 kW/m² (NIST Class IV and III, respectively) for each exposure. Additionally, the NIST Thermal Class durations generated using the heat flux data suggested that firefighters were in conditions that exceeded the Class III maximum recommendation (5 min) for each exposure. The fire behaviour (flashover simulator) unit used in the tests was fuel loaded with two wooden pallets and straw in a fire set located in the middle of the upper level; with one 19mm x 1.2 m x 2.4 m sheet of oriented strand board (OSB) against a side wall of the upper level; and one 19mm x 1.2 m x 2.4 m sheet of medium density fibreboard (MDF) against the adjacent wall of the upper level. Each experiment began by igniting the middle fire set and allowing the fire to develop and rollover to the lower level. Ventilation conditions were varied and small amounts of water were applied to the fire several times to demonstrate the impact of changing these parameters.

The research paper reported one important, yet unexpected, outcome from the 'flashover simulator' scenario was that towards the end of one test, the SCBA face piece of the firefighter equipped with the portable measurement and data acquisition system began to form bubbles (the firefighter immediately left the structure upon noticing this damage). Although the ambient temperature data indicated an extremely mild environment was present during the 'flashover simulator' exercise and the firefighters participating in the exercise felt no physical discomfort, the thermal environment was quite severe.

According to Putorti[122] et al, a 2 mm to 3 mm thick polycarbonate SCBA face piece lens may begin to bubble when exposed to a heat flux of 5 kW/m² for approximately 12 min or 7 kW/m² for 6 min. During this test bubbling of the face piece began to occur after roughly 9 min of Class I and Class II heat flux exposures (average of 1.1 kW/m²) followed by less than 9 min of Class III heat flux exposures containing an average of 5.0 kW/m². Preheating of the SCBA face piece during the initial Class I and II conditions likely contributed to the earlier onset of damage at this heat flux level (i.e. 9 vs. 12 min). Thus, predicting risk for damage to PPE should account for the entirety of the exposure conditions, even those that may be considered subcritical based on the NIST Thermal Classes. Interestingly, during a similar test no degradation occurred after more than 10 min of Class II exposure averaging 1.7 kW/m² and just under 10 min of Class III conditions averaging 3.1 kW/m². This outcome could suggest that conditions in this range may need to be studied more carefully in terms of risk for PPE damage. It was suggested, for example, Foster and Roberts UK research proposed slightly different exposure criteria where 'Hazardous Conditions' include heat fluxes from 1 kW/m² to 4 kW/m² while 4 kW/m² to 10 kW/m² would be considered 'Extreme Conditions'.

As shown repeatedly in these training scenarios and demonstrations, the measured heat flux suggests a more severe environment than the temperature measurements. In this case, the local temperature measurements indicate a very mild environment, yet firefighting PPE was damaged. Only by measuring the local heat flux would we have any indication that these conditions reach the Class III level and present risk for such damage. While temperature measurements can be made in a simple and cheap manner, it appears more valuable to characterise the fire environment by the relatively expensive heat flux measurements.

---

122 Putorti AD Jr, Bryner NP, Mensch A, Braga G (2013) Thermal performance of self-contained breathing apparatus face piece lenses exposed to radiant heat flux, NIST TN 1785. National Institute of Standards and Technology, Gaithersburg 36. Medtherm Corporation Bulletin 118, 64 Series heat flux transducers, Medtherm Corporation, Hunstville 2004

Firefighter Tenability – Personal alarms

There have been attempts over the past two decades to provide firefighters with personal protection alarms that enable adequate warning of an untenable environment occurring that is not immediately obvious for up to 60 seconds, to the firefighter who is encapsulated in high performance heat resistant protective clothing. By the time the heat has penetrated this clothing and reached the under clothing, it is too late and severe skin burns may occur.

The firefighters' turnout gear ensemble can only tolerate exposure to Class IV fire conditions heat fluxes of 10 – 100 kW/m^2 for less than one minute before degrading. The National Institute of Standards and Technology in the USA undertook a research project in 2005[123] and evaluated available 'PASS' alarms (Personal Alert Safety System (PASS) or Automatic Distress Signal Units ADSU) against a range of live fire conditions. The full-scale data demonstrate that current thermal sensing/PASS implementations were unlikely to provide a fire fighter with sufficient warning of an acute thermal hazard. Current PASS devices (2005) may provide information to fire fighters about their longer term exposure to thermal conditions, but there is a significant delay of 25 s to 120 s before the fire fighter receives this alarm or information. However, there is current research (2016) into innovative technology emerging that may soon be able to provide effective warnings to firefighters of dangerous and untenable environments. The 'Burn Saver Thermal Sensor'[124] is now under development and has demonstrated response times of less than ten seconds to Class IV radiant and 260°C ambient air temperatures.

11.13 KNOWLEDGE GAPS IN TRAINING AND IN TACTICS

Communication

Now you might think your own training programme accounts for all of these 'knowledge gaps' but do they really? It is not only the fire service that suffers from communication issues for it is part of human nature that 'Chinese whispers' syndrome causes a failure in the transfer of key elements of a message. As we give and receive commands on the incident ground we are further hindered by the stressful environment of what we are actually doing, or are about to do. People who are indirect in the communication of their messages tend to hint at things, give mixed messages and avoid getting to the point. It's as if they expect people to be mind-readers. Those receiving the message may be multi-tasking at the same time and this will limit their ability to listen and understand effectively. Basically, the message may not be effectively passed nor received. If that message is then passed on to a third party, its content is likely to include just ten percent or less of the original message as originally intended. This is a skill that demands far greater attention in the training field if we are to reduce communication error at fires. Also, note that its number one in the above list (page 208) as all the incidents involving firefighter life loss had communication failings as major causal factors.

Tactical Ventilation and Flow-Path control

These two are so closely linked but few firefighters ever appreciate the dynamics associated with flow-paths and the impact that tactical venting might have on flow-path velocity and direction. It's simple to explain but not so easy to impart such knowledge into the mind-

---

123 Bryner. N; Madrzykowski. D; Stroup. D; Performance of Thermal Exposure Sensors in Personal Alert Safety System (PASS) Devices; National Institute of Standards and Technology; Report NISTIR 7294:2005
124 https://www.dhs.gov/sites/default/files/publications/Burn%20Saver%20Thermal%20Sensor-508_0.pdf

set of an on-scene firefighter. This is also new language that refers to a fire-ground hazard that has always been there but has never been defined from a tactical perspective. In the past we have always taken events of rapid fire behaviour and ventilation as two separate topics whilst recognising the pros and cons of venting structures. What we should be doing is taking more time to describe how flow-paths form, develop velocity and direction and sometimes change direction, or even reverse. Then we should inform and instruct on the control measures that can be used to prevent flow-path formation, or direct flow-paths so that combustion products are moving away from advancing firefighters. This book informs how the author worked over several years to amend UK building design codes in 2015 so that installed ventilation smoke shafts in extended length corridors will force an automated flow-path away from advancing firefighters in nearly all situations. Earlier systems had been configured to oppose any firefighting advance as design engineers failed to account for firefighting operations that followed any building or fire floor evacuation. In the same way, firefighters need to understand how they can take actions in a fire to reconfigure a natural flow-path, where smoke temperature will always drive the flow-path vertically, and horizontally where openings exist. In fact, it is always high-pressure towards low-pressure, where the pressure differentials will influence the velocity of the flow-path. The hotter the smoke is, the faster it will move towards a low-pressure area within a building or an opening to the exterior. In some situations, the flow-path is wind-driven and this must be reflected in our tactical deployment decisions, depending on which side of the building the wind is hitting in relation to floor layout and fire location. As firefighters we can work to prevent this movement of combustion products where it might impact on firefighters or occupants remaining in the building. We can also utilise methods of re-directing flow-paths to advantage using tactical venting or anti-venting tactics (including fire isolation by closing interior doors and fire resetting by controlling the street or fire compartment door).

Strategic Placement of Hose-lines

This is a lost art whereby the first hose-line into the building will either take the fire directly or protect the egress route for firefighters who are moving beyond or above the fire compartment to search. Many will argue that a secondary hose-line should be taken beyond or above the fire by the search teams but this is time consuming and will slow the search. In placing the first hose-line on the fire and holding their position (or extinguishing the fire) to protect the search teams is nearly always the safest and most effective means of rescuing trapped occupants. If we apply these tactics retrospectively at the fires in Blaina (UK)[125] and Keokuk (USA)[126], both identical fires where multiple firefighters and children trapped above the fire died, if the primary action had been either to isolate the fire (closing doors to the fire compartment) or placing a hose-line on the fire to extinguish or protect search teams going above the fire, the outcomes would almost certainly have been different, at least for the firefighters who lost their lives.

125 Grimwood. P; EuroFirefighter; The Blaina Fire 1996 p115-118
126 Grimwood. P; EuroFirefighter; The Keokuk Fire 1999 p119-127

Door Entry Procedures

There are basically six parts to a traditional CFBT *door entry procedure* and they are as follows –

1. Have a charged hose-line with adequate water for the occupancies fire load
2. Briefly apply water droplets in spray form to the upper part of the door

3. Apply a burst of water droplets above your head to prevent auto-ignition at the point of door opening as hot smoke or flames exit the door
4. Force access but maintain control of the door opening
5. Direct 2-3 short bursts of narrow-cone water-fog towards the ceiling through a door opening just wide enough to place the tip of the nozzle into, observing the neutral plane and smoke conditions before and after – close the door. This sequence should take less than five seconds.
6. or; consider immediate entry based on conditions prevailing within.

Repeat the above brief bursts of water droplets after a short 10-20 second pause and again observe conditions to see if the application of water has changed anything. Enter as determined safe to do so behind a 3D water-fog application into the overhead gases.

However, is further Research needed in this area?

- What evidence exists to support a spray of water droplets on the door serve any viable purpose?
- What evidence exists that supports tactics of spraying a range of water droplets above the operator's head to avoid auto-ignition of fire gases effectively? (There may be very rare occasions when this approach is viable)
- How are these tactics applied where a smoke-blocker curtain is used?!
- I wonder if far more time is spent on training firefighters in *door entry procedure* than how to effectively force or breach door locks

In effect, are these initial phases of the door entry procedure actually supported by any evidence or is it another means of training firefighters to be 'robots'. The door entry procedure is time consuming, gets firefighter wet and is something that should be shortened, so these initial actions are, in my view, unnecessary. If you believe there is a flammable smoke layer above your head then by all means attempt to inert it with water but if that's the case, you are probably already in the fire compartment by definition.

Modern fitting fire doors in corridors rarely enable dangerous fire gases to accumulate a flammable layer. Interior doors could possibly do so and in a small hallway outside the fire compartment the accumulation of gases may be combustible. Again, modern fire resisting doors are highly unlikely to conduct sufficient heat through for you to see any water evaporate[127]. So, think where and when these tactics may be necessary but don't train your firefighters to be like robots, to do this on every occasion at every door!

Advancing the right Nozzle – using the right Techniques – for the Job at Hand

Every firefighter has a favourite nozzle and hose combination for interior firefighting, whether it be a high-flow smooth-bore tip or the versatility of a combination fog/straight stream nozzle. As much as high flows are suited to larger hose-lines some firefighters prefer the smaller diameter lightweight attack lines that may flow less water, but enable a crew of just two firefighters to advance on a fire more quickly. When deploying into working fires in warehouses, storage or industrial buildings, a minimum flow of 600 L/min for each deployed hose-line is required, according to the fire loading and potential fire intensity of such buildings. In private dwelling fires, houses, apartments and small flats a minimum

---

[127] Daniel Izydorczyk Bartłomiej Sędłak, Paweł Sulik, Thermal Insulation of Single-leaf Fire Doors; Test results comparison in standard temperature-time fire scenario for different types of door sets; Applications of Structural Fire Engineering, 15-16 October 2015, Dubrovnik, Croatia

flow-rate of 400 L/min for each deployed hose-line is recommended. A 22-25mm internal diameter high-pressure hose-reel may have equal suppressive capacity to the 400 L/min low-pressure lines. However, if 400 or 600 L/min is needed at the nozzle, the frictional losses encountered in 38mm or 45mm hose-lines (in particular) may amount to flow deficiencies. In such cases the attack hose-line should be either 65mm or 70mm right up to the nozzle length, where 38mm or 45mm can be light and effective to manoeuvre at the fire end. However, if using hose-lines directly off a dry rising fire main, or a poorly pressured wet rising main, the height of the fire floor will determine just what pressure (and flow) is available. They key factor here is what is the required pressure needed at the nozzle to flow the optimum tactical flow-rates? If the nozzle is a 6-bar automatic nozzle, then the appropriate friction losses need to be added to the required pump pressure. So, if the friction losses are 4.4 bar for a 38mm 'nozzle-end' hose-line fed off a 70mm pump line at 44 metres in total, the required pump pressure to flow the 600 L/min into a commercial building fire is 10.4 bar. As most hose-lines have a working pressure of 10 bar, such a flow cannot be achieved and the line is flow deficient.

It is clearly a common factor when fire commanders fail to match involved or potential fire load with adequate firefighting water flow-rates. To deploy crews into developing fire situations in large commercial and industrial premises with 100 L/min hose-lines is failing to recognise the hazards and places crews at risk. The following chapter will explain what 'adequate' firefighting water means, relative to fire loads in a range of occupancies, where commercial and industrial fires demand the highest of water flow-rates in order to deal with some of the highest fire loads found in buildings.

Fire Behaviour in 'Big box' Buildings

At the very same time I was writing this paragraph, three colleagues, one a friend, were killed in a warehouse fire in Peru. It seemed ironic that this particular topic carries so much relevance. It is one thing to train firefighters in how fires behave in steel shipping containers and it is common that they take this knowledge away, believing all building fires will behave the same way. When they experience a 'big box' high-ceiling fire in a commercial retail superstore or industrial warehouse, with large open floor spaces and high-stacked heavy fire loads, they will see the difference. How do the tactics, needed flow-rates and nozzle practice transfer across to big-box fires? How will the fire develop at high level, hidden in the smoke layer? How will automated roof vents open at the right time and in the right place? What is a 'smoke zone'? How will air feed in to replace smoke leaving the building at high level and how might this impact on fire development in a hot gas layer? How far is it safe to deploy and penetrate into such a building with a smoke layer existing just above head height? How safe are a fire engineer's calculations or computer model in predicting fire development and the height of the smoke layer? Would you trust your firefighter's lives on the outputs of that computer model? These are all questions that need answers. It is certain that you are already thinking where to find them! (Also refer to chapter 8).

Limited Staffing on-scene and the impact on Tactical decision-making

Many of us may be faced with limited staffing arriving on-scene at times. It may even be that the first engine has responded elsewhere and you are on your own for a period with a set number of firefighters. We may also experience situations where the second and third engines, or aerial ladders, are several minutes away on a response time-line. Limited staffing means tactical options may also become limited but that should not mean we should fail in our objectives in saving life and protecting property, although in some cases the exposure to risk will increase for firefighters. This has to be managed by our choices and decisions. I have seen fires that were attended by fifteen firefighters in the early stages but a team of three firefighters using alternative tactics would have achieved a better outcome, simply by selecting the right tactical option. This is not to support politically reduced staffing levels in any way whatsoever but it remains certain that our documented operating procedures can and should dictate those alternative tactics clearly. These options may include tactics such as Vent-Enter-Isolate-Search (VEIS); exterior street door control; interior fire isolation; transitional attack from the street or a holding position until further assistance arrives on-scene. In some situations, specialized equipment such water fog/CAFS insertion spikes can have good effects in suppressing fires and reducing fire spread and positive pressure fans (PPA) can also enable a safer working environment if used according to strict protocols and tactical guidelines. Our objectives are always to save life first but in some cases, extinguishing the fire or confining the fire to its original compartment may achieve this objective more safely and effectively.

11.14 RECENT RESEARCH FROM SWEDEN INTO GAS COOLING

Matthias Van de veire presented his research thesis[128] at Lund University in Sweden in 2016 entitled 'Studies on the performance of firefighter's gas cooling'. This research focused on methods and tactics used in applying water into the gaseous layer near the ceiling to reduce the risks of rapid fire escalation. These experiments provided several conclusions that may assist deliver a better understanding of gas cooling, as a well as an optimization of the technique.
Matthias writes:

> 'First off, all nozzles are able to provide a decent cooling if one knows their characteristics. Full cone nozzles have larger droplets but allow a better distribution in the gas layer. They are very much suited for cooling the back and for having more stable temperatures in the front. They are not suited for fast and close gas cooling though. Fine mist nozzles have small droplets that cool gases almost instantaneously. However, the nozzle provided in this work did not have a sufficient reach, resulting only in a cooling in the direct environment of the nozzle operator. A fine mist nozzle could perform well if it has a stronger momentum than the outgoing gases. Hollow cone nozzles performed the best due to their combination of a far reach and the smaller droplet size. A temperature decrease from 160°C to 100°C was achieved over the whole ISO [test] room having the nozzle at a fixed position.
>
> Secondly, the interval time in between different bursts should be kept as short as possible. In this experimental setup, a continuous water flow performed slightly better than pulses every half a second although the reheating was also slightly faster. The layer height did not change that much which was theoretically predicted at this room size and gas temperature. The gas layer goes up at the nozzle interface but goes down at the edges. The layer descent is minimal and recovers within the 3 seconds. There is a relative long buffer time after the application of the water, especially in the front. The first two seconds after the nozzle closes, the gases continue to cool for 2 seconds. The speed of reheating afterwards depends on the droplet size and interval time. This was generally slower for larger droplets and for the bursts with longer interval times. This stagnation time was only true for the tests where fast bursts and a continuous

---

128 Van de veire. M, Studies on the performance of firefighter's gas cooling, Department of Fire Safety Engineering, Lund University Sweden, 2016

flow were used. The continuous flow that performed the best cooling did not show this stagnation time but rather heated as soon as the nozzle closed.

Lastly, using a higher pressure for a same nozzle always performed a better cooling regardless of the fact that parts of the stream hit the ceiling or the walls. It is likely that an upper limit exists, but this has not been reached in this work'.

What does this mean to the Firefighter?

The Swedish methods of fire gas cooling and flame cooling provided a whole new meaning to firefighting tactics for many of us from 1984 on. The realisation that many firefighters may have been somewhat ignorant of how smoke layers could transport flammable gas clouds above our heads and into adjacent areas or rooms, or down behind us, only to ignite with some ferocity, was becoming obvious. The 'new-wave' approaches of dealing with these fire gases hanging in the smoke near the ceiling was to cool them, using short bursts of water-fog in order to maintain control of the neutral plane. The sporadic 'pulses' of water droplets caused the heated gas layers to contract on cooling and in some gases caused the smoke layer to rise.

Similarly, the training in the steel shipping containers showed us how very small amounts of water could extinguish these smoke layers where flaming existed along the interface with air, or even in the fire plume itself. This however was never close to a real fire as the base fire used to create pyrolysis of the chipboards consisted of only a couple of hundred kW in released energy. Therefore, many fire behaviour trainers were being lulled into a false sense of security, truly believing that such low flow rates <50 L/min, would be able to deal with intense room fires if firefighters were well trained.

The research of over 500 real fires in London during the 1980s taking these firefighting methods onto UK fire-grounds clearly showed the potential of the techniques, but also demonstrated clear limitations in maximum compartment sizes and fire loads where high-flow direct attack was needed to extinguish the base fire, aside from what was occurring in the fire gas layers. In fact, it was observed on many occasions that firefighters were beginning to incorrectly bias fire attack into the overhead gases when the visible base fire was clearly the priority.

The optimum approach at any fire is to use both methods (direct attack at the base fire and gas or flame cooling of the overhead gases). However, the direct high-flow attack must <u>always</u> be the starting point from which to benchmark and pre-plan your target flow-rates, based on the fire loads found in the range of low-medium-high risk occupancies (Chapter 12).

So many experienced fire behaviour instructors may never have actually gained much experience in 'real-world' firefighting involving intense building fires. Therefore, sometimes their message and training is inappropriately biased by what they so regularly see, feel and learn within the confines of a 1.5 MW *steel shipping container* training fire, or perhaps by a few videos they have seen.

11.15 THE EFFICIENCY OF GAS COOLING IN COMPARTMENTS WITH HIGH CEILINGS

It was a common factor experienced during the 549 fires faced by London firefighters from 1984-1994 that the Swedish firefighting water-fog tactics achieved greater effect when using either fire gas (hot smoke) or flame cooling methods, where fire compartments had high ceilings. Similarly, stair-shaft fires seemed to fall into the same category where the water-fog was at its most effective when applied into the rising thermal currents in the stair. In standard buildings with 2.5 to 3 metre high ceilings the majority of firefighting water was applied most effectively onto the fuel base. In chapter 12 it is demonstrated how the extinguishing efficiency of water in buildings with a standard ceiling at 2.5 metres from the floor is biased towards fuel-based cooling, over flame cooling. In fact, even where we are gas cooling, the likelihood is that in small compartments with low ceilings, the greatest proportion of your flow-rate will be directed at the fuel base (fuel phase). In Fig.12.8 this balance between fuel-phase and gas-phase suppression is graphically shown to be 35% / 15% in favour of fuel phase firefighting (at 50% efficiency), based on live fire research data. We may consider a similar efficiency gradient when applying our firefighting water in high-ceiling compartments. However, where there are vertical stacks of fire load above 2-3 metres high, the priority in firefighting water application shifts back to the fire's fuel base.

Height of ceiling (metres)	Fuel-phase %	Gas-phase %
2.5	35	15
3	34	16
4	32	18
5	30	20
6	28	22
7	26	24
8	24	26
9	22	28
10	20	30
11	18	32
12	16	34

Table 11.10: The proportions of firefighting water applied into the gas-phase in compartments with high ceilings may be biased in favour of the quantity applied onto the fuel-base, as cooling the larger volumes of heated smoke and fire gases in the overhead may become the priority. The application of water is based on a 50% efficiency factor (see chapter 12).

Chapter 12

Adequate 'Firefighting Water' – *you need this!*

'As you increase the quantity of firefighting water applied during the first ten minutes on-scene, the data demonstrates the area of subsequent fire damage to the building is reduced and the heat exposure to firefighters is diminished.'

Paul Grimwood (PhD) – 5,401 building fires

If you took a large stack of wood pallets and burned them in the open air, the fire would continue to burn freely in what is termed, a *fuel-controlled fire*. That is, the continuing development of the fire is dependent on the availability of adequate fuel. As long as the distribution of fuel is effective in the stack, the natural airflow would enable most of the fuel to be consumed by the unrestricted fire. However, if we placed a very large fire resisting box over the pallets, with a small opening, the fire would then become dependent on the amount of air supply it received, according to the size of the opening. The fire is now said to be *ventilation-controlled*. With a small opening the fire will not burn well at all and produce much light-coloured smoke, but with a larger opening the fire will begin to burn more efficiently and produce darker smoke. The rate of heat energy being released from the burning pallets is therefore dependent on the size of the openings (*the existing ventilation profile*).

If a single pallet provided heat energy equal to 300 MJ, that equates to 3,000 MJ per stack of ten pallets high. Such a stack would weigh around 160 kg. The British Standards guidance in BS PD 7974 suggests that the average fuel load in dwellings is 780 MJ/m^2. Therefore, a typical room of 16 m^2 in area would require 12,480 MJ of energy provided in pallets to equal the normal average fire load found in residential dwellings. To meet such levels of fire load we would need to provide at least four pallet stacks of ten high in such a room and set light to all of them to reach expected levels of post-flashover heat release. Such a fire would be dependent on the size and number of openings to that room in order to burn efficiently and would burn in a *ventilation controlled* state.

- *A fire with large openings will burn with greater intensity (higher MW) and will burn down to decay more quickly than a fire in a room with small openings.*
- *A fire with a higher fire load (say 50 pallets) will still only burn to the same intensity for the same opening size but will burn for longer, as it releases its energy at the same rate and is totally dependent on the amount of ventilation available.*

Figure 12.1: Burning pallets in the open – a fuel-controlled fire Photo courtesy of Terry Johnson

As the burning pallets released their energy much of this (around 30%) would be absorbed into the walls and ceiling surfaces (depending on the thermal inertia of such linings) and would issue from openings as smoke and heat, or even flaming combustion. Some of this released energy would also be stored in the fire gas layers (smoke) that remain within the fire compartment or adjacent areas.

- *The size, quantity and location of the openings will determine the rate at which the heat energy is released from the fuel load (Heat release rate (HRR) in kW or MW), this is the measure of fire intensity.*
- *As more windows fail or other openings are created, more heat energy is released from the fire.*
- *As more energy is released, temperatures near the fire and near ventilation points are most likely to increase.*
- *As more heat energy is released, <u>more firefighting water is needed.</u>*
- *As a fire enters its decay stages (fuel depletion), less heat energy is released and therefore less firefighting water will be needed.*
- *Therefore, we can sometimes control ventilation to a fire (door control) to reduce the heat release, bringing the HRR within reach of the quantity of water we have at the nozzle.*
- *However, it's always better to have an <u>adequate quantity of water</u> at the nozzle in case of 'unplanned ventilation' caused by the fire!*

As an experiment, you could try to extinguish those forty pallets burning in the open air to see the effectiveness of any particular firefighting stream. Start with a garden hose and then progress through larger fire streams, recording the time taken to extinguish the fire, if at all possible. Typically, a 100 L/min hose-reel will take longer than a 500 L/min solid stream. You may note that the lesser quantity of water you use, the longer it takes to extinguish the fire. In turn, the more time firefighters would be exposed to high levels of heat flux if the same fire is approached in a ventilated enclosure, where the open-air fire plume is turned into a ceiling jet, visibility is restricted by thick smoke, heat is building up and accessing the same fire becomes far more difficult. Therefore, the most effective measure of fire intensity (HRR) is directly proportional to the quantity of water needed to extinguish a room, compartment or building fire.

The amount of firefighting water required is calculated on the **_chemical heat release_** from the fire's fuel load, before it is lost through openings or at boundary surfaces. However, at building fires we can only ever estimate within reasonable parameters of accuracy according to existing experimental data, both the heat release rate (MW) and the needed flow-rate (L/min). A more accurate method may be to record the actual amount used by firefighters at a broad range of occupancy types and relate this to subsequent floor area of fire involvement/damage. This then offers some data that can be used in engineering terms, or for pre-planning firefighting interventions. In such cases it is more important to over-estimate needed flow than under-estimate.

12.1 THERE WERE 5,401 BUILDING FIRES – RESULTING IN GUIDANCE IN BS PD 7974-5-2014

Now there is one of the best known tactical sayings in the fire service. 'Big-fire = Big water'! But when you think of it, a big fire can sometimes be handled by a small amount of water. It all depends on the quantity of fire load involved, the level of ventilation potentially available, or the level of fire containment, the skill of the nozzle operator and the *method* and effectiveness of the *nozzle* used to apply water or foam.

Some key points of guidance – The **_amount of water_** (L/min) you have coming out of the nozzle, and your ability to **_reach or access_** the fire and effectively **_penetrate_** the involved fire load with **_an adequate percentage_** of that water, is what will get this job done – have no doubt!

But how do we determine what 'big water' is and when to apply it? If fire is issuing from one window do you need 'big water'? Let's say fire is issuing from three windows? What if it's going through the roof? How much water is required? We all have different ideas in our mind of the answers! Do you have the staffing and resources to deliver that large amount immediately on-scene? Does the type of construction or the type of occupancy affect the needed amount of firefighting water? As an on-scene **fire commander** how do you quantify how much water you need *now* … and are likely to need over the *next 90 minutes*? How many fire appliances are needed to deliver that quantity of water? If the fire spreads to involve 50 percent of the building in front of you – how do you determine and plan for the required amount of water flow-rate on-scene? If you are a **fire engineer** do you design to prescriptive codes when determining firefighting water requirements from a rising fire main or a building that is five miles from a water source, or do you determine needed flow density on a quantitative basis depending on fire load, occupancy and compartment size? In other words, should a rising fire main provide the same flow-rate density (L/min) in an apartment building as opposed to an open-plan office tower? You might save the client a great deal of money by quantifying needed flow-rate density, matched to the level of risk. A good fire engineer will not only analyse the required flow for the building but also flow-test the available water resources (hydrants) at different times of the day to gain a mean average, before presenting a fire engineered strategy.

There are various means of obtaining an adequate water supply as follows:

1. Water carried on-board the fire engine (generally around 1,500 – 1,800 litres
2. Augmented supply from street hydrants (generally 300 – 3000 L/min ea.)
3. Augmented supply from open water sources, ponds, lakes and rivers
4. Transport of firefighting water via bulk water carriers
5. Water relays (open or closed) set-up between fire engines

On arrival at a building fire, firefighters require immediate access to all parts of the building to allow for a rapid deployment of water onto the fire. In the UK, it has been the 55-metre length of high-pressure hose-reel (or sometimes two) that is used at the vast majority of building fires (76% of working fires). It is for this reason that building regulations require access to all parts of a residential structure (in particular) to be within 45 metres of the fire engine, or a rising main (or sprinklers) may compensate where this is not viable. If the distance to the furthest point is beyond 50 metres as hose-reel is laid into the building, then firefighters are forced to lay larger hose-lines. Extending the hose-reel will dramatically reduce flow-rate and should not normally be done. This will delay the deployment by several minutes and allow the fire to grow and develop further with internal search and rescue attempts also being compromised. Most fires of this nature are dealt with quickly within ten minutes. However, a further 24% of working fires may need additional firefighters to assist as larger hose-lines become necessary. Long hose-lays into buildings beyond 60 metres should be discouraged due to the physiological barriers of firefighters carrying heavy equipment and working in heat for long durations, therefore an adequate number of access points along a large percentage of the perimeter will assist firefighters by reducing needed hose-lay distances and provide less distance to an escape point from the building in emergencies.

To be able to contain a compartment/structural fire safely and effectively, firefighters must ideally be supplied with an *adequate* amount of firefighting water if they are to control a developing fire during the initial stages. This 'early' water 'at the nozzle' must take account of fire in both the gaseous and fuel phases of combustion and to direct water into the overhead gas-phase alone is not always the best way to deal with a fire. It is certainly recommended that every attempt is made to cool hot fire gases using short controlled bursts of water droplets, but it is equally important to try and extinguish the base fire as soon as possible. Research into the optimum use of firefighting streams reported by the author from earlier work showed that when London firefighters used general fog patterns to cool gases and extinguish fire in the gas-phase, and then reverted to straight streams to deal with fire at the fuel base, the measured cooling ratios of both applications combined were 36% gas-phase and 64% fuel-base fire during several experimental full-scale compartment fires.

The Glasgow Caledonian University (GCU) research (author's PhD), analysed Incident Recording System (IRS) data and real time water-smart flow metered data, collated directly from the two fire-grounds of a **county**, and a city **metro**, fire service across 5,401 serious (working) building fires in the UK occurring over a three-year period 2009–2012. This highly concentrated study determined that building fires rarely reached or exceeded a maximum of 500m^2 fire damaged floor area, where the average fire sizes of >5,000 'working' building fires[129] ranged between 16m² of floor area for residential dwellings up to 95m² for industrial units, with all other building and occupancy types falling somewhere in between. For these fires, average firefighting water flow-rate densities ranging between 8 and 12 L/min/m² of fire involvement were used to achieve control. It was noted that once fires had burned beyond 500m² of floor space, the buildings were generally lost completely. Building fires in this research were also broadly seen to spread further than annual government statistical reports would suggest as these were all 'working' fires' and not simply 'calls' to fires. However, 66% of all working building fires in the GCU research were extinguished using just the 1800 litres carried on-board. Based on this study, it is clear that the term 'adequate' firefighting water flow-rate should, in general, be matched against fire load density and compartment size to fire-resisting boundaries.

It has been observed that the accumulation and ignition of hot fire gases in an average UK five-roomed 70m² apartment can cause fire to spread from one room to five rooms in less than sixty seconds. It is also known that steady state fire spread in large open-plan office areas can travel at a rate of 22 m²/min (fast t-squared growth), whilst cellular offices are seen to reduce such fire spread rates to around 7-15 m²/min (medium t-squared growth). Similarly, vertical fire loads in storage warehouses or retail superstores can burn beyond the control of water deployed from a single hand-held firefighting hose stream in less than sixty seconds. It is estimated that a 20-25MW fire is probably the maximum fire intensity a single 500 L/min hand-held hose-line can deal with. Where sprinklers are not installed in such situations the fire service are quite often helpless in containing such rapid fire-growth.

A fire commander needs to be able to immediately estimate fire-ground water requirements with some reasonable accuracy using simple rule-of-thumb guides. A fire protection engineer takes more time in calculating with some greater accuracy, the quantity of water that is best suited to any particular structure using prescriptive or performance based design principles. It is worth pointing out that the quantity of firefighting water needed to protect an apartment building is far less than for open-plan offices or industrial risks, yet existing prescriptive design codes and standards do not allow for any variation when installing rising main (standpipes) and storage water in a range of buildings and occupancies. The determination of firefighting water requirements in large, tall and complex buildings in the UK has largely been based on data retrieved from studies into full-scale test fires in residential and commercial buildings, undertaken by the Fire Research Station (FRS) from 1955-1970. This scientific research effectively formed the basis for the design and configuration of rising fire mains in tall buildings, as well as water storage provisions for town centre and infrastructure planning. The resulting guidance is also applied to individually isolated buildings that are abnormally distanced from the nearest available water supply. However, it is suggested the UK has since fallen behind many international standards and codes, where firefighting water provisions are more reflective of modern building design as movable fire loads, compartment dimensions and window sizes (ventilation factors) have increased over the years.

All fire and rescue authorities in the UK are required by law[130] *to take all reasonable measures to ensure the provision of an **adequate** supply of water in the event of fire and to secure its availability for use in firefighting*. It is further noted that adequate firefighting facilities, adequate hydrants or storage water provisions, and close access to buildings must also be provided to enable firefighting water to be effectively and promptly deployed.

---

129 A 'working' building fire in the GCU research was defined as one where breathing apparatus was worn and a flow-rate of at least 100 L/min (hose-reel or main-line hose) was used to achieve control.
130 Fire and Rescue Services Act 2004; http://www.legislation.gov.uk/ukpga/2004/21/contents

12.2 HOW MUCH WATER IS 'ADEQUATE' FOR TACKLING BUILDING FIRES?

To provide 'adequate' firefighting water means that there must be enough water available for firefighters to control fire development during the growth and steady state periods of fire development, in order to protect the structural frame and ensure stability within the building's design limit state. In general, international building design guidance is reflective of firefighting needs in respect of *minimum* flow-rates (L/min) and maximum hose lay distance (m) from the fire main (stand-pipe) outlets to the furthest point on a floor plate (usually a 45–60 metre maximum distance).

However, the **flow-rate density (L/min/m²)** of applied firefighting water is rarely if ever, something that is considered or prescribed. A performance based approach takes the flow-rate density factor into account and uses this key measure as a means of providing adequate firefighting water across all parts of a floor plate, whilst maintaining economy in the overall water storage provisions. It also accounts for compensatory reductions in flow and storage, where sprinklers or other protection facilities are provided.

The influence of the fire service intervention depends on the time when the fire service arrives at the fire scene, the resources available and the time when ready to start firefighting (applying water). These issues are thus dependent on the relative distance the fire service must travel to an incident, their availability and the allocation of resources at community level. There is another issue that may become important especially in fires where the performance of automatic suppression systems is inadequate or unavailable. It is the ability of the fire service to extinguish the fire. This issue has been quantified in the NFPA Fire Protection Handbook[131] as curves on the ability of different kinds of firefighting units to extinguish a fire of a given size. The units considered are –

- An average person with an extinguisher is likely to be able to extinguish a fire of ~ 2 m² floor area
- A trained in-house fire brigade. is likely to be able to extinguish a fire of size of ~ 5m² floor area
- An average strength fire department [volunteer, part time or lesser equipped] is likely to be able to extinguish a fire of size of ~ 50m² floor area, where provided with adequate firefighting water
- The strongest fire department [full-time professional or well-equipped] is likely to be able to extinguish a fire of size of ~ 100m² floor area, where provided with adequate firefighting water

The recent publication of BS PD 7974:5:2014[132], based on the author's research published through GCU in 2015, goes some way to fulfilling the need for up-to-date performance based design guidance in meeting the fire service's requirement for an 'adequate' supply of water in large, tall or complex buildings (s.8.5). This guidance was recently called for by representative bodies in both the UK[133] and USA[134]. However, whilst there is prescriptive guidance in place detailing the required firefighting water provisions, it

---

131 Bush, S. E. & McDaniel, D. L. 1997. Systems Approaches to Property Classes; Fire Protection Handbook. 18th Edition. Quincy, MA: National Fire Protection Association. Pp. 9–92.9–109.
132 BS PD 7974:5:2014, Application of fire safety engineering principles to the design of buildings (Fire Service Intervention), British Standards Institution, 2014
133 Fire Hydrants and Firefighting Supplies; UK Water Industry Research Limited 2010
134 Evaluation of Fire Flow Methodologies; Hughes Associates; The Fire Protection Research Foundation; NFPA January 2014

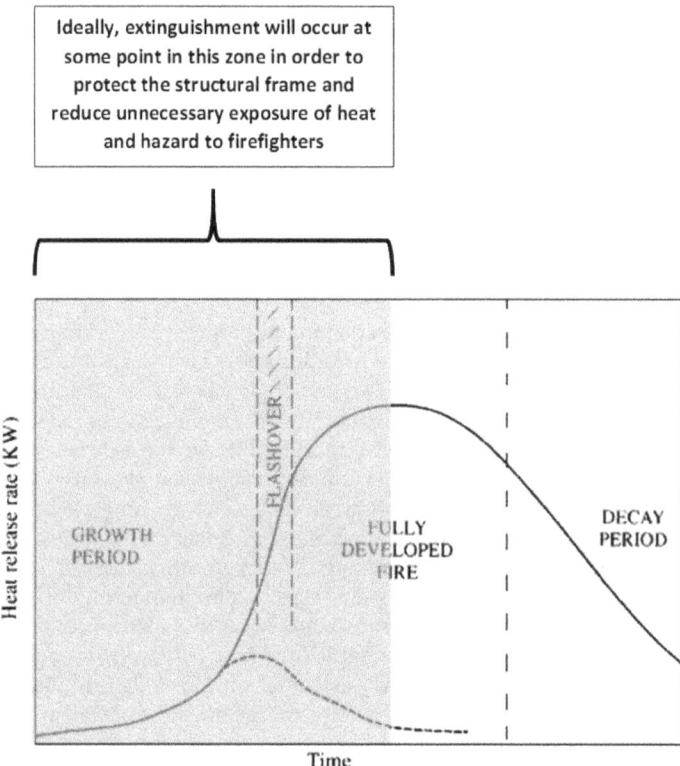

Figure 12.2: The objective must always be to respond, deploy and extinguish the fire within the optimum time frame, to protect the structural frame and reduce unnecessary exposure of heat and hazard to firefighters

is logical that any design engineer or building developer is going to ask why the more onerous recommendations apparent in PD 7974 part 5 should form a part of any fire design strategy. It is important to realise that the base calculations in '7974-5' are for *non-sprinklered* buildings and reductions in firefighting water can be achieved through the addition of a sprinkler compensatory coefficient to the calculations. There are several options open to the fire engineer to include a coefficient in the base calculations as compensation for sprinkler protection, where provided. An objective should be to deliver adequate firefighting water provisions integrated within an active/passive fire protection and management strategy that remains cost effective but is effectively risk based.

Critical firefighting water flow-rates:

In the GCU research used to form the fire engineering design methodology in BS PD 7974:5, the following benchmarks were demonstrated against fast developing fires in non-sprinklered compartments:

- **Critical** flow-rates, below which a developing fire is unlikely to be controlled during the growth or steady state periods. 2.0 L/min/m²

- **Minimum** flow-rates where suppression is achievable but firefighters may be exposed to longer duration fires and more punishing conditions. 3.7 L/min/m²

- **Optimum** (adequate) flow-rates where control of the fire is achievable without unnecessary punishment to firefighters. 6.0 L/min/m² (two dwelling rooms totalling 32m²)

  Note: 'Optimum' means the absolute <u>minimum</u> amount required to extinguish a certain sized fire <u>effectively and safely</u>. A secondary safety (back-up) line of at least equal flows should always be provided in addition, in support.

  6.5 L/min/m² (commercial building fire 50-100m²)

  0.407 L/s/MW (Grimwood)

  0.385 L/s/MW (Barnett)

Table 12.1: Benchmarks for firefighting water flow-rates can be determined in order to establish the level of risk to firefighters and the structure, caused by extended duration fires.

What does this mean to the Firefighter?

When deploying the initial attack hose-line into a fire compartment, it needs to be able to deal with the potential fire load existing in that room. The ideal water application will control fire before flashover or where travelling fire spread occurs. In some cases, fires will be fought internally as they burn at the peak of their heat release, which may last a few seconds or minutes. That peak in heat release can increase further still where windows fail or are vented.

During the initial attack in a compartment fire, a high-flow of water may extinguish the fire faster or a low flow may not be effective, or take longer to achieve control. In some circumstances a low flow rate will achieve control over a longer period of time, particularly where the rate of burning starts to slow down and decay as the fire load is depleted, but firefighters may be exposed to far greater thermal stresses over this longer duration.

In practice, we should aim to deploy the optimum amount of water in anticipation that a working fire may burn beyond the compartment of origin (47% of working fires) and apply enough water at an early stage to avoid longer duration fires, the potential for structural collapse and elements of increased building fire and smoke damage. A key statistic informs us that a room fires burn beyond the compartment of origin at one in every two 'working fires' in buildings. That's a fifty percent chance it will. A third of these *working fires* will also spread to involve other floor levels.

It would therefore be wise to consider if a 19mm hose-reel is sufficient for primary deployment into working building fires? It is also worth considering here that 22mm high-pressure hose-reels can be twice as effective as 19m hose-lines in taking control of developing fires and the author, along with the Kent Fire and Rescue Service in the UK, has been leading the background research here since 1999.

Note: A 'working' building fire is defined as one where breathing apparatus is worn and firefighting water from a low or high-pressure hose-line is necessary to achieve control.

What does this mean to the Firefighter?

Fire-ground 'rule of thumb' fire-ground estimates –

When on-scene at a fire there are so many things a fire officer has to consider during the dynamics of a very fast-moving and stressful environment, such as that presented by a developing building fire. The calculation process must be easy to apply but be relevant and useful. As a Deputy Chief of the New York City Fire Department during the 1980/90s, Vince Dunn was experienced in commanding several serious high-rise fires in the city. His respected view was that a 2,500ft^2 (232m^2) fire on an open-plan office floor plate, was the largest size fire that his Manhattan firefighters could effectively deal with before control was lost and it would take at least one 63mm attack hose-line delivering 300galls (US)/min (1,134 L/min or 4.9 L/min/m^2 of floor area fire) in order to do so. Such a high flow-rate demands a crew of 3-4 firefighters on the line to be able to deliver this fire stream into the fire floor.

Another well-known US fire chief of the same era Bill Peterson (Plano Fire Department) further stated that based on his extensive practical firefighting experience, there would likely be a 50% failure rate where firefighters are deployed internally to control a developing fire, once the fire development had surpassed 925ft^2 (86m^2) in floor space. These empirical observations from well-respected fire commanders of the time come very close indeed to the recent GCU research data outputs that are based on a large body of real fire experience in the UK.

Based on the author's earlier research from 100 fires in London in 1989 and the GCU research described here, a series of rough fire ground rule of thumb guides were developed for UK national operational guidance (NOG) as follows:

- Area of fire (m^2) multiplied by 5 (for fires involving between 100-500 m^2 of floor area)
  A x 5 = required flow-rate (L/min) - (A = Area of floor in m2)
- One low flow hand held fire stream (350 L/min) (say 100 galls/min) per 75 m^2 of floor area fire involvement
- One medium flow hand held fire stream (500 L/min) (say 125 galls/min) per 100 m^2 of floor area fire involvement
- One high flow hand held fire stream (750 L/min) (say 200 galls/min) per 150 m^2 of floor area fire involvement

- These flow-rates must be deployed as 'water on the fire', as compartment fires are most likely to develop and spread rapidly, at a rate of 22m²/min in an open-plan office (for example) – it may take several minutes following deployment into a large fire before water actually reaches the fire
- It has often been seen that a fast-growing fire in an open-plan floor area, or a vertically stacked fire load, can burn beyond the control of a single firefighting jet (hose-line) in less than 10 minutes (or far less on a vertical stack), once the flame reaches around a metre high from point of ignition.

In figures 12.3 to 12.6, the outputs and engineering formulae for required flow-rates from the GCU research are displayed. The three finely dotted lines represent dwelling (houses and apartments) fires (lower); industrial and storage buildings (upper); and 'all other buildings' adjacent to the central line. Within the charts, the dashed line represents the flows calculated by the A x 5 rough fire-ground formula. It can be seen where that fits in with the more complex engineering formulae proposed by the research. Where the fire area is less than 120m² the fire-ground rule-of-thumb is inadequate and where fire areas >600m² it begins to estimate very high flow-rates. Therefore, the fire-ground formula A x 5 is only suited for use within the parameters as discussed above. Where deployment is for a fire area less than 120m², the minimum <u>target</u> flow-rate on any primary attack hose-line should be 200-500 L/min (1 or 2 x 22mm hose-reels) on high pressure or 350-500 L/min on low pressure. (Also, see guidance in chapter 3 – Fire Dynamics).

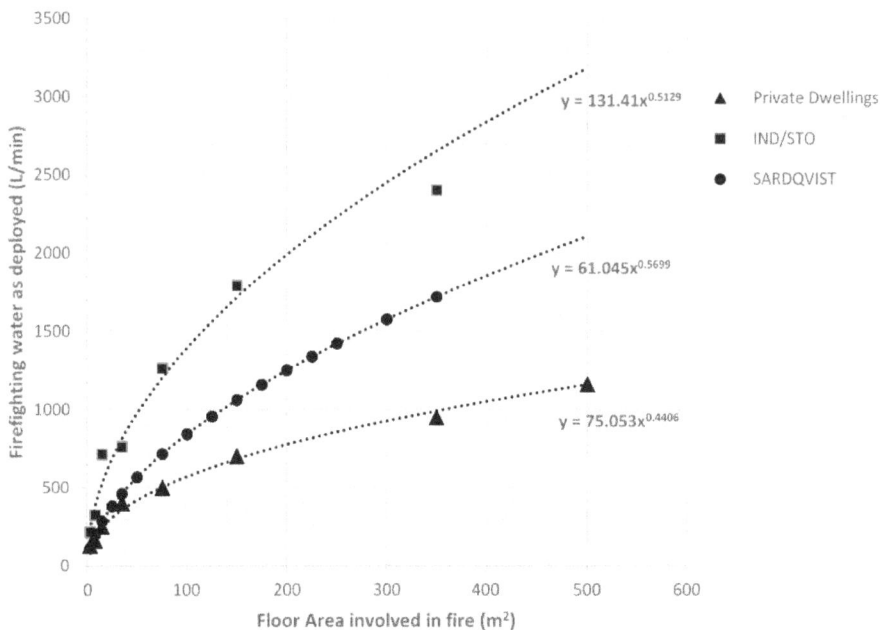

Figure12.3: Measured flow-rate from 5,401 building fires 2009-2012 - GCU Research

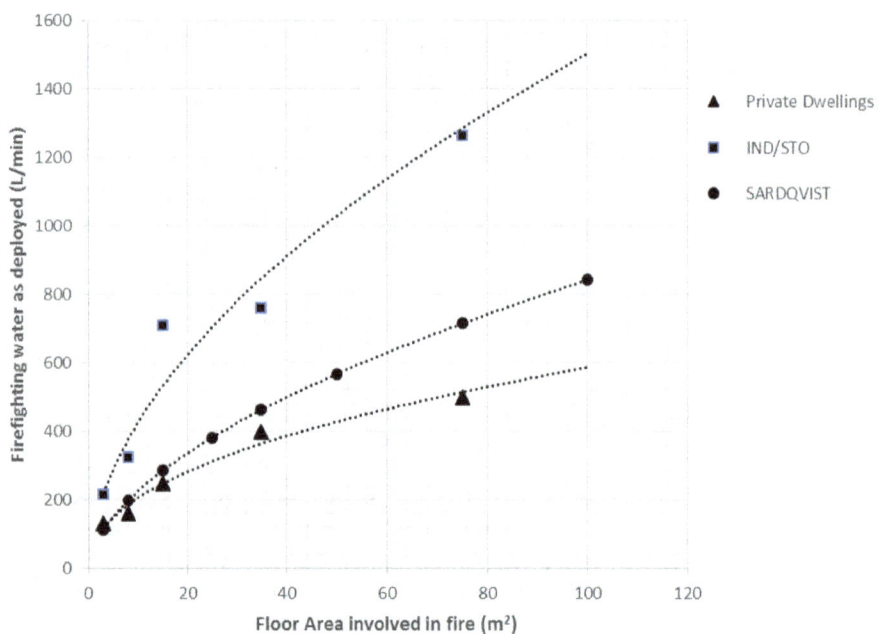

Figure 12.4: Measured flow-rate from 5,401 building fires 2009-2012 - GCU Research

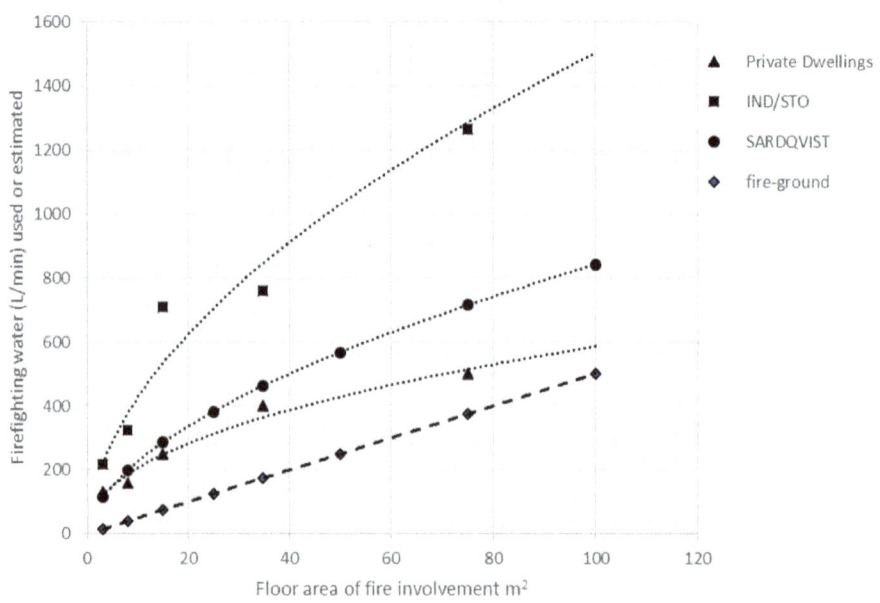

Figure 12.5: Fire-ground 'rule of thumb' flow-rate in comparison - GCU Research

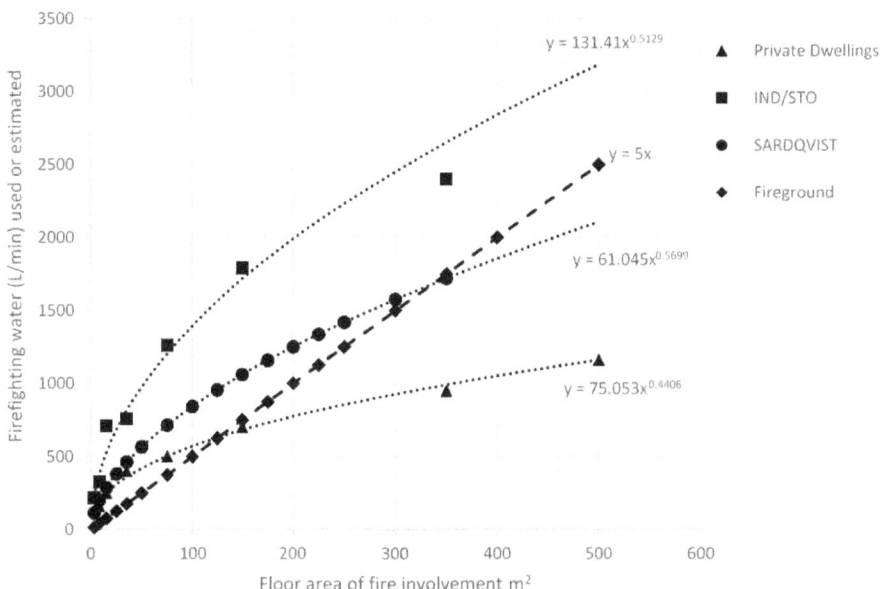

Figure 12.6: Fireground 'rule of thumb' flow-rate in comparison - GCU Research

12.3 THE THEORY OF FIRE SUPPRESSION USING WATER

Cooling Ratios

The cooling ratio of firefighting water is defined by the comparative percentages of applied water used or required to deal with fire in the gas phase (the volume of burning gases generally in the overhead but occasionally throughout a compartment) and fire in the fuel phase (generally fire across the surfaces of involved fuel at the fire's base).

It is natural that fire behaviour instructors will generally direct the greater proportion of their training in dealing with under-ventilated smoke, gases and fire in the gas-phase. It is rare that all firefighters are able to practice with fires in the fuel-phase or post-flashover, for the costs of repeatedly burning large amounts of carbonaceous fuels. Some will utilise LPG gas -fired simulators but these only present a false representation of controlling large amounts of fire and do not serve to teach firefighters how to fight fires. In turn, these methods of training create a dangerous bias in the mind-set of operational firefighters who generalise primarily in applying more water to the gas-phase than the fuel base in the belief that this will achieve greater suppressive effect in all instances.

Research (reported in 1979-1984) from several full-scale ventilation controlled fire tests at Karlsruhe University (Fire Research Station) in Germany revealed some commonality during the overall extinguishing process, where 36 percent of applied water was seen to suppress active (flaming) combustion, with the remaining 64 percent cooling the fuel base surface fire. This was noted in the live fire tests and then validated using a complex mathematical model developed to support the test process. An undefined amount of applied water may be observed as 'run-off' at building fires and an amount of warm or hot water may remain on the floor or even flow out from the involved fire compartment, having already extracted much of the heat from the fuel base. It is this division in actual firefighting water absorptive capacity that may determine the true practical 'efficiency' of how firefighting water is actually applied.

Where fuel mass was lost at around 10.9 kg/min during a series of post-flashover fire tests heat release peaked at 3 MW in a room of 4 x 3m x 2.4m high, where the predicted energy release of 3 MJ/s (185 MJ/min) was only able to reach 92.5 MJ/min in the reaction zone due to ventilation limits (vent limited fire). The firefighters were able to control this fire in under two minutes using just 100 litres, with 35.7 litres extracting all the heat from the reaction zone at a rate of 2.59 MJ/litre (2.6 MJ/kg). The remaining 64.3 litres extracted around 246 MJ from the seat of the fire. With a fuel surface of about 40m^2, this represents an application of about 1.6 L/m^2 to cool the fuel base below its ignition temperature.

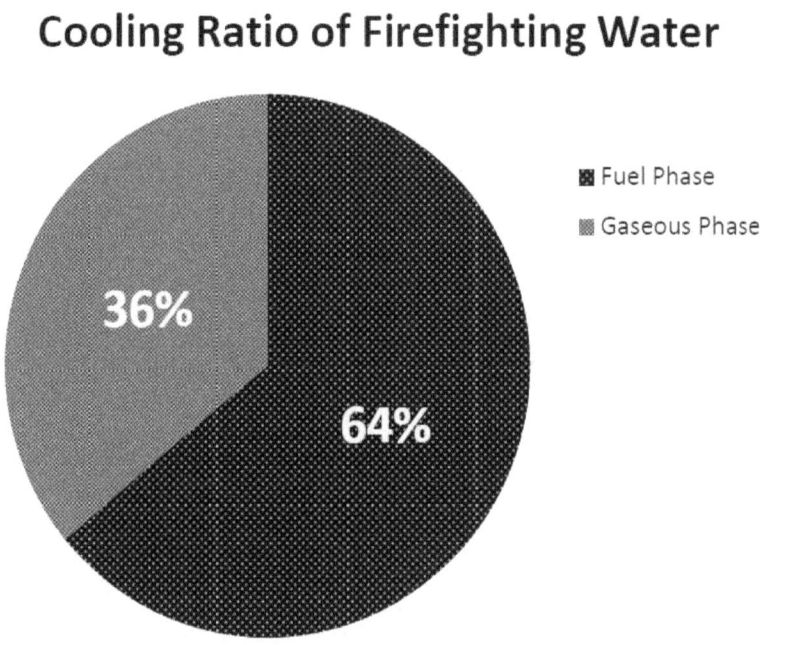

Figure 12.7: The ratio of cooling when using water applied into a fully-involved room fire is generically seen as 64%/36% fuel-phase/gaseous-phase, where neither approach is biased over the other.

In 1998, shortly following the deaths of three firefighters, UK firefighters began to receive 'fire behaviour driven' gas-phase suppression training, as taught in Sweden, beginning a twenty-year learning phase where the gas phase fire received far more attention than fire in the fuel phase. In retrospect and in reviewing the two fires where life losses occurred, it was certainly not a lack of knowledge of fire behaviour that killed those firefighters but more a lack of *tactical awareness*. Firefighting tactics play a greater part in firefighter life losses than fire behaviour knowledge, although admittedly the two are intrinsically linked.

If the reader believes that is not the case I suggest taking a close look at the prime casual factors in nearly all firefighter line of duty death (LODD) reports where:

- failure to apply early water on the fire base
- failure to effectively brief and debrief crews and act on information
- failure to communicate effectively with relevant parties on the fire-ground
- allowing moral pressure from bystanders to cloud situational awareness and prevent making logical tactical decisions
- disorganised self-deployments without directives (freelancing)
- failure to isolate fires by closing doors
- failure to carry out wind impact assessments prior to deployment
- mistimed and inappropriate tactical ventilation actions
- failures to maintain physical contact with crew members;

All of these clearly led the way in tactical omissions that contributed as prime causal factors in the majority of LODDs.

This is a most relevant point to bear in mind. It is also important to point out that the gas phase is generally of greater relevance to the firefighter in its *unburned state* as flammable fire gases will accumulate at the ceiling, in spaces and hidden voids. However, the fuel phase should always remain a primary concern when actually extinguishing a fire. Fire behaviour training should meet the objectives of practical fire dynamics by preparing firefighters to face the hazards of the gas phase in being able to identify dangerous conditions in a fire compartment or building. This training should also create situational awareness of the impact of venting a fire compartment or building. Finally, effective cooling of the gas layer to prevent a fire progressing to flashover should be taught.

At this point, there is some greater emphasis urgently needed in imparting adequate knowledge transfer of fire suppression in the fuel-phase. If this does not take place, then the bias in fire suppression will continue to prioritise gas phase over fuel base.

There are several examples of this:

1. At a high-rise fire in London the first crew in with a 45mm hose-line attempted to fight the fire by 'pulsing' water droplets into the gas layer but were unable to gain headway on an office fire involving a single work station. The actual work station was actually within reach of a high-flow smooth-bore stream applied from the stair but this was not available. The fire eventually spread to five floors and several firefighters had narrow escapes as they were overcome by the heat, as the fire rapidly developed.
2. At a residential high-rise fire in the UK, a team of firefighters attempting to reach trapped colleagues applied gas-phase flame cooling in an effort to make headway into the apartment. At this point one firefighter was believed to be still alive but the fire continued its development as the gas-phase attack took priority. A high-flow smooth-bore stream may have enabled progress to be made into the hallway. Tragically, there were two firefighter life losses at this incident.
3. Another residential high-rise fire south on the UK south coast saw a rescue team of firefighters trying to rescue four trapped firefighters by directing a 'pulsing' fog pattern into the burning gas layers when a high-flow straight stream pattern directed at the fuel base may have been far more effective. Two more firefighters tragically lost their lives at this incident.

It was engrained in the mind set of these firefighters through *fire behaviour training* to prioritise fire in the gas-phase over the fuel base fire during suppression operations. In both cases (2 and 3) the Coroner's investigations raised discussion points concerning the relevancy of training firefighters in this way.

> **What does this mean to the Firefighter?**
>
> Over a long period of time European firefighters have been taught quite rightly to be aware of smoke, heat and flaming combustion, particularly if transporting above their heads and behind them. This critical information can and does save lives. However, at the same time whilst working in fire behaviour training units, CFBTI's (compartment fire behaviour training instructors) have observed how the 1.5 Mega-Watt gas-phase training environment can be easily and most effectively controlled using very low flow-rates (even below 50 L/min). This has led to the development by some manufacturers to respond to a request for low flow 'fine droplet' hand-line nozzles.
>
> <u>This can be dangerous!</u>
>
> It will be demonstrated later where the heat release rates and speed of fire development that can be expected in small room to corridor fires, and larger open-plan compartments, can rapidly outpace low flow nozzles at an early stage, most particularly as the 'energy efficient' building fire environment evolves.
>
> Where a fire's fuel base is not immediately accessible or observed, shielded by walls, furniture or in voids, the gas phase becomes the priority and low-flow fine droplets will perform best. However, where the fire's fuel base may be reached by bouncing a mid-high flow stream off the ceiling, a wall, or more importantly with a direct application, then this approach should be used.
>
> Prioritising training and equipment and then procedural protocols, to bias one method of firefighting over the other (gas phase/fuel phase), is ineffective and potentially negates a risk based approach to interior firefighting.

Note: This evidence based guidance does not apply to the use of fog insertion tools from external locations to the fire compartment.

Extinguishing Efficiency

The theory of fire suppression has been studied and researched by practitioners, scientists and academics for several decades. However, in practice the efficiency of water as a heat sink is usually determined by the application technique, as water that fails to reach the seat of the fire cannot contribute to its ultimate extinguishment. In typical firefighting sprays, for example, only a small fraction of the relatively large droplets in the delivery will realise their maximum heat extraction potential through evaporation, while the majority will remain in the liquid phase and form runoff. Conversely, if the water is delivered in the form of very fine droplets with the aim of promoting rapid evaporation, the horizontal application of the spray may not possess the momentum required to penetrate the flame. The net result is that water is wasted and firefighting efficiency is compromised. If ultra-high-pressure (UHP) water droplets are used, they may have some greater cooling effect as the momentum will carry the finely divided droplets further.

So too is there generally some major water run-off when firefighting water is delivered directly onto a burning fuel base. Estimates in research have placed this efficiency of applied firefighting water at around 30-50 percent. That is, for every 100 litres applied, only 30-50 will take part in the suppressive and cooling phase, with the remainder possibly finding its way onto the floor and out of the structure. Researchers have broken this down to 35 percent efficiency when applied into the fuel base and 15 percent efficiency when applied into the gas-phase (total 50 percent). Research by Rasbash suggested primary efficiency factors that conform to later work by Barnett in producing a cooling efficiency factor.

> **What does this mean to the Firefighter?**
>
> As you apply water to a confined room fire, probably only half of the water from a firefighting stream will reach the fire and evaporate. The rest will 'run-off'. Where a fog pattern is applied effectively there will be less run off as most water will evaporate. However, getting a fog pattern to strike the base fire is not always easy or effective and a straight or solid stream is more often needed to penetrate into, and cool, the burning items. Pump operators must learn to optimise their water supply, working in flow-rate (L/min) as well as pressure in providing adequate water to the hose-lines in operation. Too much pressure at the nozzle may create a hose-line that is difficult to handle but too little water may place firefighters at great risk. It is important to get the balance right. Smooth-bore nozzles can flow more water for less pressure and should be considered in some circumstances, particularly where heavy fire is likely.

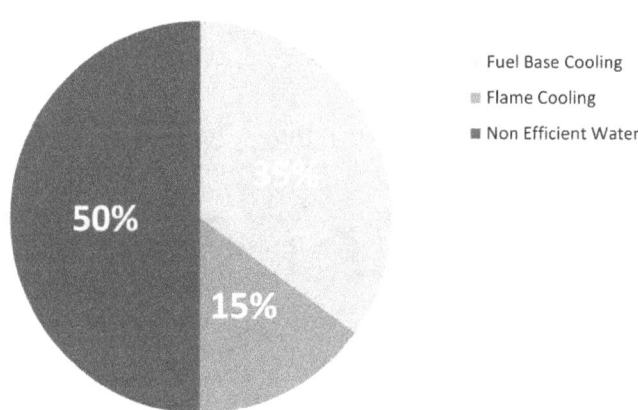

Figure 12.8: The widely accepted extinguishing efficiency of water applied into building fires.

The theoretical cooling effect of water can be calculated using specific heat capacity and latent heat values. Assume that when one litre of water is applied to a fire, it increases in temperature, turns to water vapour and then the water vapour increases in temperature until it reaches the temperature of the fire gases.

- To heat water from 10°C to 100°C, the energy input required is 90°C x 0.00418 MJ/kg °C = **0.38 MJ/kg**
- To vaporise water at 100°C requires **2.26 MJ/kg.**
- To heat the steam further requires an energy input that equals **(T - 100) x 0.002 [MJ/kg]** (specific heat of steam*), where T [°C] is the actual steam temperature

This means that to transform 1 kg (1 litre) of water at 10°C to steam at 600°C, an energy input of **(0.38 + 2.26) + (600-100 x 0.002) = 3.6** MJ is needed. The heat absorption capacity is therefore 3.6 MJ per kg of water, used to its maximum at 600°C.

Note*: *The specific heat of steam is 0.002 MJ/kg at 100°C and 0.006 MJ/kg at 300°C, so an average could be used, based on steam tables. If water were applied into a 300°C gas layer as opposed to 600°C, according to the above calculations an energy input of 3.04 MJ is needed. The heat absorption capacity would then be just 3 MW.*

Cliff Barnett's research 2004:
(A computer analysis based on data from Grimwood's 100 fire study in 1989)

It is also pointed out that the combustion efficiency of a confined fire almost never reaches 100 percent inside the compartment. Fig.12.9 illustrates the changes a building fire goes through as it grows in size. Barnett[135] informs us that 'when a small fire first ignites in a building, the air supply will be in excess and the airflow ratio will be between 2 and 3 (fuel control), at say near point B. As the fire grows in size, the relative airflow ratio falls to the point where the air supply is controlled by the size of the available openings (ventilation control). The airflow ratio decreases with increasing fire size from B to C to D and finally finishes up somewhere between D and E. Only for a small part of the growth will the fire efficiency be close to the suggested conservative design value of $k_F = 0.50$. After growth is completed, the fuel efficiency factor could well lie between E and D, or between 0.15 and 0.45 for the remainder of the fire duration'.

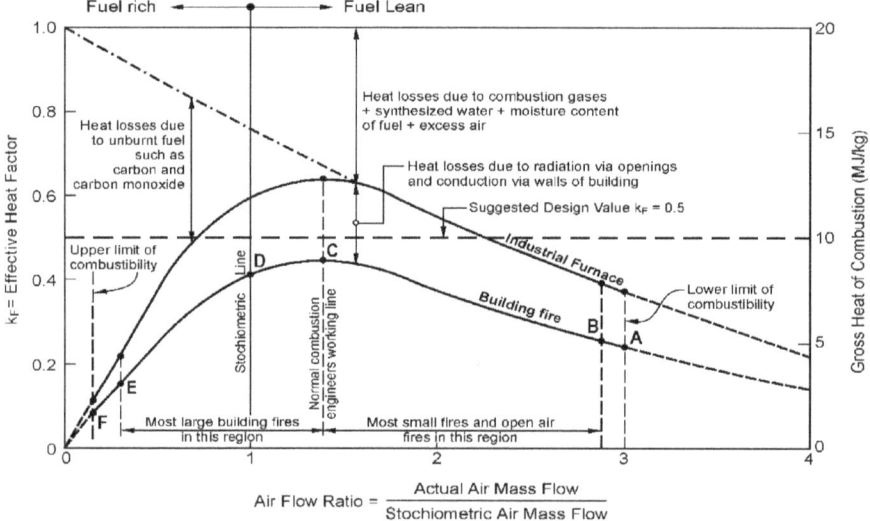

Figure 12.9: Combustion diagram for a typical air-dry wood fuel showing relationship between gross heat of combustion and air-flow. For a building fire, the effective heat coefficient k_F rarely exceeds 0.45 thus a k_F value of 0.50 can be considered a conservative and upper limiting value for design purposes. (**Reproduced from sfpe-TP2004-1, Barnett.C**)

135 Barnett.C, SFPE (NZ) Technical Publication – TP 2004/1

In combining the efficiencies of firefighting water applications with combustion efficiency Barnett uses the following formula:

$$F = \frac{k_F * Q_{max}}{k_w * Q_w} \quad Eq.12.1$$

> Where
> F   required firefighting water flow in l/s
> k_F heating efficiency of fire (conservatively 0.50)
> k_W practical cooling efficiency of applied water (conservatively 0.50)
> Q_{max} maximum heat output of fire in MW
> Q_W theoretical absorptive capacity of water at 100°C = 2.6 MW/l/s

Or alternatively a re-arrangement of Equation 12.1, to calculate the heat absorption (MW) of the water used for suppression:

$$Q_s = \frac{F * (k_w * Q_w)}{k_F} \quad Eq.12.2$$

> Where
> Qs   The heat absorption (MW) of the water used directly on the fire for suppression
> F   delivered firefighting water flow in l/s

In simple terms this means that for each MW of Q_{max} in a real fire, according to Barnett the firefighting water flow will need to be 0.50 / (0.50 x 2.6 MJ/kg) = 0.38 L/s/MW or 23 L/min/MW of Q_{max} (the point of peak heat release).

Paul Grimwood's Research[136] (based on 5,401 'working' Building Fires 2009–2012)
Glasgow Caledonian Fire Engineering PhD research 2015

Benchmark research by Rasbash (the 'fire point' theory), supported in later work by Beyler, demonstrated that for diffusion flames of dimension 5-60cm, removal of about 30-35% of the heat of combustion from within the reaction zone would cause the flame to be extinguished. This indicates that for diffusion flames in the combustion zone (gas-phase), the theoretical absorption capacity (3.6 MJ/kg) should be multiplied by a factor of 3.3 (3.6 x 3.3 = 11.9 MJ/kg) to obtain the extinction capacity. Therefore, in theory one L/s of water may absorb 11.9 MW of heat release in the combustion zone.

The following approach by Grimwood (Equation 12.3) demonstrates very close correlation with Barnett's equations given above and is further validated by the data outputs from the 5,401 building fires in the GCU research at. It is observed that by combining Rasbash's fire point theory with the full-scale fire tests undertaken by Karlsruhe University (above) – for a flow of 1 L/s the total **heat absorption** capacity (per Mega Watts) of firefighting water is –

---

136 Grimwood.P, A study of 5401 UK building fires 2009-2012 comparing firefighting water deployments against resulting building fire damage, Ph.D thesis 2015, School of Engineering and Built Environment, Fire Risk Engineering, Glasgow Caledonian University Scotland, United Kingdom 2015

MW (heat absorbtion) = (W_{gas} * E_{gas} * 3.3 * L/s * 0.15) + (W_{fuel} * E_{fuel} * L/s * 0.35)   Eq.12.3

Where

W_{gas} = Percentage of water applied into the gas phase
W_{fuel} = Percentage of water applied into the fuel phase
E_{fuel} = An energy input of 2.6 MJ/kg is required to vaporise 1 kg of water at 100°C when applied onto the fuel-phase
E_{gas} = An energy input of 3.6 MJ/kg is required to transform 1 kg of water at 100°C to steam at 600°C when applied into the gas-phase – *If water were applied into a 300°C gas layer as opposed to 600°C, according to the above calculations a lesser energy input of 3.04 MJ is needed. The heat absorption capacity would then be just 3 MW.*
3.3 = An efficiency factor based on the Rasbash 'firepoint' theory
0.15 = The practical efficiency factor of water applied in the gas phase   } **50% Total**
0.35 = The practical efficiency factor of water applied in the fuel phase   } **efficiency**
L/s = The quantity of water applied into the fuel or gas phase in Litres/sec.

When advancing hose-lines internally firefighters will utilise a range of fire stream applications and techniques that include narrow and wide angled spray patterns to protect themselves from heat and smoke and direct combustion products away from their position. Locating the fire base in limited visibility and accessing the fire when it is shielded also create difficulties. Short bursts of water-fog may be used to suppress fire and cool heat in the hot gas layer and then eventually water will be directed at the fire's base to complete suppression. It is therefore virtually impossible to apply water in an optimal way during real firefighting operations and this means that one litre of water will rarely achieve its theoretical cooling potential.

There have been several attempts to generate a range of formulae by which generic firefighting water demands at building fires can be calculated, based on a theoretical analysis of water's cooling properties, or through empirical research of real fire data. In some models, both of these approaches have been combined.

However, these models or building design guides often promote a wide variance in recommended firefighting flow-rates needed to control and extinguish fires. In some of the models the methodology may be open to question where source input data is derived from small scale laboratory fires, or where specific firefighting techniques limit the use of resulting formulae. In other cases, the source data is inappropriate or the methodology itself is incomplete, or is structured without any in-depth validation.

Many of these models base their formulae on a 0.3 water application efficiency factor which supports laboratory or experimental research, whereas others utilise empirical fire flow data obtained from actual building fires. The methodology below is based on a 0.5 water application efficiency factor (k_W at 50 percent effective), derived from an existing computer analysis of 100 building fires in London (1989)[137] and further validated by the GCU County/Metro research of over 5000 UK building fires occurring between 2009-2012.

In terms of application efficiency, Grimwood can attest to a range of situations and variable dynamics affecting flow-rate performance where access to the fire base was difficult or the distribution of the fuel load led to more energy being released into the gas-phase. In some instances firefighters would apply large amounts of water into smoke because they were guided by sounds as to where the base of a fire is likely to be. In these situations the water run-off was high and the efficiency was low. On other occasions the 'air' appeared to be burning, visibility was good and spray patterns of water droplets were directed into the burning gas-

[137] Grimwood. P, Fog Attack, FMJ Publications, Redhill Surrey, UK 1992

phase with great efficiency and less run-off. In some stair-shaft fires where the base and head of the stairs were vented, the chimney effect was ideal for approaching the fire from below. Using short bursts of finely divided droplets in a spray pattern, the natural buoyancy of the fire transported the water droplets up higher into the stairs and vast quantities of steam exited the high-level vent, extinguishing fire all the way. In such cases the water run-off was almost zero and the efficiency was close to 100%. However, a generic approach is used to calculating water application efficiency overall.

The GCU research was very pragmatic in its approach to estimating the actual amounts of firefighting water used to control building fires but recognised the need to establish a theoretical foundation in support of the data.

Calculating 'adequate' firefighting water flow-rate (Grimwood Eq.12.3)

- According to Grimwood (Eq.12.3), the ('real fire') heat absorption of water at 1 Litre/second = 2.46 MW per L/s – (0.407 L/s/per MW at peak HRR)
- Or; 24.42 L/min/MW of Qmax (the point of peak heat release) is required to achieve control and suppression of building fires, which closely correlates with Barnett's calculation of 23 L/min/MW.

Flame Suppression (36%)   0.36 x 3.6 x 3.3 x 1 x 0.15   = 0.64 MW

Fuel Base (64%)   0.64 x 2.6 x 1 x 0.35   = 0.58 MW

Total	= **1.23 MW**
Q_s	= 1.23 / 0.5(k_F)
	= **2.46 MW** *per L/s*
	= 1 / 2.46
	= **0.407 L/s/MW**

Note: Q_s = the heat absorption (MW) of the water used directly on the fire)
K_f = the combustion efficiency taken as 0.5 (50 percent efficiency)

Example One:
19mm HP Hose-reel tubing x 54m at 1.83 L/s
Flame suppression 0.36% x 3.6 MJ/kg x 3.3 x 1.83 L/s x 0.15   = 1.18 MW
Fuel base cooling 0.64% x 2.6 MJ/kg x 9.16 L/s x 0.35   = 1.06 MW

Total	= 2.24 MW
Q_s	= **2.24/0.5(k_F)**
Total heat absorption capacity	= **4.48 MW**

Example Two:
Smooth-bore 22mm nozzle at 9.16 L/s
Flame suppression 0.36% x 3.6 MJ/kg x 3.3 x 9.16 L/s x 0.15   = 5.93 MW
Fuel base cooling 0.64% x 2.6 MJ/kg x 9.16 L/s x 0.35   = 5.33 MW

Total	= 11.26 MW
Q_s	= 11.26/0.5(k_F)
Total heat absorption capacity	= **22.5 MW**

What does this mean to the Firefighter?

All these calculations lead to one conclusion. There is always going to be a large amount of water from your firefighting stream that ends up on the floor, particularly if the fire is predominantly involving dense fire loads, furniture, items and heavy solid combustibles. Where the water is applied into large amounts of gaseous flaming combustion, the drop-out may be far less, depending on droplet size and velocity as it exits the nozzle. In general, we apply a fifty percent overall efficiency factor to your applied firefighting water stream, which means half the water applied goes to waste.

Couple this with a calculated heat absorption capacity (Mega-watts or MW) and we can see how effective your hose-line flow-rate is likely to be on a certain sized fire:

Nozzle	Water Flow-rate (L/s)	Water Flow-rate (L/min)	Heat Absorption Capacity (MW)
Smooth-bore 14mm	4.83	290	11.9
Smooth-bore 22mm	9.16	550	22.5
Smooth-bore 22mm	15	900	36.9
Automatic Combination Nozzle	5.83	350	14.3
Automatic Combination Nozzle	1.66	100	4.0
Fog Nail	1.2	72	2.9
LP Fog Nozzle <1mm water droplets	5.0	300	12.3
LP Fog Nozzle <1mm water droplets	7.91	475	19.4
19mm HP Hose-reel tubing x 54m**	0.75	45	1.84
19mm HP hose-reel tubing x 54m**	1.83	110	4.5
22mm HP hose-reel tubing x 54m**	3.3	200	8.1
25mm HP hose-reel tubing x 54m**	4.16	250	10.2

Table 12.2: Minimum needed flow-rates for suppression based on heat release rate (HRR) at **2.46 MW per L/s** (ref: Grimwood Eq. 12.3).

** The use of UHP pumps may increase stream velocity, reduce water droplet size, and can increase suppressive capacity in the gas-phase by up to 3.5 times, when compared to low pressure streams at the same flow. However, where the fire growth rate is rapid, or where the fire load is heavy (or the structure becomes the fire load), the low flow-rate may then expose firefighters to dangerous levels of thermal exposure

An important point to note is, that even though water applied into the flaming gaseous phase might appear an effective use of the limited amount of water

> available at the nozzle, its low overall efficiency of just 15 percent is founded in the fact that although far less water is required to cool flaming fire gas layers, somewhere between two and six times more water will be needed at that same nozzle to finish the job by controlling the fuel base fire. Therefore, in practice, a 40 L/min nozzle that is effective in the gas-phase of a particular sized fire is likely to require at least 80-250 L/min in order to deal safely and effectively with the fuel base fire.

12.4 THE AMOUNT OF FIREFIGHTING WATER USED BY FIREFIGHTERS

A detailed analysis of 100 serious building fires in London was undertaken by the author in 1989 where, as a London Fire Brigade fire investigator, attendance was made at each fire to record the extent of fire damage and the flow-rates needed to control the fires. The primary objective of this research was in support of finding a practical method that fire commanders could use at the scene of fires to estimate the quantities of water needed to control fire spread. The outcome from this limited research demonstrated that target flow-rates of 4 L/min/m² of fire involvement (fire damage) would normally suffice for building fires where the fire was vent controlled and vent openings were limited. Where fires were well vented and approaching fuel-controlled conditions, a minimum flow-rate of 6 L/min/m² was recommended. Although this research covered a limited number of working fires in London, the recommended flow-rates were very close indeed to later research, again carried out by the author between 2009-2012, where 5,401 fires were again studied for the amounts of firefighting water used by firefighters to control developing fires in buildings, where the fires were so serious that breathing apparatus was required.

This research data was drawn from fires in two fire brigade areas, one a county brigade of 3,544 km² and a Population Density of 413 people/km² where 1,152 working Fires occurred over a three-year period; and a city metro brigade of 1,276 km² with a Population Density of 2,105 people/km² that had 4,249 working Fires over the same period. To achieve greater accuracy, much of the data was retrieved direct from the fire-ground to computer terminals monitoring actual flows used with water-smart technology vehicle pump flow-meters.

These fires involved all building types ranging from residential dwellings, high-rise apartments, hotels, shops, restaurants, schools and care homes to very large industrial units and warehouses. Six hundred of these fires (11 percent of the total) involved fire damage exceeding 50 square metres and a further 1.3 percent of fires involved fire damaged floor areas exceeding 500 square metres. The average fire sizes ranged between 16m² of floor area for residential dwellings up to 95m² for industrial units, with all other building and occupancy types somewhere in between. For these fires, firefighting water flow-rate densities ranging between 8-12 L/min/m² of fire involvement were used to achieve control. In total, 65% of all building fires in the GCU research (86% of fires that just used high-pressure hose-reels) were extinguished using just the 1800 litres carried on-board.

Using data from the UK National Incident Recording System (IRS) the data for 5,401 building fires occurring in two UK fire authorities between 2009–2012 are presented below under a list of 'purpose groups', according to the UK building regulations (table D1 2007 version). The data is also listed under 'County' and Metro' fire service areas for comparison.

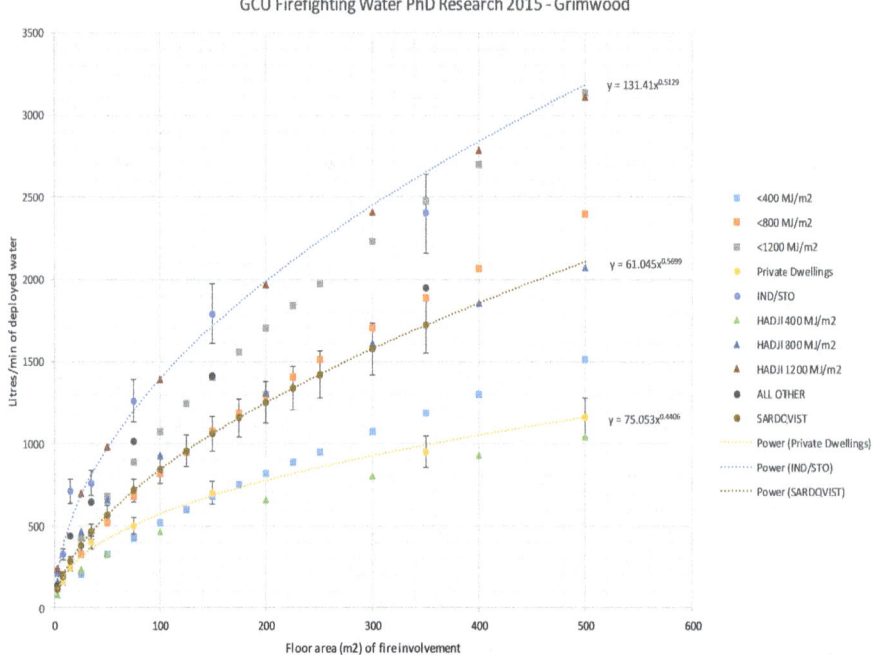

Figure 12.10: 10% error bars are added to the author's GCU research for consideration. Also shown in the chart are the firefighting water flow-rate data taken from other notable research by Cliff Barnett (squares), Stefan Sardqvist, and Hadjisophocleous, G.V. & Richardson, J.K. (triangles) (at 50% applied water efficiency).

The building fires were all considered 'working fires' where a recordable area of fire damage occurred and water was flowed internally by firefighters wearing breathing apparatus. The research did not include exterior fires, exterior roof fires, derelicts buildings or chimney fires. The data is validated in part by water-smart flow meter technology that records actual flows used by pumping appliances (fire engines) at fires.

The purpose groups listed are as follows:

- **Dwellings** (private houses, apartments and flats)
- **Residential (institutional)** (Hospitals; Homes; schools; prisons)
- **Residential (other)** (Hotels; colleges; hostels; halls of residence)
- **Assembly & Recreation** (Public entertainment; conference; museums; churches; law courts; health centres; day centres; clinics; passenger terminals)
- **Offices**
- **Shop & Commercial** (retail)
- **Industrial** (Factories)
- **Storage and other non-residential** (Storage warehouses and car parks)

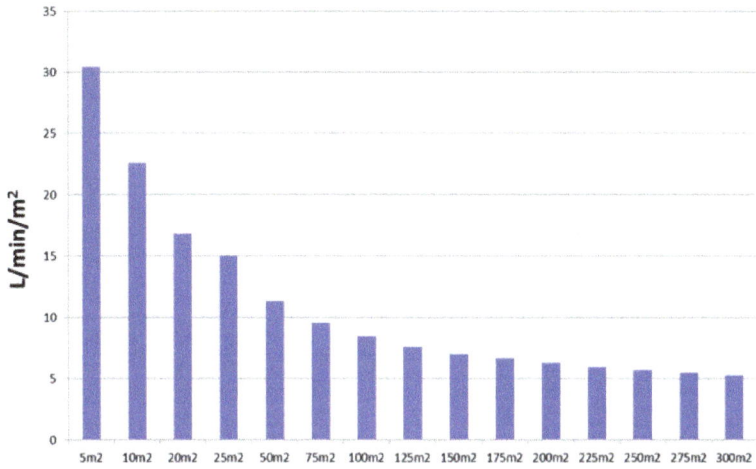

Figure 12.11: Typical flow-rate densities recorded for fires in occupancies demonstrating medium fire loads

12.5 BUILDING FIRE DAMAGE AND FIREFIGHTING WATER FLOW-RATE – THE LINK

The GCU data demonstrated average fire sizes of 38.78m^2 in County and 30.33m^2 in Metro areas and showed that fires in the Metro area were less likely to spread from the room or floor of origin and were less likely to result in multi-floor fires.

In comparison to UK fire spread statistics (2001-2011) from the Department of Communities and Local Government (DCLG), this research data of 5,401 working fires (2009-2012) presents a different picture from the DCLG representation of fire containment where it is seen that 86 percent of fires are contained within compartment of fire origin. According to the County/Metro research, when a building fire requires firefighters to deploy hose-lines and wear breathing apparatus, only 53 percent of fires are confined to the compartment of fire origin and fire spread to other floors within a building occurred at a quarter of fires in the Metro area and more than a third of the County fires.

It was noted that once fires had burned beyond 500m^2 of floor space, the buildings were generally lost completely. Building fires in this research were also broadly seen to spread beyond the compartment of origin more often (48%) than annual government statistical reports would suggest (8%), as these were all 'working' fires' and not simply just 'calls' to fires.

It was a further observation of the research that the building fire damage (fire containment) in the metro fire area was less than in the county area. This was interesting as both fire services applied exactly the same total firefighting flow-rate density at 12 L/min/m^2 of fire damage across the total number of fires. However, the metro fire service delivered a higher flow-rate during the very early stages following arrival on-scene than the county fire service did. This caused the county fire service to apply greater amounts of water during the later stages of firefighting to achieve control. It is suggested that this is why the building fire damage was greater in the county area. It appears that greater amounts of resources, staffing and flow capacity is needed later in a firefighting operation if the water flow-rate is inadequate from the outset.

The County fire service deployed nearly twice as many lay-flat attack hose-lines later in the fires to deliver the same final quantity of water as Metro, requiring greater resources overall.

Glasgow Caledonian University Research of 5,332 UK Building Fires 2009–2012

COUNTY	Occupancy	Incidents	L/min/m²	Average Fire Size m²	ROOM %	FLOOR %	MULTI-FLOOR %	19mm HP H/REELS %
	Dwellings	839	13.84	16.82	61.74	16.69	21.57	92.37
	RESI INST	24	8.34	36.7	70.83	8.33	20.83	84.5
	RESI OTHER	43	8.92	29.3	39.53	18.6	41.86	81.4
	ASSEMBLY	72	11.93	33.38	49.32	16.44	32.88	82.33
	OFFICES	23	19.54	15.13	60.87	4.35	34.78	81.96
	SHOPS	107	17.55	22.86	56.07	17.76	26.17	77.31
	INDUSTRIAL	35	9.59	62.68	54.29	8.57	37.14	50.29
	STORAGE	9	6.3	93.44	11.1	22.2	66.67	33
	TOTALS	**1152**	**12.00**	**38.78**	**50.46**	**14.11**	**35.23**	**72.89**

METRO	Occupancy	Incidents	L/min/m²	Average Fire Size m²	ROOM %	FLOOR %	MULTI-FLOOR %	19mm HP H/REELS %
	Dwellings	2939	12.18	15.72	67.68	15.75	16.57	93.81
	RESI INST	27	18.71	6.92	74.07	18.52	7.41	92.88
	RESI OTHER	72	18.74	12.59	54.17	11.11	34.72	87.83
	ASSEMBLY	300	12.22	19.24	64.67	17.33	18	89.67
	OFFICES	66	7.96	39.77	54.55	10.61	34.85	89.79
	SHOPS	459	10.48	30.17	44.66	27.45	27.67	84.39
	INDUSTRIAL	288	9.61	51.00	46.53	26.39	27.08	59.39
	STORAGE	98	9.72	67.27	42.86	31.63	25.51	45.9
	TOTALS	**4249**	**12.45**	**30.33**	**56.14**	**19.84**	**23.97**	**80.45**

Table 12.3: The response data from two UK fire services (County & Metro) were analysed over a three-year period for building fires where breathing apparatus was used and water was flowed to achieve fire control. All fires in the table involved less than 500m² of fire area – there were 69 further fires in the research that were larger than 500m².

Key	
Group	Purpose Groups according to the UK building regulations
m²	area of fire damaged floor space in square metres
L/min	Litres/minute flow-rate as deployed to achieve control/suppression
Room	percentage of 'working' building fires confined to room of origin
Floor	percentage of 'working' building fires confined to floor of origin
Multi-floor	percentage of 'working' building fires spread beyond floor of origin
H/reels	percentage of fires dealt with using 60m x 19mm rubber tubing first-strike firefighting hose @ 100 LPM

Note: Hose-reel data displayed separately – overall 65% of **total fires** were dealt with using tank water without augmenting the supply (87% of **hose-reel fires** used tank water only) 13% of fires required additional firefighting water augmented from other sources.

During this period, there were 70 additional large building fires involving 500 – 10,000 sq/m of fire damage

> **What does this mean to the Firefighter?**
>
> It pays to pre-plan the firefighting water flows that can be immediately available to firefighters on arrival and therefore 'adequate' firefighting water becomes a key factor in the entire response and deployment model. It has also become clear that it is often those in a position to harness such changes within the fire service response model who fail to understand or appreciate the technical aspects associated in meeting an adequate flow-rate provision (L/min) and matching this against the building risk profile (fire load) in the area served.
>
> The evidence show, quite simply, that less water applied from the outset of operations usually means more water, hose, staffing and resources are needed later on in an incident. Add to this the suggestion that building fire damage increases at 48 percent of working fires, as fires spread beyond the room and even floor of origin, wherever first response water deployments are inadequate.

The data also demonstrated that fires in the Metro area were less likely to spread from the room or floor of origin and were less likely to result in multi-floor fires. In comparison to UK fire spread statistics (2001–2011) from the Department of Communities and Local Government (DCLG), this research data of 5,401 working fires (2009–2012) presents a different picture from the DCLG representation of fire containment where it is seen that 86 percent of fires are contained within compartment of fire origin. According to the GCU research, when a building fire requires firefighters to deploy hose-lines and wear breathing apparatus, only 53 percent of fires are confined to the compartment of fire origin. It was further noted that fire spread to other floors within a building at a quarter of fires in the Metro area and at more than a third of the County fires.

The primary means for distribution of delivered water flow-rate (Fig.12.15) varied between the two fire services in the research where the county utilised main lay-flat hoses to deliver 57 percent of their firefighting water but only 43 percent when using high-pressure 19mm hose-reels. In contrast, the metro used less lay-flat hose but more hose-reels to deliver the same average quantity of water density as the county across all building fires. This research data from GCU suggests that applying greater flows (L/min/m^2) from the outset may lead to less building damage and a reduced staffing/resource requirement during the latter stages of firefighting.

Continued overleaf.

	County	%	Metro	%
Fire Incidents	1116		4573	
Total reels & jets laid	1822		6459	
Jets/reels	466/1356		1069/5390	
Fires just Hose-reels	813		3682	
Reels laid	915		5058	
Reels off tank supply	746@702 fires	86.35	4170@3205 fires	87.05
Reels Augmented	164@107 fires	13.16	888 @477 fires	12.95
Just Jets	78 Jets @ 35 fires		745 Jets @366 fires	

County	Metro
1116 fires/1822 hose-lays – 1.63 per fire	4573 fires/6459 hose-lays – 1.41 per fire
1116 fires/1356 reels laid – 1.21 reels per fire	4573 fires/5390 reels laid – 1.17 reels per fire
1116 fires/466 jets laid – 1 jet per 2.4 fires	4573 fires/1069 jets laid – 1 jet per 4.27 fires
Just reels dealt with 72.85% fires	Just reels dealt with 80.5% fires
Just reels – 1.12 reels laid per fire	Just reels – 1.37 reels laid per fire
1.06 reels off tank supply only – per fire	1.3 reels off tank supply only – per fire
1.53 reels of augmented supply – per fire	1.86 reels of augmented supply – per fire
At 3.14% fires 'just jets' were used	At 8% fires 'just jets' were used

813 FIRES JUST REELS		3682 FIRES JUST REELS
7.72m^2	Average Fire Size	11.14m^2
3.2MW	Estimated HRR @ 15–20% vent – one door	3.9MW
14.73 litres/min/m^2		12.36 litres/min/m^2
Room of origin – 591 (72.7%) Floor of origin – 118 (14.5%) Spread to other floors – 114 (14%)	Fire Confinement and Fire Spread	Room of origin – 2621 (71.2%) Floor of origin – 601 (16.3%) Spread to other floors – 460 (12.5%)

Table 12.4: Fire suppression data – GCU research (includes fires in derelict buildings)

12.6 OTHER MOST RECENT RESEARCH INTO FIREFIGHTING WATER FLOW-RATES

Fire Protection Research Foundation (NFPA)
A review of nineteen international firefighting water flow methodologies 2014

A review of nineteen international fire flow methodologies, including the author's 1989

Adequate 'Firefighting Water' – you need this! • 255

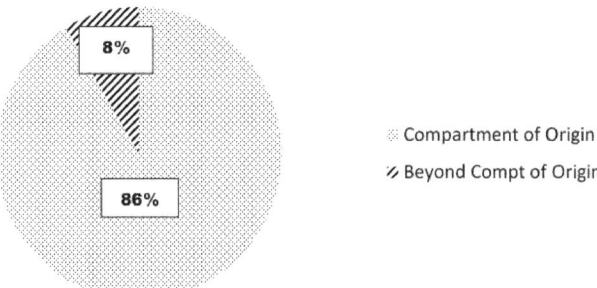

Figure 12.12: Nationally recorded (UK) levels of fire spread or fire containment for all building fires (Note: 6% of these fires spread beyond the building of origin) (Many of these fires were recorded in the national incident recording system as 'no action' by fire service) **Source: Table 4 of UK Fire Statistics (DCLG) 2001-2011.**

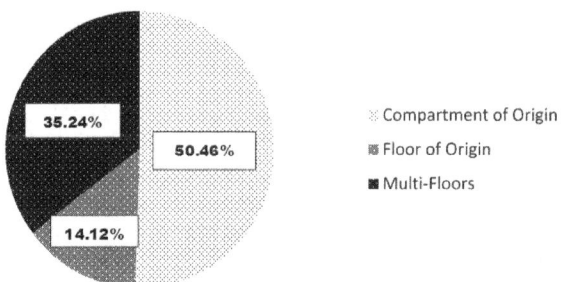

Figure 12.13: Levels of fire spread or fire containment in the County Area (1,152 working fires)

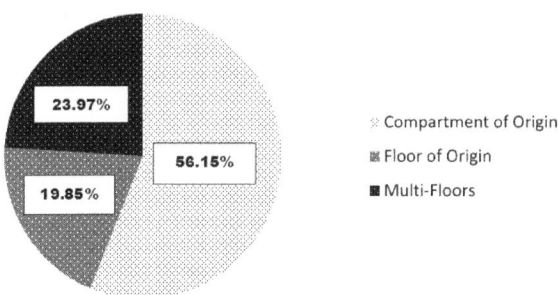

Figure 12.14: Levels of fire spread or fire containment in the Metro Area (4,249 working fires)

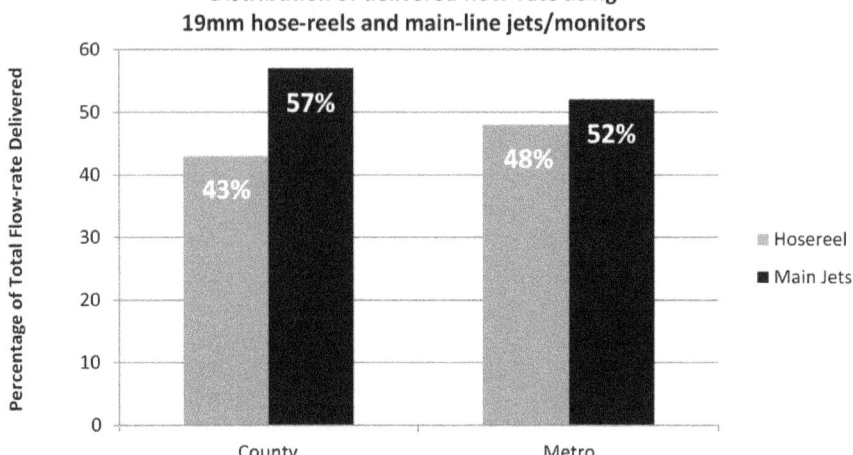

Figure 12.15: How the two fire brigades (County and Metro) delivered their firefighting water in the GCU 5,401 building fires (hose-reel versus main jets).

research (described as the 3D method), was undertaken on behalf of the NFPA's research foundation in 2014. The review demonstrated a wide variance in calculated fire flow requirements. This was visibly apparent, particularly where calculation processes were aimed at (a) infrastructure planning, (b) building design or (c) 'easy to use' on-scene firefighting calculation in support of fire service intervention. Planning methodologies supporting piped hydrant grids based on applied risk profiles resulted in the highest flow demands, followed by individual building design needs and then on-scene fire attack methods.

Of the sixteen methods used for comparative purpose, only four utilised the heat release rate as a means of assessing needed flow-rates and only three methods linked an *efficiency factor* of applied water. The SFPE (NZ) TP 2004/1 (Barnett) methodology (detailed at 12.3 above) was the only one to link both HRR, and water application efficiency factors derived from empirical research.

Stefan Sardqvist's Research 1998[138]:
(A study of Research into 307 Non-residential Fires in London 1994-1997)

Stefan Sardqvist undertook some detailed research into 307 fires in non-residential buildings in London from 1994 to 1997 and this research has since been widely referenced for flow-rate input data in subsequent firefighting water models globally. The fires in this study were relatively limited in size where just seven of the incidents exceeded 100m^2 in fire damaged floor area. The data refers solely to fires in public and commercial premises, schools and hospitals, industrial premises, and hotels and boarding houses. The data suggested that the upper limits of fire service intervention appeared to be about 50m^2 for a two-pump incident (2 fire engines) and 100m^2 for a four-pump incident (4 fire engines). The resulting Sardqvist formula to calculate needed flow-rate produces quite reasonable correlations for the applicable range of buildings when compared to the GCU County/Metro research in a single line flow-curve.

---

138 Särdqvist. S; Real Fire Data; Fires in non-residential premises in London 1994-1997; Report 7003 Lund University

Hadjisophocleous & Richardson 2005[139]
A validated methodology for calculating adequate flow-rate

A further theoretical methodology that appears to enhance the existing Canadian NRC FIERA model is presented by **Hadjisophocleous & Richardson** and allows the fire engineer to input any particular chosen extinguishing efficiency factor (kw) in the calculation process, (as does Barnett's). However, when Barnett's 0.5 extinguishing efficiency factor is used as an input (kw) to Eq.12.5, the outputs from this method compare well with that proposed by Barnett and also show reasonable correlation with the GCU County/Metro research real fire data, although not quite so where the floor plate is large.

In the development of Eq.12.5 the 50 percent combustion efficiency input proposed by Barnett is already included. Also, included in Eq.12.5 is work by P H Thomas in 1959 (Eq. 12.4) where an attempt to draw a relationship between horizontal fire area and time to fire extinguishment is integrated within the calculation process –

$$t_{ext} = 3.3 * \sqrt{A_{fire}} \qquad Eq.12.4$$

Where:

- t_{ext} time to extinguish the fire (mins)
- A_{fire} Area of fire involvement (m2)
- 3.3   A coefficient based on the use of what is deemed adequate water (L/min/m^2)

The formula proposed for 'offensive' firefighting is therefore:

$$Fire_{off} = \frac{0.058 * q_k * \sqrt{A_f}}{K_w} \qquad Eq.12.5$$

Where:

- $Fire_{off}$ Required water flow-rate to achieve extinguishment (L/min)
- k_W practical cooling efficiency of applied water (user's choice)
- q_k fire load density for the compartment (MJ/m^2)
- A_f Total internal floor area of undivided compartment (m^2)

Note: 'Offensive firefighting' here is defined as the process of extinguishing a building fire generally from interior positions.

---

[139] Hadjisophocleous G.V. and *Richardson J.K.; Water Flow Demands for Firefighting; Fire Technology*, 41, 173–191, 2005

12.7 FIRE PROTECTION ENGINEER'S DESIGN GUIDANCE[140]

Prescriptive guidance in the UK

The existing guidance in the UK for providing firefighting water supplies to buildings and residential/industrial estates is as follows[141]:

	Occupancy	Minimum flow-rate
1	**Housing** to 2 Levels	480 L/min
	Housing above 2 levels	1200–2100 L/min
2	**Lorry/Coach Parks; Multi-Storey car parks or Service Stations**	1500 L/min
3	**Industrial** to 10,000 m^2	1200 L/min
	Industrial to 20,000 m^2	2100 L/min
	Industrial to 30,000 m^2	3000 L/min
	Industrial over 30,000 m^2	4500 L/min
4	**Shopping; Offices; Recreation and Tourism**	1200–4500 L/min
5	**Education; Health; and Community Facilities**	
	Village halls	900 L/min
	Primary schools and single storey health centres	1200 L/min
	Secondary schools; colleges; large health & Community centres	2100 L/min

Table 12.5: UK Water/LGA/CFOA approved water provisions guidance

The above guidelines are only performance based as far as being based on research data provided in the 1950s-1970s, calculated on older traditional construction and fire loads. The existing prescriptive criteria in the UK for rising main and water storage provisions in tall buildings are as follows:

1	Fire mains should have a **minimum** nominal bore of 100 mm and the system should be designed to withstand a pressure of one and half times its predicted maximum operating pressure (BS 9990:2015).
2	At least two rising fire mains are to be provided where floor area exceeds 900m^2. (Building Regulations (Fire Safety) part B)
3	A **minimum** flow-rate from each wet riser of 1.67 L/min/m^2 (1500/900m^2) must be achievable (calculated from BS 9990:2015 and Building Regulations (Fire Safety) part B).
4	A maximum hose-lay distance (45 or 60m – depending on the existence of a firefighting shaft and sprinklers) is stipulated from each fire main, according to BS 9999:2008 and BS 9991:2011.

Continued overleaf.

---

140 A BS 7974 flow-rate design calculator tool, with added sprinkler compensations, can be downloaded from www.eurofirefighter.com

141 National guidance document on the provision of water for firefighting; UK Water & Local Government Association; with approval from the UK Chief Fire Officers Association. 2007

5	**Building Regulations (Fire Safety) part B; Section 15:** Where a building, which has a compartment of 280m² or more in area, is being erected more than 100m from an existing fire-hydrant, additional hydrants should be provided as follows; • **Buildings provided with fire mains** – hydrants should be provided within 90m of dry fire main inlets. • **Buildings not provided with fire mains** – hydrants should be provided within 90m of an entry point to the building and not more than 90m apart. • Each fire hydrant should be clearly indicated by a plate, affixed nearby in a conspicuous position, in accordance with BS 3251 Where no piped water supply is available, or there is insufficient pressure **and flow**** in the water main, or an alternative arrangement is proposed, the alternative source of supply should be provided in accordance with the following recommendations: • A charged static water tank of at least 45,000 litre capacity; or • A spring, river, canal or pond capable of providing or storing at least 45,000 litres of water at all times of the year, to which access, space and a hard standing are available for a pumping appliance; or • Any other means of providing a water supply for firefighting operations considered appropriate by the fire and rescue authority. **To achieve 'adequate' firefighting water (flow) a dry fire main is expected to achieve 7 bars at the highest outlet (50 metres) if pumped to 12 bars at the inlet, or 4 bars at the highest outlet if only pumped to 10 bars at the inlet (60 metres) (pre-2015 code). The quantity of water needed is not prescribed in the codes but an adequate amount for 2-3 firefighting hose-lines is considered to be between 1,000 and 1,500 L/min. The required flow-provision for residential premises may be less than that required for offices or commercial premises.
6	Pressure-reducing valves (wet-rising fire mains) should be provided to regulate the flow and pressure to (750 ±75) L/min at (8 ±0.5) bar per outlet (BS 9990:2015).
7	Where more than one wet fire main is installed in a building, **the potential need for additional water storage and/or pumping capacity should be taken into account** (Building Regulations (Fire Safety) part B).

Table 12.6: Prescriptive requirements for rising fire mains and firefighting water storage in the UK

The final point in Table 12.6 (7) suggests that for each additional fire main above 900 m² floor area, the additional water required should be considered, based on an 'adequate' provision for the purposes of extinguishing a developing or fully developed fire. A previous UK building regulation (1991 ADB) recommended two fire mains to every 2,000m² of floor area and an additional fire main was required after 2,000m² for each additional 1,500m², or part thereof. So, for a 3,500m² floor area above 60m in height, a flow provision of 1.29 L/min/m² was required. However, no consideration was given to occupancy type or compartment dimensions and the same provisions were applicable to small apartments, as well as large open-plan office floors.

Modern performance based codes in the UK

The GCU research has clearly established that the severity of building fires is dependent on several factors (listed below). It is further demonstrated that the quantity of firefighting water that may be deemed as 'adequate' may be effectively linked to the estimated ongoing heat release rate for any specific occupancy fire and can be presented on a gradient, ranging from the lesser amount (L/min) required in residential buildings (low-flow), upwards through offices or commercial mixed use buildings (mid-flow), to industrial and storage facilities (high-flow). However, this finding is not reflected in current design codes.

It is demonstrated that **adequate** firefighting water provisions are dependent on six main factors:

- Occupancy type (residential fires require far less water than industrial/storage occupancies, or even office buildings)
- Fire load energy density (MJ/m²)
- Floor area (m²)
- Existing passive or active fire protection measures
- Potential vent openings combined with floor space ratios (A_v/A_f)
- An adequate water source being available in the first place

The actual flow-rates as used by firefighters to control over 5,000 serious building fires in the UK (2009-2012), in a wide range of occupancy types, is used to form equations for building and infrastructure design purposes. The formulae resulting from the Glasgow Caledonian University (GCU) research for non-sprinklered buildings are included in BS PD 7974:5:2014 and can be used to calculate the required flow (L/min), by applying flow-rate density per m² of floor area (L/min/m²), for various occupancies based on the above six factors.

The following two charts demonstrate the flow-rate density outputs from the GCU research for building fires below 500m² in floor area (Fig.12.16) and large fires involving 500-10,000m² (Fig.12.17). It is this data from the firefighting flows used by the fire service to control and extinguish such building fires that was used to derive equations for performance based design purposes.

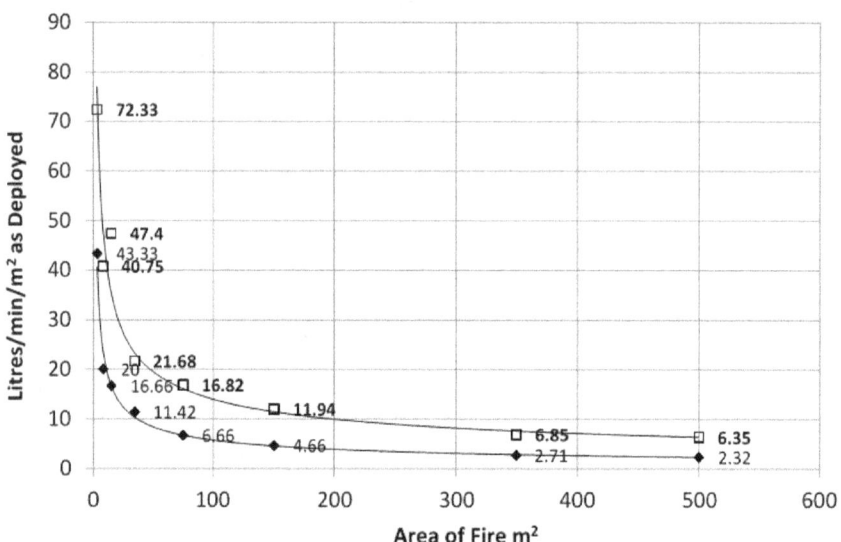

- Private Dwellings $75*A_f^{-0.56}$

- IND/STO $131*A_f^{-0.487}$

Figure 12.16: County/Metro fire fire-flow research of 5,401 building Fires 2009-2012 showing deployed water flow-rates (L/min/m²) for Private Dwellings and Industrial/Storage (INDO/STO) building fires less than 500m² fire damage.

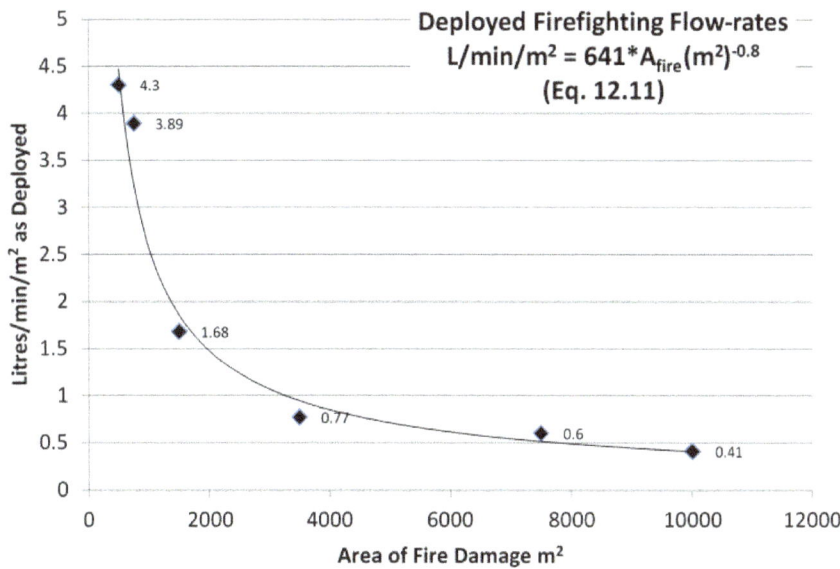

Figure 12.17: County/Metro building fire fire-flow research of 70 large Fires 2009-2012 in excess of 500m² (A_f) fire involvement.

As the fire engineer starts to apply the following formulae to projects, it soon becomes apparent that in some cases, greater amounts of water flow-rate (hence storage water) are calculated than prescriptive codes recommend. In other cases, it will be less. However, it's important to ensure that in any performance based design an adequate amount of firefighting water is provided to enable firefighters to contain fire spread. In some situations, this might require one hose-line using a flow-rate of just 500 L/min for 10 minutes, but in other situations it may take several hose-lines applying several tens of thousands of litres to achieve final extinguishment. Large office floors (for example) where sprinklers are not installed, or have failed, can lead to extended firefighting operations over several hours.

Ideally a fire should be handled by a single hose-line, with an additional safety hose-line in support, and this should be the objective in any design. As a rule of thumb, one hose-line has the suppressive capacity to handle anything up to a 20 MW fire. If the design fire, combined with the expected fire service response and deployment time (water applied to the fire), exceeds a 20 MW growth period then additional passive or active measures will be needed if the building is to be effectively protected. Clearly, early liaison with the local fire service is required for response and deployment analyses. Whilst no time-line can ever be guaranteed, the fire service should be able to provide data for expected response and arrival times for a varied percentage of circumstances.

The following equations for each occupancy group represent the starting point for designing firefighting water strategies:

The GCU – BS PD 7974:5:2014 Equations **12.6, and 12.7 (and 12.8 from Sardqvist's research)** below utilise occupancy type and floor area fire (m²) inputs whereas Equation **12.9** incorporates fire load density (MJ/m²), which can also be applied to high-stacked vertical fire loads (**Eq.12.10**).

For fires in dwellings (house, flats, maisonettes and apartments) – $$F_{dwe} = 75 * A_{fire}^{0.44}$$	Eq. 12.6
For fires in factories, industrial units and storage warehouses - $$F_{ind} = 131 * A_{fire}^{0.51}$$	Eq.12.7
For fires in public, office, commercial, schools, hospitals, hotels and smaller industrial buildings – $$F_{other} = 61 * A_{fire}^{0.57}$$ (based on Sardqvist's 1998 research – see section 12.6) Where: F_{dwe} Deployed Flow-rate (L/min) for dwellings F_{ind} Deployed Flow-rate (L/min) for factories, industrial and storage warehouses F_{other} Deployed Flow-rate (L/min) for 'all other' buildings (Sardqvist) A_{fire} Floor area of fire involvement (m2)	Eq. 12.8
Or alternatively using the estimated fire load energy density (MJ/m²) for all building occupancies, from BS PD 7974:5:2014 - $$F_{design} = 0.00741 * (q_k * A_f)^{0.666}$$ Where: F_{design} design flow-rate for a fully involved fire compartment burning at maximum intensity (L/s) q_k fire load density for the compartment (MJ/m²) A_f total internal floor area of the fire resisting compartment (m²)	Eq. 12.9
Or for high-stacked fire loads (retail; industrial or storage etc) $$F_{design} = 0.0074 \left((q_k * h) A_f \right)^{0.666}$$ Where: h height of stacked storage (metres)	Eq. 12.10
For 1-2 storey, large floor area buildings, (infrastructure and estate planning) based on the GCU 70 large building fire research data 2009-2012 (fires were >500m² to 10,000m²) $$F_{500\ design} = 641 * (A_f)^{-0.8} = L/s$$	Eq.12.11

There are global benchmarks that can also be used to determine the amount of firefighting water that is required to extinguish real fires in tall buildings and these have influenced international building codes when designing rising fire main installations and water storage provisions in support.

What is currently used in the UK is a prescriptive wet rising fire main design specification, based on small residential sized compartments, which would fail to meet most international standards for tall buildings where commercial or mixed use buildings are concerned.

Obtaining Sprinkler compensations

One of the most common designs for tall office buildings is based around a central or offset core housing lifts, stairs and services. The core of the building generally occupies 15-25% of the overall floor-plate, so a typical floor area of 3,000m^2 might consist of a 750m^2 central core and a 2,250m^2 open-plan office floorplate. With this design, there should be maximum hose runs from the centre of the core to the corners of the floor of around 40m. The calculations provided above, and in BS PD 7974:5:2014, do not take into account a sprinkler controlled fire. Where sprinklers are installed according to relevant design standards the fire engineer should be looking for a compensatory coefficient to be included in the equations, acknowledging that although sprinklers are generally 97% effective, there is a 3% chance (or greater) they will not control a fire. At the heart of most sprinkler failures are management, maintenance and inappropriate system design issues. There are several methods open to the fire engineer to account for sprinklers in the calculation process that will reduce both firefighting water storage and supply provisions. (A firefighting *water flow-rate design* calculator spreadsheet can be downloaded from www.eurofirefighter.com).

Coefficients used in the calculations, that might be considered to compensate for approved fire suppression provisions, are available in:

- BS PD 7974
- NFPA 14
- Eurocode EN 1991-1-2 (Annex E)
- Points scale (GCU research proposal – below)

A BS PD 7974 probability model[142] has been applied to office buildings, retail premises and hotels in order to evaluate a suitable sprinkler coefficient, defined as the ratio between the design fire load densities for sprinklered and non-sprinklered compartments. The value of the coefficient, which depends on the distributions of fire load density for three occupancies, ranged from 0.53 to 0.68. These results showed that the fire resistance of sprinklered compartments of these occupancies might be around 60 percent of the resistance specified for non-sprinklered compartments, suggesting a possible 40 percent reduction coefficient for sprinkler provisions. The impact of sprinkler protection on fire resistance may be referenced in other ways[143]. In the case of BS EN 1991-1-2 and associated guidance in the form of BS PD 6688-1-2, the inclusion of fire load density in the calculation procedure may be reduced by 39%. This is on the proviso that life safety

---

142 British Standard PD 7974:7:2003, Application of fire safety engineering principles to the design of buildings, Probabilistic risk assessment
143 Hopkin. D; A Review of Fire Resistance Expectations for High-Rise UK Apartment Buildings; Fire Technology 2016

sprinklers (taken as 'additional measures to improve system reliability and availability' from 2015), in accordance with BS EN 12845, are provided. However, this may have the impact of distorting the resulting fire dynamics as the principle of using a straightforward post-flashover fire model, in practice, would not reach such severity should the sprinklers operate successfully.

Alternatively, more contemporary approaches seek to distinguish the structural reliability from the overall reliability of the fire resistance system through consideration of sprinkler reliability. The structural (or passive) reliability governing the fire resistance requirements of structural elements will depend upon the contribution (if any) from any proposed suppression systems of a given reliability.

As a further example of sprinkler compensatory coefficients, NFPA 14[144] may account for a reduced supply of firefighting water for sprinklered buildings by reducing the rising fire main (standpipe) and water storage requirements by 20 percent. Therefore, by adding a 0.8 coefficient to the calculations we can account for a sprinkler protected building by adapting NFPA guidance in the following way, in a 2,250m2 open-plan office floorplate:

$F_{other} = 61 * 2250^{0.57} * 0.8 =$ **3,973 L/min** using **Eq.12.8**

$F_{design} = 0.00741 * (570 * 2250)^{0.666} * 0.8 =$ **4,159 L/min** using **Eq.12.9**

In each case the practical design solution may be to provide two 150mm rising fire mains, each with at least two outlets per floor, providing a minimum total flow-rate of 4,000 L/min.

Two fire main outlets per floor assists the laying of an attack hose-line additional to a safety hose-line in support, from the same floor (preferably the fire floor, to reduce hose-lay distances and also to protect the stairs from smoke infiltration) – (confirm with local fire service procedures).

Using the above method for a sprinkler protected environment provides a minimum flow density of 1.8 L/min/m², which is considered a reasonable flow-rate for a sprinkler protected open-plan office floor plate. In comparison, a UK prescriptive design would require at least two 100mm fire mains providing 3,000 L/min in total, with or without sprinklers. This results in a flow density of 1.3 L/min/m², which is **below the critical flow rate** determined by GCU research.

The provision and effect of sprinklers to a reduction in the design value of the fire load [MJ/m²] is also taken into account in Eurocode EN 1991-1-2 (Annex E) using factors considering different active firefighting measures. Two factors deal with sprinklers and take into account: automatic water extinguishing system (0.61) and independent water supplies (**0** = 1.0; **1** = 0.87; **2** = 0.7).

A simple but most effective system proposed within the GCU research incorporates a point scale where a building occupancy is risk-graded, based on the following factors:

---

144 National Fire Protection Association, NFPA 14:2013, Standard for the installation of standpipe and hose systems.

	• **11-12 pts** Can achieve a <u>40% reduction</u> in riser flow-rate and water storage provisions • **7-10 pts** achieves <u>30% reduction</u> • **1-6 pts** achieves <u>20% reduction</u>	**Compensatory points**
A	Provision of 'additional measures' *(additional measures to improve sprinkler system reliability and availability)* from BS-EN 12845 2015)	1
B	The provision of zoned maintenance with <u>additional live fire watch</u> during system 'down' times	1
C	The provision of a BS 9999 level one management structure, meeting the requirements of PAS 7*	1
D	A structure designed mainly from concrete or steel framed construction (adds no additional fire loading in cases of intense fire spread) therefore less firefighting water would be required	2
E	Detailed analysis demonstrates an effective fire service intervention should occur (adequate water applied to the fire) prior to the fire reaching a stage beyond the control of a primary attack hose-line	2
F	A simultaneous or effectively phased evacuation strategy has been documented within the building fire safety plan	1
G	The potential for fire spreading to additional floor levels, above or below the floor of origin, has been countered by design (e.g.: combustible cladding, curtain walls or large window expanses avoided)	2
H	The existence of rigid cellular partitioning may slow the development of fire to a medium t-squared growth rate (no open-plan floor areas)	2

Figure 12.18: Glasgow Caledonian University Fire Engineering – sprinkler reduction points scale

As examples in using the GCU points scale, a building that is fitted with an ordinary hazard sprinkler system when only a light hazard system is required, or perhaps where additional measures to improve system reliability and availability are provided, will qualify for the two points under 'A'.

Taking the 2,250m² open-plan office floor-plate described above, by applying the GCU points scale effectively, a total of 11 points can achieve a 40 percent reduction in sprinkler flow, pump-set capacity and water storage, equating to the following:

$F_{other} = 61 * 2250^{0.57} * 0.6 =$ **2,980 L/min** *using Eq.12.8*

$F_{design} = 0.00741 * (570 * 2250)^{0.666} * 0.6 =$ **3,120 L/min** *using Eq.12.9*

So by using a performance based methodology that is founded upon the core principles of modern day fire attack data, we have still only matched a prescriptive flow-rate of just 1.3 L/min/m², but we have ensured in return, a very high standard of structural design and building management (PAS 7) as well as coordinating and gaining agreement with the fire service on their potential (a fair assumption but never a guarantee) to mount an early time-lined attack on the fire, based on standard response and deployment data for the area served.

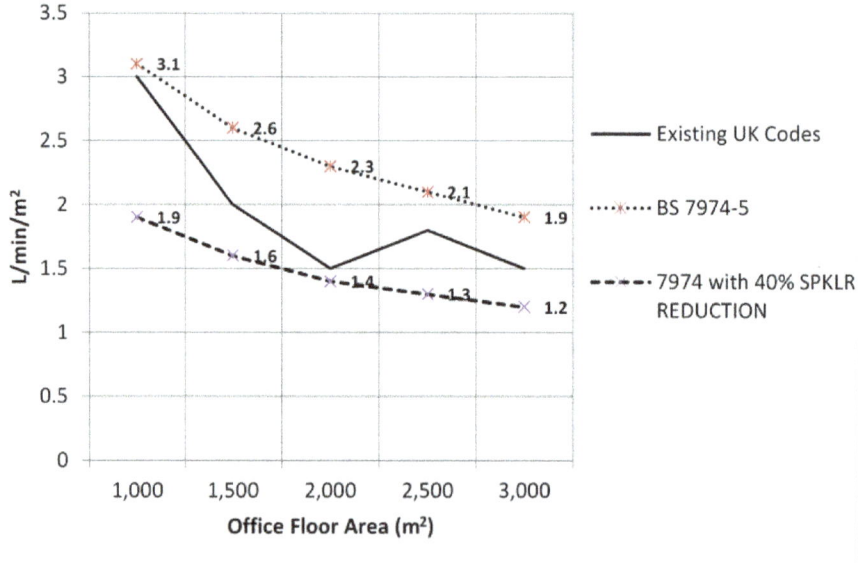

Figure 12.19: A comparison of flow-rate density provisions across occupied floor space, between prescriptive UK regulations and the performance based BS PD 7974:5:2014 (excluding core areas) where a 40% sprinkler reduction has been applied, in accordance with figure 12.17.

Equally, a 40% sprinkler coefficient might be used, based on the GCU points scale, when calculating fire protection for high-rise flats or 100m² apartments as follows:

$F_{dwe} = 75 * 100^{0.44} * 0.6 =$ **340 L/min** *using Eq.12.6*

So, for a sprinkler protected 100m² apartment fire, with maximum hose distances within prescriptive limits from the fire main outlets, two hose-lines flowing 340 L/min (one serves as a safety line, as needed in high-rise) would be adequate. Therefore, the engineered flow provision is rounded to 700 L/min, with a much-reduced water storage and pump set requirement in buildings >30m in height, compared to a code compliant provision. Again though, we have traded needed firefighting water flow-rate and storage provisions for sprinklers, coupled with reliable construction under fire, 'PAS 7' management levels and effective firefighting access, systems and facilities maintenance, and alarm/evacuation arrangements (possibly public address voice alarms) that have all been considered, even if it's just the fire floor and local zones above and below that are initially evacuated.

Isolated Buildings – no immediate Water Supply

In some rural situations, there is coded guidance for establishing water storage provisions where the nearest water supply is several miles away from a building. In the UK building regulations (ADB) the provision of a large static tank of at least 45,000 litre capacity is sometimes used as a reference point. This enables a 45-minute supply at 1,000 L/min firefighting water that may need to be topped up from an alternate source. However,

---

* PAS 7: 2013 Fire risk management system, Specification, British Standards Institute

taking this need for an isolated water supply, it makes sense to calculate a needed flow-rate density per m² of floor area, also taking sprinkler compensatory coefficients into account. Therefore, based on 7974-5 flow-rate density guidance, a two-storey house isolated from a firefighting water supply by some great distance could be provided with a calculated water storage provision as follows –

Two storey non-sprinklered house with a total (2-floors) floor plate of 600m² requires the following storage water for firefighting:

$F_{dwe} = 75 * 600^{0.44} *= $ **1,252 L/min** using **Eq.12.6**

Then, 1,252 L/min x 30-minutes supply = 37,560 Litres
1,252 L/min x 45-minutes supply = 56,340 Litres

With sprinklers installed, the same water supply serving sprinklers could be combined to provide 750 L/min of firefighting water (22,500 Litres for 30-minutes supply at 40% compensatory reduction for sprinklers). A smaller house of just 300m² total floor space across two floors would require a far lesser firefighting water provision.

Note: Based on UK Water guidance 2007, a flow of only 480 L/min is required, totalling 21,600 L/min for a 45-minute supply. Such a supply may be adequate for a single firefighting jet only, but would not enable an additional safety jet to be deployed to optimise firefighter safety.

12.8 FIREFIGHTING WATER CALCULATOR TOOL (BASED ON BS 7974-5:2014)

A firefighting water calculator spreadsheet tool, based on the BS 7974-5:2014 (Fire Service Intervention) guidance has been developed. The tool offers the user the means to compensate the design if sprinklers are fitted and the 8 points on the GCU scale (Fig.12.18) have been addressed. This tool can be downloaded for free from **www.eurofirefighter.com**.

12.9 TRAVELLING FIRES

A review of fires in large open-plan compartments (typically with floorplates greater than 200 m²) reveals that in general, these fires do not conform to normal *flashover* fire development where fire involves the entire enclosure at one time. Instead, these fires tend to move across floor plates, reaching peak levels of heat release across a limited or zoned area at any one time. These fires have been labelled *travelling fires* and in some texts this process of fire development has been referred to as *progressive burning*.

Fire spreading across a large open-plan office floor is typical of this type of behaviour where 'fast and even ultra-fast ***t-squared*** growth rates during the initial few minutes may well be reached, with average fire spread rates in excess of 22m²/min. It may appear that the entire floorplate is burning; however, the fire will likely be at different levels of intensity across the floor space. In some areas, the fire may still be in its growth stages whilst elsewhere on the floor-plate the fire may have reached its peak intensity burning at steady state, with distant far field areas burning into the decay stages.

A fire involving an open-plan 1,400 m² office floor plate located around a central core is more likely to demonstrate peak heat release rates across three zones of 470m² or four zones of 353m², depending on fuel distribution and ventilation parameters. The concept of travelling fire spread makes reference to Alpert's ceiling jet correlations and burning time calculations. In terms of meeting an adequate firefighting water provision the concept of a 'travelling fire' in large enclosures should be looked at primarily using a growing fire analysis. If fire service intervention is unlikely to control the fire at the point the primary hose-line is deployed then additional active or passive fire protection features, or the effective siting of additional firefighting shafts with 150mm fire mains to support multiple secondary hose-line deployments at the fire floor, should be considered and discussed with the fire service at the Qualitative Design Review (QDR) stage.

A good example of travelling fire spread across large open-plan office floor space occurred in Los Angeles in 1988 at the First Interstate Bank building fire. The author visited the site of the fire a few weeks after it occurred and interviewed LAFD fire chiefs and firefighters who fought the fire. The tower has a structural steel frame with lightweight concrete slab on profiled steel deck. The external cladding system consisted of glass and aluminium. The fire started on floor 12 of the 62-storey office tower at 2225 hours and spread up to floor 16 before it was brought under control by the fire service some four hours later. An automatic sprinkler system, although installed, had been temporarily shut down awaiting installation of water flow alarms.

This analysis is of the fire spread on floor 12 where the fire originated. The open-plan office floor space was located around a central core that contained lifts, stairs and service shafts. The office space surrounding the core measured 188 metres by 7.5 metres, totalling 1,410 square metres. The analysis looks at the actual spread of fire across the floor plate as recorded in real time (video) and utilises the actual amounts of ventilation available on the fire floor as windows were seen to fail at various locations and times. The fire itself took 65 minutes to wrap around the core and involve all the available open-plan floor space.

Although a 'medium' t^2 fire growth curve is normally applied to office accommodation, the growth stages in this fire very soon developed a 'fast' rate of growth, spreading through 22 square metres of floor space per minute. The fire has been analysed here using three and four zone models to represent how a travelling fire could spread across the floor plate, demonstrating near field and far field temperatures that may impact on structural elements in a way not accounted for in current codes or standards. To evaluate the requirements for 'adequate' firefighting water, the energy release has been calculated in the 3 and 4 zone models as the entire compartment does not burn to maximum energy release at any one time. In both models, there are only brief periods of burning where two zones are burning at Q_{max} at the same time.

As the fire developed, a sufficient number of windows were broken by heat to enable the fire to burn in a fuel controlled state throughout the entire duration of the fire, with 20-25% A_v/A_f vent opening to floor space ratios fairly constant. Had the fire floor been modelled as a single compartment fully involved in fire the estimated energy release, without suppression activity, would have been around 218 MW, based on a fire load energy density of 570 MJ/m^2 burning at 154 kW/m^2. When modelled as a travelling fire advancing through three zonal areas of floor space the peak energy release (Q_{max}) in each zone is estimated at 104 MW, although there remains an element of additional heat release in adjacent zone as they pre-heat in growth or cool in decay in the far field zones. However, from a firefighting approach, Q_{max} can be used on a zonal basis to calculate the required water flow-rate.

Even though earlier attempts were made, Los Angeles firefighters were unable to deploy adequate water until 34 minutes into the fire due to a delay in calling the fire department and ineffective fire pump settings serving the wet rising fire mains (standpipes). At this

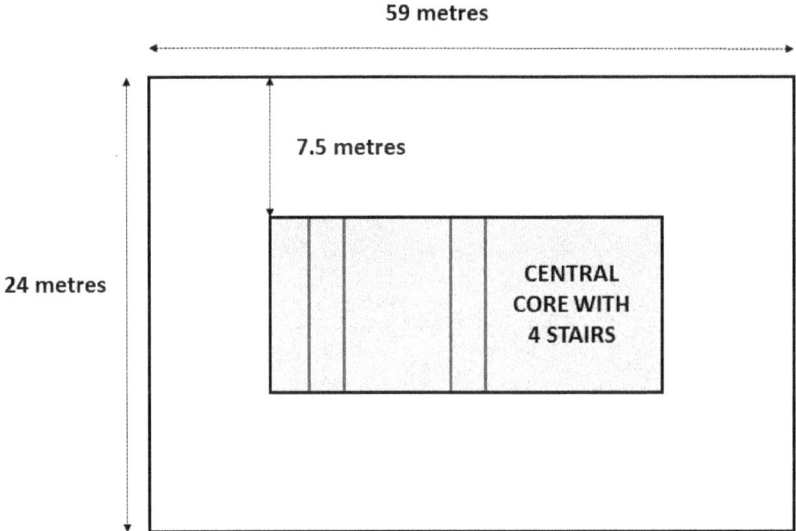

Figure 12.20: Plan of the 12th floor at the First Interstate Bank 1988

point two thirds of the 12th floor was fully involved in fire although on a zonal basis, part of the floor space would be in decay. Even so, as the firefighters reached the 12th level at 21 minutes into the fires 'fast' growth rate, any attempt at intervention would have been unlikely to succeed as the fire had already spread to a quarter of the floor plate with an estimated heat release of 86 MW.

Los Angeles Fire Department Operations

At 2237, the Fire Department received three separate 911 calls from people outside of the First Interstate building reporting a fire on the upper floors[145]. At 2238, a Category 'B' assignment was dispatched consisting of Task Forces 9 and 10, Engine 3, Squad 4, and Battalion 1 -- a total of 30 personnel. (A Task Force in Los Angeles consists of 10 personnel operating two pumpers and one ladder truck.) The first report of the fire from inside the building was received at 2241, as the first due companies were arriving at the scene. While responding, Battalion 1 had observed and reported a large 'loom-up' in the general area of the building. As he arrived on the scene, the Battalion Chief observed the entire east side and three-fourths of the south side of the 12th floor fully involved with fire. Battalion Chief Don Cate immediately called for five additional Task Forces, five Engine companies, and five Battalion Chiefs. This was followed quickly by a request for an additional five Task Forces, five Engine companies and five Battalion Chiefs, providing a total response of over 200 personnel within five minutes of the first alarm. Two Fire Department helicopters were also dispatched. The High-rise Incident Command System was initiated with companies assigned to fire attack and to logistics and support functions from the outset.

In accordance with Los Angeles City Fire Department policy, elevators were not used and all personnel climbed the stairs to the fire area. The first companies to reach the fire

145 Extract from United States Fire Administration (USFA) report 1988

floor found smoke entering all four stair-shafts from around the exit doors. Hose-lines were connected to the standpipe risers and the initial attack began at approximately 2310. Due to the magnitude of the fire on the 12th floor, attack was initiated from all four stairways. The crews had great difficulty advancing lines through the doors and onto the floor. As the doors were opened, heat and smoke pushed into the stairways and rose rapidly to the upper levels of the building.

First Interstate Bank Fire 12th Floor analysis	3 Zone Fire	4 Zone Fire	Entire 12th Floor
Floor Area (m^2)	470	353	1410
Heat of combustion (MJ/kg)	20	20	20
Fire load energy density MJ/m^2	570	570	570
Estimated fire load (kg)	13,253	10,046	40,185
Fire growth rate	Fast	Fast	Fast
Ventilation or fuel controlled	Fuel	Fuel	Fuel
Ventilation opening ratio (%)	25	25	25
Maximum burning rate (kg/s)	5.2	4.3	10.9
Zonal Qmax (MW)	104	86	218
Time to uncontrolled burnout (t) (mins)	160	154	185
Time to extinguish (Fire Service) (mins)	124	124	124

Table 12.7 Travelling fire analysis at the First Interstate Bank Fire in 1988 (Firesys)

'Firesys' is a detailed suite of fire modelling spreadsheets, developed by New Zealand structural and fire engineer Cliff Barnett. In 2004 the author combined his previous research with Barnett's work, providing flow-rate input to Firesys model 8-E, using data outputs from the 100 fire study in London (1989). The foundation of the Firesys model is based on the BFD Curve, an alternative "natural" fire curve that, according to Barnett, fits the results of large scale fire tests closer than any previously known fire modelling methods (2007) including Eurocode parametric curves.

A Firesys spreadsheet calculator can be downloaded from **www.eurofirefighter.com**

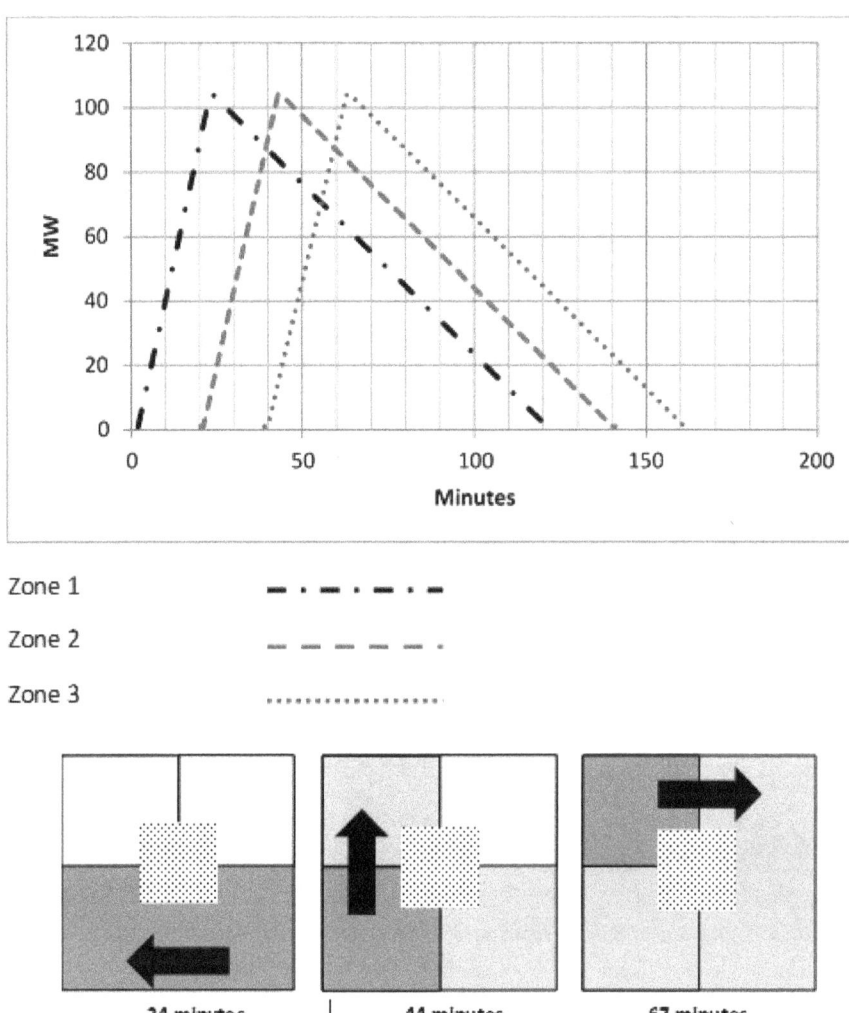

Figures 12.21 & 12.22 – Three Zone model (Interstate Bank fire)

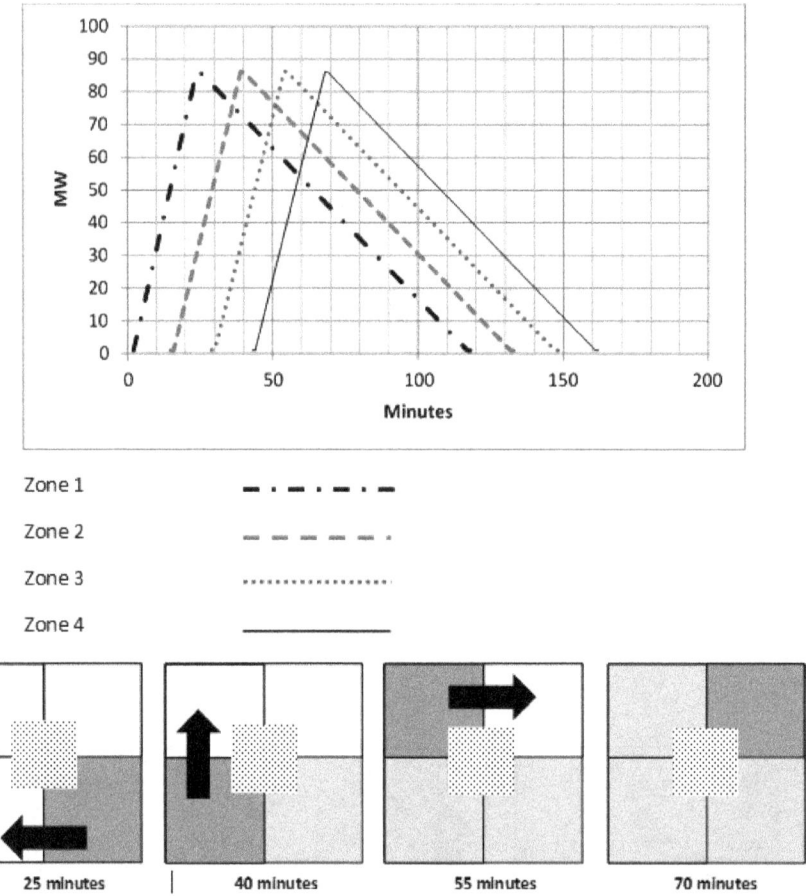

Figure 12.23: Four-zone model of heat-release rates on the 12th floor at the First Interstate Bank fire 1988 / Figure 12.24 – Four Zone model (Interstate Bank fire)

Adequate 'Firefighting Water' – you need this! • 273

First Interstate Bank Fire (1,410m² 12ᵗʰ Floor) (Fuel control Fire)		
Mass of fuel	40,185	kg
Calorific value used	20	MJ/kg
Fire Load Density	570	MJ/m²
Energy contained in fuel	803,700	MJ
Growth factor (fast fire)	150	s
Peak Heat Release Rate (PHRR)	217.8	MW Q_{max}
Time to reach PHRR (growth phase)	36.9	mins
Energy released over time	160,736	MJ
Decay factor	600	secs
Time to end of decay phase	147.6	mins
Energy released in time	642,959	MJ
Total duration of burning	<185	min
Heating efficiency	0.50	k_F
Cooling efficiency	0.63**	k_W
Firefighting water flow-rate required L/s	66 (3,960 L/min)	L/s

Table 12.8: The design fire and flow-rate calculations used in 'Firesys' for the Interstate Bank fire:

Note: A 63% efficiency factor** is applied due to a fuel-controlled fire mostly likely extinguished in the decay phase (based on data from 100 large fires) – this is not a design fire calculation.

Calculating Adequate Water for the Interstate Bank Fire

The design flow-rate for the 1,410-square metre open-plan office floor space at the building can be calculated using Eq.12.8 or Eq.12.9 as follows:

$F_{design} = 0.00741 * (570 * 1410)^{0.666}$ = **3,808 L/min using Eq.12.9**

$F_{other} = 61 * 1410^{0.57}$ = **3,805 L/min using Eq.12.8**

If applying a travelling fire analysis to estimate firefighting water demands it would be necessary to calculate as follows:

3 Zone model:

465m² fire area at 104 MW Q_{max}

$$F = \frac{0.5 * 104}{0.5 * 2.6} = 2,400 \; L/min \; using \; Eq.12.1$$

Added to this will be a requirement to deal with adjacent zones, mostly during decay stage burning, where 2 l/min/m² may be generically applied as follows –

Two additional zones of 465m² totalling a potential 930m² of decay stage burning leading to an additional requirement of 930 x 2 = **1,860 L/min**

2,400 + 1,860 = **4,260 L/min**

Alternatively, where additional zonal floor space calculated as *decay stage burning* is greater than 1,500m² then the calculation given in figure 12.16 (chart) may be used in addition to Eq.12.1 requirements.

4 Zone model:

353m² fire area at 86MW Q_{max}

$$F = \frac{0.5 * 86}{0.5 * 2.6} = 1,984 \; L/min \; using \; Eq.12.1$$

Added to this will be a requirement to deal with adjacent zones, mostly during decay stage burning, where 2 L/min/m² may be generically applied as follows –

Three additional zones of 353m² totalling a potential 1,059m² of decay stage burning leading to an additional requirement of 1059 x 2 = **2,118 L/min**

1,984 + 2,118 = **4,102 L/min**

With fire spreading at a 'fast' t² rate across the 12th floor as firefighters arrived on-scene and deployed to the 12 floor, an effective attack on the fire was initially unlikely due to the energy release at this point. As the fire progressed, between five and eight 500 L/min fire streams totalling 4,000 L/min (2.83 L/min/m²) were deployed internally on the 12th floor (totalling 9,000 L/min to all fire involved floors from four 150mm fire main stand-pipes). To ensure this water continued to flow, firefighters were relieved at approximately 15 minute intervals on each hose-line. The staffing and resource demand was extremely high.

The fire extended at a rate estimated at 45 minutes per floor and burned intensely for approximately 90 minutes on each level before entering the decay stages of burning. This resulted in two floors being heavily involved at any point during the fire. The upward extension was stopped at the 16th floor level, after completely destroying four and one-half floors of the building.

Using the following equation, the time needed to effectively extinguish a fully involved fire on floor 12 can be estimated as follows:

$$t_{ext} = 3.3 * \sqrt{1,410} = 124 \; min \; using \; Eq.12.4$$

THE FIRE	Interstate Bank LA 1988 – 4 Zone Fire
Q_{max} PHRR (MW) also (MJ/s)	86.4 MW as 4-zone travelling fire
PHRR at 50% Comb Efficient k_F	43.2 MW
Duration of Steady State Fire	11.6 min
Mass Loss (kg) at Steady State	2727 kg
Maximum rate of burn kg/s	3.9 kg/s average (peak at 4.3 kg/s accounted for)
Maximum rate of burn kg/min	258 kg/min
Energy Release at Q_{max} MJ	54733 MJ
Energy Release at Q_{max} MJ/s	86.4 MJ/s
Energy Release at Q_{max} MJ/min	5184 MJ/min
Energy Release at 50% Comb Efficient k_F	2592 MJ/min at 1.5 MW
THE WATER	
ADEQUATE WATER (Barnett) 0.38 L/s/MW (Grimwood) 0.40 L/s/MW	0.38 x 86.4 MW = 32.8 L/s At Q_{max} = 1968 L/min

RASBASH/GRIMWOOD			
Note: According to several studies, extinguishing efficiency (Fig.12.8) is estimated to be between - 32-35% directly at the fuel base and 15-18% when water is applied into the flaming reaction zone. The remaining 50% of water applied is generally seen to 'run-off' with minor cooling effects.	0.36 x 3.6 x (1/0.3) x 32.8 x 0.18%	=	25.48 MW
	0.64 x 2.6 x 32.8 x 0.32%	=	17.46 MW
	Total	=	43 MW
	Q_s = 43 MW / 0.5 k_F	=	86 MW per 32.8 L/s
	86 MW / 32.8 L/s	=	2.62 MW/L/s
	32.8 L/s x 2.62 MW	=	86 MW Q_{max}
	32.8 / 86	=	0.38 L/s/MW

THE HEAT EXTRACTION	
Reaction Zone (RZ) @ 50% Combustion efficiency Gaseous Phase	43 MW 43 MJ/s 2580 MJ/min
Water reaching RZ	36% of 1968 L/min = 708 L/min (12 L/s)
Water need in Suppression of RZ	2580 MJ to be extracted
	Therefore at least 717 L/min is needed; (2580 MJ / 3.6 MJ/kg);
	The remaining 64% of applied water (1,259 L/min) is used to cool the fuel base – an application rate of around 3.6 L/min per m2 of floor surface coverage is needed.

Table 12.9 - Summary of the Interstate Bank firefighting water flow requirements

At the Interstate Bank fire the water was actually flowing for just over 120 minutes on floor 12 before the fire was finally controlled and the remaining fires up to floor 16 were extinguished within 4 hours from time of arrival. It is estimated that *uncontrolled burning* would have exceeded 165 minutes on the 12th floor. During such a period, the structural elements and steel frame connections would have been subjected to an extended heating and cooling phase, with a much greater potential for failure.

> **What does this mean to the Firefighter?**
>
> As a fire begins to spread across open-plan floor space with low ceilings, such as in an office building, the speed of deployment to get initial water on the fire must be rapid if the fire is to be controlled. Past experience has shown us that such fires spread at an average rate of around 22 square metres every minute in these buildings and once a fire reaches one metre in height, the suppressive capacity of a hose-line is likely to be surpassed within the next 10 minutes (500 L/min) or 13 minutes (750 L/min) or 15 minutes (1,000 L/min), as the fire takes hold on the floorplate and develops into its fast growth curve.
>
> Where there are high ceilings (storage warehouse), such rapid fire spread will depend on the vertical fire load and the distances between stacks of goods, where un-sprinklered or unvented fire development may be equally dramatic as burning brands quickly spread fire around and internal racking collapse becomes the primary hazard. In these instances, the smoke layer may drop rapidly upon firefighters, sending them into darkness and increasing their vulnerability to hidden fire spread in the overhead. Definitely something that should be anticipated and a planned evacuation from the structure may need to be made ahead of any such conditions occurring.

12.10 THE IMPACT OF INADEQUATE FIRE MAIN CAPACITY ON FIRE SERVICE INTERVENTION

Some evidence based analysis should be applied to the predicted point in time at which the effectiveness of fire service intervention begins to fail and where additional active or passive fire protection measures might be considered as part of the overall fire strategy for a building. Depending on the design for any particular building objectives (including life safety, business resilience and environmental impact), the most extreme fire may be expected to be one that remains within the largest single fire resisting compartment. Where fixed fire suppression exists, the largest fire is estimated according to the design capabilities of such systems. It is this basis upon which most methodologies relating to firefighting water provisions and needed flow-rate are founded, although consideration should be given to the reliability, probability and consequences of such systems or barriers failing to meet design needs effectively when required. Where a fire spreads beyond the largest fire resisting compartment upon which firefighting water design calculations are based (GCU statistics suggest 47% of 'working fires' will do this), building design objectives are generally seen to have failed and the structural elements themselves as well as adjacent exposures may be at risk. In the case of some low-rise buildings this may be considered a reasonable and acceptable worst case scenario.

There are targeted time frames on how quickly and how effectively the fire service are able to apply water (or other suppressive agents) into a fire involved area before the elements of structure are compromised or before the fire spreads unsafely and unplanned, beyond the fire resisting compartment of origin. At some stage the fire will move into the decay phase of fire development where the maximum rate of heat (energy) release (Q_{max}) is surpassed and therefore a lesser amount of water becomes necessary to control the fire. The general trend suggests that effective fire service intervention operations began the transition into 'defensive mode' as primary flow rates into large floorplates fell below

3.75 L/s/m² and/or the fire involvement exceeded 600m² in floor area. It should be re-emphasised here that flow-rates below 4 L/s/m² are seen to create an element of added thermal exposure to firefighters and therefore their working times on the floorplate are decreased and additional staffing and resources are needed in support, as fire floor relief times are reduced.

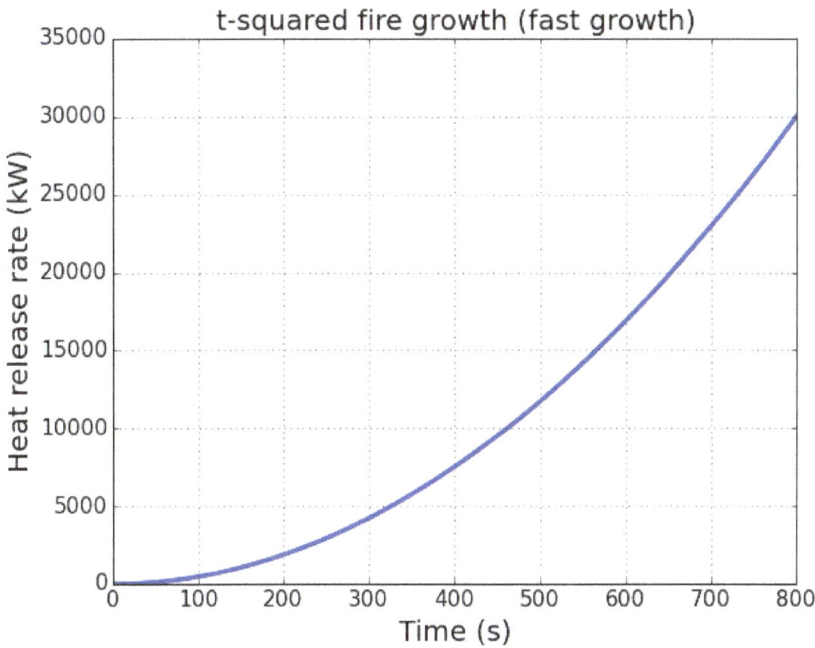

Figure 12.25: A t-squared fire growth curve demonstrating a 'fast' growth rate.

At this juncture, the ability of any specific fire service intervention to be able to contain, control and extinguish a large developing fire will depend on a 'window of opportunity' related to the fire size and speed in deploying 'adequate' water; the staffing resources and equipment available in support of maintaining an attack on the fire, and viable access to the burning fire load. Also, consider the subjective statement of US fire chief Bill Peterson that there is likely to be a 50 percent failure rate in interior fire service control once fire development has surpassed 86m² in floor area on first deployment into large floor space. Similarly, 'the rule of thumb' guidance that a primary attack hose-line flowing 500 L/min offers a suppressive capability of up to 20 MW of heat release should be considered. These two practical firefighting estimates appear very close in their recommendations.

As large fires develop beyond the initial control of fire service intervention, a lesser amount of water is then required once the fire burns into decay stages and this will be a time dependent turning point. The example of the Interstate Bank fire demonstrates adequate water was achieved at 2.8 L/min/m² just before the stability of the structural frame was threatened. The calculated design requirements for adequate water in this building (without sprinklers functioning) ranged from 2.7 to 3.0 L/min/m². To achieve these firefighting water flow-rates the First Interstate Bank had four 150mm wet risers serving the 1,400m² floor plates, with two 2,800 L/min pumps and water storage that would allow 81 minutes of continuous firefighting at a 4,000 L/min flow-rate. In comparison, the current UK regulatory guidance for rising mains in this building would recommend just 2.1 L/min/m² flow coverage from two 100mm rising mains. However, if maximum hose-run distances conformed to 45-60m from the central core, then the UK flow requirement could be halved to 1.0 L/min/m² (single 100mm main) which would be grossly inadequate to sustain an extended firefighting operation on a single fire floor.

A good example of an engineered rising main strategy can be seen in Singapore, where the vast majority of high-rise office buildings provide adequate firefighting water with a minimum flow in the first riser in use of 2,280 L/min (2x300gpm (US) hose-lines) and 1,140 L/min in subsequent risers brought into operation, providing a minimum flow coverage of 2.45 L/min/m² on a 1,400m² floor plate. These flows are achieved via 150mm wet risers with two floor outlets per rising main in all office buildings over 45m in height.

12.11 THEN THERE WERE 249 MORE FIRES ... 2015-16 (POST-SCRIPT)

As a post-script to his PhD research, the author undertook further analysis of an additional 249 'working' building fires that occurred in Kent, UK from April 2015 to April 2016. In order to ensure a comparable analysis, again all fires were considered as 'working' fires with derelict building fires in both sets of data excluded. A new response model had been introduced across the service's 56 fire stations, covering an area of 3,544 km² with a Population Density of 413 people/km². This new fire service response and deployment model was the outcome of a two-year service delivery capability review combined with financial funding restrictions following UK government requirements to optimise the public service delivery. It led to changes in response, attendance times, training, equipment, firefighting tactics and intervention strategies. The ultimate optimal performance objectives were to reduce life losses through fire, and from this particular research perspective the amount of fire damage resulting from building fires, or the ability to contain fires to smaller areas, is considered as a logical method in measuring service delivery and performance at building fires (**property protection**).

In the three-year period 2009-2012 in Kent there were 4,430 building fires that resulted in 1,152 working fires and from 2015-2016 (after the changes to the response and deployment model had been made) there were 291 fires, that fell into the category of *working fires in occupied buildings*. The new equipment being introduced on the fire-ground included –

- A wider use of Positive Pressure Ventilation (PPV) as an attack tool (PPA)
- The introduction of 22mm high pressure hose-reels flowing 200-250 L/min as a first-strike tool compared with 19mm reels (110 L/min) that were most commonly utilised in the 2009-12 research period
- The provision of 22mm diameter smooth-bore nozzles across the 56 stations (complementing the automatic nozzles already provisioned) to deal with high-rise fires above the 10th level and also rapidly developing fires elsewhere with a higher flow-rate being achievable at lower nozzle pressures
- Portable attack monitors on all fire engines, again to provide greater flow-rates where needed
- Larger diameter attack hose-lines such as 51mm replacing 45mm attack hose
- 90mm supply hose-lines replacing some 65mm for use with portable monitors or

as hydrant supply hose where one 90mm line can equal the flow of two 65mm lines, freeing up staffing on the fire-ground to initiate fire attack more quickly;
- 'Fog-spike' (fog-nail) water-fog insertion nozzles, issued to all fire engines, supported by five COBRA units, used where fire is contained to small compartments or is spreading through structural voids and attic spaces making access difficult.

The changing dynamics of our building fire environment warranted some updating of the fire response model and these changes have clearly been justified when comparing the fire data both from before and after changes were made (Figs. 12.26 and 12.27) – an increased target flow-rate and tackling roof-space and void fires more effectively being central to the changes.

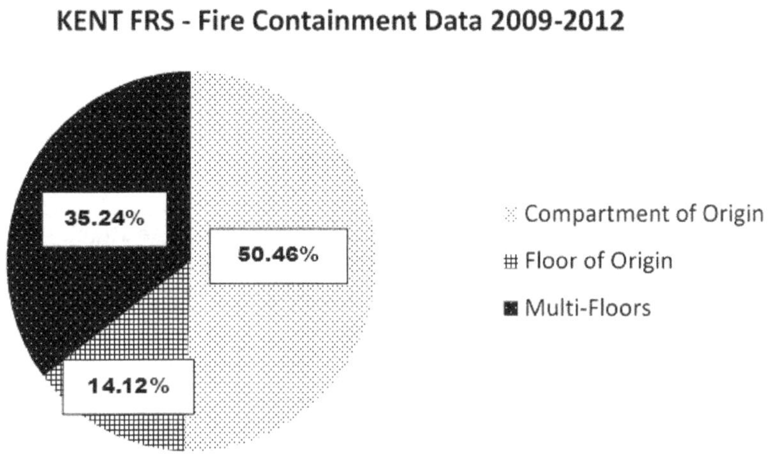

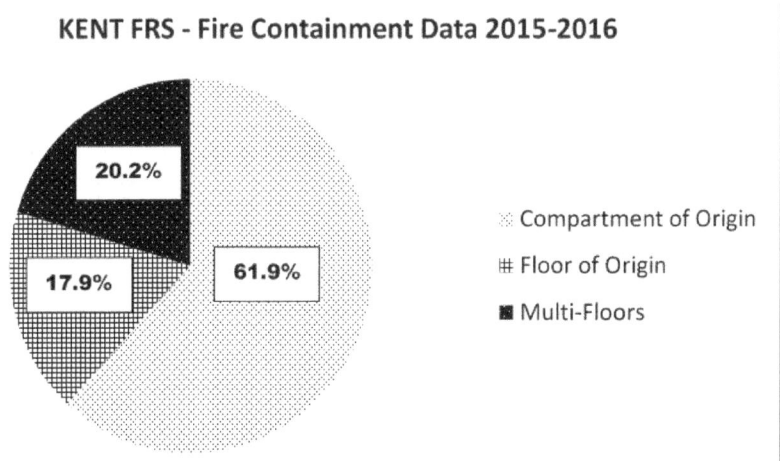

Figure 12.26: The data from the Kent (County) fires in 2009-2012

Figure 12.27: The data from the Kent (County) fires in 2015-16

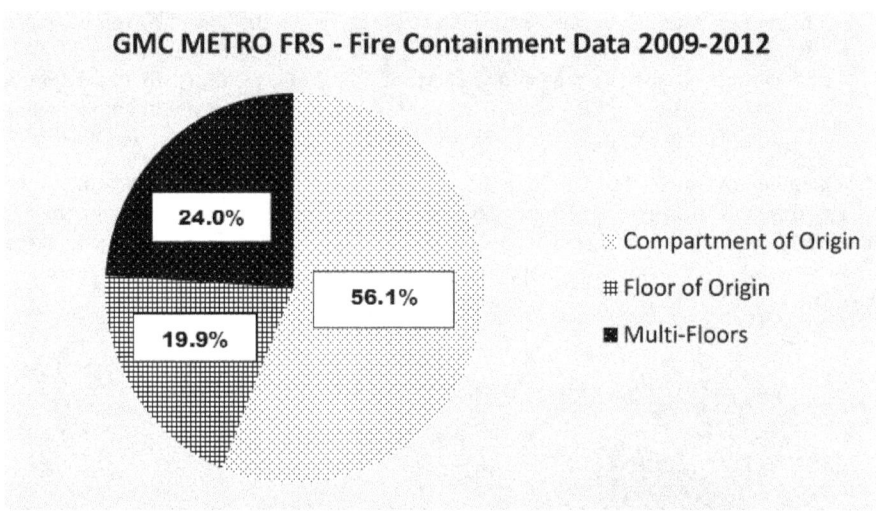

Figure 12.28: The data from the GMC (Metro) fires in 2009-2012

High Pressure (HP) versus Low Pressure main Hose-line attack

HP is certainly nothing new, using small diameter rubber hose-tubing (19mm-25mm diameter bore) to discharge small amounts of water (100-300 L/min) at high pressure (around 30-50 bar at the pump) into the fire can create some impressive knockdowns. Similarly, fog insertion tools such as COBRA (UHP) can provide very finely divided water-fog streams. The firefighting method can be traced back as far as the Royer/Nelson IOWA and Lloyd Layman research in the 1940–50s. It is a strategy that has been well used by London Fire Brigade and throughout Europe for several decades with some great success.

2 x 20m x 22mm hose-reel (40m total length)			
20 bar @ Pump	25 bar @ Pump	30 bar @ Pump	35 bar @ Pump
170 L/min	250 L/min	270 L/min	300 L/min

*Table 12.10: Even higher flow-rates are achievable by reducing the total length of 22mm hose-reels to just 40 metres, although this length **does not integrate with building regulation design specifications** for firefighting access to apartment blocks (45m from fire appliance/engine to furthest apartment).*

Adequate 'Firefighting Water' – you need this! • 281

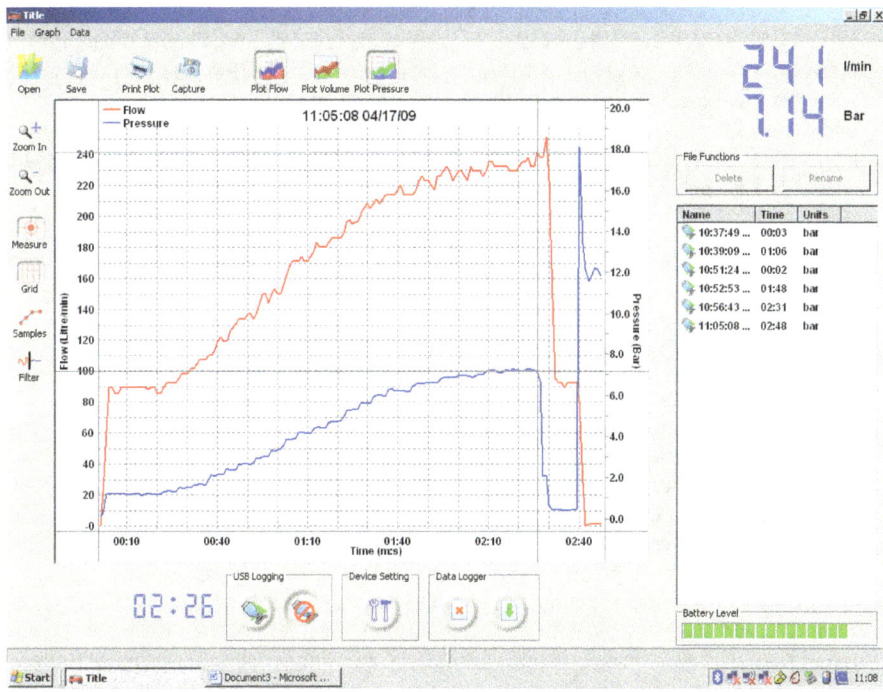

Figure 12.29: To achieve 7 bar and 240 L/min at the nozzle of a 55-metre run of 22mm high-pressure hose-reel you may need to pump to >40 bar at the pump (3,200 RPM) – a 100% improvement in flow compared to a 19mm hose-reel.

What does this mean to the Firefighter?

Advantages of High-pressure
- Very quick to deploy water on fire
- Firefighting water tank lasts longer (lower flow than main-line hose)
- Less staffing required
- Less reliance on hydrant support
- Can easily be deployed to 4-5th storeys by rope line, hauled aloft
- A 120 L/min HP hose-line may be equal in gas-phase suppressive capacity to a 350 L/min low-pressure line (three times more effective)
- Great tool to access fires in voids
- 76% of 5,401 working UK building fires** were dealt with using 19mm HP hose-reels

Disadvantages of High-pressure
- Crews can become over-confident in HP's suppressive ability
- The low flows have distinct suppressive limits in the fuel-phase
- Firefighters can get into trouble quickly if caught by sudden fire spread
- In timber frame properties, the fire load may be too much for UHP to handle
- May increase levels of thermal exposure experienced by firefighters (some fires take longer to extinguish)

** *Glasgow Caledonian University (authors) research 2015 (described in this chapter)*

The use of HP pumps and small diameter hose-line may increase stream velocity, reduce water droplet size and can increase suppressive capacity in the gas-phase

Continued overleaf.

> by up to 3 times, when compared to low pressure streams at the same flow-rate. However, where the fire growth rate is rapid, or where the fire load is heavy (or the structure becomes the fire load), the low flow-rate may then expose firefighters to dangerous levels of thermal exposure for an increased duration. However, the innovative use of UHP COBRA/PYROLANCE or FOGNAIL/FOGSPIKE tools (piercing nozzles) applied from external positions is increasing cross Europe.

12.12 19mm DIAMETER HOSE-REELS

The use of 120 L/min 19mm hose-reels in the UK is finally being acknowledged by many as an outdated and potentially dangerous practice when matched against the higher fire loads and the more intense and faster compartment fire spread being seen today. The author has been campaigning now for nearly two decades for European fire services to recognise that the time for change has come. His work with London Fire Brigade in 2006 has finally resulted in the introduction of 22mm high-pressure reels flowing up to 250 L/min and carrying near equal suppressive capability as 45mm low-pressure hose-lines. Similarly, Kent Fire and Rescue Service, Greater Manchester FRS and several other fire services have implemented the same transition to 22mm reels. We have seen recent fires in London, Warwickshire and Manchester where firefighters have lost their lives after 19mm reels had been deployed on arrival into commercial and industrial premises for periods in excess of 30 minutes before higher flow hose-lines were decided on. The question needs to be asked, is it viable to deploy eight firefighters with four hose-reels into a fire or is this a misuse of resources. Are four low-flow tactical vantage points equal to one or two higher flow options? Is a total of 500 L/min applied from four locations better than a total of 1300 L/min applied from two vantage points? Should we be looking to deploy greater numbers of firefighters to manhandle larger hose-lines earlier on in commercial buildings to achieve better effect?

These considerations are of course dependent on various strategic factors and may vary according to the type of fire or occupancy profile. Greater Manchester FRS achieved good effect in the flow-rate research (County v Metro) above in flowing two or more hose-reels during the initial stages at many of their fires but most of these were in domestic properties. However, I am a strong believer that we should deal with 'working fires' involving commercial and industrial fire loads from the outset, using minimum flows of 500-700 L/min per hose-line. Such hose-lines demand an appropriate staffing deployment of 2-4 per line. At any stage that firefighters are advancing hose-lines into such occupancies, safety hose-lines of *equal flow* or *suppressive capability* are also a primary need at an early stage.

The Reaction Zone – Cooling Mechanisms and Heat Extraction

- Cooling the fuel surface, which reduces the pyrolysis rate and so the rate of fuel supply to the flaming reaction zone, thus reducing the heat release rate and the radiative feedback from the flame to the fuel surface.
- Cooling the flame zone directly, which disrupts the chemical reactions responsible for combustion. Some portion of the heat of reaction is abstracted in heating and evaporating the liquid water; therefore, less thermal energy is available in the vicinity of the reaction zone.

	Room Fire	CCAB Chicago Office	Interstate Bank**
HEAT PROFILE			
Floor Area m²	12 m²	244 m²	353 m² (Zone 1)
Fire Load Mass kg	520 kg	6,977 kg	10,046 kg
MJ/kg Calorific Value Heat of Combustion	18 MJ/kg	20 MJ/kg	20 MJ/kg
FLED MJ/m²	780 MJ/m²	570 MJ/m²	570 MJ/m²
Fire Load Energy MJ	9,360 MJ	139,536 MJ	200,925 MJ
Peak Mass Burn Rate kg/s	0.3 kg/s	3.3 kg/s	4.3 kg/s
Peak Mass Burn Rate kg/min	18 kg/min	198 kg/min	258 kg/min
Heat of reaction MJ/s	5.1 MJ/s (5.1 MW)	65.1 MJ/s	86 MJ/s (86 MW)
Heat of reaction MJ/min	306 MJ/min	3906 MJ/min	5160 MJ/min
Steady state duration (min)	11 min	2 min	1 min
Peak HRR (Qmax) MW	5.1 MW	65.1 MW	86.4 MW
Steady State Fire	306 MJ/min x 11 = 3366 MJ	3906 MJ/min x 2 = 7812 MJ	5160 MJ/min x 1= 5160 MJ
FF Water required L/min	5.1 MW x 0.38 = 1.9 L/s x 60 = 116 L/min	65.1 MW x 0.38 = 24.7 L/s x 60 = 1484 L/min	86.4 MW x 0.38 = 32.8 L/s x 60 = 1968 L/min
23 L/MW =	23 x 5.1 = 117 L/min	23 x 65.1 = 1497 L/min	23 x 86.4 = 1987 L/min
FireSys Calculates	2 L/s (120 L/min)	25 L/s (1500 L/min)	33 L/s (1980 L/min)
HEAT EXTRACTION			
Total Reaction Zone (RZ) @ 50% Overall Peak Combustion Efficiency	2.55 MW 2.55 MJ/s 153 MJ/min	32.5 MW 32.5 MJ/s 1950 MJ/min	43.2 MW 43.2 MJ/s 2592 MJ/min
36% Water reaching Flaming reaction zone Gaseous Phase	36% of 116 L/min = 42 L/min (0.7 L/s)	36% of 1484 L/min = 534 L/min (9 L/s)	36% of 1968 L/min = 708 L/min (12 L/s)
64% Cooling efficiency on the flaming fuel surfaces	64% of 116 L/min = 74 L/min (1.23 L/s)	64% of 1484 L/min = 950 L/min (16 L/s)	64% of 1968 L/min = 1260 L/min (21 L/s)
Heat Extraction from reaction zone	153 MJ to be extracted/minute	1950 MJ to be extracted/minute	2592 MJ to be extracted/minute
Water required for RZ = MJ / 3.6 MJ/kg	153 / 3.6 = 42.5 L/min	1950 / 3.6 = 542 L/min	2592 / 3.6 = 720 L/min
The remaining water is used to cool and penetrate the fuel base	74 L/min	942 L/min	1248 L/min

** *This is taken as one single zone (353 m2) of a four-zone (1410m2) travelling fire*

Table 12.11: Typical heat extraction analysis from the flaming reaction zone (gas-phase) of three fires (demonstration examples) based on Karlsruhe University (Germany) data[146].

146 Fuchs. P; On the Extinguishing Effect of Various Extinguishing Agents and Extinguishing Methods with Different Fuels, Karlsruhe University (Germany), *Fire Safety Journal, 7* (1984) 165–175

12.13 A SLIDING-SCALE FLOW-RATE ANALYSIS FOR FIRE COMMANDERS

It is important for fire commanders to be able to roughly estimate the limitations of the amount of firefighting water available on-scene, against a large growing fire. Where a fire in a large floor space with a high fuel load continues to grow, a time may come where the speed of fire development outpaces the quantity of water being applied to the fire and a transition to *defensive mode* of attack may become necessary. Aside from obvious triggers such as unstable structures, inadequate staffing or hazardous storage, a quick analysis of the flow-rate available can determine at what point the fire may begin to outpace safe and effective interior offensive operations.

It can be seen in the upper graph's flow-line (L/min) that as floor area fire involvement increases, so too does the needed (applied) flow-rate increase. This is nothing new to firefighters! Where non-residential buildings are involved the required flow-rate is even greater. Such information presented this way would suggest that as a fire grows larger, we simply apply more water. However, it does not necessarily present a picture of when defensive operations might become necessary. The lower graph (**L/min/m²**) demonstrates how the ratio of applied flow-rate decreases as the fire size increases.

There is evidence from the research presented in this chapter that suggests when applied flow-rates fall below 4 L/min/m² in non-sprinklered compartments, we are entering the '*inadequate water*' phase of fire attack where *total burnout* may ensue. This is the point where the fire is growing faster than we can apply adequate water. This may be because we do not have enough staffing or resources to apply the required amount of water or because we are unable to access the fire area, or perhaps we are unable to transport sufficient quantities of water to the fire scene quickly enough. A rough fire-ground calculation can be made here by taking the total floor area (m²) involved by fire and multiplying it by 4. If the total floor area of the building is 600 m² and 200 m² is involved in fire, multiplying 200 x 4 = 800 L/min. This is the minimum flow-rate where we can safely continue offensive firefighting operations, based on flow-rate alone, without placing firefighters in compromising situations (also see table 12.16). However, in reality that 200 m² of fire may soon become 300 m² if we aren't able to apply water with good effect.

Note: A minimum flow-rate deployment of 200 L/min (High-Pressure) is recommended in UK dwelling fires.

The reader should take note that this research and recommended flow-rates for firefighting or design purposes are the result of 5,401 working fires across all occupancy groups and building types in England, but would generally represent the UK and many parts of Europe in general. In other countries the construction materials, floor space and ventilation factors may vary from those common to this research data and therefore any guidance here should be considered on the basis that both fixed and movable fire loads may present fire intensities that are greater and are therefore likely to affect calculated outcomes.

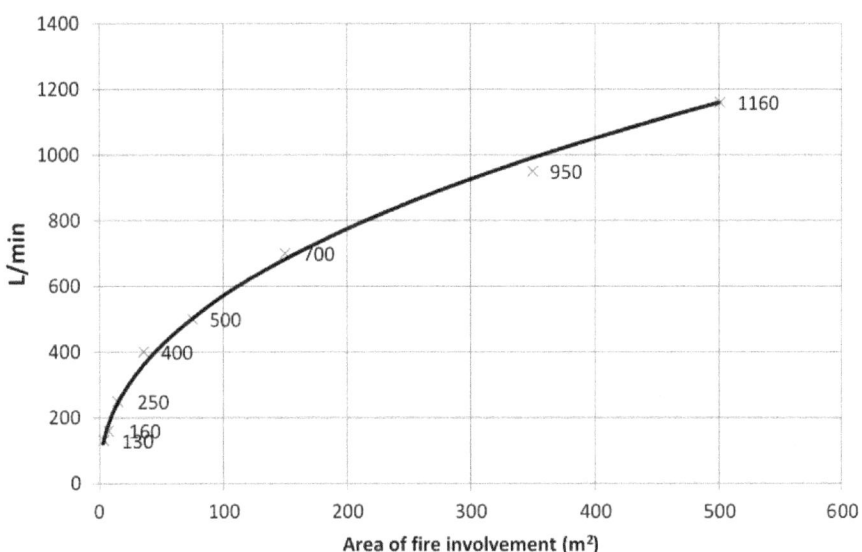

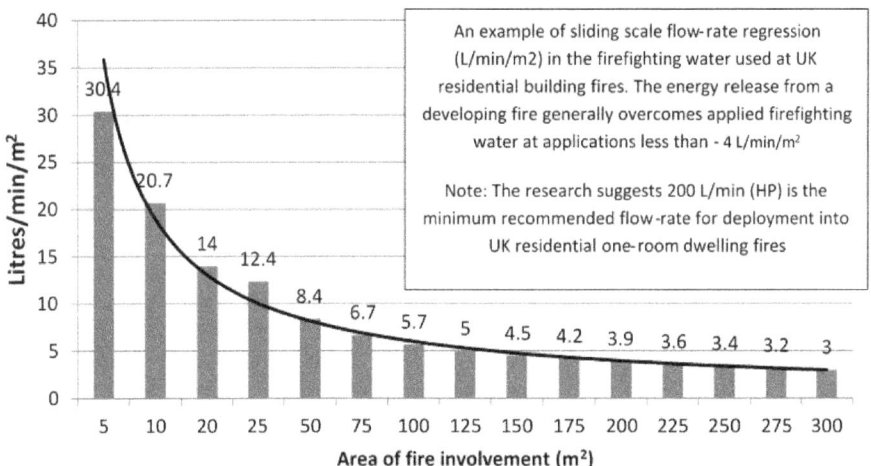

Figure 12.30: Sliding scale deployments and regression graphs of firefighting water (UK dwelling fires)

Chapter 13

Prescriptive & Performance based Building Designs

13.1 PRESCRIPTIVE CODES AND REGULATIONS

Regulations covering fire safety requirements for buildings in the UK are set out in Approved Document B (ADB) of the Building Regulations (DCLG, 2006), and provide prescriptive rules for a variety of measures in relation to building characteristics. A developing and updated standard, BS 9999 (BSI, 2017), has recently been published (originally in 2008), providing an alternative approach based on risk factors associated with the nature of the occupants, the building itself, and the factors likely to influence the severity of a fire such as ventilation and fire load. It incorporates compensatory allowance for sprinklers where appropriate. In general, this leads to a reduction in the structural fire resistance requirements compared with ADB, and in particular 2-storey offices less than 1000 square metres per floor, require only 15-minutes fire resistance. The standard is based on fire engineering principles and allows more flexibility than is possible using conventional prescriptive guidance. Guidance is also provided for aspects not previously covered in depth such as atria, fire service access, and post occupancy safety management. It therefore represents a significant advance, but compliance does not, in itself, represent performance based design.

13.2 PERFORMANCE BASED FIRE ENGINEERING

Even in the UK, where performance based fire engineering design is probably as well developed as anywhere in the world, a full fire engineering approach is still relatively rare, although it is becoming gradually more common. The extensive suite of documents contained in BS 7974 offer guidance to UK fire engineers in their quest for fire engineered approaches although they are equally able to utilise other guides (such as NFPA or IBC) should they so desire. The term fire safety engineering is now generally taken to imply a performance based approach and is firmly established in the UK as an alternative to simple compliance with the prescriptive rules of ADB. Indeed, in some types of large or complex buildings such as airport terminals, it may be the only practical way to achieve an appropriate standard of fire safety. The value of a more fundamental approach is also beginning to be recognised by building developers and owners, who are seeking efficient, reliable and consistent levels of fire safety. The basis for performance based design is founded in much experimental work, full-scale fire tests and computer modelling to validate particular process. The accuracy and reliability of any computer model is only as good as the person/s using it as well as the quality of inputs applied. The most searching questions should be asked about the limitations of any computer model and the accompanying fire strategy should be detailed in how the model itself can support each specific objective.

Main objectives of performance-based fire engineering design:

- Greater design flexibility
- Innovation in design, construction, and materials
- Equal or better fire safety
- Maximisation of cost/benefit

13.3 STRUCTURAL FIRE ENGINEERING

The prescriptive functional requirement B3(1) of the Building Regulations for England states that 'The building shall be designed and constructed so that, in the event of a fire, its stability will be maintained for a reasonable period'. The equivalent requirement (2.3) in the Scottish Building Regulations states that 'Every building must be designed and constructed in such a way that in the event of an outbreak of fire within the building, the load-bearing capacity of the building will continue to function until all occupants have escaped, or been assisted to escape, from the building and any fire containment measures have been initiated'.

In the context of structural fire engineering[147], the most common approach is simply to use an alternative to the standard fire *to justify a relaxation in the required fire resistance time*. This might typically be based on the time equivalent method or the Eurocode parametric fire curves. In many cases the structure is then simply protected according to the requirements for this reduced fire resistance time. This is largely because the cost of more detailed studies on the response of the structure to fire is often not justified by the possible cost savings, since in the UK there has been a significant reduction of the cost of fire protection. With a very competitive and effective fire protection industry efficiency has been improved dramatically, and costs have consequently been driven down, with a fall of as much as 60% in some types of protection in the past decade or so.

For buildings higher than about 7 storeys, there is a clearer potential economic benefit from a detailed study of the structural response. In such cases designers will consider realistic fire scenarios using one of the approaches outlined above, and model the structure using a finite element model such as Vulcan or ANSYS. This not only enables a more precise specification of the fire protection requirements for individual members, but also allows close inspection of critical details to ensure that the desired overall performance for both structure and compartmentation can be achieved.

Prescriptive codes and standards have the benefit that they are easy to apply and enforce. Additionally, buildings designed to prescriptive codes and standards have a good history of performance in fires. However, they do not result in uniform levels of safety or cost-benefit. Consider, for example, stores classified as mercantile occupancies. A store that sells greeting cards would fall under this occupancy classification, as would a store that sold liquor in bottles. Although the protection that would be required in these stores would be similar, the fire hazard presented by these stores would be different.

---

147 Plank. R.J, Performance Based Fire Engineering in the UK, Council on Tall Buildings and Urban Habitat, International Journal of High-Rise Buildings Volume 2 Number 1, 2013

The Shard Tower, London

At 310 m, this 70-storey mixed use structure is presently the tallest building in Western Europe, and for a building of this size and nature a performance based approach is very important to achieve the required level of fire safety. The structure changes at different levels, but is steel-framed from levels two to 40, which accommodate retail and commercial floors.

The fire engineering study[148] consisted of a consideration of worst case post-flashover fire scenarios using the parametric time temperature curve and taking account of a variety of realistic ventilation conditions corresponding to different degrees of glazing failure ranging from 25% to 100%. Thermal modelling was then used to determine member temperatures and whole frame modelling to determine overall performance. The fire rating of the concrete-filled steel box section external columns, fabricated from 100 to 125mm-thick plate, was determined using external flaming calculations to BSEN 1991-1-2 (BSI, 2002), and partly because of their size, which provided a significant degree of inherent fire resistance, only a thin layer of intumescent protection was required.

Special consideration was given to the main transfer structures over the backpack/tower interface under the effects of severe and highly localised fires to determine a worst case limiting temperature. Additionally, a qualitative assessment of the likely performance of the concrete construction was undertaken by reference to historical performance and recent research.

As a result of these studies, considerable savings were achieved for the fire protection required for the steelwork, whilst achieving a clear and consistent level of safety.

Figure 13.1: Some of the main transfer structures over the backpack/tower interface can be seen above

148 WSP Fire Engineering, London

Chapter 14

Fire Safety Regulation

'If you change the way you look at things, the things you look at change.'
                                                                    Dr. Wayne W. Dyer

In the case of fire, the Building Regulations set 'functional requirements' which must be equalled or surpassed in design. A series of 'Approved Documents' have been produced in support of these functional requirements and one of these, Approved Document B, deals with fire safety.

The main sections within Approved Document B cover such things as:

B1 Means of Warning and Escape
B2 Internal Fire Spread (linings)
B3 Internal Fire Spread (structure)
B4 External Fire Spread
B5 Access and Facilities for the Fire Service.

14.1 THE FUNCTIONAL REQUIREMENTS OF ADB (ENGLAND AND WALES)

ADB offers guidance in England and Wales on a prescriptive basis and lists several 'Functional Requirements' as Benchmark Solutions. Whichever way the design for a new or renovated building goes, the functional requirements must be met or achieved, even where performance based solutions are produced.

- **B1** (1) The building shall be designed and constructed so that there are appropriate provisions for the early warning of fire, and appropriate means of escape in case of fire from the building to a place of safety outside the building capable of being safely and effectively used at all material times. (Not prisons).
- **B2** (1) To inhibit the spread of fire within the building, the internal linings shall;

    (a) adequately resist the spread of flame over their surfaces; and
    (b) have, if ignited, a rate of heat release or a rate fire growth, which is reasonable in the circumstances.

- **B3** (1) The building shall be designed and constructed so that, in the event of fire, its stability will be maintained for a reasonable period.
    (2) A wall common to two or more buildings shall be designed and constructed so that it adequately resists the spread of fire between those buildings. For the purposes

of this sub paragraph a house in a terrace and a semi-detached house are each to be treated as a separate building.

(3) Where reasonably necessary to inhibit the spread of fire within the building, measures shall be taken, to an extent appropriate to the size and intended use of the building, comprising either or both of the following;

    (a) Sub-division of the building with fire resisting construction;
    (b) installation of suitable automatic fire suppression systems.

(4) The building shall be designed and constructed so that unseen spread of fire and smoke within concealed spaces in its structure and fabric is inhibited.

B4 (1) The external walls of the building shall adequately resist the spread of fire over the walls and from one building to another, having regard to the height, use and position of the building.

(2) The roof of the building shall adequately resist the spread of fire over the roof and from one building to another, having regard to the use and position of the building.

B5 (1) The building shall be designed and constructed so as to provide reasonable facilities to assist firefighters in the protection of life.

(2) Reasonable provisions shall be made within the site of the building to enable fire appliances to gain access to the building.

It is the responsibility of anyone carrying out building work to comply with the relevant requirements of the Building Regulations. The Approved Documents associated with the Regulations give practical guidance on how to comply. The design and construction of building work is subject to checks by a building control body. Applicants can decide whether to apply to the local authority for building control or to appoint an approved inspector. The two systems of building control are detailed in the Manual to the Building Regulations. When a building is likely to be used for a purpose that is subject to the fire safety requirements of other legislation, consultation between the relevant bodies is an essential part of the building control procedure.

Whilst there are clear distinctions of jurisdiction between building work and an occupied building it is often impossible when considering the overall level of safety in a building to separate physical fire safety measures and the way in which the building will be managed when occupied. Relevant guidance[149] describes the statutory consultations that fire authorities and building control bodies are obliged to carry out. It also indicates where discussions may be needed to consider fire safety proposals at an earlier stage than those required for statutory consultation so as to keep all parties informed and avoid delays. These discussions may be between the applicant and the building control body only, or they may also involve the fire authority.

Early discussions with the fire authority prior to any consultation process are welcomed where any of the functional requirements of B1-B5 (above), or firefighting water provisions (according to 12.7) are likely to be non-code-compliant or are subject to a fire engineered approach. In fact, where alternative measures or solutions are under consideration, a QDR[150] approach is most useful in achieving regulatory acceptance.

---

149 Building Regulations and fire safety: procedural guidance
150 Qualitative Design Review or QDR is defined in PD 7974: 2001 as the initial stage of any fire engineering design in which the basic design parameters for a project are established and the scope and objectives of the fire strategy can be defined

14.2 THE HOUSING ACT 2004

This legislation places duties and responsibilities on Local Housing Authorities to risk assess and bring about improvements in the housing stock through the HHSRS (Housing Health and Safety Rating System). HHSRS is a risk assessment system and requires 29 categories of 'hazard' to be considered when deciding whether any residential property is suitable for occupation – by virtue of being free from hazards that could harm the health of the occupiers. Hazards are divided into bands according to the seriousness of the risk to health. The highest risk hazards are called category 1 hazards (bands A – C), category 2 hazards then range from Band D down to the lowest banding where there is little or no risk and action to remedy the problem is not warranted. '**Fire**' is, for the first time, included as a general hazard that affects all occupiers of residential property. The vulnerable group – the people most likely to be seriously affected – are the over 60's. Therefore, fire risks must be assessed in all housing types and tenures, but with a focus on the types of housing with the highest likely risks – HMOs, blocks of flats, hostels, housing for vulnerable groups etc.

'If a local housing authority consider that a category 1 hazard exists on any residential premise then they must take the appropriate enforcement action in relation to the hazard.' – *Section 5, Housing Act 2004*. For modern post 1991 housing, the Building Regulations include appropriate fire safety measures including mains operated smoke detection to protect occupiers. The Housing Acts give Local Housing Authorities powers and duties to address the faults or defects (including fire safety defects) in relation to **existing housing** to minimise harm to occupiers.

14.3 THE REGULATORY REFORM (FIRE SAFETY) ORDER (RRO) 2005

This is the main legislation enforced by Fire and Rescue Authorities. There are areas of overlapping duties in relation to the Housing Act 2004 provisions. The Fire Safety Order 2005 concentrates on the communal areas of higher risk residential properties such as houses in bedsits (HMOs), high rise flats and buildings with vulnerable occupiers. The legislation also applies to workplaces, hotels, factories and the majority of occupied premises other than domestic premises occupied as a single dwelling. The central feature of the Order is that fire safety responsibility is now the duty of the owners and occupiers of buildings so there is a similarity with health and safety legislation. The Order identifies the responsible person as the individual who must take responsibility for fire safety – this could be the owner, agent or manager depending on who has the powers/duties and the finance to manage fire safety within the building. The responsible person must undertake a fire risk assessment of the building and then act upon it.

The RRO came into force on 1st October 2006. The RRO replaced over 70 pieces of separate fire legislation. The approach of the RRO is that those best placed to identify fire risks and address fire safety should be responsible for doing so. Therefore, it places duties on the 'responsible person' to conduct risk assessments, implement appropriate fire safety measures to minimise those risks and to keep the assessments under review and up to date. The RRO is specifically directed towards 'the safety of people in relation to the operation and use of certain buildings once occupied'

'Responsible person':

Article 3 RRO defines the term 'responsible person'. Pursuant to Article 3 there are three categories of people that can be the 'responsible person':

(a) Employers: If the premises are a workplace the employer will be the responsible person, to any extent that the workplace is under his control. An employer will not be responsible in respect of parts of the premises which are not under his control.
(b) Persons in control in connection with a trade, business or undertaking: If the premises are not a workplace, such as shared parts of a block of flats, the person in control of the premises in connection with a trade, business or undertaking, will be the responsible person.
(c) Owners: If the person in control of the premises does not do so in connection with a trade, business or undertaking, the owner of the premises will be the responsible person.

There may be a number of people with shared responsibility in relation to premises. For example, an owner/landlord may be responsible for fire safety for the parts of the premises over which he retains control, whereas his tenant who uses the premises as part of his business may be responsible as an employer in respect of those parts of the premises over which he has control. The enforcing authority can therefore bring a prosecution against more than one person in respect of the same premises.

In summary, responsible persons are expected to:

- take general fire precautions to ensure the safety of any employees and non-employees on the premises;
- carry out a risk assessment to identify what general fire precautions are required for the particular premises and keep this under review;
- keep records, which must include the significant findings of the fire risk assessment (and any review); the fire precautions that have been, or will be put in place to address the significant findings; and any group of persons identified by the assessment as being especially at risk;
- make and give effect to fire safety arrangements, as appropriate;
- take measures for firefighting and fire detection, including the provision of appropriate equipment at the premises;
- ensure that the premises, any fire safety equipment, and all emergency routes and exits are properly maintained and kept in working order;
- cooperate with other people who have fire safety responsibilities and coordinate with each other; provide information and safety training for employees; and
- provide information about fire safety risks for third parties who may be affected, and those who are responsible for such people.

Maintenance of measures provided for the protection of firefighters
Section 38

(1) Where necessary in order to safeguard the safety of firefighters in the event of a fire, the responsible person must ensure that the premises and any facilities, equipment and devices provided in respect of the premises for the use by or protection of firefighters under this Order or under any other enactment, including any enactment repealed or revoked by this Order, are subject to a suitable system of maintenance and are maintained in an efficient state, in efficient working order and in good repair.
(2) Where the premises form part of a building, the responsible person may make arrangements with the occupier of any premises forming part of the building for the purpose of ensuring that the requirements of paragraph (1) are met.

(3) Paragraph (2) applies even if the other premises are not premises to which this Order applies.
(4) The occupier of the other premises must co-operate with the responsible person for the purposes of paragraph (2).
(5) Where the occupier of the other premises is not also the owner of those premises, the reference to the occupier in paragraphs (2) and (4) are to be taken to be references to both the occupier and the owner.

14.4 THE FIRE RISK ASSESSMENT

The Fire Risk Assessment (FRA) means that each building which is subject to the Fire Safety Order is individually assessed in detail and all necessary works are identified and prioritised. FRA also takes into account of any vulnerable groups – those with impaired mobility or psychological issues which could affect behaviour in the event of a fire.

The greatest risks for occupiers of residential properties are to be found in multi-occupied properties where there are 3 or more storeys. This may include houses that are converted into flats, hostels, managed or sheltered accommodation, purpose built multi-storey buildings and flats above shops. The risk rises with increased occupancy, multiple ignition sources (cookers, heaters, fires, smoking), vulnerable occupants, poor construction and lack of fire prevention measures. Analysis of national fire statistics have concluded that you are six times more likely to die in a fire if you live in any house in multiple occupation (HMO), compared with a single-family house.

14.5 THE LICENSING ACT

The Licensing Act 2003 came into force on 24 November 2004 and introduced a single licensing scheme for premises which:

- Supply alcohol
- Provide regulated entertainment
- Provide late night refreshment

The Act has four principle objectives:

- The prevention of crime and disorder
- Public safety
- The prevention of public nuisance
- The protection of children from harm

Fire safety

The Fire Service are required to comment on the fire safety provisions and notify the licensing authority on the suitability of premises. To enable us to do this the fire service will normally require:

- A copy of the application
- A copy of the operating schedule
- A plan of the premises in the prescribed form

It would also be beneficial to provide a copy of the *Fire Risk Assessment*, although there is no legal requirement to do this.

Following receipt of an application the Fire Service may decide to audit premises to determine suitability for licensing. This audit will be carried out under the Regulatory Reform (Fire Safety) Order 2005. The Government has produced sector specific guidance for small, medium and large places of assembly which provides valuable information on your responsibilities and technical guidance for the majority of premises.

Chapter 15

Structural stability under Fire Attack

'Don't let us look back tomorrow and say what we did today, could have been done better.'
Prepare – Plan – Train

The United States Fire Administration reported that 1,230 fire fighters died between 2000 and 2012. Structural collapse caused 142 of these fire fighter deaths (11.5%). Structural collapse often results in multiple fire fighter injuries and fatalities. The number of fatalities includes firefighters killed by collapse inside and outside a structure. How is this relevant to firefighters in the UK and Europe? Simply because our residential building construction is now trending towards fast-build lightweight modular framing systems and the working environment is likely to follow similar hazards as those experienced by US firefighters over the past 25 years where the firefighter life loss has demonstrated dramatic increases as a ratio of the number of building fires (see Ch.17.2). The life losses amongst firefighters both inside and outside residential buildings has proportionally been increasing for 25 years and various guidance has been issued to fire departments

Although structural collapse is a significant cause of injury and death to fire fighters, the potential for a structural collapse is one of the most difficult situations to predict. A collapse zone is defined as the area around the perimeter of a structure that could contain debris if the building collapses. This area is sometimes referenced in establishing a perimeter at a distance from the building that is equal to 1½ times the height of the structure. Once established, a collapse zone should be identified by a fire-ground transmission, coloured tape, signage cones, flashing beacons, fences, or other appropriate means. A 'No Entry' policy should be enforced by the Incident Commander and defensive modes of attack should be activated from outside those areas identified for potential collapse.

When it comes to fire performance of building materials, detailed heavy timber structures for instance may perform well by slowly forming a char layer. However, slender timber in lightweight form is generally protected by fire-rated board and great care is required in construction and ongoing maintenance across its life-span if fire resistance is to remain effective. Meanwhile, the fire performance of concrete, with its variety of mixes and formulations, may be difficult to determine precisely from the street. Steel's great advantages are that it is non-combustible and that there is a large body of research into its performance in fires. The strength of hot rolled structural steel decreases with temperature. Following an extensive series of standard fire tests, that strength reduction has been quantified. Recent international research has also shown that the limiting (failure) temperature of a structural steel member is not fixed but varies according to two factors, the temperature profile and the applied load. That load can change as the building comes under fire.

Lightweight steel framing is far less likely to demonstrate any real stability under fire attack unless well protected. For small, fully loaded hot rolled sections, exposed on all four sides, the inherent fire resistance without added protection can be as little as 12 minutes. For very large, hot rolled sections, lightly loaded and with some partial protection from concrete floor slabs on the upper flange, this can be as high as 50 minutes. Where the heated perimeter is further reduced by the method of the construction, up to 60 minutes inherent fire resistance can be achieved. The best known form of construction which uses this principle is 'Slimdek'.

Quite often we may see a combination of materials forming main structural elements and their performance in fire generally depends on their cohesion within the structural frame and most importantly, how effectively they are connected to each other.

15.1 A 'REASONABLE' PERIOD

The building regulations for England and Wales, for instance, set out the legal obligations for building designers. These provide 'functional requirements' that outline what must be achieved through a series of 'approved documents' that explain how such requirements may be met. In terms of fire safety, the relevant publication is Approved Document B (ADB). An example of a functional requirement that covers structural stability in a fire is that 'the building shall be designed and constructed so that, in the event of a fire, its stability will be maintained for a reasonable period'.

The regulations do not actually define what a 'reasonable period' is, but ADB does offer some guidance in terms of minimum required structural fire resistance periods. Structural fire resistance periods vary between 30–120 minutes. In some instances of smaller buildings or industrial units, no fire resistance to the structural elements may be required at all and in some codes (BS 9999) only 15 minutes is required in small office buildings. The overriding objective of 'reasonable period' is aligned with a relevant time frame that enables occupants to escape, firefighters to enter and access the fire, and remain therein for a reasonable period to control and extinguish the fire. In some cases, fire suppression is achieved in minutes but in others, an interior firefighting operation may be extended to several hours as fire spreads to adjacent areas or upper floor levels. It is critical that firefighters are trained in fire resistance expectations of various occupancies, building sizes and structural designs.

15.2 DISPROPORTIONATE COLLAPSE

UK guidance for design against disproportionate collapse has its origins in the measures implemented in the Building Regulations shortly after the Ronan Point collapse in 1968 and remains largely unchanged in current Codes of Practice [Cormie][151]. Over the intervening 40 years, the design of structures has advanced substantially, with structural spans increasing substantially over the period from typical 6×6m structural grids up to 12x13.5m and even up to 13.5x18m now being common. More efficient design has been

---

151 Cormie. D (Arup), Review of international research on structural robustness and disproportionate collapse; Department for Communities and Local Government; DCLG, Centre for the Protection of National Infrastructure (CPNI), 2011

led by technological advances and by advances in methods of analysis and computing power. The drive for faster erection times on site due to labour costs and site mobilisation costs has led to lighter methods of construction and modularisation. The densification of housing has resulted in substantial multi-storey dwellings, with timber frame construction becoming commonplace for up to six storeys and in some cases up to nine storeys. Structural steel construction has become lighter, with longer spans and lighter connection types. Significant advances have been made in cold-formed lightweight steel construction, and precast concrete is enjoying renewed popularity. Curtain walling is the cladding type of choice for many buildings, whereas masonry infill panels once so prevalent have all but disappeared. The move to open-plan offices and the demand for more flexible fit-out design has made the masonry partition wall largely extinct.

All the above factors have indisputably diminished the robustness of our buildings. The document referenced above [Cormie] is well worth reading in terms of its broad international definitions of building stability and collapse and various approaches on mechanisms to resist collapse, including redistribution of loads, where columns and beams in particular may fail.

A *progressive collapse* is one which develops in a *progressive* manner akin to the collapse of a row of dominos. A collapse may be progressive horizontally – successively from one structural bay to those adjacent to it and propagating through the structural frame. A collapse may also be progressive vertically – e.g. the collapse of the columns supporting a floor slab due to the dynamic shock load caused by the collapse onto it of the storey above it, or the successive collapse of the columns supporting a number of floors due to the dynamic shock load as the block of mass is brought to rest as it impacts with more rigid structure below. These examples of vertical progressive collapse are often termed 'pancaking' (downward and upward respectively). The term 'progressive' refers to a characteristic of the behaviour of the structural collapse.

A *disproportionate collapse* is one which is judged (by some measure defined by the observer) to be disproportionate to the initial cause. This is merely a judgement made on observations of the consequences of the damage which results from the initiating events and does not describe the characteristics of the structural behaviour.

A collapse may be *progressive* in nature but not necessarily *disproportionate* in its extents, for example if arrested after it progresses through a number of structural bays. Vice versa, a collapse may be *disproportionate* but not necessarily *progressive* if, for example, the collapse is limited in its extents to a single structural bay but the structural bays are large.

The terms *structural robustness* or *robustness* are used to describe a quality in a structure of insensitivity to local failure, in which modest damage (whether due to accidental or malicious action) causes only a similarly modest change in the structural behaviour. More specifically, a robust structure has the ability to redistribute load in the event that a loadbearing member suffers a loss of strength or stiffness, and characteristically exhibits ductile rather than brittle global failure modes. A robust structure does not mean one that is over-designed: the ability to resist damage is achieved through consideration of the global structural behaviour and failure modes so that the effects of a localised structural failure can be mitigated by the ability of the structure to redistribute the load elsewhere, and so that the effects of the initial failure are gradual in onset.

Eurocode 1 (BS EN 1991-1-7) describe robustness as 'the ability of a structure to withstand events like fire, explosions, impact or the consequences of human error without being damaged to an extent disproportionate to the original cause', thereby linking it explicitly to the concept of disproportionate collapse while recognising that total collapse is an acceptable outcome from a gross hazard.

So far as reasonably practicable – SFARP

The tolerability of risk and the probability of collapse is detailed and complex and the Building Regulations (ADA) cover such issues, along with other guidance documents. Safety legislation in construction talks of elimination of risk or its reduction 'so far as reasonably practicable' (SFARP). For the last 43 years, under the requirements of the Health and Safety at Work Act 1974, those in control of the premises or work activity have been obliged by law to reduce risk SFARP.

Structural Fire Engineering

Significant work has been undertaken in the structural fire engineering arena to develop analytical methods for designing structures to be resistant to fire-induced collapse. There is substantial similarity in the analytical methods employed, although there are also some important differences:

Structural fire engineering is generally focussed on demonstrating structural stability within a single floor or a small group of floors. This is in contrast to damage-induced structural collapse, for which the whole building frame is likely to be employed in resisting the collapse.

Fire-induced structural collapse takes place over a longer timescale and modelling of effects such as the softening of the structural steelwork and heat-induced buckling of members is necessary. Damage-induced structural collapse takes place over a fundamentally shorter timescale and the dynamic effects of the problem are therefore much amplified.

Notwithstanding the above differences, it remains that there is substantial similarity between these two areas and that the knowledge available in the structural fire engineering community should be exploited in the development of knowledge in damage-induced structural collapse. The University of Edinburgh is a centre of excellence in the structural fire engineering arena (Flint, Lane et al, var.).

15.3 TOTAL 'BURNOUT' COMPARTMENTS

Refer also to s 17.6

Many international design codes are now stipulating a requirement to design a structure to compartment '**burn-out**', maintaining structural stability throughout the duration of a fire rather than designing to a 'reasonable period'. I respect anyone who has such design ambitions but I'm not sure if this is practically viable or even within reach at this point in time. However, designing to 'total burnout' – what actually does that mean? In the New Zealand design code[152] it means 'full burn out without intervention'. The definition it gives is as follows:

*'**Burnout**' Means exposure to fire for a time that includes fire growth, full development, and decay in the absence of intervention or automatic suppression, beyond which the fire is no longer a threat to building elements intended to perform loadbearing or fire separation functions, or both'.*

It follows with section 2.4 in the NZ code:

---

152 C/VM2 Verification Method: Framework for Fire Safety Design; For New Zealand Building Code Clauses **C1-C6 Protection from Fire; 2012**

Section 2.4 Full Burnout Design Fires
NZ C/VM2 Comment:
Design fire characteristics include parameters for Fire Load Energy Density *(FLED)*, *fire growth* rate and heat of combustion. This means a post-*flashover* 'full *burnout design fire* can be defined. The 'full *burnout design fire*' for structural design and for assessing *fire* resistance of *separating elements* shall be based on complete *burnout* of the *firecell* with no intervention. However, the maximum *fire resistance rating* for a sprinklered *firecell* need not exceed 240/240/240 (4-hour fire resistance) determined using AS 1530.4.
There are three choices for modelling the full *burnout design fire*:

- Use a time-equivalent formula to calculate the equivalent *fire* severity and specify *building elements* with a *fire resistance rating* not less than the calculated *fire* severity. In this case, an equivalent *fire* severity of 20 minutes shall be used, if the calculated value is less.
- Use a parametric time versus gas temperature formula to calculate the thermal boundary conditions (time/ temperature) for input to a structural response model, or
- Construct an *HRR* versus time structural *design fire* as described in Paragraph 2.3.3. Then, taking into account the ventilation conditions use a *fire* model or energy conservation equations to determine suitable thermal boundary conditions (time/temperature/flux) for input to a structural response model.

Further comment is provided in **Practice Advisory 18 Note**: Fire safety design for tall buildings – *The protracted time usually required for occupant evacuation and firefighter operations requires the structure of tall buildings to remain stable for the full duration of a fire. Overall global structure instability is not an acceptable performance outcome while occupants or firefighters are in the building, or where structural collapse due to the effects of fire causes damage to other property. A <u>cautious assessment of the fire severity associated with complete fire burnout and design strategies for maintaining structure stability is needed.</u> This structural fire performance requirement applies to tall buildings in order to comply with Building Code Clauses B1 and C6. Therefore, structural systems in tall buildings need to be designed to remain stable for the complete duration of a fire (for complete burnout and cooling to ambient temperature) to maintain safe access and egress for firefighters and building occupants. <u>Consequential collapse of elements not required for structural stability must not compromise occupant safety or firefighting access.</u> This usually requires particular attention to fire resilience of shafts containing stairs and lifts used for fire egress and firefighter access, and areas adjacent to final exits. The likelihood and consequence of fire in podium spaces, car parks, transportation hubs, retail areas and similar spaces at lower levels of tall buildings need to be considered to the extent it affects overall structural stability.*

Time equivalence
A common approach to use with this Verification Method is the 'equivalent fire severity' method described in Eurocode 1 Actions on structures, Part 2-2. This allows the equivalent time of exposure to the *standard test* for *fire* resistance to be estimated based on the compartment properties, *FLED* and available ventilation given complete *burnout* of the *fire-cell* with no intervention.

In 2014 the building research association of New Zealand (BRANZ) reported on the various methodologies that may be used to quantify the necessary fire resistance to meet full burn out conditions. The report concluded that 'while the equivalent time of exposure could be used as a basis for quantifying the destructive potential of the fire on a given construction element and as a basis for setting fire resistance requirements, it cannot be relied upon to provide an accurate prediction of the expected failure time in a compartment fire'. The report went on, 'Existing time-equivalent methods do not explicitly allow for materials that burn or char, during or following consumption of the primary fuel load within the compartment', and 'the current methods may not be appropriate for evaluating the performance of assemblies with combustible materials because once these materials burn or char they may continue to be consumed after the fire intensity decays. This continued self-sustaining degradation of the fire-rated elements may be significant and result in eventual failure of the element or assembly, even after the contents of the compartment have been consumed. Thus, applying the concept of 'withstanding burnout' may not be appropriate unless it can be assured that either the degrading material does not get involved or else continued degradation of the material ceases at some point or the residual structure continues to maintain stability or prevent fire spread depending on the design objective. For example, for gypsum plaster-lined and timber-framed assemblies, a temperature-based time-equivalence method could be based on preventing the onset of char. The specification of fire resistance ratings for combustible construction to ensure structural adequacy or prevent fire spread requires much additional research.

The concept of withstanding burnout is liberally used within the New Zealand Building Code 'Protection from Fire Clauses and Verification Method', and can be taken to mean the construction continues to perform its function as a barrier or load-bearing structure following exposure to a compartment fire for its full duration including any decay period. Using time-equivalence methods to specify a fire resistance rating sufficient to withstand burnout does not provide certainty that the functional requirements of the Building Code will be achieved (i.e.no collapse or fire spread prevented).

In 2010 an article[153] in the Structural Engineer gave this view on Time Equivalence – 'The standard furnace test does not take any account of any ventilation being provided and works on the worst-case assumption that no venting is provided from the fire compartment. Increased levels of ventilation allow greater amounts of heat and smoke to be vented from the fire compartment, reducing the temperature to which the structure is subjected, allowing it to maintain its performance for longer. The time equivalent calculation quantifies this reduction in required structural fire resistance. The time equivalent method provides a way of calculating the fire resistance period needed in a building, expressed as a time equivalent to fire resistance duration as tested in a standard furnace test. The standard furnace test does not take account of several real building characteristics which affect the conditions to which the structure and other fire rated elements are subjected, and therefore the duration they will maintain stability during a real fire. Perhaps the most obvious factor is that it tests elements in isolation without reflecting their interaction with other elements of structure. By assessing the effect of each of these variables it is possible to calculate the time equivalent period. The variables considered include ventilation, fire loading, and rate of heat absorption to linings and the presence of sprinklers. The method has been experimentally derived through large scale fire testing. When applying the time equivalent method, it is important to check that the building being assessed falls within the validity range of the original experiments, as expressed by values for the variables given in the guidance, and therefore that the results are valid'.

---

153 Marshall. W; The Structural Engineer; 88: November 2010

The **National Institute of Standards and Technology** (NIST) in the USA undertook a review[154] of fire resistance in timber buildings in 2014. Their report states:

*'The most common way of **designing for burnout** is to use a time-equivalent formula to estimate the equivalent fire severity (exposure to a standard fire) for the complete process of an uncontrolled fire from ignition through fire growth, flashover, burning period and decay to final extinguishment. Such time-equivalent formulae assume that the fire severity is a function of the fire load, the available ventilation, and the thermal properties of the surrounding materials of the fire compartment. This has some problems, especially if we do not know the worst case scenario for fuel load and ventilation. These values should be determined on a probabilistic basis, with higher safety factors for increasingly tall buildings. More research is required to assess the applicability of current time-equivalent formulae for use in multi-storey timber buildings. The fire severity, hence the time-equivalent formula, will depend on whether the wood structure has no protection, limited encapsulation or complete encapsulation'.*

There are two important factors that need acknowledgement here. The concept of 'full or total' compartment burnout appears to mean designing to support firefighters to undertake extended firefighting operations during a severe compartment fire, even into the decay stages. Therefore 'total burnout' means that the fire duration and heat release will be reduced by firefighting actions and that at least part of the fuel load mass would remain unburned. To define this any other way, meaning that the building will support a total compartment burnout because the structure design will prevent upwards fire spread and the structural elements will likely withstand 30 minutes of PHRR and another 60-120 minutes of far field temperature exposures without any cooling applied, is I think uneconomical and beyond reach at this point in time. The importance and effectiveness of automatic fire suppression systems is of course so highly relevant in this respect and the uppermost levels of fire safety management within a building, to include regular and effective system maintenance, are key.

Total Burnout

154 Buchanan. A; Ostman. B; Frangi. A; The Fire Resistance of Timber Structures; NIST USA 2014

15.4 FURNACE TESTING DOES NOT REPLICATE 'REAL FIRE' CONDITIONS

In a paper[155] by *Bisby et al.* reporting that after about a century with the standard fire resistance test being the predominant means to characterise the response of structural elements in fires, both research and regulatory communities are confronting the many inherent problems associated with using simplified single element tests, on isolated structural members subjected to unrealistic temperature-time curves, to demonstrate adequate structural performance in fires. As a consequence, a shift in testing philosophy to large-scale non-standard fire testing, using real rather than standard fires, is growing in momentum. A number of custom made, non-standard testing facilities have recently been constructed or are nearing completion. Non-standard fire tests performed around the world during the past three decades have identified numerous shortcomings in our understanding of real building behaviour during real fires; in most cases these shortcomings could not have been observed through standard furnace tests.

A report commissioned by NIST[156] in 2008 contains descriptions of 22 cases where multi-storey buildings experienced fire-induced collapses between 1970 and 2002, with approximately equal distribution between steel, concrete, and masonry buildings, and with the majority of fires in office or commercial buildings. Specific details of the various collapses are omitted here. However, the following statements can be made on the basis of the information presented:

1. in most cases the critical failure modes observed in real buildings could not have been predicted on the basis of standard fire resistance (furnace) testing;
2. structural interactions and connection response played important roles in all cases; and
3. in several cases the collapses occurred during the cooling phase of the fire.

The report notes that 'connections are generally recognized as the critical link in the collapse vulnerability of all structural framing systems, whether or not fire is involved'. Also presented is a review of high-rise building fires without collapses, but with major structural damage, which leads to the suggestion that further work on the structural fire response of entire building frames should be conducted for both steel and concrete construction. The final contribution of this report is to present a global survey of structural fire resistance testing capability, essentially by surveying fire research and testing laboratories around the world as to their capabilities with respect to both vertical and horizontal structural elements. The focus appears to be on structural fire testing capabilities involving furnace testing, as opposed to tests using real fires. Based on the responses, the authors conclude that many laboratories are able to perform standard furnace fire resistance tests of various sizes, types of fire exposure (heating but not necessarily controlled cooling), loading, and measurements. They note that several unusually large furnaces exist and that these could be used to evaluate structural connections or combinations of building elements, however they also note that no single laboratory is currently able to test large-scale structural assemblies under the full range of applicable loading and fire exposure conditions.

The paper by Bisby et al. went on to describe in some detail the outcomes of the full-scale Cardington tests in 1996 (seven fire tests in an eight-storey steel-framed building).

---

155 Bisby. L; Gales. J; Maluk. C; A contemporary review of large-scale non-standard structural fire testing; Fire Science Reviews 2013, 2:1 http://www.firesciencereviews.com/content/2/1/1
156 Beitel and Iwankiw (2008) - NIST GCR 02-843-1 (Revision)

This test program surely represents the most comprehensive and realistic test series that has ever been performed, and is a key reason why the steel industry has been able to aggressively promote performance-based structural fire design in the subsequent decades, with significant economic and sustainability benefits in steel-framed buildings. The 21 m × 45 m building was three bays by five bays, and had a total height of 33 m. All beams were designed as simply supported, acting compositely with a floor slab on steel decking. Bisby et al. – *'Taken together, these seven tests demonstrated many important aspects of the full-structure response of composite steel-framed buildings during fire. In particular, they shed light on the secondary load carrying mechanisms which can be activated during fire to prevent collapse, the potential importance of restraint to thermal expansion on heating (and thermal contraction on cooling) on localized buckling and/or connection failures, and the fact that full-structure response in fire is markedly different than that observed in standard fire resistance tests performed in furnaces. In the case of regular grid plan composite steel-framed buildings such as the one tested at Cardington, the fire resistance appears to be far greater than is normally assumed on the basis of furnace tests. This conclusion has been used to great advantage in recent years by the Steel Construction Industry (in particular in Europe) to justify removal of passive fire protection from secondary steel beams and steel decking in composite framed steel buildings, with considerable aesthetic, economic, sustainability, and constructability benefits'.*

The paper goes on to provide interesting details of a number of similar full-scale fire tests in both steel and concrete buildings that have taken place around the world, with variable results. The authors conclude:

'Taken together, the tests described in this report highlight a number of important construction details and potential construction errors which may appear inconsequential to a building contractor, but which may have a profound impact on the structural fire response and integrity of a building during fire. Examples of this include integrity of fire stopping during large deformations, lapping of steel reinforcing mesh, anchorage of steel reinforcing mesh over shear studs on protected perimeter beams, use of deformed versus smooth bars for reinforcement (potentially leading to strain localization and tensile failure of deformed steel bars during fire), proper anchorage and grouting of hollow core slabs, use of specific types of bolted steel connections to promote connection ductility and rotational capacity during fire, quality, uniformity, and robustness of structural fire protection materials (either passive or intumescent), and so on. Serious unknowns continue to surround many, if not all, of these issues, and there is a need for testing to support the development of best practice guidance which can be used to provide quality assurance programs on construction sites of so-called 'fire engineered' buildings'.

In particular[157], Bisby et.al. report on Test #6 of Cardington's 1996 tests series that was meant to assess the global structural behaviour of a large rectangular corner compartment. The fire load was given by representative office furniture, which, based on previous fuel surveys, resulted in more fuel than the 80% fractile fire load recommended by the European standards. In this test, as with most compartment fire tests performed during the Cardington test program, the columns were protected to avoid local buckling as was experienced in one of the early tests. A subsequent Cardington test was performed in 2003. All of the compartment fire tests executed during the Cardington tests series were done exposing the structure to both a heating and cooling phase; in some occasions this resulted in failures (local buckling near the connections and cracking of the composite

---

157 Bisby. L; Gales. J; Maluk. C; Structural fire testing – Where are we, how did we get here, and where are we going? 15th International Conference on Experimental Mechanics, 2012

slabs) during the cooling phase. This behaviour has also been seen in real fires and in other non-standard large scale structural fire tests. Taken together, the seven Cardington tests demonstrated many important aspects of the full-structure response of composite steel-framed buildings during fire. In particular, they shed light on the secondary load carrying mechanisms which can be activated during fire to prevent collapse, the potential importance of restraint to thermal expansion on heating (and thermal contraction on cooling) on localized buckling and/or connection failures, and the fact that full-structure response in fire is markedly different than that observed in standard fire resistance tests performed in furnaces.

Bisby et al. further report that in 2001, as part of the European Concrete Building Project (Bailey, 2002[158]), a test on a seven storey reinforced concrete building was performed, also at the Cardington UK test site. The full-scale building represented a typical commercial office building. The three bays by four bays building had total dimensions of 22.5 × 30 m, with two core areas, which included steel cross-bracing to resist lateral loads. The test took place inside a 15m × 15m ground floor compartment. The main purpose of the test was study the global behaviour of the reinforced concrete building, with special attention in assessing the influence of restraint from the surrounding cold structure. Also, the impact of spalling on the structure's load-bearing capacity was a focus of the analysis. Vertical loads were applied, throughout the whole building, by means of evenly distributed sand bags, replicating design imposed loads, partitions, raised floor, ceiling and service loads. The fire load was given by timber cribs evenly distributed (40 kg/m2) to simulate one of Eurocode's parametric fires (CEN, 2004). 40 kg/m2 is known as a representative fuel loading of an office structure (see Lennon and Moore, 2003[159]). Structural stability was maintained during testing. However, significant deformation of perimeter columns was observed. This was attributed to lateral thermal expansion of the slab undergoing compressive membrane action. Severe spalling of the underside of the floor slab was also observed, starting six minutes from the start of the test. This was attributed to the high in-plane compressive stresses within the slab undergoing compressive membrane action. This test again demonstrated that the performance of real structures in real fires is markedly different than the response of isolated elements tested in standard furnaces, both in terms of structural response and potential failure modes.

As noted by authors going back to (at least) 1981, the standard temperature-time curve is not representative of a real fire in a real building. In order to truly understand the response of real buildings in real fires, tests of structures and structural elements are required under credible worst case natural fire exposures. Depending on the type of structure and the occupancy under consideration, this may require experimental consideration of localized, compartmentalized, horizontally and/or vertically travelling, smouldering, or hydrocarbon fires, all of which have the potential to introduce structural actions or interactions which are not captured by the standard fires. Furthermore, given that many structural fire engineers already have serious concerns about the quality of installed fire stopping between floors in multi-storey buildings, large-scale nonstandard fire tests should perhaps be considered in which vertical fire spread is explicitly simulated using natural fires to evaluate the structural impacts of credible worst case fires burning simultaneously on more than one floor of a structure. Taken together, the large-scale non-standard fire tests reported in the literature highlight a number of important construction

---

158 Bailey CG, 2002. Holistic behaviour of concrete buildings in fire. *Proceedings of the Institution of Civil Engineers, Structures and Buildings*, 152 (3), pp. 199-212.
159 Lennon T, 2003. The Natural fire safety concept- full scale tests at Cardington. *Fire Safety Journal*, 38. 623-643.

details and potential construction errors which may appear inconsequential to a building contractor, but which may have a profound impact on the structural fire response and integrity of a building during fire.

15.5 COOLING PHASE AND RESIDUAL CAPACITY

Bisby et al further reports that a number of localized structural failures or adverse structural responses of steel connections, concrete flat plate slabs, and hollow core slabs have been observed during the cooling phase of both real fires in real buildings (e.g. Firehouse. com 2004, Bamonte[160] et al. 2009) and non-standard heating regimes in large-scale structural fire experiments (e.g. Bailey and Lennon 2008, British Steel 1999). Structural actions resulting from creep, localised and/or global plastic deformation, local buckling, and thermal contraction and restraint, <u>all need to be better understood for all types of structures if designers are to realistically be expected to design for full burnout of a fire compartment without structural collapse.</u>

Furthermore, the residual structural capacity of fire damaged structures which have undergone large deformations is not well known, meaning that many fire damaged structures will need to be demolished after a fire (e.g. New York Times 1997). This is particularly true for so-called fire engineered composite steel frames, which explicitly rely on large deformation behaviour to mobilize the tensile membrane actions which are necessary to support gravity loads and prevent collapse during fire (British Steel 1999).

A critical issue for the structural fire protection of timber structures appears to be the integrity of fire proofing materials such as gypsum plasterboard (Lennon et al. 2000). Additional research is required to better understand the factors leading to 'fall-off' of plasterboard and other fire stopping materials and systems during fire. It is important to recognise the difference between light timber frame buildings (where the fire resistance of wood must be protected by gypsum lining materials) and heavy timber buildings which rely on the predictable charring rate of large timber members.

Another critical issue for timber structures is the performance after the fire has gone out. Designers are often concerned that charred timber will continue to char after all flaming has ceased, which makes it difficult or impossible to design for a complete burnout as is done for concrete or steel structures. More research is needed in this area, and this can only be done with large scale structural fire testing.

15.6 FAILURE CRITERIA IN COMPOSITE FRAME STRUCTURES (LAMONT)

In order to assess the data provided from a finite element analysis some means of defining failure criteria must be established. There is currently in the UK no Regulatory definition of failure. The term 'failure' is not straightforward to define in the context of this type of analysis on the basis that, although a compartment fire may lead to large deflections of main and secondary beams, this is unlikely to cause structural collapse i.e. stability requirements can be met. However, for compartmentation large deflections could cause a breach of the separating function of the element such as the floor, or the escape cores.

---

160 Bamonte P, Felicetti R, and Gambarova PG, 2009. Punching Shear in Fire-Damaged Reinforced Concrete Slabs. *ACI Special Publication 265, American Concrete Institute*, Farmington Hills, USA, pp. 345-366.

On this basis, the following were proposed in a fire engineered strategy as acceptance criteria:

- Stability of structure maintained throughout the design fire. This was primarily assessed by looking at the rate of deflections during the fire. Runaway deflections (a rapid increase in the rate of deflection) were assumed to indicate failure of the floor system and pulling in of the columns.
- Horizontal compartmentation was also assessed by monitoring the rate of deflection of the composite floor. A rapid increase in deflection in any region of the floor plate was assumed to imply compartmentation failure.
- Vertical compartmentation via the vertical firefighting shafts (a fire rated shaft required in the UK in buildings with a floor 18m above fire service access level to provide a place of relative safety on each floor for fire fighters) was assessed by monitoring the connections at the shaft wall to ensure that they maintained their capacity for the fire period.

15.7 GRANARY WAREHOUSE WALL COLLAPSE, LONDON 1978

[In large warehouses], 'Cast iron gives way from so many different causes that it is impossible to calculate when it will give way. The castings may have flaws in them; or they may be too weak for the weight they have to support, being sometimes within 10 percent or less of the breaking weight. The expansion of the girders may thrust out the side walls. For instance in a warehouse 120 x 75 x 80 feet, there are three continuous rows of girders on each floor, with butt joints; the expansion in this case may be twelve inches. The [cast iron] tie rods to take the strain of the flat arches must expand and become useless, and the whole of the lateral strain be thrown be thrown on the girders and side walls, perhaps weak enough already. Again, throwing cold water on the heated iron may cause an immediate fracture. For these reasons, firemen are not permitted to go into warehouses supported by iron, when once fairly on fire'.

Sir James Braidwood
Chief Fire Officer Edinburgh 1824-1833
Chief Fire Officer London 1833-1861 (Killed by a warehouse wall collapse at the Tooley Street fire)
'Fire Prevention and Fire Extinction 1866' (Memoirs published after his death in 1861)

There is a tragic irony that London's Chief Fire Officer Sir James Braidwood might have predicted his own death some years before in published papers, where a Tooley Street warehouse wall would collapse and bury him in at a major fire in 1861.

Despite the decline in the UK's Victorian industrial heritage, many of these transit sheds or warehouses still remain in use. Some are original although most have been redeveloped to retain external walls, floors, roofs and facades whilst housing some of the city's most illustrious and glamourous apartment complexes. In some cases they have been turned into office buildings, museums, restaurants, shops or universities.

The 1st October 1978 saw one of London's largest post war fires occur at the Granary warehouse, Camley Street NW1. It was here, 117 years after the tragic death of Braidwood that further tragedy occurred, as another London firefighter was killed and several others were seriously injured by a masonry wall collapse. The scale of the blaze is evidenced by the rapid development of the LFB's mobilization: 2.58am first call to incident; **3.05am** make pumps 4; **3.07am** make pumps 6; **3.12am** make pumps 10; **3.19am** make pumps 15 and turntable ladders 2; **3.39am** make turntable ladders 4; **3.51am** make pumps 20 and turntable ladders 6; **4.19am** make pumps 25; **4.30am** make hose layers 2; **5.13am** make pumps 35.

Structural stability under Fire Attack • 307

Figures 15.1; 15.2 and 15.3: The corner of masonry warehouse wall collapse that killed a London firefighter in 1978 and critically injured several other firefighters. **Photos courtesy Paul Wood**

This Victorian era warehouse had been constructed using an elaborate framework of cast-iron columns and beams, heavy timber floors consisting of 50m thick floor boards sitting on 300m deep timber beams, substantial brick arches and walls constructed with a lime mortar mix of 2 to 1 sand/lime showing Flemish bond to the exterior but English bond to the interior face. The walls were supporting lines of heavy stone cornices that cantilevered over the brick parapets. These were tied to each other with giant cast-iron ties set into their top faces, all in a severe state of corrosion, expanding and splitting the stone off the parapets.

The five-storey building was approximately 100 x 100metres and at the time of the fire the building had lay disused for some years but still housed a partly stocked paint and thinners store. Over 200 firefighters attended this fire and after an internal firefighting operation during the first 30 minutes following arrival, crews were withdrawn to the exterior after some interior structural collapses occurred. A total of 18 hand-held fire streams and 6 aerial ladder streams supported with a further 6 ground monitors were then deployed from external defensive positions.

At 0450 hours several firefighters were caught by the wall collapse where one died. I spoke with some of these firefighters based at my fire station (Soho) a few hours later from their hospital beds. One firefighter told me he had been directing his stream through window openings from his location directly below the wall face. He suddenly started to see flames emerging through the brick mortar joints and immediately became concerned.

Within a few short seconds it became clear the wall was disintegrating and collapsing. With no time to run, he turned and dived underneath a fire engine behind him and this undoubtedly saved his life as the wall came down on top of them. The very same fate that befell to Braidwood at the Tooley Street fire had repeated itself 117 years on.

Granary Warehouse Construction

In 2011 a new development was completed just a few hundred yards from the location of the Granary warehouse fire, forming the cultural centrepiece of the mixed use commercial scheme at London's Kings Cross was the striking new Central Saint Martins College of Art and Design.

'The most interesting thing about this new development is that approaching from three sides, you would be hard pressed to notice there was in fact a new building there. This is because the whole college is flanked not only by two Victorian 'east' and 'west transit shed, but to the south by a 55m x 31m six storey brick, cast iron and timber granary warehouse'.[161] In fact, being so close to the fire damaged warehouse this range of grade ll listed buildings were almost certainly designed and constructed in the same way by renowned architect Lewis Cubitt in 1851.

The structural engineer's analysis and works in the college development demonstrated how the fire building itself was perhaps already structurally weakened through natural deterioration over time, even before the fire occurred. According to structural engineer Michael Beare of AKS Ward Lister Beare, the Granary warehouse restoration was a fascinating project due to its composite construction. Its exterior brick walls taper from 1m thick to 600mm at the top, but within is an elaborate cast iron structure of columns and beams supporting a heavy timber floor. Historically designed to take huge loads, some of the floorboards are 50mm thick, sitting on 300mm deep timber beams. 'It was a sophisticated structure,' says Beare. 'There were even metal strips between every floorboard, stopping dust transferring through floors. We also discovered fine threaded rods running across the building below the floors, acting as ties for the walls, preventing them from bowing outwards.' But although the building was designed to take high loads, change of use requirements meant engineers had to analyse the structure for disproportionate collapse. 'It was difficult to do, as modern codes of practice are not really written to take account of the composite nature of structural materials, such as we had here, so it took time to prove that the building could meet them,' says Beare. 'In the end, we proved that the structure was so sophisticated, that each system was independent. Column heads went through floors, meaning the steel frame was an independent structure – whole sections of timber floor could be taken up without any negative effect'.

Beare seeks to increase awareness of the use of traditional lime mortars in new construction. The industry has been quick to dismiss the use of lime mortar, claiming it is weak and can't take tension, but that is purely a curing time issue, says Beare: 'The main thing is that it is a cement-free mortar, with the sustainability benefits that that brings with it.' He adds: 'The fact that the free lime within it dissolves and resets means that it 'micro-cracks' at close centres. This is unlike cement mortar, which can take tension better, but which, as a result, will crack more significantly but at wider centres. So in a sense you want the intrinsic flexibility that lime mortar brings.'

However, from a firefighter's perspective, the type and quality of mortar used has some relevance on the stability of a wall under prolonged fire attack. Older constructions have

---

161 Magazine of the Chartered Institute of Building 9/2011, http://www.construction manager.co.uk//features/first-class-return/

been known to fail and collapse as the lime mortar was washed out by the firefighting water streams directed at the wall itself whilst targeting streams at openings within, destabilizing the walls ability to stand strong without roof and floor supports in place.

Chapter 16

Wind Driven building fires

The dangers of wind driven fires are nothing new to the fire service and have always existed. However, there have been countless building fires over past decades where firefighter injuries and fatalities have resulted where an incoming wind wreaked havoc with interior fire conditions. What would have been a routine one-room fire suddenly became a heavy resource dependent inferno. Innovation and development over the past twenty years has seen a changing environment in which we live and work, where plastics now form such a major part of the fire load in our buildings, which in turn creates a fire that produces greater amounts of flammable fire gases. It is in the smoke that these dangerous fire gases transport into adjacent areas surrounding the fire compartment and the introduction of an exterior wind, as windows fail (or are vented by firefighters), causes horizontal flaming combustion to travel further into the building and ignite these gases. It as if a 'blowtorch' was being directed at advancing firefighters. This effect is worsened still as larger window openings and long narrow internal corridors are now becoming a major feature of modern building designs. The net result is that far greater volumes of air (m^3) travelling at higher velocities (m/s) enter the windows causing the worst fire conditions ever experienced, heading into areas where a blow-torching effect is enhanced even further.

This evolving problem has, over the past twenty years, impacted on firefighter injuries and life loss in ways that had never been predicted. So much so that far greater emphasis is now being placed on how firefighters must;

- Recognize that wind driven fires may impact all low, medium and high-rise buildings
- Size-up to determine fire location
- Immediately recognize wind direction
- Immediately estimate wind speed
- Establish the safest primary entry point and direction of advance (sector, stair or side of a building etc.)
- Emphasize fire isolation by closing interior doors
- Determine any ventilation requirements to prevent over-pressurization of fire compartment
- Consider control measures if forced to deploy against a headwind

Further considerations of wind driven fires should be made surrounding the impact on structural stability as such severe blow-torching effects can cause structural elements and fire resisting partitions to fail much earlier than might normally be experienced.

16.1 THE 360° RECONNAISSANCE AND SIZE-UP

When firefighters are arriving on-scene it is rare that external wind speed and direction considerations influence their primary deployment options. Experience has demonstrated to us that such considerations are normally secondary to the prime concerns surrounding occupant status, fire location and crew deployments to gain the optimum vantage points for speedy firefighting and rescue. When firefighters are involved in training exercises or given table-top scenarios, rarely do they place wind conditions at the top of the list of tactical considerations. This has to change!

When undertaking a brief 360° survey, or internal building reconnaissance, all firefighters and commanders must place existing wind related considerations right up there with those of occupant status and fire location. Experience has shown us that if this does not occur prior to deployment, then it may be too late to prevent firefighter injury. It is useful for fire commanders to ask themselves, how many times to we deploy primarily by the front door on the A side of a structure fire? This is clearly determined by street to structure geometry and so often the shortest hose-lay and most direct access route to the fire or search options are via the front door. However, think again how wind direction might impact on this standard approach and at what wind speed might there be a real danger? Also, consider that there may be low wind velocity, but high gusting speeds that may be concealing the dangers.

16.2 OVER-PRESSURIZING THE FIRE COMPARTMENT

If you are deploying a positive pressure fan to ventilate a fire building during the attack phase would you do this without an outlet vent? Almost certainly not! Therefore, if a good tailwind is at your back as you enter the street entry door to fight a fire, or undertake search and rescue, would it be prudent to create that opening? This situation presents a serious risk of over-pressurizing the fire compartment if an opening is not in existence or is not created by firefighters. Such an over-pressurization of the fire compartment may dramatically intensify fire development and cause high heat, smoke and possibly flaming combustion to backflow along the approach route. Therefore either consider creating a vent outlet on the opposite side of the fire or control the street entry door (door control assignment).

16.3 WORKING AGAINST A HEAD-WIND

This situation is perhaps where most injuries occur and firefighters should be aware if they are advancing towards a side of the building that is facing a headwind (from the leeward side of the building). If a window on the windward side of the building should fail through heat, or even be inappropriately vented by firefighters, the intensification of the fire may create untenable conditions for firefighters (and of course occupants) within a few seconds. If firefighters are committed too far beyond a safe egress point they will feel the full force of the blow-torching fire effects and may not survive. Even though the fire may be located on the far side of the building and the intention is to fight the fire from the unburned side, in these situations, it is far safer to deploy and approach the fire from the windward side of the building. This may even mean that longer hose-runs are necessary, engines and fire apparatus may need re-positioning, and a greater amount of fire damage may result within the building. This is a tactical challenge for it is possible that as an opening is made (entry point) on the windward side with a fire located on that side of the building, the fire might be pushed further into the structure and occupants therein may be subjected to intensifying fire conditions. Therefore, door control at the entry point becomes essential and a partially closed door (door control assignment), or an effective smoke blocking device[162], becomes a priority.

16.4 THE BEACON STREET FIRE, BOSTON 2014

A tragic fire occurred in Boston in March 2014 where two firefighters lost their lives in a wind-driven fire scenario. As with so many other fires of a wind driven nature, the firefighters entered the building taking a hose-line from the front 'A' side with the fire being located on the far (C-side) of the building. The weather at the time of the fire was overcast skies, temperature 34 degrees F, humidity 27%, wind was from the NW sustained at 32.2 mph with gusts to 67mph[163].

The structure involved in this incident was called a 'brownstone.' Brownstone is the common name for a variety of brown, red, and pink sandstone widely used as building materials from the mid-1800s until the early 1900s [New York Landmarks Conservancy 2003][164]. The property was designed as an attached four story single family Victorian townhouse but at the time of the fire was occupied as a multi-family eight apartment building (including two studio apartments in the basement) (known in the UK as an HMO). This period design included what was known as a 'garden floor' at the basement level characterised by a front entrance underneath the front exterior staircase and a walk out rear exit that often led to a garden at the back of the house. This arrangement effectively created four identifiable floors when viewed from the front 'A' side, and five floors when viewed from rear 'C' side. The front basement apartment was entered by a street doorway sited just under the first floor entrance steps whilst the rear basement apartment was entered by in interior stairs from the first floor. This was the entry route taken by the two deceased firefighters when laying an uncharged hose-line into the basement towards the rear apartment.

The first floor apartment had fixed decorative security bars in place over the windows in both the front and rear of the building. Both basement apartments had security grates on all windows that were designed to hinge when opened. These grates were likely installed for security purposes and are very common in this area of Boston. After the fire, keyed padlocks were noted on the front basement apartment window grates. The rear apartment window grates were forcibly removed from the rear windows during fire attack and rescue operations. It was undetermined whether there were padlocks on the rear basement apartment window grates.

Within two and half minutes of entering the basement Engine 33 crew transmitted a mayday. It is reasonable to assume they were trapped in the basement and there means of egress back up the stairs was cut off by fire that had traveled across the ceiling above their heads. Following on from this they were radioing for urgent water to charge their 45mm hose-line on several occasions. These radio contacts went on for eight minutes following the mayday before contact with E33 was lost, where the two engine 33 firefighters were repeatedly asking for their hose-line to be charged with water. In all instances like this the communication barriers created by the rapidly changing dynamics of a challenged

---

162 http://www.rauchverschluss.de/index_e.htm
163 Board of inquiry report; Boston Fire Department, USA 2014
164 Lieutenant and Fire Fighter Die and 13 Fire Fighters injured in a Wind-driven Fire in a Brownstone, Massachusetts; National Institute for Occupational Safety and Health (NIOSH) report March 2, 2016

fire-ground are never as clear as with any hindsight provided by debrief and conclusions, where building occupants are being reported at the time as trapped on upper floors and engines and command are arriving and rapidly being assigned roles. The E33 hose-line had actually been charged but had burned through at the top of the stairs. At the same time a second crew with a hose-line were attempting to gain access behind them in support but were faced with a rapidly changing wind-driven environment. command at 14:59:07) that the companies were unable to get into the building to rescue the trapped members due to the extreme conditions. One fire commander stated that it was not a spectacular fire as he arrived on Beacon Street but the conditions changed drastically by the time he had donned his SCBA. He stated that there were very distinctly different types of smoke issuing from the building: heavy brown smoke from the first floor entryway and thick black smoke from the basement.

The E33 crews urgent call for water coincided with a sudden change in fire conditions forcing firefighters on the first floor at the top of the basement stairs to quickly evacuate the building, with some retreating firefighters suffering second degree burn injuries. This sudden fire escalation motivated the Incident Commander to strike a second alarm, calling on additional engines and ladders. Based on interviews with firefighters positioned at the top of the stairs, conditions there suddenly became untenable. A sudden increase in the first floor hallway ceiling layer temperatures also caused charring of Engine 7's 65mm uncharged line of hose and melting of the nozzle's elastomeric covering in the short time it took to escape from the hallway to the front stairs six metres away. The most likely explanation for the sudden fire escalation in the basement hallway and stairway is, according to the Boston Board of Inquiry's report, probably the failure of the hallway door window due to the glass cracking and falling out or being blown out by pressurization of the rear entry foyer. This allowed higher flow rates of the hot gas and smoke from the rear entry foyer into the hallway and up the stairway, caused by heavy gusting winds (to 67mph) directly hitting the rear face of the building.

There was subsequent evidence that a number of items had been stored in a cabinet at the top of the basement stairs. Investigators noted a number of ruptured paint, aerosol and flammable or combustible liquid containers within the debris found at the base and within the stairwell. Radiant heat flux exposure tests were conducted on exemplar Boston Fire Department hose samples in order to determine how long a section of dry hose can withstand the radiant heat flux from a flame or hot combustion products under a ceiling. Again, according to the Boston Board of Inquiry's report, the magnitude of the radiant heat flux impinging on the hose depends on the temperature, thickness, and thermal properties of the flame or hot gas/soot layer as well as the distance (height) above the hose. In the case of a 0.6 metre deep layer of combustion gas and soot from a typical residential furniture flame (temperature of 927°C to 1038°C), the radiant flux at the floor would be about 9 kW/m^2 for 2.7 metre high ceiling and about 20 kW/m^2 for 2.1 metre high ceiling. In the case of a 0.9 metre deep flame under the 2.7 metre high ceiling, the radiant flux at the floor would be about 15 kW/m^2.

Tests in the Worcester Polytechnic Institute (WPI) cone calorimeter were conducted using hose samples representative of the 45mm attack hose used by Engine 33. Per the December 2013 BFD Attack Hose Specifications, the inner and outer hose jackets are made of filament polyester yarn. The waterway is a synthetic rubber. Radiant heat fluxes of 10 kW/m^2 and 15 kW/m^2 were used for the hose heat flux exposure tests. Two hose samples were used for each of the two radiant heat fluxes. Observations were made of the exposure time to the initiation of smoking, charring, and melting. Results are shown in 16.1. Melting of the jackets occurred after about two minutes of exposure to the 15 kW/

m² heat flux, and after about four minutes of exposure to the 10 kW/m² heat flux. Smoking and charring of the hose, indicating the initial degradation of hose strength and high pressure water carrying capability, occurred 20 to 30 seconds prior to the actual melting of the hose samples.

Hose Sample	Heat Flux (kW/m²)	Time to smoking (s)	Time to charring (s)	Time to Melting (s)
1	15	84	94	125
2	15	91	98	118
1	10	200	222	251
2	10	175	197	230

Table 16.1: Cone calorimeter heat tests of fire-hose samples used at Beacon Street fire in Boston.

Firefighting hose in the UK is tested (type 3 hose as used by local authority fire brigades) according to British Standard 6391:2009, where charged hose-lines filled with water are subjected to conduction temperatures at specific points on the hose of 400°C for 120 seconds and 600°C until burst occurs (which should be at least 15 seconds before any signs of leakage or burst occurs). There is no requirement to test hose that is not charged with water for heat exposures. The Boston fire has led to researchers at the fire protection engineering division of Worcester Polytechnic Institute (WPI), one of the three fire safety research centres in the United States, to research and create a more fire-resistant fire hose that is far less likely to fail through exposure to high heat conditions. The hope is that they can create a hose that is even more fire-resistant, while still being affordable. The general feeling is that a combination of sound tactics combined with improved hose-lines that are more resistant to fire are needed. By early 2016 the WPI had identified over 170 similar failures to fire hose caused by exposure to heat throughout the USA. This problem is exacerbated by using the lightweight hoses for routine house and apartment fires, since these hoses are supposed to only be used for high rises. This was not the case in the Boston fire, but it's a common issue seen elsewhere.

What does this mean to the Firefighter?

The hazards associated with flow-path reversal under exterior wind conditions are not solely reserved for high-rise fires. A wind-driven fire can occur at any time and in any building. Know the wind speed and direction when you book on duty and take account of these two critical features in tactical size-up and deployment at every fire as you dismount from the fire engine.

A safe strategy for basement fires has always been –

- A firefighting crew should not descend into a basement fire without wearing BA and taking a charged hose-line of at least 45mm diameter (or one that provides adequate water).

- Water supply must be adequate for purpose and augmented immediately.
- A second crew with at least a 45mm diameter (or larger) <u>charged</u> line should be in support as soon as staffing assignments allow.
- The second hose-line can be sited at the top of the stairs to protect the attack teams egress.
- Remember – where hose-lines are taken down stairs into basements, the hose behind you may now be at **ceiling height** and could be exposed to ceiling fire temperatures.
- Where conditions at the top of the basement stairs suggest any descent by the primary attack team is likely to expose firefighters to untenable conditions, a holding position at the top of the stairs may protect fire from spreading upwards to trap occupants or firefighters undertaking search and rescue on upper floors.
- Do not implement any low level tactical venting actions where firefighters are located above the fire or are descending into a basement.
- At the earliest opportunity, consider alternative and optimum entry points to a basement fire and also consider the viability of an exterior stream to gain some reset of the fire.
- Ongoing communications with interior crews are essential.
- Where communication is lost with firefighting crews in a basement fire, the second team in support should descend into the basement to locate them and a third charged hose-line should be placed at the top of the stairs as cover.

NIOSH concluded in their report on the incident that contributing factors to the firefighter fatalities at Beacon Street were:

- *Delayed notification to the fire department*
- *Uncontrolled ventilation by a civilian*
- *Occupied residential building with immediate life safety concerns*
- *Staffing*
- *Scene size-up*
- *Lack of fire hydrants on Side Charlie (a private street)*
- *Lack of training regarding wind-driven fires*
- *Unrestricted flow path of the fire*
- *Lack of fire sprinkler system*

NIOSH Key Recommendations
Fire departments should define fire-ground strategy and tactics for an occupancy that are based upon the organization's standard operating procedures. As part of the incident action plan, the incident commander should ensure a detailed scene size-up and risk assessment occurs during initial fire-ground operations, including the deployment of resources to Side Charlie. Scene size-up and risk assessment should occur throughout the incident.

16.5 WIND DRIVEN FIRE RESEARCH (NIST USA)

During the previous three decades in particular, there have been a large number of fires in the USA, the UK and elsewhere, where multiple firefighter deaths were caused by extremely intense wind driven fire conditions. The sudden reversal of the flow-path, causing floor to ceiling temperatures that are untenable for firefighters, must be anticipated by commanders and countered by effective firefighting tactics that may include the use of exterior streams, deployment into the windward side of the building, wind control devices, floor below nozzles in high-rise buildings and adequate resources on scene that enable rapid deployment of support crews with back-up safety lines.

In February 2008, a series of 14 experiments[165] were conducted in a 7-story building in New York City to evaluate the ability of positive pressure ventilation fans, wind control devices (WCDs) and external water application with floor below nozzles (high-rise nozzles) to mitigate the hazards of a wind driven fire in a structure. Each of the 14 experiments started with a fire in a furnished room. The air flow for 12 of the 14 experiments was intensified by a natural or mechanical wind. Each of the tactics were evaluated individually and in conjunction with each other to assess the benefit to fire fighters, as well as occupants in the structure. The results of the experiments provide a baseline for the hazards associated with a wind driven fire and the impact of pressure, ventilation and flow paths within a structure. Wind created conditions that rapidly caused the environment in the structure to deteriorate by forcing fire gases through the apartment of origin and into the public corridor and stairwell. These conditions would be untenable for advancing fire fighters. Each of the tactics were able to reduce the thermal hazard created by the wind driven fire. Multiple tactics used in conjunction with each other were very effective at improving conditions for fire fighter operations and occupant egress. The experiments were conducted by the National Institute of standards and Technology (NIST), the Fire Department of New York City (FDNY), and the Polytechnic Institute of New York University with the support of the Department of Homeland security (DHS)/ Federal Emergency Management Agency (FEMA) Assistance to Firefighters Research and Development Grant Program and the United States Fire Administration.

Figure 16.1: The Governors Island wind-driven fire tests undertaken by the National Institute of Standards and Technology (NIST), Fire Department New York (FDNY) and New York University 2008 **Image courtesy of the National Institute of Standards and Technology (NIST) USA**

165 Stephen Kerber; Daniel Madrzykowski; Fire Fighting Tactics Under Wind Driven Fire Conditions: 7-Story Building Experiments; NIST Technical Note 1629; 2009

With an imposed wind of 9 m/s to 11 m/s (20 mph to 25 mph) directed mostly into 3 m^2 windows and a flow path through the fire floor and exiting out of the door onto the roof, temperatures in excess of 400°C and velocities in the order of 10 m/s (22 mph) were measured in the corridor outside the fire apartment and in the stairwell above the fire floor. These extreme thermal conditions are not tenable, even for a firefighter in full protective gear. These experiments demonstrated the 'extreme' thermal conditions that can be generated by a 'simple room and contents' fire and how these conditions can be extended along a flow path within a real structure when wind and vent openings are present. Potential tactics which could be implemented to interrupt and control the flow path are door control from inside the structure and or the placement of a WCD. From the floor below the fire, external water application was demonstrated to be effective in reducing the thermal hazard in the corridor and stairwell. The use of PPV in the stair shaft had some limited success in controlling smoke in the stair and cooling the hallway environment. PPV fans alone could not overcome the effects of a wind driven condition. However when used in conjunction with door control, WCDs, and FBNs, the PPV fans were able to maintain tenable and clear conditions in the stairwell. The key to successful use of PPV fans was to mitigate the wind driven fire condition via door control or other tactics. Then the PPVs can be used to clear the stair and then pressurise the stairwell to provide a safe working environment. Although the PPV fans, when used alone, could not reverse the flow of a wind driven fire, PPV fans always improved conditions in the stairwell.

The wind driven condition can be described as hot gases or flames flowing horizontally out of the room of fire origin. The wind driven fire condition has been described as a 'blow torch' by firefighters. In these experiments the inlet to the flow path was the windward side window in the room of fire origin. The flow path then went through the apartment, into the corridor, and exited out of the door onto the roof, via the stairwell. Without a flow path, the wind driven fire condition inside the structure cannot occur.

The Fires

The bedrooms and living rooms had a conservative average fuel load of 21 kg/m^2 and 15 kg/m^2 respectively. The room fire experiments were run until the fire burned down and smoke production was minimal. The experimental duration varied between 23 min and 53 min depending on the growth of the fire and the impact of the tactics used.

- A fire in the bedroom from the first experiment (**no wind**) did not extend beyond the room of origin despite reaching flashover and served as a benchmark for comparison.
- The second test involved two rooms of fire (living room and bedroom) with **no wind**. Temperatures in the hall of the apartment reached 500°C and 200°C in the apartment service corridor. However, with stair and roof door opened a through air flow-path was created and temperatures in the corridor reached an untenable 300°C and as high as 250°C on the stair landing. The use of the PPV fan to pressurise the stairs was able to clear the stair of smoke and return temperatures in the corridor and stair to tenable levels for firefighters. A maximum air flow of 2 m/s (4 mph) as recorded leaving the fire apartment door into the corridor.
- The third test involved a **simulated wind** of 9m/s to 11 m/s (20 mph to 25 mph) directly into a bedroom fire. In this test, flames extended throughout the apartment and along the corridor and into the stair. An initial fire room temperature of 500°C reduced to 200°C as the window started to fail (25%). As the entire window failed the temperature exceeded 600°C and flashover was achieved in the bedroom. The three temperature measurement locations in the corridor behaved similarly in reaction to the

events with the higher temperatures at positions closer to the fire apartment. Corridor temperatures reached 400°C before they became ventilation limited between 700s and 1100s. Temperatures in the corridor remained very steady until a ventilation path was created. The stair door was opened which caused the corridor temperature outside the fire apartment to increase from approximately 400°C to 550°C. The temperatures declined again due to the lack of oxygen until the roof door was opened. Once opened a wind driven condition developed increasing the corridor temperatures to above 600°C. Closing the roof door and allowing the fan to pressurise the stair caused the stair temperature to decrease. Corridor temperatures reached approximately 300°C to 600°C before they became ventilation limited between 700s and 1100s. Temperatures in the corridor remained very steady until a ventilation path was created. The stair door was opened 0.08 m (3 in) at 1100 s which caused the corridor temperature outside the fire apartment to increase from 350°C to more than 600°C. The temperatures decreased again with the lack of oxygen until the roof door was opened. Once opened a wind driven condition developed increasing the corridor temperatures to above 800°C (1472 °F). The temperatures decreased when the roof door was closed and the PPV fan was flowing. The flow of hot fire gases and flaming combustion out of the fire apartment with the stairwell door opened was approximately 5 m/s and increased to 20 m/s when the roof door was opened. Once the wind driven flow-path was fully created it forced flames all the way from the ignition bedroom to the stairwell door in the corridor. These conditions would not be conducive to an interior direct frontal fire fighter attack. When the ground floor was open and the fire floor stair door and roof door were closed the gases forced around the fire floor door travelled down the stairs and exited the street entrance, seven floors below.
- The remainder of the 14 tests demonstrated very similar results in terms of fire development, temperature and velocity scales, flow-path movements and untenable conditions that were only relieved by closing doors and isolating the fire flow-path. In some cases stairwell temperatures exceeded 700°C in the open flow-path. Various tools and strategies were successfully demonstrated during the latter series of tests that included combinations of door control; wind control devices (WCDs) (fire service deployment of portable fire curtains covering windows); floor below nozzles utilising an exterior attack at height, and positive pressure ventilation (PPV) of the stairs and corridors.

What does this mean to the Firefighter?

Opening doors at the stair in a high-rise can cause you big problems, as can the automatic opening of a smoke shaft or vent (AOV) during a wind driven fire**. As a flow-path is created (air inlet to smoke and heat outlet) the temperatures created by a wind driven fire will increase. As a door or automatic opening vent (AOV), or a smoke shaft opens, the flow-path is initiated. In a wind driven fire these additional openings will intensify the fire development and 'pull' fire and heat into the paths of firefighters and escaping occupants, placing firefighters directly in the hot zone of a flow-path. The pressure differentials that can be set-up by AOVs or smoke shafts opening behind firefighters with a developing fire in front of them can be hazardous and all means of placing barriers, such as closed doors, between the fire and the vent outlets should be utilised at all costs.

***The firefighter should be aware of some smoke control systems that are designed to open the stair door automatically (specifically smoke extract systems) where a corridor smoke shaft creates a negative pressure in the corridor, to draw smoke into the smoke shaft and enable a flow-path of make-up air to enter the corridor via a partially open stair door. This effect can be overridden by firefighters via a control panel should the engineered smoke control system fail to prevent smoke infiltration into the stair. If the system is overridden in this way, the stair door should automatically close.*

Figure 16.2: Exterior wind driven fire with AOV Closed

Figure 16.3: Exterior wind driven flow-path with AOV opened

Wind Driven building fires • 321

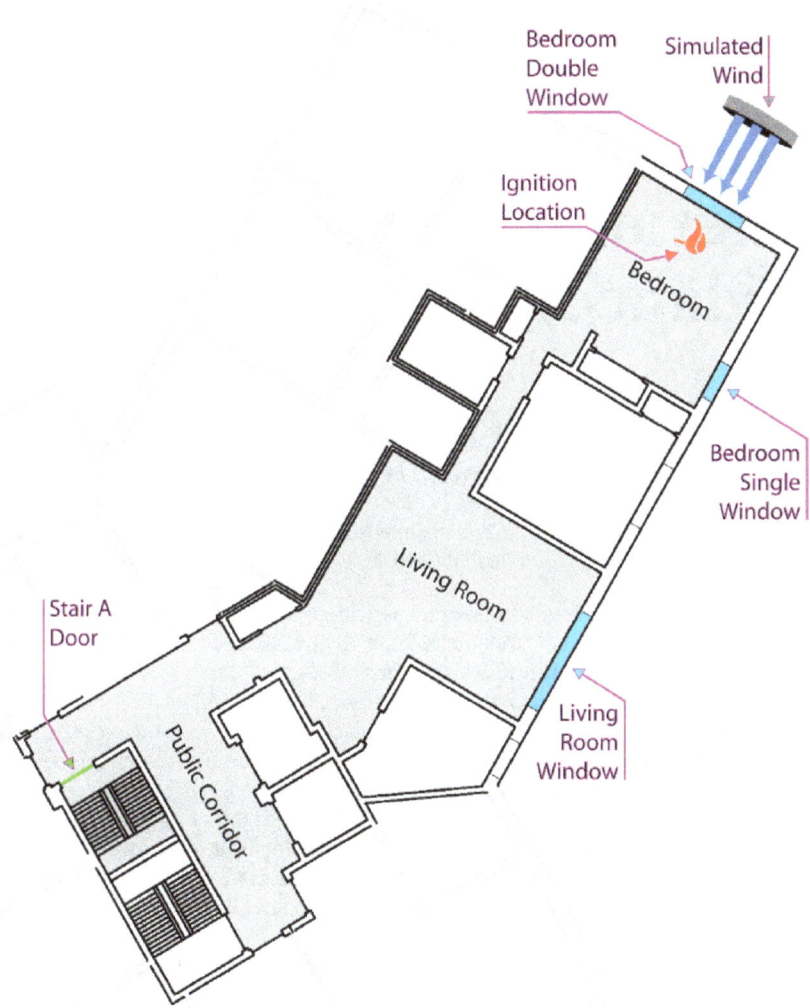

*Figure 16.4 A typical apartment layout used in the 14-experimental wind-driven fire tests by NIST **Image courtesy of the National Institute of Standards and Technology (NIST) USA***

Chapter 17

Rules of Engagement (interior deployment guidance)

'*Rules of engagement are (ROE) determined by your department SOPs and implemented by the on-scene Incident Commander (IC). In all cases, ROE should be followed on a 'risk versus benefit' profile, but operational discretion may allow the IC to vary from SOP guidance, although only with reasoned and justified purpose'.*

There are *five golden minutes* following arrival on-scene that should represent the very foundation upon which firefighters base their actions at fires, their decisions as commanders, and how we must formulate our SOPs:

Research data[166] has demonstrated that at least 25% of building fires become worse, following fire service arrival, based on fire service actions alone. Other research (GCU) of 5,401 UK fires shows that 47 percent of all 'working' building fires spread beyond the compartment of origin before control is achieved. Guidance further suggests also there is only a 50 percent chance of a fire being controlled once it reaches an 86m^2 perimeter and a 500 L/min hose-line will become outpaced by increasing heat release much beyond that sized floor area. The largest *medium fire load* floor area fire a high-flow (>1,000 L/min) hose-line can deal with is around 230m^2, providing an adequate amount of staffing (3-5) is handling the hose-line.

With this in mind, maybe we need to look more closely at our tactical approaches within the first few minutes of arrival on-scene. In effect, pre-planning not only our response and weight of attack, but also our primary deployment operations becomes critical. 'Quick water' certainly but reasoned deployments become important here.

On many occasions, we may have created [vent] openings to a building, possibly opening an entry door or venting a window, prior to being in a position to immediately get adequate water onto or into a fire. Take note here of the term 'adequate' because its' relevant in terms of suppressive capacity at the time we deploy. In some situations it may be wiser to delay opening the building, or indeed closing it down, until we have enough staffing and water on-scene.

However, there are situations where the possibility of rescue from an external position may be presenting themselves and this should become a priority. In this scenario closing

---

166 Sardqvist.S; Fires in non-residential premises in London 1994-1997 – Report 7003, Lund University 1998

the fire down may slow the fire's growth and buy some valuable time to take a quick look into any room venting smoke, from ladder positions if necessary. In these situations, is Vent-Enter-Isolate-Search (VEIS) a viable option? Is it something your fire department practices? It can carry some level of risk to firefighters but if following strict protocols, such a strategy may reap high rewards.

New building designs, coupled with limited staffing and water availability within the first few minutes, dictate changing tactics. This chapter therefore offers a detailed analysis of what are termed 'rules of engagement' for firefighters faced with challenging situations at building fires, ensuring safe but effective tactical approaches are considered under a structured approach according to risk and hazard analysis.

- **Primary search** – a rapid search of the areas of most threat to occupants, usually at locations that are nearest the fire. This might occur ahead of or above the firefighting hose-line's placement (or where the fire is effectively isolated), according to predetermined 'triggers'. Control measures should normally be put in place
- **Secondary search** – A follow-up search in all areas and often after the extinguishment process has taken place, or when the fire is confirmed as *under control* (such fires should never be left alone to re-establish a further growth curve).

*A good SOP uses 'triggers' where **primary search** is actioned by -*

1. Persons are reported trapped or missing by an escaped occupant or reliable source.
2. A person was seen or heard within by firefighters.
3. A person is on the phone from inside the property to fire control.
4. The IC has operational discretion to deploy firefighters into a structure for any other reason or purpose with reasonable justification to achieve a viable objective. But there **must** be justification. If an entire family are huddled together at the front of the property and all assure that there is absolutely no chance anyone else are in the house/apartment, the tactical approach should be based on the IC's assessment of risk versus benefit for *saveable property*. This should then be determined based on fire involvement/conditions, staffing, resources and an established and augmented water-supply.

If a commander deploys firefighters to search an abandoned building ahead of or above the firefighting hose-line, where the fire is demonstrating serious fast spreading fire conditions, I would question his/her justification without sound evidence or reasoning. If there was gut feeling based on knowledge that the building was likely to be occupied, then that may be just enough reason but this is a borderline decision. Had the crews attended fires earlier in the shift, or even the day before and noted occupants living in this building then that may be justification enough. However, without any pre-existing knowledge that there are likely to be occupants, any interior search must follow extinguishment.

Similarly, in a known occupied house or apartment, any search above or ahead of the firefighting hose-line can only be justified according to the above 'triggers'. If there is reasoned belief, due to *time of day, car on the drive, lights on in the house* etc., any primary search above or ahead of the firefighting hose-line should only occur where the fire compartment can be effectively isolated (closed door) or where the fire is sprinkler-controlled etc.

Note: 'Viable victims'? Unless occupants are in a fully involved fire compartment we can never determine any realistic viability of occupants until we have either reached or extracted them.

But remember, ROE are determined by your department SOP (if you have one) – so follow it!

> 'Rules of Engagement' are proposed here as a series of tactical actions and deployment procedures at building fires that take into account weight of attack, response times, firefighter safety, search and rescue and structural stability expectations in a changing and more hazardous working environment.

17.1 RULES OF ENGAGEMENT

In a speech entitled 'US Forces: The Challenges Ahead', Colin Powell stated, 'We owe it to the men and women who go in harm's way to make sure that ... their lives are not squandered for unclear purposes'. He was challenging leaders to make strategic decisions based on a core ethic: **Don't waste human life**. Implicit in his speech and in the Powell Doctrine is that committing troops to combat should neither be an easy nor an automatic decision. In fact, such a decision should be made only if there is a significant advantage to be gained[167.]

Fire Service veteran Eric Lamar writes[168];

> 'The 21st century battleground is dynamic, chaotic and complex, and so is the fire-ground. As with the military, we have gone to great lengths to employ organizational systems and technology, to instil a degree of order and predictability to the working fire environment. Both line firefighters and infantry soldiers now have an array of modern protective gear, surveillance equipment and offensive tools to achieve rapid victory. The uniform application of command and control systems is designed to ensure coordinated and effective action and to strictly limit casualties. In reality, our systems, protocols and technology often fail us with disastrous results. Why?'

He continues:

> 'Almost without exception, our firefighting forces are most vulnerable during interior structural firefighting. This operational environment most closely resembles the combat setting to which Colin Powell refers in his famous Doctrine. In his view, committing forces requires four imperative strategic considerations':

- Committing troops [firefighters] must be an absolute necessity
- There must be a compelling risk posed by not acting
- Overwhelming resources must be applied
- A clear exit strategy must be in place

He continues:

> 'Do fire officers and firefighters routinely commit to interior operations where the objectives are fuzzy and the strategy is unclear? Are firefighters routinely killed in

---

167 Grimwood.P.; Eurofirefighter p154; Jeremy Mills Publishing Ltd West Yorkshire UK; 2008
168 http://firehouse.com – Firefighter Safety, August 2007

interior environments where their responses to these four strategic considerations should suggest completely different tactics?'

A database of premises, based on fire resistance periods, from where developing some rules of engagement provides an opportunity to create a database of premises, based on fire resistance periods, from where developing some rules of engagement.

17.2 FIREFIGHTER INJURIES CAUSED BY STRUCTURAL COLLAPSE

During the Victorian era it was quite common for firefighters to be injured or killed whilst caught in masonry collapses of burning buildings, similar to those firefighters serving during the WWII Blitz. During the post war years the life losses and injuries gradually reduced as the built environment became modernised to some greater extent. Even so, there were still some life losses and serious injuries of firefighters during the 1970–80s where fires occurred behind listed and graded Victorian facades that disguised internal renovations with separating walls removed and additional structural loads distributed ineffectively.

Currently, the number of firefighters who are reportedly killed or injured in the UK by structural collapse is reducing, but will the ever-increasing transition to lightweight modular construction, now becoming more commonplace across the UK, see any increase on firefighter injury statistics in the coming decades? What seems certain is that an ever-evolving built environment will present a wider range of fire-ground hazards and the fire service must address their strategic approaches in future and encourage greater situational awareness, according to the four considerations listed by Lamar (above).

It must be anticipated that the fire resisting protection of key structural elements will, in some buildings, be non-existent and that such buildings under fire attack are likely to collapse much earlier than previously experienced, often without warning. It is also more likely that 'falls' will become a more common fire-ground injury as floors may be weakened and collapse much earlier during interior firefighting operations.

The experience in the USA has seen a similar transition to modern lightweight and pre-engineered modular constructional methods since the 1970s. It is of great concern that a continuing trend towards lightweight modular construction in the U.S. may have unwittingly exposed firefighters to increasing operational hazards, resulting in a higher life loss ratio (per 100,000 building fires) both during and after a 23 year transitional period that saw the introduction of these fast-build modern methods of construction (MMC). It was established in a U.S government sponsored 'structural collapse' study[169] related to firefighter life losses in the USA during interior firefighting operations, that whilst a general reduction (by a third per 100,000 fires) in traumatic or fire related firefighter fatalities was achieved over a 23 year period, the life loss ratio in cases of 'structural collapse' remained steady at 1.5 per 100,000 building fires. Additionally, it was shown that the percentage of firefighters killed by structural collapse in residential buildings had actually increased during this time period. In the UK firefighters are now adapting their tactics where lightweight construction is involved in fire, utilising such tools as water-fog insertion equipment, high-flow portable attack monitors or compressed air foam streams for rapid knockdown from exterior positions to control rapid fire spread, reduce heat release and internal temperature prior to entry.

---

169 Brassell. L. and Evans D.; Trends in Firefighter Fatalities Due to Structural Collapse, 1979-2002; National Institute of Standards and Technology (NIST/FEMA); NISTIR 7069:2003

What does this mean to the Firefighter?

The UK and many parts of Europe are following design and construction trends, using lightweight timber in residential buildings, that became most popular in the USA over 25 years ago. Therefore a look at North American firefighter injury statistics from fires that occurred during recent periods might serve as some sort of warning to us. According to data from the United States Fire Administration (USFA), during the period 1977 through 2002, there was an average of 115 fire fighter fatalities per year, not including the 343 fire fighters who died at the World Trade Centre on 11 September 2001. Over the same period, there have also been approximately 45,000 non-fatal firefighter injuries annually, occurring at a fire scene. Structural fires pose particular hazards to fire fighters from structural collapse or becoming lost or disoriented and unable to escape before running out of oxygen. For the period 1977 through 2000, the number of fire fighter deaths in structure fires declined 59%; however, the rate of fire fighter deaths (per 100,000 fires) from traumatic injuries during structural firefighting has actually increased over this same time period.

- Although the number of fire fighter deaths in structure fires has declined over the last 20 years, the rate of fire fighter deaths from traumatic injuries during structural firefighting has actually increased over this same time period.
- Structural fires pose particular hazards to fire fighters; fatalities while fighting structural fires typically involve structural collapse, rapid fire progression, and trapped firefighters.
- Assessing the integrity of a burning lightweight building is a complex task, and presents a difficult problem for the fire service in determining when a catastrophic, life threatening event, such as a major collapse, will occur; at present, there are no uniform assessment tools to assist incident commanders and fire fighters in accurately determining when to exit a burning structure before it collapses.
- In only 20% of the NIOSH (USA) investigations of fatalities due to structural firefighting had fire fighters begun an evacuation before the structure collapsed.
- To reduce fire fighter fatalities during structural firefighting, additional efforts must be directed to more effectively use what we have learned through past experience.
- Additional training for incident commanders and firefighters may enhance awareness of the hazards of structural fires and be useful in guiding critical decision making on the fire ground.

17.3 FIREFIGHTER TRAINING IN BUILDING CONSTRUCTION

A report by the Fire Brigades Union in 2008[170] informed that a poll amongst firefighters showed that half (48%) of those polled said they had not received any building construction training in the last 12 months. Only 5% said they had received specialist building construction training and 15% said their training had consisted of a DVD or similar method.

---

170 'In the Line of Duty'; Fire Brigades Union UK 2008

The regional picture as reported was highly uneven. In the North East, three in five firefighters (60%) had not received building construction training, followed by 59% in the North West and 53% in both the South West and Northern Ireland. The West Midlands (51%) and the South East (50%) were also poor on this score. The quality and duration of training were also of concern. Nearly two-thirds (65%) had received half a day or less of building construction training, while fewer than one in ten had had more than two days.

At a time where fire service staffing and resources are repeatedly under pressure to optimise their service response and delivery, it is absolutely imperative that fire commanders, firefighters, fire safety officers and building inspectors are effectively trained in order to identify construction types, assess risk factors associated with each building and its contents, evaluate the stability of a building whilst under fire attack and recognise key warning signs of potential collapse.

17.4 NATIONAL OPERATIONAL GUIDANCE (NOG)

In 2014 the UK fire service, under the lead of London Fire Brigade, established an authority controlled Wikipedia style educational website[171] for the purposes of upgrading the training materials used to teach, inform and update on firefighting knowledge. The Programme works with fire and rescue services and experts from a wide range of organisations developing 'best practice' online guidance that helps UK fire and rescue services to respond to incidents safely and effectively.

This website hosts an ever-increasing catalogue of national policy and tactical operational guidance. It also contains industry-leading research performed during the course of guidance development and, for the first time, a catalogue with all of the existing guidance in one place. The UK fire service rely on fire service training manuals and national operational guidance (NOG) to inform firefighters of the hazards and risks associated with a range of incident ground scenarios. Within that mass of information there is guidance on traditional building construction as well as MMC and typical warning signs of building collapse when under fire attack provide critical indicators to firefighters and fire scene commanders.

A 198 page UK NOG guidance manual[172] includes the following information for firefighters:

General considerations

Modern timber frame buildings utilise traditional or lightweight and/or engineered timber along with wood based sheathing board such as oriented strand board (OSB) or chipboard which imparts rigidity to the frame. Timber is combustible and its performance is somewhat reliant on fire protection measures (e.g. plasterboard). Lightweight timber and traditional timbers are usually covered or protected by linings either externally or internally. Traditional or lightweight timber frames along with the sheathing board interact with air filled cavities and provide a combustible fuel should a fire enter this cavity. This is particularly the case if the cavity has not been adequately compartmented with cavity barriers and the fire is allowed to spread unhindered in the cavity.

---

171 National Operational Guidance Programme; http://www.ukfrs.com/
172 Fires in the built environment and Fires in buildings under construction or demolition; Knowledge and information; Building Research Establishment (BRE) National Operational Guidance DCLG 2016

In some properties traditional or lightweight timber can also interact with composite cladding systems, which if inadequately installed can allow for concealed fire spread.

Fire and rescue service considerations

- Consider the presence of timber framed construction at any newly built property or development
- Consider the potential for rapid undetected fire spread
- Investigate any potential fire spread (cutting away of cladding or coverings may be required)
- Consider the use of thermal imaging cameras
- Consider defensive firefighting tactics for timber framed buildings under construction due to rapid fire spread and potential early collapse

The guidance points out that it is sometimes difficult to identify lightweight frame construction and many modern buildings are camouflaged with false brick or stone appearance cladding. It is a critical failing of design standards that the fire service may not be able to ascertain the type of construction involved until either they cut into it, or the fire spreads outwards, or some structural collapse occurs.

17.5 RULES OF ENGAGEMENT (ROE) – B-SAHF

In the context of the building itself, the term 'Rules of Engagement' (ROE) should refer to clear guidance based protocols that determine when interior firefighting is viable and relatively safe, based on a recognised analysis of what impact the existing fire conditions have on structural stability. B-SAHF (building, smoke, air track, heat, and flame) indicators. In addition to *Lamar's four strategic considerations*, an immediate and ongoing assessment of fire conditions and structural stability must be undertaken. This will necessitate a primary size-up by the first on-scene incident commander, followed by continual review by the IC and assigned sector commanders or safety officers, as well as firefighters working within the hazard zone. It is absolutely critical that relevant information and observations are communicated through the appropriate channels. If a firefighter notes any of the established warning signs at any location within and around the structure, this information must be passed on to the nearest safety officer or sector commander.

In the simplest of definitions, ROE means do we deploy internally or not? Do we work within a hazard zone or not? Do we implement additional control measures or not?

B-SAHF

Shan Raffel (Queensland Fire and Rescue) and Ed Hartin (Central Whidbey Island Fire and Rescue) originated and developed the B-SAHF concepts of key indicators, upon which ROE can be based.

B	Building	(B) **Building** – Structural stability (observed; and anticipated based on fire conditions)
S	Smoke	(S) **Smoke conditions** – Colour, volume, velocity, opacity, location (may not indicate the fire compartment), stability (bouncing or turbulent), and position of the neutral plane at an opening (window or door), or within the fire compartment itself.
A	Air-track (flow-path)	(A) **Air-track (flow-path)** – Where are the air inlets and where are the smoke outlets? Are they accompanied by noise (whistling or roaring)?

H	Heat	(H) Heat – How hot is the fire? This may be determined by indicators such as glass cracking, or smoke colour, thermal image readings or the velocity in the smoke.
F	Flame	(F) Flame – What indicators are there around the intensity of the flaming, the shape of the flame from a window, or the colour of the flames?

Table 17.1: B-SAHF indicators that can be used for guiding 'rules of engagement' decisions

In the earliest firefighter training manuals (Manuals of Firemanship) a list of 'principal signs of masonry wall and timber floor collapse' appeared as follows:

- Cracked or dropping arches over doors, windows and other openings; [*horizontal cracks were always more indicative of collapse than vertical cracks although both offer warnings*];
- Spalling of stonework, falling of cornices, etc., particularly on buildings with heavy facings of ornamental stone;
- Sagging floors or beams or gaps between the edges of floors and the walls;
- Displacement of steel or cast-iron pillars supporting joists or beams.

Further information was provided;

'Heavy machinery on the upper floors of a building can prove dangerous in the event of weakening of the walls or supports and water should be removed from upper floors as soon as possible. A few inches of water spread over a large floor area may weigh several tons'. In general, much of the water used for firefighting will either evaporate or run out of the building. If we estimate conservatively that 30 percent in flow of a single high-flow exterior firefighting stream remains within a large building, adding to imposed loads on the structure, we might estimate that in excess of 36,000 litres could accumulate within an hour. This would add 35 tons to the imposed structural load in a building which is possibly already weakened by fire! Sometimes this water is absorbed by stock and contents and this in itself can cause expansion of the materials, pushing against walls and columns to create further added loads.

This author has attended several fires where the weight of firefighting water from aerial water towers contributed towards the collapse of floors in large buildings. On some occasions, firefighters were still undertaking interior firefighting operations below the collapse. Therefore where such large amounts of water are being directed at or into a fire building, the possibility of collapsing floors should determine where and if firefighters should remain within the hazard zone. In such cases, warning signs of collapse should not be relied upon to substantiate their position within the building as a catastrophic collapse can occur suddenly, without any warning whatsoever.

The manuals go on to discuss further warning signs of masonry wall collapse, beginning with pieces of mortar or stone falling out of walls from above, followed by visual signs of leaning, cracking or bulging. In some cases, masonry walls can be pushed out through the expansion of heated steel joists and beams. As these beams are cooled, they can sometimes shorten enough to fall from the masonry walls, initiating floor collapse.

Modern construction methods utilising lightweight timber frames and roof/floor trusses, Glulam, engineered Cross-Laminated Timber (CLT), Laminated Veneer Lumber (LVT), Structural Insulation Panels (SIPS), light gauge steel framing and lightweight cladding have changed the way the fire service deploy at fire scenes. Although never 100 percent reliable, the most common method now for pre-determining buildings for firefighting deployment strategies is to determine the level of fire resistance within the main elements of construction. The structural fire resistance period is based on the buildings use, compartment size and the height of the building. These parameters reflect the anticipated relative fire sizes and the difficulties of fighting fire in larger and taller buildings. A larger compartment can generally be expected to have a greater fire load and therefore a more severe fire. However, some town centre shops in particular may demonstrate much higher fire loads per floor area ratios and this can lead to very intense, long duration fires that are difficult to extinguish. In these cases the structural integrity is most likely to fail early as the level of fire resistance is low, possibly allowing the fire to spread into adjacent compartments, roof spaces and adjoining buildings.

17.6 FIRE RESISTANCE RATINGS (FRRS) AS A 'RULE OF ENGAGEMENT'

Refer also to Chapter 15

Can firefighters use assigned fire resistance ratings (FRRs) for individual buildings as a 'rule of engagement' (building interior deployment) for what is termed, 'a reasonable time period' to undertake search and rescue? Is firefighting deemed under 'search and rescue' operations, or according to undefined regulatory guidance once determined all occupants are accounted for, are firefighters expected to end operations there? How are fire resistance ratings for a building determined and do they offer reliable guides? Experimental research and computer aided analysis forms part of it as does the certification process surrounding furnace tests. However, there are distinct knowledge gaps relating to how a fire impacts limited amounts of construction material assemblies, or structural framework in the test house or on a computer analysis, compared to actual building fires. **Time equivalence** methods appear most commonly used for determining the structural fire resistance ratings required, when little is known about specific materials and the type of construction to be used. However, the usefulness of time-equivalent methods is limited by the accuracy to which the compartment fire temperatures or heat fluxes can be predicted. Even so, there is much more research to be undertaken in ensuring FRRs provide greater accuracy within adequate safety margins, whilst still remaining commercially viable in their application.

The building codes and standards recommend structural fire resistance as a period in minutes, most generally varying from 15 min for small and low rise buildings to 120 min for tall buildings or large compartment areas. Single storey buildings do not normally require structural fire protection. The common exceptions to this are where the building is in a boundary condition (i.e. where there is a danger of fire spread to adjoining buildings should a wall collapse in a fire) or where the insurance company or owner deems that it is necessary. A component or element of construction that achieves 120 min fire resistance is one that maintains its integrity for 120 min when tested in a standard furnace test[173]. This does not necessarily mean it will maintain integrity in a real fire to an equal duration as the test fire, when forming part of the structural frame. However, this does offer a reasonable method of *pre-determining* an estimated 'maximum time of operational engagement'. The structural frame and individual elements could maintain integrity for longer than the test results suggest, or fail at an earlier point, depending on ventilation parameters, fire load, fire intensity, fire duration and locality of the hottest part of the fire. An intense fire in a room involving a fully loaded timber two-metre-high wardrobe, could cause a BS476 ceiling to fail within a time period much less than a designated test fire resistance standard.

This provides an opportunity to create a fire response database of premises, based on fire

---

173 BS 476 Fire tests on building materials and structures Part 20 Method for determination of the fire resistance of elements of construction (general principles); or BS EN 1365-1:2012; as examples.

resistance periods, where developing some *rules of engagement* may begin. An example protocol may be that all unprotected buildings with less than 30 min fire resistance (FRR) should not be entered if a severe fire is in progress on arrival. Another might be to evacuate firefighters from lightweight buildings where the fire has spread beyond the floor of origin (unless limited to minor cavity or void spread) However, *Lamar's four strategic considerations* should always be considered before any on-scene decision is made and local risk-based protocols take priority.

Lamar's four strategic considerations (Repeated here) –

- Committing troops [firefighters] must be an absolute necessity
- There must be a compelling risk posed by not acting
- Overwhelming resources must be applied
- A clear exit strategy must be in place

Building Regulations (England & Wales): *Use building height, sprinklers and occupancy type as the primary factors for assessing the FRR of structural elements.*

AD-B Occupancy Type	Building height to highest floor above access level (metres)			
	to 5m	to 18m	to 30m	> 30m
AD-B Dwelling	30	60	90	N/P (120)
AD-B Hospital	30	60	90	120
AD-B Hotel	30	60	90	120
AD-B Office	30	60 (30)	90 (60)	N/P (120)
AD-B Retail	60 (30)	60	90 (60)	N/P (120)
AD-B Assembly	60 (30)	60	90 (60)	N/P (120)
AD-B Manufacturing & Storage	60 (30)	90 (60)	120 (90)	N/P (120)

NP – Not Permitted

Table 17.2: Typical fire resistance ratings (in mins) – (bracketed for sprinkler-protected buildings) (Note: 30 mins increased to 60 mins for compartment walls separating buildings) also note special situation notes in BR-ADB.

British Standard 9999:2017 (UK Code of Practice)

The British Standard code of practice 9999 offers designers a way of navigating beyond prescriptive regulatory guidance in a way that allows greater design freedoms. The COP uses a *risk and consequence analysis* based around a time equivalent approach to determine fire resistance ratings.

A COP task force made up of fire safety engineers consultancies and research organisations in the UK, developed a methodology that encompasses engineering calculations used in a probabilistic manner as a basis for achieving satisfactory levels of fire safety. Recommendations for specifying the fire resistance requirements for different occupancy purpose groups have been derived which are linked to the occupancy and building height. One of the benefits of the '9999' approach proposed that should changes in the levels of safety be required in the future, the methodology within the COP enables these changes to be quantified in terms of risk to life safety from structural failure.

In order to predict the fire behaviour in a compartment, four main design parameters are considered:

- Fire load density.
- Ventilation characteristics
- Compartment geometry
- Thermal properties of the materials used in the construction of the compartment.

Some of these parameters are variables, whose values will randomly fluctuate with time during the design life of the building. In an attempt to eliminate the uncertainty for variables such as fire load, current design guidance often recommends values that represent the 80th fractile.

British Standards Codes of Practice BS 9999 *Use building height, sprinklers, ventilation and occupancy type as the primary factors for assessing the FRR of structural elements.*

Building height categories limits

The height categories beyond which current fire resistance periods increase in Approved Document B (ADB) are:

- Less than 5m
- Between 5m and 18m
- Between 18m and 30m
- Over 30m.

The development task group involved in BS 9999 considered that two new height criteria should be introduced (11m and 60m) and these were linked to changes in either fire escape requirements or firefighting operations. For simplicity, the acceptable risk was assumed constant and to provide a benchmark common value for all occupancies, a height of 18m corresponding to the 80% fractile was adopted. Using a series of risk, frequency, maximum dimensions of compartments, height, consequence and probability charts and equations, the acceptance criteria for the relative risks were then calculated. The fire resistance periods given in the '9999' COP are based upon the ***elements of structure surviving a burn-out***. For property protection purposes, the construction separating one compartment from another is usually expected to withstand the burn-out of the contents of the compartment. This applies to the integrity, insulation, and if relevant the load-bearing functions of the separating elements. A sprinkler system, suitably designed and installed for the hazard to be protected, can be expected to prevent the rate of heat release from significantly exceeding that at the time of sprinkler operation. In most instances, it will assist in controlling the fire. The fire resistance of the compartment walls and floors can therefore be reduced in a sprinklered building or compartment.

When determining the structural fire resistance for a building using BS 9999, two methods are available. The first method is independent from ventilation conditions, assuming with and without sprinkler protection with the fire load density incorporated into the design. This approach varies slightly from the guidance in Approved Document B. This is due to the risk categories and split of several occupancies into more discrete groups in the engineered BS 9999 document. However, it largely follows the guidance in Approved Document B, with the assumption that the fire is based in an unventilated environment.

A second method is based on fundamental fire safety engineering i.e. a combination of real fire behaviour and evaluation of risk and consequences. There are limitations and this method is only applicable if the ventilation conditions in BS 9999 are met. The ventilation conditions are:

Minimum periods of fire resistance in minutes (independent of ventilation conditions)					
Fire Growth Risk Profile	Risk Characteristics	Height[A] of top occupied storey above access level			
		< 5 metres	< 18 metres	< 30 metres	> 30 metres
A1 Slow	Occupants who are awake and familiar with the building	30	60[B]	90[C]	120
A2 Medium		30	60	90	120
A3 Fast		60	90	90	120
B1 Slow	Occupants who are awake and unfamiliar with the building	30	60	90[C]	120
B2 Medium		30	60	90	120
B3 Fast		60	90	90	120
C1** Slow	Occupants who are likely to be asleep Long term individual occupancy	30	60	90[C]	120
C2** Medium	Long term managed occupancy	30	60	90[C]	120
C3** Fast	Short term occupancy	30	60	90[C]	120

** not individual residential

Note: 15 min fire resistance may be used for open-side car parks above ground level and with a top occupied storey of not more than 18m above access level *(increased to 30 min protecting vertical means of escape)*.

A) Buildings above 30 m are not permitted unless they have sprinklers in accordance with BS 5306-2 or BS EN 12845
B) 30 min if sprinklers conforming to BS EN 12845 (new systems) or BS 5306-2 (existing systems) are fitted.
C) 60 min if sprinklers conforming to BS EN 12845 (new systems) or BS 5306-2 (existing systems) are fitted.
D) 90 min if sprinklers conforming to BS EN 12845 (new systems) or BS 5306-2 (existing systems) are fitted.

Minimum fire resistance periods for elements of structure (based on specified ventilation conditions*)						
Fire Growth Risk Profile	Risk Characteristics	Height of top occupied storey above access level				
		< 5 m	< 11m	< 18m	< 30m	< 60m	> 60m
A1 Slow	Occupants who are awake and familiar with the building	15	30	30	60	75	90
A2 Medium		30[E]	30	60	90	120	150
A3 Fast		60	60	90	120	300	300
B1 Slow	Occupants who are awake and unfamiliar with the building	30	30	30	60	60	75
B2 Medium		30	30	60	75	90	120
B3 Fast		30	45	75	105	135	180
C1 Slow	Occupants who are likely to be asleep	30	30	30	45	60	60
C2 Medium	Occupants who are likely to be asleep	30	45	60	75	90	105

Risk profile C3 is unacceptable under many circumstances unless special precautions are taken.
The provision of sprinklers may enable an increase in risk profile and a reduction in fire resistance requirements.
E) Reduced to 15 min when ground floor area is less than 1000 m2.

*Table 17.3: Examples of fire resistance ratings for structural elements taken from BS 9999:2017 – For precise guidance of fire resistance requirements and **ventilation conditions used*** (with openings ranging between 2.5% and 10% as a percentage of floor area, with opening height further referenced as a percentage of storey height), the reader is advised to consult the British Standard*.*

* BS 9999:2017, Fire safety in the design, management and use of buildings – Code of practice, The British Standards Institution, Published by BSI Standards Limited 2017

- Minimum ventilation opening area (i.e. glazing breaking during fire) as a percentage of floor area. This typically ranges from 2.5% to 10% depending on the occupancy characteristic
- Height of opening as a percentage of floor to ceiling height in the compartment. This typically ranges from 30% to 100% depending on the occupancy characteristic.

In a well-ventilated compartment the duration and/or the severity of the time-temperature environment is generally less than in a Standard Fire Test. Therefore, the methodology proposed in BS 9999 with ventilated conditions suggests that taking the ventilation profile into account can lead to a reduction in FRR (fire resistance rating).

The compartments used in the tests were small by modern standards but the results are indicative of the influence of fire load and ventilation on the time-temperature environment generated within fire compartments.

However, somewhat in contrast to the 9999 approach –

New Zealand Building Regulations (Based on time equivalent, performance based, BRANZ research) – *Using fire load and ventilation opening factors as the primary references for an assessment of FRR*

	FIRE LOAD											
	400 MJ/m²				800 MJ/m²				1200 MJ/m²			
	Actual Fire Resistance Rating in Minutes											
VENT	15	30	45	60	15	30	45	60	15	30	45	60
0.01	22	41	59	NF	22	40	60	80	22	40	60	80
0.02	12	24	38	NF	12	24	36	49	12	24	37	50
0.03	9	18	31	NF	9	19	28	38	9	18	28	37
0.04	7	15	28	NF	7	15	23	32	7	15	23	31
0.05	6	13	26	NF	7	13	20	28	6	13	20	27
0.06	5	12	25	NF	6	12	18	26	6	12	18	24
0.07	5	11	23	NF	5	11	16	24	5	11	16	22
0.08	5	10	23	NF	5	10	15	22	5	10	15	21
0.09	4	9	23	NF	4	9	14	22	4	9	14	19
0.1	4	9	14	NF	4	9	14	22	4	9	13	18

Table 17.4: Predicted time to failure (t_{fail}) in minutes for assemblies with a fire resistance rating (FRR) from a standard fire test, based on ventilation opening factors. (NZ BRANZ)

NF – No structural failure (it is estimated that fuel will deplete before element failure occurs)

In order to calculate the ventilation opening factor (O) 'Branz' used the following method:

$$O = \frac{A_v \sqrt{H_v}}{A_t} \quad (m^{0.5}) \qquad \text{Eq. 17.1}$$

Where

O = opening factor

A_v = is the total area of vent openings

H_v = is the height of the vent openings

A_t = is the total area of internal bounding surfaces (including openings)

It can be seen from Table 17.4 above that the times to failure are much less than the specified FRR, for opening factors of 0.02 or more.

Conversely, to assist building code requirements in New Zealand –

Actual Fire Resistance Rating in Minutes	FIRE LOAD											
	400 MJ/m²				800 MJ/m²				1200 MJ/m²			
	Time required for safe evacuation or intervention											
VENT	15	30	45	60	15	30	45	60	15	30	45	60
0.01	10	22	34	45	10	22	34	45	10	22	34	45
0.02	19	38	50	55	19	38	56	75	19	37	55	73
0.03	25	44	52	55	25	49	71	83	25	49	72	95
0.04	30	47	52	54	30	58	75	83	30	58	85	103
0.05	33	48	52	52	35	64	76	81	35	66	92	104
0.06	35	48	52	52	38	66	76	80	39	74	94	102
0.07	37	48	52	52	42	68	75	76	42	78	93	98
0.08	38	48	50	50	48	68	74	75	45	80	92	96
0.09	39	48	50	50	48	68	73	73	48	80	92	93
0.1	40	48	49	49	50	68	72	72	51	80	88	90

Table 17.5: Required FRR in minutes required to ensure safe evacuation times are achieved.

An example can be provided as follows -

Office floor area	20 x 20m (3m high)	400m²
Vent opening area in fire	8 x 2m high	16m²
Fire load density	570 MJ/m²	570 MJ/m²
Bounding surfaces	(20 x 4 x 3m) + (400 x 2)	1,040 m²
Opening ratio (A_v / A_f)	16 / 400	0.04
Opening Factor (O)	$O = \dfrac{16\sqrt{2}}{1040}$	0.02 m$^{0.5}$

Table 17.6: Data used as an example demonstrating the NZ code based on ventilation opening factors

There is no doubt that *ventilation profiles* will have a major impact upon the expected duration of fire resistance and both UK and NZ codes of practice provide solutions that relate this. Taking a look at earlier research, where heavy steel assemblies were tested under full-scale realistic fire conditions in the Cardington research laboratory in 1997-98 and again in 2004, we can assess ventilation impacts.

Cardington full-scale office fire test 1997[174]

Office area 13.5 x 10m (135.12 m² floor area) with a full office furniture fire loading equal to 45.6 kg/m² (wood equivalent) at 820 MJ/m². This fire reportedly went to 58 MW maximum HRR at temperatures in excess of 1000°C with an opening factor of 0.03 m$^{0.5}$. Main columns and column to beam connections were protected but all other elements were exposed.

Cardington full-scale fire test 2003[175]

11 x 7m (77m² floor area) with a wood equivalent fire load of 40 kg/m² at 720 MJ/m². This fire had an opening factor of 0.043 m$^{0.5}$. The heavy steel members, supporting a weight-loaded reinforced concrete slab floor, had a sprayed protection fire resistance rating of 90 minutes. The fire reached a maximum temperature of 1108°C after 54 minutes and was still in decay at >150 minutes. Maximum beam and column temperatures exceeded 1000°C during the test. The time equivalent design FRR for this structural arrangement was 72 minutes.

Although large deflections were measured during and after the tests, the steel frame was structurally stable and still carried the loads following the fire. However as a result of this deformation the integrity of the floor slab around the internal column had suffered and it was also evident that during cooling, very high tensile stresses developed at the connections due to thermal contraction of the deflected beams. Similar levels of deformation without collapse have been observed at several fires by the author. However, local collapse of secondary steel beams, masonry, cast-iron and lightweight non-load-bearing steel framework has been commonly noted.

---

174 Kirby. B.R, Large Scale Fire Tests: the British Steel European Collaborative Research Programme on the BRE 8-Storey Frame; British Steel PLC UK 1997
175 Lennon. T, Full-scale tests at Cardington, BRE 2003

Time Equivalence (also ref. 15.3)

The most common approach to quantifying fire severity, when determining fire resistance, is to equate the performance of a load-bearing structural element exposed to a real fire in terms of an equivalent exposure to the standard heating regime applied in the standard fire test. This approach to quantifying fire severity is known as determining the 'time equivalence' of a fire. The thermal inertia of the compartment linings also has some effect on the intensity and the duration of a fire. Linings that have more insulation, such as plaster, reduce heat transfer from the compartment to the walls and ceilings, so the temperatures and the fire duration in a well-insulated compartment are greater. The time-equivalent analysis method is appropriate for compartments where the fuel load can be characterised, being offices, retail areas, schools, hospitals, residential apartments and hotels, but the method is not generally applicable to extensive storage areas with high fire loads or special hazard areas, such as those with significant quantities of flammable liquids. Also, the method is not generally appropriate for compartments with minimal or isolated fuel load, such as baggage check in and concourse areas at airports and railway stations, where fully developed fires (flash-over conditions) could not be possible.

The application of 'time equivalence' in assessing fire resistance requirements of structural elements was brought into question by Dr Gordon Cooke in an article[176] where he wrote;

'The measured time equivalence appears to depend on the construction [type] being heated. In the large compartment ($138m^2$) tests reported here [BRE FRS] the measured time equivalents used to validate calculated values were determined using protected steel members. The large compartment tests showed that the individual time equivalents for identical protected steel beams at the front, centre and rear of the compartment can vary by 15 minutes. The use of the average measured time equivalent will lead to under-estimates of time equivalence. The tests showed that fires spread from front towards the back of a compartment, [based on the amount of ventilation provided at the 'front']. If the test compartment had been as wide as it was deep and if there had been equal ventilation openings in two adjacent walls, we might expect that a fire ignited in the corner furthest away from the window walls would spread in two directions, and the effect on the variation and magnitude of time equivalence is a matter of conjecture'.

Consistency and 15 minute increments in FRRs

In a paper exploring the inconsistencies of fire rating systems O'Loughlin and Lay[177] suggest that three fundamental problems arise from the current system of 15-min increments:

- Inconsistent, ill-defined levels of safety are delivered. Such crude increments between fire resistance classes often require significant rounding-up of engineering output to the next grade, which many in the industry misinterpret as a comforting degree of contingency or safety factor. However, the variations in rounding-up (from 1 min to 14 min in theory) yields highly inconsistent levels of contingency, meaning that future changes in building fabric or use, or in the event that a design assumption transpires to be un-conservative or a more severe fire scenario than

---

176 Cooke. G; Time Equivalence – is it a good measure of compartment fire severity; Fire Safety Engineering Journal.
177 O'Loughlin. E, Lay. S; Structural fire resistance: Rating system manifests crude, inconsistent design; Case Studies in Fire Safety 3 (2015) 36–43; Open access Elsevier Ltd

that considered takes place, can result in an un-conservative fire resistance design.
- Discussions with design teams, approvers and stakeholders tend to focus on the outcomes of any performance-based study and what it means for project costs and the perceived standards of safety, as opposed to the particular design and analysis in question. Margaret Law highlighted this problem 20 years ago [4] and her views remain just as relevant today: departures from prescriptive recommendations attract undue scrutiny, to levels which would never be applied to solutions that comply with the "magic numbers", at the expense of rational, risk-based and engineering-based review. The implications of a 15-min or 30-min change in fire resistance are often viewed as too much reduction in safety for an approver to accept (or indeed an unnecessary increase in the eyes of a client or project team if, in the unlikely event that, a fire engineer was to suggest a fire resistance higher than the prescriptive guideline was appropriate) or too risky for a project team to pursue without confirmation of approval, regardless of what an engineering assessment demonstrates.
- The efficiency of the final solution is compromised, sophistication in the design methodology is negated and benefits of an all-inclusive, integrated design approach are limited if not wasted. If structural fire engineering is to be incorporated as a core design discipline to the overall advantage of the design, it needs to be quantifiable in a manner that allows accurate engineering-based assessment, metrics for incorporating various stakeholder aspirations (for example the differences between designing for a structural reliability purely intended to prevent disproportionate collapse and one intended to meet higher standard, life safety-based requirements) and is less dependent on subjective regulatory approvals, as per the previous point above. "Rounding-up" factors, owing to the 15-min increments, which can differ by an order of magnitude do not lend themselves to accuracy. The consistency of crudeness sought after in any engineering process is therefore not attained.

The authors provided an example of inconsistency in fire resistance ratings –

Two hypothetical cases were assessed in order to explore the differences between the current 15-min interval system for fire resistance ratings and a system with ratings refined to 1-min grades. For the comparison, two 28 m tall, un-sprinklered office buildings in the UK were chosen. As such, from a prescriptive perspective (for example, in accordance with AD B or BS 9999 UK codes of practice), both are identical and should have a fire resistance of 90 minutes. However, the buildings differ in their form of construction, amounts of potential ventilation openings (windows) on their façades and their compartments sizes. As such, when risk-based and / or performance-based fire engineering methods are applied to establish the appropriate fire resistance for each building, two very different answers are obtained: **74 min** is determined for Building A and **61 min** is determined for Building B. In practice, currently when specifying the fire resistance for these two buildings, the fire engineer would need to round up both of these figures to the nearest commercially useable fire resistance rating, namely 75 minutes for both. The paper goes on to offer proposals resulting in more accurate and appropriate fire resistance ratings, according to the authors methodology.

> **What does this mean to the Firefighter?**
>
> Fire resistance ratings (FRRs), the time allotted to a building's minimum fire resistance before structural elements may begin to fail, cannot be relied on to inform firefighters of a safe firefighting occupation time. Just because a building has a minimum 90-minute FRR this cannot guarantee the structural elements will maintain stability for that period if under constant and severe fire attack. The ventilation profile and layout of the involved fuel load distributed across the floors may reduce FRRs substantially. The only way to ensure firefighters are working within safe time-frames are to establish safe working zones (and collapse zones), to monitor structural warning signs of collapse and prepare for an evacuation well in advance, where this might be needed. The FRRs can only be considered as a rough guideline for safe occupation times. In some situations, such as two storey offices with floor-plates less than 1,000 m^2, industrial portal frame sheds or some lightweight structural elements, a safe firefighting occupation time may be just a few short minutes following arrival on-scene where fire is taking hold.

17.7 RULES OF ENGAGEMENT – THERMAL IMAGING

By Chief Andrew Starnes (Project 'Kill the flashover' USA)

For many firefighters, their basic training on fire behavior phenomenon's such as rollover, flashover, and backdraft has occurred in the classroom and in a well ventilated burn building where they could see these indicators they were taught. With visibility being limited and almost zero in the fires they see today, what inherent cues or clues are they to rely to prevent rapid fire progression? Consider the following photograph that places a side by side view of the fire environment and compares the thermal imaging view vs our naked eye. Without the aid of a thermal imaging camera in the hands of a properly trained firefighter the end result could be a line of duty death.

According to NFPA 1971 'The minimum TPP rating for NFPA 1971-compliant coats and trousers is 35, which provides 17.5 seconds of protection until a second-degree burn is caused, in dry gear.' The TPP of firefighter PPE is of such high caliber in most cases that by the time they feel the heat their skin is nearing 130 degrees Fahrenheit. Then at 140 degrees Fahrenheit a firefighters skin becomes numb, thus their skin is no longer a valid indicator for extreme fire behavior. Even armed with this information, many firefighters have not considered the cumulative heating effect high heat has on their gear as they progress towards their objective. The built in safety factor (or thermal protective performance) is often exhausted without any sign or recognition as firefighters crawl in superheated convective heat currents that gradually absorbs into our PPE. Then when a rapid fire progression event occurs, firefighters do not have the 17.5 seconds or more of protection due to their P.P.E. being saturated with heat. This is exemplified by the thermal signatures noted on this firefighters P.P.E in Fig.17.2.

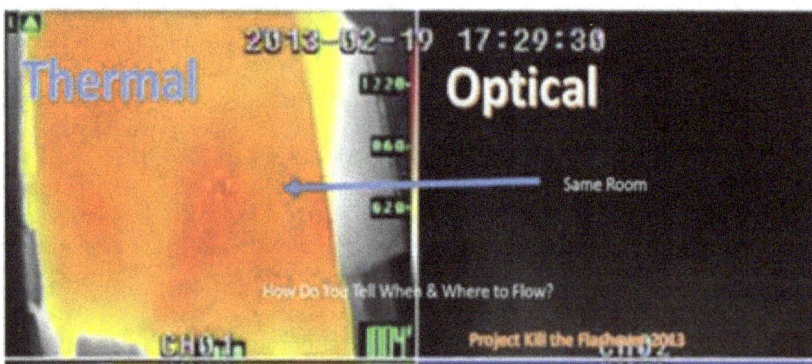

Figure 17.1: A side by side view of the fire environment, comparing the thermal imaging view against normal vision in heavy smoke. **Photo courtesy Project Kill the Flashover**

> The Thermal Protect Performance (TPP) of Personal Protective Equipment (PPE) has been a point of much research for several decades, as clothing fibres and standards are constantly developing. However, very little scientific research has evaluated how the broad range of firefighting environments may impact on firefighter's PPE. Firefighters can receive burns from radiant heat energy (heat flux) or in direct contact with flames. The high levels of protection afforded by modern PPE can also increase a firefighter's risk of heat exhaustion when exposed to even medium levels of heat flux for long durations (*see Fig.11.7*). Some injuries may occur where the clothing is compressed against under-garments, or skin. Garments that are wet may exhibit significantly higher heat transfer rates than garments that are dry. TPP testing measures the amount of heat transfer through a firefighter's clothing composite (all layers included) from an 84 kW/m^2 heat flux. This level of flux is chosen in order to replicate a flash-fire, or mid-range post-flashover environment. A minimum TPP rating of 35 is required according to the NFPA standard, providing a firefighter with 17.5 theoretical seconds to escape, prior to sustaining second degree burns. However, some research (Krasny) has suggested that TPP 35 garments may only offer protection for around 10 seconds in such a thermal environment.

European firefighter clothing standards are addressed by EN 469 (loves, helmets and other ancillary items come under different standards). In general, European standards do not meet such stringent requirements as those under NFPA standards. *(Further detailed information is available in EuroFirefighter p249-50) (Jeremy Mills Publishing UK).*

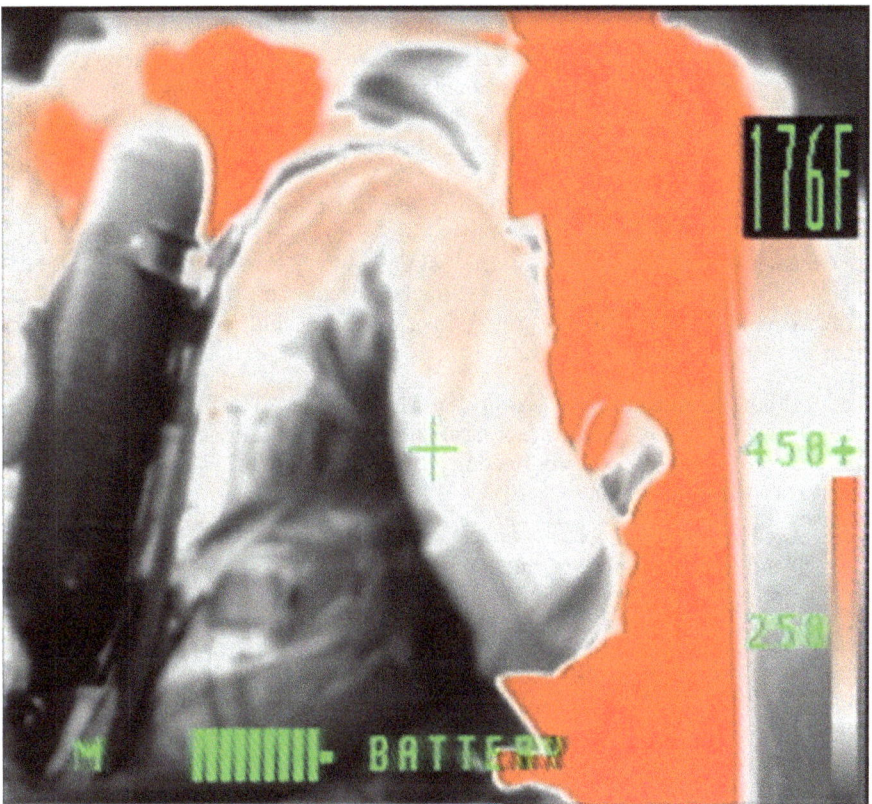

Figure 17.2: Thermal signatures noted on this firefighters PPE denote high levels of heat flux is being absorbed into the clothing. **Photo courtesy Project Kill the Flashover**

Over recent decades, we have seen our working environment as firefighters change due to experiencing fewer fires overall but rapidly developing fire progression combined with light weight construction, and higher TPP of our P.P.E has actually created a great need to thermally diagnose the severity of the conditions firefighters face. We should not dismiss nor fail to recognize that a number of firefighters die each year due to rapid fire progression: 'Lieutenant Jeff Parker pointed out in a recent article for Fire Engineering that not having a TIC at a structure fire was the 6th ranked risk factor in firefighter fatalities.

NIOSH reports had identified that they had failed to be deployed in 38% of fatalities.[178]

If we study our past we can also see where firefighter LODD reports reveal numerous situations where if only the firefighters were equipped with a thermal imaging camera (TIC), and were properly trained to use it, the TIC might have saved their lives. In addition to these line of duty deaths, if we read the reports further we will see another disturbing phrase repeated under contributing factors which is:

'The failure to recognize, understand, and react to deteriorating conditions'.

Firefighters are continuing to die in rapidly progressing fire environments, that collapse on them quickly, and they cannot 'feel the heat' to diagnose the true severity of the environment thus they are at a true disadvantage without proper training on fire behavior, tactical thermal imaging, and the 'know how' to mitigate such conditions. The need for a paradigm shift in our understanding of the fire environment has never been greater or more necessary.

The following photo has the definition of flashover overlaid in it for a profound reason. This room is nearing flashover but to the naked eye this room was had zero visibility. A firefighter entering this room without the assistance of a TIC would fail to recognize the indicators (which are not visible with the naked eye), would not feel the heat until it's too late (due to the thermal protective protection of their PPE), and could be injured or killed in the impending flashover.

Tactical Thermal Imaging can be defined as the usage of a TIC to better diagnose the *IDLH (immediately dangerous to life or health)* environment we are in for the purposes of strategic decision making. The information gleaned from Tactical TIC use can assist us in all aspects on the fire-ground such as:

- Size-up (Tactical 360)
- Identifying and managing the flow path
- More efficient stream placement and interior fire attack efforts
- Enhanced search methods and more.

The first area that can be greatly improved through the implementation of tactical thermal imaging is the 360 degree survey. Numerous LODD reports have cited that the failure to complete a 360 has been a contributing factor towards their demise. Firefighters are becoming well versed in this tactic and it is becoming more common place in its application yet many firefighters aren't carrying the TIC or viewing the structural thermally during their walk-around. Firefighters who choose to carry their TIC and enhance their 360 degree survey will find that they not only have more information but can be more efficient in the execution of their tasks. The following photo clearly explains this phenomenon as the Incident Commander is starting his/her Tactical 360. Notice as the Incident Commander completes the Tactical 360 the difference in the second photo optically and thermally –

With a TIC the Incident Commander can look at the temperature of an opening exhausting smoke. It is important to note that a Thermal Imager reads surface temperatures and not the temperatures of the gases (TIC's can only read the temperatures of three gases that fit within the preset emissivity range of the camera, (such as ethane, ethylene, and hydrogen cyanide). Firefighting Thermal Imaging Cameras read a specific range of light

---

[178] Whitty, Michael (2010). Maximizing Thermal Imaging Use in the Emergency Services. (pp.5-). retrieved from: http://www.esf.com.au/documents/reports/MichaelWhitty.2010.MFB.pdf

Rules of Engagement (interior deployment guidance) • 343

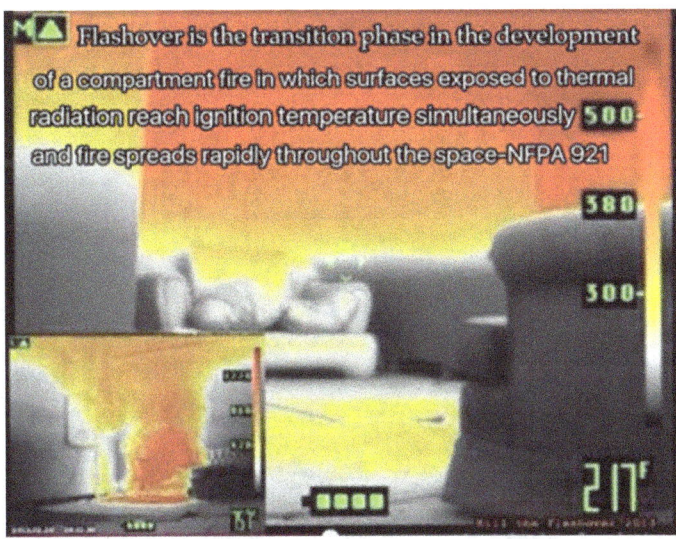

Figure 17.3: This room may present zero visibility to the naked eye but seen through a thermal imager, the room is approaching flashover. **Photo courtesy Project Kill the Flashover**

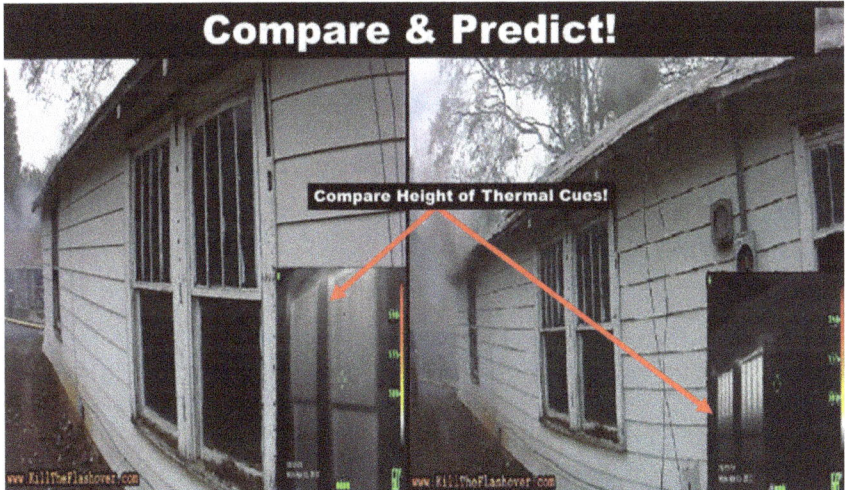

Figure 17.4: Thermal cues provided by a thermal imager from outside the building – such as the height of the heat layer and possibly the neutral plane **Photos courtesy Project Kill the Flashover**

between 7-14 microns at a set emissivity (.95). Thus an exhaust opening showing over 260°C (500 degrees Fahrenheit) at the top is the temperature of the immediate environment and not necessarily the gases. Firefighters need to be aware that those superheated gases may be much hotter! Also the Incident Commander needs to be aware that the focal point of the TIC is typically a measurement of a 300mm² (12 inch) area only (dependent upon

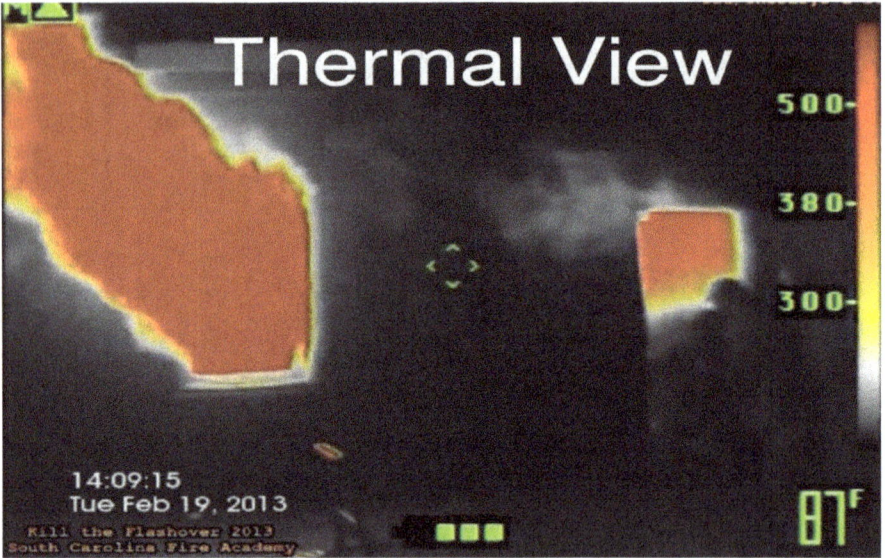

Figure 17.5: Reading the whole picture and not just the spot temperature is critical to gaining awareness of the fire environment. **Photo courtesy Project Kill the Flashover**

the distance to spot ratio). The focal point of the TIC produces the direct temperature measurement (DTM) in the lower right hand corner of the screen. Firefighters who read just this temperature can be lulled into a sense of complacency if they fail to read the palette or the big picture as you see the photo shows a spot temperature of 87 degrees Fahrenheit (30°C) (Which is typically located in the bottom right hand corner of the TIC screen) whereas the color palette indicates temperatures of 500 degrees Fahrenheit (260°C) or more.

Tactical Thermal Imaging is not simply using a TIC for search & overhaul. It is the understanding of fire behavior, higher heat release rate fuels, and proper interpretation of the thermal information in the hands of a well-trained and educated firefighter. These examples mentioned in this brief overview only are a mere sampling of the uses how Tactical Thermal Imaging can enhance overall effectiveness on the fire ground. Firefighters must be diligent not only to stay current on their fundamental skills but the ever improving technologies that are impacting our service delivery. My hope is that this section of the book inspires other firefighters to learn to enhance their overall fire ground effectiveness far beyond my own efforts.

Instructor Andy J. Starnes

Chapter 18

Tactical Ventilation – survive in the 'Flow-path'

'It's dangerous being alive – you can die!'
                              David Eberhard, Chief Psychiatric Doctor, 2005

'The building in front of us was breathing like a gigantic dragon. It was taking in air and breathing smoke out in ever increasing gulps. Suddenly there was an eerie silence for around a minute. Things didn't look good and we had five firefighters inside that monster's stomach! The order went out over the air-waves to evacuate the building immediately. Within what seemed like seconds there was a strong wind heading into the doors. It was so strong it swept passed our ears like an arctic squall, with street rubbish being picked up in its path and hurled at the wide open doorways into the super store. A wheelie bin started to move at quite a pace towards the building. [The flow-path was forming in a big way]. Then, as three of our firefighters came out of the doorways in front of us, there was a groaning creaking sound, followed by a whistling sound. The backdraft occurred just a second later …. It was immense, like nothing we had ever seen before'!

As recounted by Firefighter Alf Gardener
Kent Fire and Rescue Service
Dover B&Q Fire 1995

The above incident was also reported by the Dover Historian[179] as follows – *'On the evening of 16 November that year [1995], at about 19.30hrs, the Service was called to the B&Q superstore on Charlton Green. The blaze was relatively confined but while five firemen were searching the building a flashover cause a fireball to envelop it. Of the five firemen who went in only three came out. Eventually the other two appeared, their uniforms smouldering. All five were taken to hospital with fire fighter Colin Cox, being admitted into the East Grinstead burns unit for injuries to his shoulder'.*

On return from a research detachment with fire departments in New York during the 1970s I had learned a lot about the fire-ground strategies associated with vertical and horizontal ventilation. The firefighters in New York were trained, equipped and effectively staffed and resourced to position themselves on roofs, on balconies and external stair access points, ready to create openings and vent windows. These actions were generally pre-determined by SOPs but were dependent on building type, occupancy and prevailing fire and wind conditions. I observed that not all firefighters were so well practiced in the critical aspects of 'timing' the vents and optimizing vent location and on occasions, fires would spread uncontrollably beyond their capacity to extinguish due to mistimed or

---

[179] https://doverhistorian.com/2013/07/27/dover-fire-service-part-2-from-1939/

inappropriate venting actions. However, there were clear situations where venting actions saved lives and provided safer working conditions for firefighters. All this in contrast to London firefighters who were more likely to lessen any venting actions, reduce the heat release from the fire load and control fire spread by closing down structure fires (closing doors). However, it was clear that what was needed was some middle-ground approach, harnessing the best of both approaches.

In the 1980s I developed and introduced, through the national journals, an integrated (UK/US) fire-ground strategy to UK firefighters termed 'tactical ventilation'. Since then the term 'tactical ventilation' has become widely used in many texts and training programs around the world but the original definition has become somewhat clouded and diverse and the objectives are now widely varied. So in this chapter we go back to the original source definition and examine the objectives of *tactical ventilation* as a fire-ground strategy, demonstrating how essential it is for firefighters to familiarize with 'tac-vent' protocols if they are to survive whilst working in the *flow-path*.

> *As the fire load in buildings gradually evolved through the 1970-90s to include a higher hydrocarbon/plastic content, and as buildings became more air-tight and energy efficient, we began to lose more firefighter lives due to rapid fire phenomena. This raised a training need in how to manage the hazards associated with an increasing ability for combustible fire loads to transform and then transport in the fire gas layers, leading to flashover, backdrafts and fire gas ignitions (including smoke explosions). What in fact is needed is a training programme aimed at teaching firefighters how to 'survive whilst working in the flow-path' if future life losses are to be prevented. Therefore a firefighter must be able to identify how a flow-path forms, what controls it, how and where it may be located in a building fire, where the most dangerous part of the flow-path is and what impact it is likely to have on fire development. Further to this, a far greater understanding is required of how tactical or unplanned venting will either worsen or improve conditions in the flow-path.*

18.1 TACTICAL VENTILATION - FLOW-PATH CONTROL

So, let's examine that definition more closely. Tactical ventilation is used primarily to take control of the fire's burning regime to tactical advantage (in fact you might even interchange 'burning regime' with 'flow-path' to achieve tactical advantage). To be able to do this the firefighter must understand basic practical fire dynamics and be aware of how flow-paths form, transport, reverse and expose firefighters to great risk. The strategy begins with 'containment' and then *may* be followed by venting actions.

> *'Tactical ventilation is the containment or venting actions by firefighters, used to take control from the outset of a fire's burning regime, in an effort to gain tactical advantage during interior structural firefighting operations'.*
>
> Paul Grimwood
> **London Fire Brigade 1987**

This means we start by controlling the fire spread by cutting off its air supply, wherever we can. If the street door is open we partially close it until our hose-line is charged and firefighters are ready to advance. On a '360' primary size-up we may find opportunities to close doors to the rear also, particularly at the lowest points in the building. As we make entry we have created our first vent opening! Do we need to control that opening? Now the Incident Commander (IC) may consider horizontal cross ventilation or even vertical ventilation. If the strategy of Positive Pressure Ventilation (PPV) or PP Attack (PPA) is a consideration, we must follow strict protocols to lessen the chances of worsening the situation. As firefighters advance in, they should take every opportunity to close doors as they pass them. This is especially important if they are heading to upper floors. This will reduce further the possibility of air feeding in to the fire, creating an ideal flow path for PPA to take optimum effect, and may isolate any fire spread into the stairs or other areas where the fire may not yet have been located.

In 1996 the UK Home Office (Fire Department) published the author's tactical ventilation strategy in a fire service manual[180] where they failed to address the original tactical objectives of fire isolation and confinement; the manual stated:

'Ventilation at fires is, the removal of heated air, smoke and other airborne contaminants from a structure, and their replacement with a supply of fresher air; whereas Tactical Ventilation requires the intervention of the fire service to open up the building, releasing the products of combustion and allowing fresher air to enter'.

This guidance came following a home office sponsored report by Adrian Hay[181] in 1994 that incorporated the *tactical ventilation* strategy as a basis for his report, following discussions with the author. This was shortly followed by a series of fire ventilation tests in 1995 by BRE[182]. However, again one of the key principles of Grimwood's integrated strategy (fire isolation and confinement) was missing. It is important to take into account the entire integrated US-UK strategy when venting (or not) structures.

The traditional approach to cutting holes in roofs and breaking windows, in an effort to relieve conditions inside a fire-involved structure has long been under question. As fire loads increase in content and density and buildings become better insulated with energy efficient heat retaining walls, the fire dynamics we see now are not suited to venting without clear purpose and forethought as to what the effects on fire development may be. There are direct conflicts in removing combustion products and replacing them with a fresh air supply. What is needed is clearer guidance to avoid those unknown moments where venting a window or roof might 'seem' a good idea at the time; or where things are getting hot and breaking a window out from inside will hopefully provide instant relief from heat and possibly prevent a flashover. It is most certain that where firefighters 'vent as they go' from the inside, the temporary relief obtained by escaping heat and a rising smoke layer will last only a few short seconds. As additional air feeds in through the openings created, unless a firefighting stream is immediately and effectively applied at the fire the fire development that follows may be rapid and severe, possibly trapping or burning firefighters and any remaining occupants.

---

180 The Behaviour of Fire; Tactical Ventilation of Buildings and Structures; Manual of Firemanship (supplement), Home Office UK 1996
181 Hay. A; A Survey of Fire Ventilation; Warrington Fire Research. 5/94; 1994
182 Fire ventilation trials; FRDG 15/96; UK Fire Research and Development Group 1996

What venting actions might lead to an event of extreme fire behaviour?

- Incorrect location of a vent opening
- Mistimed vent opening (uncoordinated)
- Inappropriate vent opening (no objective or purpose)
- Inappropriate entry point, or procedure, for gaining access to a ventilation controlled fire
- Creating vent openings without confining the fire into zones or siting a charged attack hose-line at a critical point in the structure (e.g.; vertical access routes)

In 1980 the author began to question some of the ventilation (or anti-ventilation) tactics being used in the USA and across Europe and twenty five years later the NIST and UL research in the USA were producing the science that established what happens inside a building on fire as vent openings are made. The changing fire environment with modern fire loads and energy efficient buildings, where more smoke is generated and greater amounts of flammable fire gases are produced, creates greater risk in the gas-phase. As fires burn hotter and develop faster, the strategy of creating openings to release these products of combustion only speeds the burning process up and in some cases, this occurs faster than firefighting water can be applied effectively.

Even so, there are still fires where tactical ventilation by firefighters will optimise the approach and the author continued to analyse building fire outcomes by comparing US .v. European tactical approaches. Would New York City (FDNY) firefighters have achieved a better outcome in specific fires we had in London and vice versa, would London firefighting tactics have been better in some situations in New York fires? The general conclusion was yes to both! As an example, there was an occupant trapped in an attic fire in the city that took firefighters wearing breathing apparatus several long minutes to reach by internal stairs due to the excessive heat build-up in the stair-shaft. The fire was five floors below the attic and had been fully controlled prior to rescue attempts. Would FDNY firefighters have been able to reach this occupant earlier with their SOP'd 'roof access to skylight approach'? One could argue that this would be the case. Would the occupant have stood a greater chance of surviving? Possibly yes. But then there were fires elsewhere where the creation of vent openings may have placed occupants at greater risk as the fire developed more rapidly beyond the original compartment of origin.

A look at fire spread data between the US and the UK showed that US building fires were far more likely to spread out of and beyond the compartment of origin and whilst there may be many reasons for this, the excessive use of ventilation tactics was perhaps a primary cause for such an outcome. Another comparison was made with the roof venting tactics used by firefighters. In NYC it was common to utilise roof and trenching cuts to prevent fire spreading through common cocklofts and this was a very necessary and successful strategy. However, the author frequently observed firefighters *outside the city* undertaking venting operations of pitched roofs over domestic dwellings where the fire had not yet entered the roof space (note: FDNY firefighters themselves rarely vent pitched dwelling roofs). This was considered an unnecessary tactical approach that nearly always resulted in additional damage to the structure, above that caused by the fire itself! What may have been seen as a principal strategy in the pre-1970s period is now being considered as a strategy of less importance as the firefighting environment demonstrating faster fire spread rates continues to evolve.

Communicate & Coordinate Venting

Whilst communication and coordination have always been key objectives during any ventilation actions, experience has shown that omissions of these two necessary phases are frequently the cause of tactical failings or errors resulting in near misses, increased property damage or even life loss. It is useful to discuss the important stages and aspects surrounding successful and safe venting strategies with crews and a bullet list of key points for discussion can be used to start a simple debate for guidance, as follows -

Key points in the Tac-Vent strategy: *(points for round-table discussion with crews)*

- A 360 size-up provides key information
- Which direction is the wind heading? What is the relevance and possible impacts?
- Containment actions by partially closing doors prior to deployment with a charged hose-line
- The entry door may be our first venting action!
- How is wind likely to affect our deployment? Is in at our rear or in front of us?
- Do we need to create an opening ahead of the fire if the wind is at our rear?
- Do we need to take actions to close any openings if the wind is in front of us?
- Do we control this entry door?
- Isolation and containment tactics by closing all interior doors on passing them and why?
- Has the lowest point of fire been located? Basement?
- If firefighters change levels, or advance towards the windward side of the building, the importance of interior door control may be critical
- When and why would we use horizontal ventilation?
- When and where should we use vertical ventilation?
- Do we have a **reason** or **directive** to vent?
- Do we have a reason NOT to vent?
- Who gives the order for tactical ventilation to take place?
- Are ALL interior crews aware and in support of a venting action?

The words of Chief Fire Officer Braidwood are just as important now as they were then (1830-1866)

'On the first discovery of fire, it is of the utmost consequence to shut, and keep shut all doors, windows, or other openings. It may often be observed, after a house has been on fire, that one floor is comparatively untouched, while those above and below are nearly burned out, this arises from a door on that particular floor having been shut, and the draught (air-track or flow-path) directed elsewhere'.

Superintendent James Braidwood
London Fire Engine Establishment
On the Construction of Fire engines and Apparatus, the Training of Firemen and the Methods of Proceeding in Cases of Fire (Memoirs dated 1830)

The men of the fire brigade were taught to prevent, as much as possible, the access of air to the burning materials. What the open door of the ash-pit is to the furnace of a steam boiler, the open street door is to the house on fire. In both cases, the door gives vital air to the flames'.

Master of Fire Engines James Braidwood
Edinburgh Fire Engine Establishment
Fire Prevention and Fire Extinction (1866)

Tactical Horizontal Ventilation
- The Incident Commander (IC) should be the person who initiates any such action
- This should be coordinated with requests by an interior crew
- Before any venting the IC should consider how it may impact on the flow-path and the fire's intensity
- Before any venting the IC should ensure all crews working inside are informed what is about to occur
- The timing of horizontal ventilation should not normally occur until water is being applied to the fire
- The suppressive capacity of the nozzle in use should be adequate for any increase in heat release that may occur. Where a vent is to be created a main-line hose (not hose-reel) should be in position at the fire compartment.
- Where even a light wind exists on the windward side of the building and firefighters may be working on the leeward side (of the fire) a main hose-line (not hose-reel) should be deployed.
- Where even a light wind exists and firefighters are entering from the windward side, consideration should be given to the creation of a cross-vent opening on a leeward facing fire compartment, to relieve pressure and force fire away from the advancing hose-line.
- If the fire compartment is not leeward facing in such a situation a range of options are available but will depend on building size, layout and occupancy type (discuss options with crews)
- Use horizontal ventilation to release fire pressure when advancing a hose-line on the fire
- Use horizontal ventilation to redirect heat and smoke away from firefighters (reverse the flow-path) as combustion products will head for an air supply and if the compartment door is partially closed this will reverse the flow-path to advantage

Tactical Vertical Ventilation
- Consider what you are achieving by doing this and analyze risk versus benefit
- The IC will dictate when vertical ventilation should take place – timing and location are everything
- Vertical ventilation is a life safety strategy
- To place firefighters above a fire to cut holes in the roof should only be undertaken by the most experienced of city fire departments. It is a strategy fraught with hazards and is not recommended.
- To access stable rooftops and utilize fixed openings such as roof hatches, roof access doors and other permanent openings above stair-shafts or occupied rooms may be a life-saving strategy if timed correctly.
- To open windows or other vent outlets on upper floors above the fire could escalate the fire dramatically by creating the ideal flow-path, unless the fire has already been effectively suppressed. Venting the top of a stair-shaft could draw fire into the stairs, particularly if the windows to the fire compartment have vented
- Smoke logged stairs can be cleared using this strategy, but be aware that smoke control systems (pressurization) may be compromised by such actions.

Venting 'Hot-wall' or 'Cold-wall' Fire Compartments
The compartment boundaries (walls, ceiling and floor) can influence the conditions inside a fire compartment in three ways:

1. There is a large heat loss through non-insulated walls to the exterior. In this situation the wall is cold and the hot layer is losing heat.
2. The wall may be insulated and prevent heat escaping from the hot layer. In this case the heat is retained in the hot layer and radiated heat-flux increases dramatically.
3. The wall itself may act to absorb large amounts of heat (ie: brick or concrete walls etc.), and this hot wall condition will serve to increase the heat in the hot layer further still.

As radiation becomes the dominant mechanism of heat transfer in a compartment fire, the situation may be termed **'hot-wall'**, where the walls are able to retain heat. Where walls are not able to retain heat the heat is either lost through the wall to the exterior or bounced back into the hot layer (insulated wall).

In **cold-wall** situations where convection is the dominant mechanism of heat transfer, venting leads to heat loss from the hot layer and a reduction in compartment temperature (this may only be temporary as air feeds in through the vent and increases fire development).

In **hot-wall** situations where radiation is the dominant mechanism of heat transfer, venting generally leads to some heat loss from the hot layer, but the radiation from the hot walls may quickly overwhelm these losses as the combustion process accelerates (resulting in ***thermal runaway*** – leading to room flashover caused by venting).

18.2 TACTICAL (VERTICAL) VENTILATION

A colleague of mine [*Chief Ed Hartin*] provided an interesting *set of conclusions* on his firefighter training website[183], in relation to the UL research into tactical vertical ventilation. The following tactical considerations related to *vertical ventilation* are based in part on the research results and tactical considerations developed by the UL Firefighter Safety Research Institute, ongoing study of practical fire dynamics, and fire-ground operations.

- The air track from vertical ventilation openings in or directly connected to the involved area of the building is most likely to be unidirectional, and outward [*author's note: sometimes the air-track can be bi-directional (both air in and smoke out) or oscillatory (pulsing in and out), depending on the size of the opening*[184] *(larger openings in particular)*].
- The air track from horizontal ventilation openings above the fire is likely to be unidirectional, outward, may be bidirectional (out at the top and in at the bottom), or may be pulsing (in and out).
- The air track from horizontal openings on the same level as the fire is likely to be bidirectional, but may be unidirectional, outward or inward or it may be pulsing (in and out).
- The air track from horizontal openings below the level of the fire is likely to be unidirectional, inward, but may present differently depending on conditions.
- Air track is influenced by the location and size of openings, the distance of the opening from the fire, wind conditions, the burning regime (fuel or ventilation controlled), and if ventilation controlled the extent to which ventilation is limited.

---

183 http://cfbt-us.com/wordpress/?tag=b-sahf
184 Karlsson. B, Quintiere. J, Enclosure Fire Dynamics, p111, CLC Press LLC, Florida, USA, 2000

As with all of the B-SAHF (building, air track, heat, smoke, and flame) indicators, air track must always be considered in context.
- Larger vertical ventilation openings will release a larger amount of smoke and a correspondingly large volume of air will be introduced into the building.
- With natural tactical ventilation, if the area of the inlet or inlets is small in relation to the exhaust opening, the movement of both smoke and air will be constrained and ventilation will be less efficient. Correspondingly if the area of the inlet or inlets is large movement of smoke and air will be more efficient.
- When using natural tactical ventilation, the inlet area should whenever possible be two or three times the size of the exhaust opening (note that this is reversed when using positive pressure ventilation).
- If the fire is in a fuel controlled burning regime, effective vertical tactical ventilation will provide a lift in the smoke level and slow fire development even if fire attack is delayed. This was commonly seen in the legacy fire environment, but is unlikely in the modern fire environment due to the high heat release rate of modern fuels and fuel loads found in today's buildings.
- If the fire is ventilation controlled (most likely in the modern fire environment) and either horizontal or vertical tactical ventilation is performed absent fire attack, the lift (if it occurs) will be momentary as increased heat release rate and smoke production will likely overwhelm the size of the ventilation opening.
- If the fire is ventilation controlled, the effectiveness of vertical tactical ventilation on improving conditions is dependent on concurrent application of water onto the fire. Note that this requires effective fire attack, not simply a charged line at the door or being advanced into the building. Once ventilation openings are created, the clock is ticking on increased heat release rate.
- Coordinating fire attack and vertical tactical ventilation requires close communication between companies assigned to fire attack and those assigned to ventilation. Communication when water is being applied to the fire is critical. However, it is also important to evaluate observed conditions in conjunction with reports from the interior.
- If using existing vertical openings such as skylights, scuttles, or roof bulkheads, it may be necessary to delay opening until the hose-line is in place and operating.
- Vertical ventilation through cut openings takes longer than using existing openings and as such hose-lines may be in place and operating before the hole is completely cut. However, it is important for company or team performing ventilation to verify that this is the case before opening the cut hole.
- Effective coordination between fire attack and ventilation requires that command and company officers have a good idea of how long specific tactical operations take in different types of buildings and with varied construction types. If you don't know, it is time to get dirty and find out!

18.3 THE 'FOUR TENETS' AND 'SIX COMMANDMENTS' OF TACTICAL VENTILATION

(Lt. Nick Papa – City of New Britain (CT) Fire Department)

The 'Four Tenets' and 'Six Commandments' of Tactical Ventilation, **developed by Nick Papa,** provide firefighters with a set of guiding principles and procedures to be utilized when engaging in ventilation activities. By identifying the most fundamental operational considerations and parameters, they collectively serve as a universal framework. These concepts are a direct input from an original program that he created and delivered at the 2017 Fire Department Instructor's Conference (FDIC) in Indianapolis, USA. Nick is a Lieutenant with the City of New Britain (CT) Fire Department in the US. I feel strongly that Nick's interpretation and message are not only relevant to today's fire environment, but have also remained aligned with the original concept of *Tactical Ventilation*, as it was originally intended, upon its inception by Paul Grimwood in the 1980s. The following section is his writing on the subject; detailing both the *'Four Tenets'* and *'Six Commandments'* of Tactical Ventilation.

The Four Tenets of Tactical Ventilation:

1. Informed

On the fire-ground, information comes at a premium. Because every situation we face is inherently unique, intelligence gathering is always a top priority. From the moment an alarm is received, we are inundated with information – varying in degree of detail and accuracy. That information must be immediately processed, utilizing our knowledge base and intuition, and translated into an appropriate action plan (also known as 'Recognition-Primed Decision Making' or RPD). In the absence of a one-size-fits-all approach, the conditions will always dictate our actions. With so much at stake, it is incumbent upon us to, not only collect as much information as possible (without delaying action), but to disseminate that which is *pertinent* to the other members operating on scene and en route. Given the infinite number of potential variables, and the uncertainty of a rapidly changing, hostile environment, we are faced with the daunting challenge of *'attempting to make perfect decisions with imperfect information.'*

To minimize that differential, we must engage our senses and maintain a heightened state of (situational) awareness. By always being acutely cognizant of our surroundings, we can identify critical pieces of information, as they present themselves, which will guide us to effective decision-making. Upon arrival, this (physical) process begins by 'sizing-up' the incident. We must keenly observe the building as we approach; if we pull past or to the far corner of the building, we are granted a visual of three sides. Upon doing so, the officer of the first-due company should provide a 'brief initial report (BIR).' A model which is commonly utilized for such instances is the 'CAN' report: condition; actions; and needs. Following this transmission, a rapid survey of the exterior (and interior) should be made whenever feasible. Often referred to as the '360,' it is more than simply 'taking a lap' to check the proverbial box, it provides us with a more complete picture of the incident. In regards to ventilation, particular attention must be placed upon building construction (layout, existing openings, and egress points), presentation of the smoke and fire, as well as the wind condition (direction and speed). We must remember that size-up is not a single act, it is a continuous process that must be strictly employed throughout the duration of the incident.

Not all of the information we obtain on the fire-ground is acquired first-hand. In fact, a great deal is received through radio transmissions (and word-of-mouth). By always keeping an ear to the radio, we can ascertain the whereabouts of other crews operating within the structure, their assigned tasks, their progress, as well as the conditions being encountered. Lt. Rob Brown, FDNY (2016) states, 'One of the best ways to do this is to listen to all radio transmissions on the fire-ground to paint a mental picture, gauge the flow of the operation and be able to anticipate potential problems in other areas that may affect the objective.' Amidst the fast-paced, high stress, and hostile environment that is the fire-ground, succumbing to the 'fog of war' can happen suddenly and often without our realization. When we allow our emotional response to elevate uncontrollably, we can become so overly task-oriented that we lose sight of what is going on around us; due to the onset of 'tunnel-vision' and 'auditory exclusion.' We must practice arousal control to maintain our composure, no matter what the circumstances.

Those engaging in ventilation activities, especially, must possess a much broader focus; *as their actions can single-handedly and swiftly impact the course of an incident – potentially with severe implications.* We must possess enough information to be sure that the ventilation plan has a defined objective that will facilitate a specific firefighting operation and the achievement of the mission – preserving life and property. Brown (2016) asserts that 'Smart, aggressive firefighters never lose sight of where they fit into the operation (accountability) and how quickly their objective can change by monitoring their situation awareness...' Possessing (and sharing) vital information is often the determining factor in the success of any operation. The need for vigilant situational awareness, coupled with accurate and concise communication then, becomes paramount. Those behaviors are just two of the fundamental components necessary for exercising fire ground discipline and the very basis for the concept of tactical ventilation.

2. Deliberate
The very basis for the act of ventilation is to improve tenability and facilitate firefighting operations. Being a supportive function, ventilation must be executed purposefully, with the resultant effect serving as an expedient towards achieving a specific objective. Historically, ventilation has been employed as a means to address the two foremost incident priorities of life safety and fire extinguishment; precipitating the common expression 'vent for life and vent for fire.' To further assert the tactical nature of its application, however, that old adage has been more aptly rephrased to as 'vent for search' and 'vent for extinguishment.' This seemingly trivial, semantical alteration serves as a blatant, yet necessary reminder that ventilation is task-oriented; requiring sound judgement to ensure that any actions taken are calculated and aid in the success of a particular operation. Given the continual changes (detriments) to the fire environment, the need for such prudence has never been more crucial.

The effect of ventilation has been commonly misinterpreted, prompting its propensity for malpractice. This deficiency has tragically resulted in numerous close calls and line of duty deaths. Operational competency is more than merely possessing the physical ability or technical proficiency necessary to execute a given tactic (the 'how'), it is equally dependent on cognitive aptitude (the 'where, when and why'). As Tom Brennan (2007) once wrote, 'It is not simply a cut-and-bust operation. We are professional and select primary and secondary channels to improve conditions and positively impact the fire behavior based on an ongoing size-up.' To be truly effective, we must marry knowledge, obtained through research and education (the 'science'), with skill, developed through training and experience (the 'art'). Thankfully, due to the extensive work of both UL and NIST, we now have substantial empirical data detailing the dynamics of fire and the impact of firefighting operations, most notably, ventilation. With this information, we can develop the comprehensive understanding necessary to close those gaps; creating firefighters who are not only aggressive, but '*smart* aggressive' (Brown, 2016).

Even with the best intentions, ventilation can have negative implications *if* certain variables are not accounted for and the subsequent precautions are not taken. The fact is often lost that ventilation, by its very design, *increases* air flow. If ventilation is not appropriately managed, *controlling the intake of air to the fire area until extinguishment efforts are (effectively) underway*, the fire will intensify. Creating an uncontrolled, low-pressure opening will result in greater air exchange and draw the fire to the new source of air supply (the path of least resistance); deteriorating conditions, rather than improving them, in the process. As William Clark (1991) stated, 'Constantly consider that ventilation will cause the fire to grow rapidly and will increase the temperature in the fire compartment. Even ventilation above the fire will accelerate the burning rate by allowing fresh air to replace the oxygen deficient atmosphere that was slowing combustion.' To prevent such an occurrence, the decision to ventilate must be predicated on the following: the conditions present (building construction and layout, fire location and extent, wind speed and direction); life hazard; progress of the interior crews; and the desired effect. Only when these factors are properly evaluated, can the appropriate actions be taken.

- **Venting for search** has traditionally been a rather loosely applied tactic. There were rarely any standardized protocols established for its use, *with the exception of Vent-Enter-Search (VES)*. The only real guideline set forth was simply that if the (interior) search effort was inhibited, due to high heat and/or low visibility, windows could be taken to improve conditions; with the intent of enhancing orientation and tenability. Little to no insight or consideration, however, was commonly given as to its potential impact on the fire and the necessary precautions that must be taken to ensure its effectiveness. Without this critical understanding, the 'vent-as-you-go' concept was often misguided and subsequently, improperly executed. As Vincent Dunn (2007) warned '...if venting is carried out before the hose team is ready, the fire may spread or firefighters inside a superheated room conducting a search may be burned by a flashover.' Venting for search, *when conducted appropriately* (communicating intentions, coordinating efforts and controlling the door), is highly effective and can reap the ultimate reward – saving a life. Executing this tactic in an area that potentially cannot be isolated, *prior to the fire being in check*, however, comes with tremendous risk and should be reserved for situations where actionable intelligence of a life hazard is received – potential benefits outweighing the consequences. Members engaging in these operations are fighting the clock and must enlist the highest degree of situational awareness and operational proficiency to limit their exposure (and that of the victim) to the increasingly unstable environment.
- **Venting for extinguishment** requires an equal degree of diligence, as it carries the same inherent risks if not applied correctly. The use of this tactic is intended to provide an outlet for the (by) products of combustion and steam generated by the stream of the hose line; relieving conditions and aiding in the advance of the engine company. Another old saying that is dangerously misleading, taken merely at face-value (and without the essential background knowledge), is 'vent early, vent often.' Successfully venting for extinguishment is contingent upon the location, timing and size of the opening created and the ability to effectively control the *horizontal* openings (doors and windows) within the area of involvement *until a charged hose line is at least in position*. Clark (1991) asserted that 'Ventilation should begin, ordinarily, as soon as water is up to the nozzle, not before. The procedure will allow more air to get to the fire, which will accelerate the burning, and a charged hose line must therefore be ready to be moved into action as soon as ventilation begins.' When *effective* extinguishment efforts are initiated, more energy will be absorbed by the water than generated by the fire and the subsequent cooling of the fuel packages will inhibit the combustion process. Upon which, the inlet of fresh air and exhaust of the residual heat and smoke will substantially increase visibility and decrease temperatures – enhancing tenability.

As we conduct ourselves on the fire-ground, we must not forget that every decision we make will, in some way, directly impact the course of the incident *and the livelihood of all those involved*. To paraphrase Newton's Third Law of Motion, 'for every action there is a reaction.' Understanding the impact of our tactics is essential to the decision-making process and ensuring the appropriate interventions are selected and implemented to address the situation at hand. As a stark reminder, the FDNY has a sign posted at their training academy (known as 'The Rock') that reads, '*Let no man's ghost come back to say,* '*My training let me down.*''

3. Coordinated
Coordination is arguably the most influential and critical of the four tenets. There are three elements of ventilation that determine its impact on the fire-ground: timing; location; and amount. While there is a set of fundamental parameters for which they follow, each of these variables is inherently unique to every incident. *Communication with the interior crews is essential to properly addressing those factors and ensuring that any action taken will facilitate a specific objective and provides a distinct tactical advantage* (Grimwood, 2008). As Lt. Rob Brown, FDNY (2016) states, 'We must make decisions based on the operational needs, not the individual needs. Doing so, he explains, requires the utmost 'positional discipline.' The decision to ventilate, therefore, must be based upon the progress of the engine company, the ventilation profile and how additional ventilation will affect conditions and operations.

The most prominent misconception, regarding ventilation, is its thermal impact. The belief, within the fire service, has been that ventilation reduces temperatures within the area of involvement. That assumption was derived from the distinctive relief felt when vertical ventilation takes place. By creating a dedicated, low-pressure outlet, vertical ventilation alters the thermal balance and causes the neutral plane to rise. The sudden lift and redirection of heat is responsible for the perceived drop in temperature. If an opening exists below the neutral plane (open door or window), it will then become an inlet. The uni-directional flow (upwards of 20mph) will proportionately draw fresh air into the space as the contaminated air is discharged. While the evacuation of the existing smoke and fire gases provides an initial benefit, the additional supply of oxygen will begin to accelerate the combustion process; increasing the heat release rates and causing temperatures to rise.

As the fire continues to develop, it's (by) products can overwhelm the vertical opening, no matter how large it may be. The capacity of today's hydrocarbon-based, synthetic combustibles (BTU and HRR) is two to three times that of purely cellulosic materials. Because fires have become so overly fuel-rich, ventilation *alone* cannot transition them to a fuel-limited state (Kerber, 2011). This fact exposes another misconception, regarding its ability to prevent the spread of fire by localizing its growth. Vertical ventilation, *on its own*, does not adequately possess the sustainability to contend with the fire's energy output and should not be employed as a *long-term* means of containment. If water is not promptly applied to the area of involvement, cooling the fuel packages and rapidly absorbing the heat being generated, the benefits of vertical ventilation will be short-lived.

Ventilation has a finite window of effectiveness, as the fire will not immediately react to the change. The time frame is dictated by the fire conditions, the ventilation profile and the opening(s) created. The higher the interior temperatures and the ventilation point, the greater the air flow and the closer it is to the fire, the faster it will respond. As the fire environment continues to evolve, fires are reacting to ventilation at an accelerated pace; diminishing that 'grace' period (>75% reduction from the era of cellulosic contents). Kerber (2016) explained, "Our timeline today is very different to what it used to be…

[Fire] responds to your tactics faster and you have to understand it...[ventilation] is *all about timing*. You have to time it to make it work.' Once it expires (approximately two to three minutes), conditions within the ventilation path will abruptly deteriorate. The area will eventually become untenable, potentially with little to no warning, and a (ventilation-induced) flashover can be imminent – as little as ten seconds (Kerber, 2011).

In the UL report, *The Impact of Ventilation on Fire Behavior in Legacy and Contemporary Residential Construction*, Steve Kerber (2011) simply explains that if air is added to the fire and water is not applied in the appropriate time frame, the fire gets larger and the hazards to firefighters (and victims) increase. As the old saying goes, 'If you put the fire out, everything else gets better' (Norman, 2012). As a basic, guiding principle, ventilation should not occur until a charged hose line is *at least* in position or the intake of air to the fire area can be effectively managed until that time, *unless there is actionable intelligence of life hazard* (i.e. VES). The success of ventilation, consequently, is contingent upon the ability to synchronize its execution with confinement and extinguishment efforts.

4. Controlled
The fire-ground is a highly dynamic environment that is layered with varying degrees of complexity and uncertainty. Each incident we face will present its own set of variables, which must be specifically accounted for, and addressed accordingly – especially in regards to ventilation. As the late Lt. Andy Fredericks, FDNY (1996) once wrote, '... ventilation parameters are highly variable and may change many times during the course of a fire.' DC Emanuel Fried (1972) similarly stated that 'Every building has its own ventilation problems. What may be the correct ventilation technique in one building may not be proper for another.' Therefore, the size-up process must include a (continual) assessment of the fire's *ventilation profile*. Paul Grimwood (2008) defines this term as, 'the amount of air available within a compartment' – determined by the location and amount of existing ventilation points (also see **3.14**). Once it has been identified, the appropriate interventions can then be determined and properly executed. Doing so will also aid in the critical task of assessing the fire's rate of growth and its direction of travel; as fire burns proportionately to the quantity of air which it receives (Braidwood, 1866).

By its very name, ventilation has always been predominantly focused on the exhaust component, when in reality equal attention must be placed on the intake. Because today's fires are so overly fuel-rich, their development and spread are almost exclusively dictated by the availability of oxygen within the compartment. Therefore, by *temporarily* restricting its supply of air (without endangering potential occupants or inhibiting vital operations), *until the onset of fire attack*, the fire can be kept in a ventilation-controlled state – promoting decay. As Steve Kerber (2011) states, you want to 'limit the air *until* you gain the upper hand.' Any open, horizontal opening must be viewed as a source of ventilation and a means for feeding air to the starving fire. Fried (1972) asserted 'Never ventilate a fire that may intensify because of the ventilation *until charged lines are in position*;' except in matters of life safety (i.e. VES). Whenever we open a door to make (forcible) entry, we are also creating a ventilation point. We must understand that doors are actually the most effective means of horizontal ventilation, as they span nearly the entire height of the wall; allowing fresh air in and hot gases out. Simply leaving a door open (without a charged line present) can be the catalyst for rapidly reenergizing and/or spreading a vent-limited fire; as was the case in the 2008 close-call in Prince George's County, Maryland, among others.

The recent fire dynamics research has conclusively shown that the fuel load of today's fires, exacerbated by energy efficient materials and construction, is producing tremendous heat releases rates and resulting in the following detriments: faster fire propagation; shorter time to flashover; greater propensity to become ventilation-limited; and increased volatility (Kerber, 2011). Because the progress of the fire's development has been accelerated, the impact of ventilation is now even more profound. Despite popular belief, this concept is not at all new. In fact, Andy Fredericks (1996) discussed this topic almost twenty years ago in his Fire Engineering article *Thorton's Rule*, stating: '… when ventilation is increased, another factor must be considered—the rate at which different materials liberate heat … when provided with increased ventilation, hydrocarbons generally will liberate heat at a rate faster than cellulosic materials. As a consequence, we can expect that rollover and flashover conditions will be achieved in much less time.' The common denominator among the aforementioned conditions is they collectively lead to a diminished operational time frame (unless we intercede); placing greater emphasis on controlling the movement of air – managing the ventilation ('flow') path.

While 'flow path' management has seemingly taken the fire service by storm over the last five years, it is *not* a recent discovery. The terminology may be different, but the concept and its associated tactics have been around for 150 years. James Braidwood (1866) was a pioneer of his time, responsible for many significant advancements in firefighting. He published, arguably, the first document on the subject (which he referred to as 'draught') and the need for isolation procedures (i.e. door control), which read as follows:

> 'In the first discovery of fire, it is of the utmost consequence to shut doors, windows, or other openings. It may often be observed, after a house has been on fire, that one floor is comparatively untouched, while those above and below are nearly burned out, this arises from a door on that particular floor having been shut, and the 'draught' directed elsewhere… The men of the fire brigade were taught to prevent, as much as possible the access of air to the burning materials. What the open door of the ash-pit to the furnace of a steam boiler, the open street door is to the house fire. In both cases, the door gives vital air to the flames. The door should be kept shut while the water is being brought, and the air excluded as much as possible.'

Managing the ventilation path can be as simple as closing an open door or window *until the arrival of a charged hose line*. The fire service has been preaching isolation procedures (door control) for decades, when engaging in vent-enter-search (VES); so much so that some have added the letter 'I' (isolation) to the acronym (VEIS). Yet, the same cannot be said for other fire-ground operations, especially traditional (interior) search. By simply closing the door upon entry, the room becomes isolated from the area of involvement. Doing so prevents further contamination, offers protection from fire spread and allows windows to be vented, *if the situation warrants it*; enhancing tenability and orientation, as well as providing an alternative means of egress/victim removal. Fredericks (2000) described how a veteran, ladder company lieutenant from the Bronx advocated for door control during the primary search at apartment fires, stating 'The calming effect on a growing fire that results from the simple act of closing the door behind you can be quite astonishing.'

There has been an apparent reluctance to fully accepting door control and other isolation techniques. Much of this can be attributed to its association with the perceived movement against ventilation. This misconception has likely derived from the use of such terms as 'anti-ventilation.' When taken at face-value, it becomes easy to understand why; as Aaron Fields always says, 'words matter.' Anti-ventilation, developed and implemented in the UK, utilizes 'zoning' tactics in an effort to confine the fire and isolate it from the rest of the structure. It does not denounce ventilation; it simply calls for a delay in its use ('unless the incident commander has identified a viable objective or reason to create openings') until the fire has been controlled. While its application as a primary strategy has been effective for our brothers across the pond, its success, however, is largely in part to the building construction which they face (differing from that which is found in the US); as it 'relies on the stability of the compartment to keep the fire in-check' (Grimwood, 2008).

Despite the differences, Paul Grimwood of the London Fire Brigade saw that there were many components that would prove beneficial as a *supplement* to the American style of ventilation – giving birth to the concept of 'tactical ventilation' in 1987. Grimwood came to this revelation while on a study detachment with the FDNY's 7th Division (South Bronx) in the mid–70's (the 'War Years') and volunteering in West Islip, Long Island. He went on to say, '*I acknowledged that the general concept of opening up buildings under specific circumstances would reap great rewards, providing the strategy was applied with a clear purpose, or intent, and the actions of firefighters were organized, disciplined, and controlled.*' Experiencing the two approaches first-hand in their respective environments, he recognized that they both had merit and possessed their own relative pros and cons. By fusing their basic principles and adapting them accordingly, he arrived at a middle-ground, tactical ventilation: **the venting or containment actions used to take control in an effort to gain tactical advantage during firefighting operations (Grimwood, 2008).**

Ventilation is, fundamentally, a matter of control; both of ourselves and the air flow within the compartment involved in fire. The actions we take, or lack thereof, will dictate the survivability of victims, the fire's progression and the well-being of firefighters. DAC John Norman (2008) poignantly explains this dynamic metaphorically: Whoever controls the supply of oxygen wins the battle. For if it is allowed to fall into the hands of the enemy *at the wrong place*, it will use it to grow stronger, possibly with explosive speed. The commanders who master the art of warfare known as ventilation have a great force on their side (paraphrased for brevity). Ventilation, when executed tactically, is not the problem; quite the contrary, as it is *part* of the solution. The real issue at hand is a lack of fire-ground discipline. By applying the Four Tenets of Tactical Ventilation (*Informed; Deliberate; Coordinated; and Controlled*), those engaging in ventilation operations will possess the insight and means to be successful and ensure any actions taken are in support of the mission – preserving life and property.

The Six Commandments of Ventilation:

I. *Thou shall communicate with the engine (and ladder) company officer prior to venting.*

II. *Thou shall temporarily control the door(s) to the fire area (prior to venting) until a charged hose line is in position.*

III. *Thou shall not horizontally vent for extinguishment until fire attack is underway – venting in opposition of their advance.*

IV. *Thou shall not horizontally vent for search until fire attack is underway and/or the affected area can be isolated until that time, unless there is actionable intelligence of a life hazard (i.e. VES).*

V. *Thou shall not vertically vent until a charged hose line is in position or the affected area can be isolated until its arrival.*

VI. *Thou shall avoid venting/operating in a manner that may compromise egress and shall attempt to establish/locate a secondary means.*

Above courtesy of Lt. Nick Papa, NBFD

It's a worthwhile exercise to take the framework Nick Papa has created by applying it to any scenario you can imagine, or have experienced in the past, and identifying the most likely outcomes. This template can also be utilized as a debriefing tool following a building fire. Upon which to pose the following questions:

- Did the fire self-vent at any point?
- Did tactical ventilation take place, either horizontally or vertically?
- If ventilation did not take place, why not?
- Were the outcomes beneficial or detrimental to (potential) occupants and/or firefighters?

Upon obtaining those answers, analyze the actions taken, identifying that which was positive and negative, to ensure the best practices are employed in the future.

Nick Papa's References

Brennan, Tom. (2007). Random Thoughts. PenWell Corporation: Tulsa, OK.

Brown, Robert. (2016). Smart Venting Using the SA Cycle. WNYF: 2nd/2014.

Brown, Robert, Smart 4 Life Lecture, April 16, 2016.

Brown, Robert. (2016). Auditory Exclusion and How the Sympathetic Response Affects Our Operational Performance on the Fire-ground. WNYF: 2nd/2016.

Clark, William. (1991). Firefighting Principles and Practices, 2nd Ed. PennWell: Saddle Brook, NJ.

Dunn, Vincent. (2007). Strategy of Firefighting. PennWell Corporation: Tulsa, OK.

Fredericks, Andrew. (1996, November). Thorton's Rule. Fire Engineering.

Fredericks, Andrew. (2000, March). Little Droplets of Water: 50 Years Later, Part 2. Fire Engineering.

Fried, Emanuel. (1972). Fireground Tactics, 1st Ed. Chicago, IL: H. Marvin Ginn Corp.

Grimwood, Paul. (2008). Euro Firefighter. United Kingdom: Jeremy Mills Publishing.

Kerber, Steve. (2011). The Impact of Ventilation on Fire Behavior in Legacy and Contemporary Residential Construction. Retrieved from http://www.ul.com/global/documents/offerings/industries/buildingmaterials/fireservice/ventilation/DHS%202008%20Grant%20Report%20Final.pdf

Kerber, Steve. (2011). The Impact of Ventilation of Fire Behavior in Legacy and Contemporary Residential Construction. Retrieved from http://www.ul.com/global/documents/offerings/industries/buildingmaterials/fireservice/ventilation/DHS%202008%20Grant%20Report%20Final.pdf; Norman, John. (2008). Fire Officer's Handbook of Tactics, 4th Ed. Tulsa, OK: PennWell.

18.4 VENT – ENTER – ISOLATE – SEARCH (VE(I)S)

'Vent – Enter – Search' has long been an effective search and rescue strategy whereby access is gained, often by ladder to upper floors, to an exterior window that may be emitting smoke. The assumption, based on knowledge and experience of building layouts, is that the room being entered is a bedroom and may well be occupied (especially at night or by a reliable report).

V – Vent a window of the room where occupants are most likely to be based on good intelligence
E – Enter the window to search (one firefighter ideally with the second at the head of the ladder
I – Isolate any fire beyond the room by locating and closing the door and restricting the flow-path
S – Search the room as quickly and efficiently as possible

Note: where fire exists in the room to be searched, bring a hose-line and deal with it first, closing the door when possible. At all times, follow your departmental guidelines and SOPs.

This is generally seen as a high-risk/high gain strategy but as long as strict controls measures are adhered to and firefighters are trained and operate according to documented department procedure, the exposure of risks to firefighters is lessened.

- Select a window that is likely to serve as a bedroom and may contain occupants or one where occupants are believed to be (are reliably reported) within.
- The strategy may precede the placement of a hose-line if the room to be searched is not the fire compartment.
- Leave from the window you entered by and <u>do not</u> enter the hallway to continue searching beyond the entry room, unless your department policy allows this tactic. Search one room at a time from a different exterior entry point.
- Advancing the search into the hallway, beyond the primary search room, where a remote fire has not been isolated or a hose-line has not controlled or fully extinguished the fire greatly increases the risk to firefighters.
- To tactically vent the room's window, if it hasn't already vented or is open, should only be done where;
    1. A known occupant or occupants are reliably reported in that room and;
    2. The window demonstrates a low heat to touch or;
    3. Where the hose attack team have the fire under control.
    4. <u>If the room cannot be accessed as quickly and safely using an interior approach.</u>
    5. A safe interior approach demands that the fire should be isolated, extinguished or effectively controlled before searching ahead of the line.
- Firefighters should not crowd the window exit where one firefighter has already entered.
- The first firefighter (entering the room) should always attempt to locate the room door and close it before searching.
- The second firefighter should only enter if called to do so by the first firefighter. The process functions more effectively if children are handed to the ladder firefighter by the interior firefighter.
- If adults are involved, or the room is exceptionally large, the second firefighter may need to enter to assist and a third firefighter should ascend the ladder to further assist. Again, the room door should be located and closed as a priority. An escape line (rope) might be used from the top of the ladder to locate the window.
- If the room is actually the fire compartment, a hose-line should be taken into the room.

VEIS is a strategy not to be feared. It is one that can be undertaken by limited staffed crews where on arrival, a three-person crew (for example), can react to direct intelligence of missing or trapped occupants, reported to be most likely within a particular room. In entering the room the first action should be to try and isolate the fire by locating and closing the room's door. This can usually be found by tracing a wall in heavy smoke. Then a search can be made of the room in question. As long as the creation of a flow-path is avoided by maintaining the street door closed, or partially closed, the search of one room at a time that has been isolated should not worsen fire development to any great extent. However, such operations are time limited and firefighters should work fast but efficiently. If the room has effectively been isolated by closing the room door, after searching this room, the same operation can be undertaken by approaching from the exterior to search adjacent rooms.

To operate **VES** means potentially *searching beyond the primary target room* by entering the hallway and then entering other rooms. This involves a much higher exposure to risk for firefighters and should not be undertaken unless:

1. The fire has been isolated from within by other crews, or
2. The fire has been controlled by a hose-line from inside the building or has been extinguished
3. A hose-line has been placed between the fire and firefighters searching, to protect their position and means of internal egress
4. Any temporary knockdown of fire from an exterior position does not support VES unless this is maintained and is confirmed as 'fire controlled'

18.5 POSITIVE PRESSURE VENTILATION (ATTACK) – PPA

Also refer to UL research detailed in Chapter 1.

In 1987 the UK first introduced Positive Pressure Ventilation (PPV) following pioneering work by the author, together with CFO John Craig (Wiltshire Fire Brigade), following their research in the USA. Together they produced the UK's first standard operating procedure in the use of PPV. This initial experience was later used as a basis for further development by the Fire Service College and Tyne and Wear FRS. The use of PPV Ventilators to control flow-paths, remove smoke and flammable fire gases and reduce compartment temperatures preceding firefighter deployment into ventilation controlled house or flat fires, is generally considered a useful and effective tactical ventilation strategy. However, it is well established that manufacturer's fan air-flow data rarely meets the actual air flows achieved through *small domestic structures*. Even so, it is important to consider the distinct benefits in using positive pressure attack (PPA) methods in a controlled and coordinated manner. In general, where PPA is used under a protocol based approach, firefighting becomes generally safer and occupants may be located far sooner.

Consideration should always be given to the protocols that will dictate any particular service approach when writing standard operating guidelines and these should include:

- The staffing requirements needed on-scene to implement PPA
- The occupancy types applicable
- Any existing ventilation profile and fire conditions that may preclude the use of PPA
- The use of PPA will usually occur where the fire is in a ventilation controlled state
- The maximum compartment dimensions in terms of any particular PPV units in use
- The ability to vent stair shafts
- The potential conflicts with pre-engineered venting systems installed
- The level of training required to become a competent PPA operator
- Conflicts in verbal communication
- Hearing protection
- The type and level of additional control measures required

- The time-lag behind the air-flow before firefighters are deployed (30-60 seconds) (allowing the fire conditions to stabilize)

A tactical objective would be to quickly force an under-ventilated fire into a well ventilated state, removing dangerous smoke and fire gases from the building and reducing compartment temperatures in order for firefighters to undertake fire suppression, search and rescue operations in a safer and more controlled environment. When a forced draught pressure differential state is created within a fire-involved structure it is possible that the rate of burning will increase in the fire compartment. This may have the effect of increasing the rate of heat release beyond that normally expected, although this increase in HRR may only be in the region of 15%. Despite an increase in burning rate that might normally lead to a temporary increase in fire room and hallway temperatures, the effect of the PPV air-flow is to generally cool the environment at lower levels and reduce temperatures overall.

Considerations for risk control measures most particularly relate to the potential for the forced draught PPV air-flow to 'push' flaming combustion or hot fire gases/smoke into uninvolved areas of the structure (via interior or exterior routes); and/or causing an ignition of the fire gas layers, leading to some rapid fire development. These are clear and relevant concerns that should be addressed by effective training and generic risk assessment, ensuring adequate resource deployments and control measures are implemented prior to using PPV against developing ventilation controlled fires.

Air-flow – practical objectives

Practical air-flow tests were undertaken by the author using a typical sized three bedroom house, to assess the performance of a typical PPV ventilator in its ability to achieve effective air movements and meet pre-defined critical limits. The actual air-flow potential at the 1m² exhaust outlet created at the bedroom window fell far short of the manufacturer data at just 2.45m³/s (8820m³/hour). This is because air inlet/outlet ratios in typical residential buildings are rarely optimized. The manufacturer's data of the tested fan stated air flow performance of 43400m³/hr (12m³/s) is achievable although third party AMCA test data suggested 29755m³/hr (8.26m³/s). Air-flow performance of a PPV ventilator is affected by a number of variables which include the capacity of the fan; the distance of the unit from the air inlet; the size of the air inlet and outlet points; size (m³) of the air flow paths from inlet to outlet as well as natural leakage paths from the building etc.

The volume flow through an exhaust air opening can be calculated using the following equation[185]:

$Q_f = C_d \, u_F \, A_F$ **Eq. 18.1**

Where

Q_f is the volumetric air flow-rate (m³/s)

C_d Is the discharge coefficient (0.7)

u_F Is the measured or theoretical air velocity through the opening (m/s)

A_F is the geometrical area of the opening (m²)

---

185 Ingason & Fallburg; Positive Pressure Ventilation in Single Medium Sized Premises; Fire Technology, 38, 213-230, 2002

The air-flow capacity of the fan in use demonstrated that for an average sized three bedroom two storey house of 98m² total floor space, with ideal air-flow paths and no excessive leakage paths, the actual through-flow of air volume for the fan is closer to 6624m³/hr. This is taking into account a normal 1.8m² street inlet door, fully open, and a partially open[186] 1.0m² window vent outlet, with the ventilator ideally located between 2-4 metres from the entry door[187].

It was estimated that the air-flow path (inlet to outlet) in existence through the building, with all room doors closed except to the stairs and into the target room, was approximately of 110m³ volume. This volumetric space represented 47% of the total volume of the building below the highest ceiling.

Using Eq.18.1 we can calculate the volumetric airflows entering and leaving the target room as follows:

$Q_f = 0.7 \times 1.9 \times 1.4 = 1.862$ m³/s airflow into the target room

$Q_f = 0.7 \times 3.5 \times 0.75 = 1.837$ m³/s airflow leaving the target room

Air-flow into the target room	1.4m² room door	1.9m/s	1.9m³/s
Air-flow out of the window	75% of 1m² window	3.5m/s	1.8m³/s

Table 18.1: Volumetric air-flow-rates (using measured air velocities and Eq.18.1) achieved through the bedroom window in the house tests. It should be noted that natural wind speeds were recorded entering the structure and leaving the window between 0.5 to 1.3m/s prior to ventilator tests and although these minor wind gusts may have influenced the above readings, they were taken at steady flow over a period of seconds where wind gusts were not apparent in the readings. They are therefore not reflected in the above data.

This demonstrates the PPV ventilator in use offers a full flow potential of around 2.45m³/s (8832m³/hr) where the 1m² window is fully vented. It is clear to see that it is difficult to replicate manufacturer's test house data when inlet and outlet dimensions are disproportionate and non-optimised, as they would actually exist under realistic scenarios when ventilating domestic houses or apartments. Where inlet and/or outlet points are larger, the air-flow capacity is increased closer to the manufacturer's published data. For example, if a larger than average doorway and 2.0m² of window/s are fully vented from the fire compartment, this will likely increase the amount of air entering and leaving the building. UK National guidance GRA 3.6 recommends that when applying PPV the size of the outlet opening should be slightly less than the size of the inlet as this facilitates the build-up of positive pressure. The outlet size may be increased if more than one fan is in use. The FSC Moreton-in-Marsh recommends that the outlet is one half to two thirds the size of the inlet.

A theoretical view[188] would suggest that the outlet vent should be at least as large as the inlet area but preferably twice the size in order to optimise air-flow efficiency. In fact, the efficiency of PPV air-flow is rated at 90% with a 2-1 'outlet to inlet' ratio and 45% at a 2-1

---

186 The cantilevered window outlet allowed only around 75% of the volume air-flow to leave the building so this is calculated into the air-flow estimate provided above
187 Where larger floor areas (for example small commercial etc) and increased opening sizes are relevant, the distance of the ventilator might be increased effectively up to 6 metres from the air inlet, depending on fan design
188 Svensson. S; Fire Ventilation p69-71; Swedish Rescue Services Agency 2005

'inlet to outlet' ratio. Whilst it is true that volumetric air-flow is generally increased in this way, the reality of residential domestic firefighting will normally see a 2-1 ratio of 'inlet to outlet' with the inlet doorway twice as large as the outlet window.

Wind effects

Where a 5mph (2.23m/s) wind is entering a $1m^2$ - window outlet this wind pressure should be countered effectively by the air-flow characteristics of a commonly used PPV fan (as table 18.1 above) and smoke will continue to be ejected, but if the outlet vent is increased in size to $2m^2$ the single fan will then be unable to cope with the additional exterior air inflow and the wind will push fire and combustion products back into the building.

We can see that a 5mph (2.23m/s) wind entering a $1m^2$ window will create a volumetric airflow of:

$Q_f = 0.7 \times 2.23 \times 1.0 = 1.561$ m³/s airflow into the target room.

If we then match this against the 2.45 m³/s airflow leaving the same window we can see around 0.9m³/s over-pressure exiting the window will just about overcome the wind. Increasing the window size vent area to $2m^2$ in this scenario however will cause the wind to overpower any smoke exiting the window.

$Q_f = 0.7 \times 2.23 \times 2.0 = 3.122$ m³/s airflow into the target room

An academic research paper[189] (using Smart-Fire CFD) into overcoming wind velocity using a PPV ventilator concluded that for the 'typical ventilator' with a manufacturer's flow rating of 6.64 m³/s, a 'critical wind speed' of 3.3 m/s existed that represented the maximum wind velocity in the flow-path that could be reversed by this particular ventilator.

However, to this point we have not accounted for additional room fire pressure in the equation.

Air flow versus fire pressure

When one area of a structure has a different pressure than an adjacent area a flow occurs between these two areas. The greater the differential pressure, the greater the velocity (m/s) of flow. Pressure (Pa) always flows from high to low. There are several approaches that may be taken to assess the effectiveness of PPV air-flow capacity in terms of the potential to improve interior conditions where a room fire exists in an under-ventilated or steady state burning regime. The following research has specifically addressed a wide range of scenarios and likely fire conditions in order to offer guidance on the critical volumetric air-flow requirements and pressure differentials needed to successfully force vent such situations. The influence of PPV on the ability to 'hold back' conditions inside a fire room may be examined by comparing pressure differentials created between the fire room and adjacent areas.

It is established that a post-flashover room fire may generate as much as >25Pa fire pressure so the countering fan pressure should ideally exceed this. In VTT report 419 from Finland[190] we are provided with a recommendation that critical air-flow rates to achieve this are somewhere between 6.66m³/s and 8.33m³/s (24000m³/hr to 29988 m³/hr)

---

[189] Arun Mahalingam, Mayur K. Patel, and Edwin R. Galea; *Simulation Of The Flow Induced By Positive Pressure Ventilation Fan Under Wind Driven Conditions;* Fire Safety Engineering Group, University of Greenwich, London 2010

[190] Research Report 419; Technical Research Centre VTT Finland

where venting intense room fires using PPV. In fact, the actual test identified this air-flow-rate potentially as a minimum critical limit against a very intense room fire, where it took over a minute to completely reverse smoke flows issuing into the hallway at the top of the fire compartment door. In this room fire test the introduction of PPV caused the burning rate to increase within the room, with corresponding temperature rises at certain points. However, overall the cooling from the PPV airflow soon caused dramatic temperature reductions throughout. The energy release also increased from 12MW to 14MW as PPV was brought into use for around 200s. This might suggest firefighters should delay their approach until approximately 30-60 seconds after PPV is directed into the building.

Work[191] by the National Institute for Standards and Technology (NIST) in the USA provided some useful data involving typical room fire pressures and inlet/outlet velocities and reported as follows: 'This study examined gas temperatures, gas velocities and total heat release rate in a series of fires in a furnished room. The use of the PPV fan created slightly lower gas temperatures in the fire room and significantly lower gas temperatures in the adjacent corridor. The gas velocities at the window plane were much higher in the PPV case than in the naturally ventilated scenario. This higher velocity improved visibility significantly. PPV caused an increase in heat release rate for 200 seconds following initiation of ventilation but the heat release rate then declined at a faster rate than that of the naturally ventilated experiment'.

The test room used by NIST measured 16m^2 with a 2.44m high ceiling. The room was approached using a 2.29 long corridor that was 1.22m wide. The doorway openings to the corridor and the room itself both measured 2.366m^2 and the single room window opening measured 1 sq. metre at a sill height of 0.8m above the floor. The natural air leakage loss would be far less in this purpose-built test house than in a real building, similar to the structure used by the author (above). The PPV ventilator used had a manufacturer's volumetric flow rating of 6.64 m^3/s (14,060 ft^3/min). The fan was positioned 2.44 m from the open doorway to the corridor at an angle of approximately 15 degrees from horizontal to create a 'cone of air' around the doorway. The room's fuel load for the live fire tests totalled 250kg (15.6 kg/m^2). The fuel load was selected in order to represent a typical bedroom configuration. It was also intended to create a fuel rich atmosphere to make burning dependent on the available oxygen (ventilation controlled fire).

In the experiments, black smoke flow was observed in the corridor prior to 300s and flames were not observed in the corridor doorway until the window was ventilated. Within 10s of opening the window, flames extended out of the corridor doorway. The PPV fan forced all burning out of the corridor and back into the room by 516s. Once the fan was activated, it took 130s to completely reverse the flow back into the room. At that point, little or no smoke was seen coming out of the room doorway. Flames were observed in the corridor of the naturally ventilated experiment until 1200s. The flames in the PPV ventilated experiment extended at least 1.83 m from the window.

The maximum heat release rate was 14 MW for the PPV ventilated fire and close to 12 MW for the naturally ventilated fire. The peak heat release rates were reached approximately 40s after window ventilation with a spike to their respective maximum. The peak of the PPV experiment occurred 5s after that of the natural experiment. This corresponded to the 5s period before the PPV fan was started. Comparing the heat release rate between the time of peak and the time where the two curves intersect showed that the PPV created a higher burning rate by approximately 60 % for about 200 s after the fire reached its maximum output. After the heat release rate spike, the PPV output remained

191 Kerber. S and Walton. W; Effect of Positive Pressure Ventilation on a Room Fire; NISTIR 7213; 2005

4 MW above that of the naturally ventilated experiment for 70s. At the end of those 70s, the rates converged until 590s when the naturally ventilated fire had the higher heat release rate. The naturally ventilated fire remained roughly 1 MW above the output of the PPV ventilated fire until the end of the experiment (figures 28, 29). The integral of the heat release rate curve in figure 30 provided the total heat released over the duration of both experiments. The fan caused heat to be released quicker in the PPV experiment, but ultimately both experiments released approximately the same amount of heat.

		PPV air flow recorded before fire tests	Room Fire Data (No PPV) outlet window open	Room Fire Data (with PPV) outlet window open
Fire Room Pressure	Top of door	-	28 Pa	62 Pa
Fire Room Pressure	Middle of Door	-	7 Pa	41 Pa
Fire Room Pressure	Bottom of door	-	Minus 14 Pa	21 Pa
Door gas velocity	At inlet	-	5 – 3 m/s**	5 m/s
Window gas velocity	At outlet	5 m/s	12 – 0 m/s**	20 – 5 m/s

*Table 18.2: With a 0.42 outlet to inlet ratio (30% PPV air-flow efficiency) the above flow and pressure data was recorded by NIST ** gas-flows were actually in two directions, ranging between air entering and hot smoke leaving from top to bottom of openings.*

During the NIST room fire tests the PPV fan alone generated gas velocities of 5 m/s in the window while the naturally ventilated fire generated velocities of nearly 12 m/s. In the experiment with the PPV fan, window gas velocities of nearly 20 m/s were generated, approximately equal to the additive velocities from the fan and the naturally ventilated fire. The fan quickly forced a unidirectional flow out of the window but took a period of time to completely reverse the flow out of the doorway and create a flow into the room. The fan was able to create a more tenable atmosphere as soon as it was turned on by reversing the natural flow out of the corridor, where the fire fighters would be approaching the fire for extinguishment.

Gas velocities into the room through the doorway were lower than those out through the window. Prior to ventilation, there was a 4 m/s to 6 m/s flow out of the top two-thirds of the doorway into the hallway and a flow into the room in the bottom one-third of the doorway of 2 m/s. After the fan was activated the bottom two-thirds of the doorway flowed into the room and the flow in the upper third of the doorway fluctuated between in and out of the room at a doorway flow of 3-4m/s or 4.65m³/s (16740m³/hr).

NIST Test Fire data compared to author's Test House (non-fire) air-flow data

Although the NIST test involved live fire, some comparisons may be made with air-flow data through a purpose-built test house and a real-world house used for recording PPV air-flows. In the author's three bedroom test house the PPV ventilator was able to achieve just 20 percent of the manufacturer's published performance data due to natural leakage paths within the structure and also because the outlet to inlet ratios were not optimised where the entry door was larger than the window outlet by a ratio of 0.7 (around 55% air-flow efficiency). In comparison, the NIST house had almost no natural leakage paths and an outlet to inlet ratio of 0.42 (less than 40 percent air-flow efficiency). The fan used in the NIST tests had a manufacturer's volumetric flow rating of 6.64 m³/s but only 3.5 m³/s was achieved (53 percent of manufacturer's published data) in pre-fire air-flow tests through the test house due to outlet to inlet ratio inefficiency (leakage path losses would have been greater in a real-world test setting). Even so, these were typical ratios encountered when using PPV/PPA in residential settings.

A similar approach is taken by reference to an equation used in pressurization design calculations[192]

$$V_k = K_v \left(\frac{E}{W}\right)^{0.333} \qquad \text{Eq.18.2}$$

Where

V_k = critical air velocity to prevent smoke backflow (m/s)

E = energy release rate into corridor (watts)

W = corridor width (m)

K_v = coefficient (0.092)

Intensity of Room Fire	Minimum Air Volume flow-rate required to Overcome Fire Pressure (Pa) at 1.4m2 internal door
1 MW	4.6m³/s
3 MW	6.4m³/s
5 MW	7.3m³/s
7 MW	8.1m³/s
10 MW	9.1m³/s
15 MW	10.5m³/s

Table 18.3: Based on work by Thomas (1970) in calculating the critical air velocity (m/s) needed to hold back combustion products in a corridor or room with an open door, then by converting to m³/s through an internal 1.4m² door, it is suggested that a 3MW post-flashover room fire (typical in a 12m² residential bedroom for example) requires a volume flow of at least 6.4m³/s to prevent combustion products entering the hallway outside the room.

Now if we refer to VTT Finland report 326[193] there are further recommendations suggesting critical air-flow requirements, when using PPV to ventilate ventilation controlled room fires. In this detailed research report the authors propose that the required air-flow should be based upon the volumetric space of the area to be vented, including the fire compartment as well as the entire air-flow path. A critical air-flow rate is recommended as 96m3/hr with optimum performance achieved at 144m³/hr.

What this is telling us is that the 110m³ flow-path, open to PPV air-flow between the inlet and outlet points in the author's test house, may approach critical limits where applications below 2.93m³/s (10560 m³/hr) are used or where applications in excess of 4.4m³/s (15840 m³/hr) are used. In other words, either rate of air-flow may either *under* or *over* pressurise the fire compartment. Such considerations are of course also dependant on the fire size, rate of fire growth and geometric factors associated with inlet and outlet vents. However, this approach offers a useful way to determine a benchmark for basic fan performance needs.

192 Klote. J; An Overview of Smoke Control Research; ASHRAE 2007
193 Research report 326; Technical Research Centre VTT Finland

The air-flow tests undertaken by the author demonstrated the following air flow velocities and volume flow-rates from the ventilator through the 98m² house, from entry point, through inner doors, and out of the bedroom window at first floor level.

Location	Fan at 1m	Fan at 2m	Fan at 3m	Fan at 4m
Main inlet door	4.0m/s (5.04m³/s)	5.4m/s (6.8m³/s)	4.76m/s (6.0m³/s)	4.16m/s (5.24m³/s)
Inner door ground floor stairs	1.4m/s (1.37m³/s)	1.9m/s (1.86³/s)	1.7m/s (1.66m³/s)	1.7m/s (1.66m³/s)
Inner door to 1st Floor bedroom	1.65m/s (1.6m³/s)	1.9m/s (1.86m³/s)	1.9m/s (1.86m³/s)	1.9m/s (1.86m³/s)
Window outlet vent 1m² used at approximately 75% efficiency	2.3m/s (1.2m³/s)	3.5m/s (1.83m³/s)	3.4m/s (1.78m³/s)	3.25m/s (1.7m³/s)

*Table 18.4: Averaged air-flow velocities and volume flow-rates at various doors in the test house used to record PPV data. It should be noted that a gusting wind of 0.5 to 0.9m/s was sometimes recorded at the entry door as a state of bench-mark prior to the tests and may affect the accuracy of data. The above data may include these possible wind gusts. However, the timed duration of each flow test was set to record a 'steady' air-flow and this was used in each case. Also, the window was cantilevered and open-able to around 75% of its full 1m² area. This would reduce volume flow-rate at all points. Taking this into account, >**2.45m3/s** (**>8820m³/hr**) is more representative of typical fan potential (test fan used) in this situation with a fully vented/removed window (fan at 2m from entry door).*

Manufacturer's brochures or AMCA test data are only representative of a particular PPV ventilator's air flow power under ideal conditions. In reality, it may be difficult to achieve anywhere near these air flow rates as inlet and exhaust vent ratios rarely achieve the ideal combination and natural leakage paths inside fire buildings may often reduce the actual amount of air flow reaching the fire room/s. The author undertook several non-fire flow tests of typical PPV fans in an average sized three bedroom two storey house of 98m² total floor space, with ideal air-flow paths and no excessive leakage paths, observing at best that they were commonly achieving only a third of the AMCA rated airflow and one fifth of the manufacturers rated airflow. The street inlet was a typical 1.8m² doorway with the ventilator ideally located between 2-4 metres from the entry door and the final outlet was a 1m² window, providing an outlet to inlet ratio of 1 to 1.8.

National UK guidance GRA 3.6 currently recommends that when applying PPV the size of the outlet opening should be slightly less than the size of the inlet as this facilitates the build-up of positive pressure. The outlet size may be increased if more than one fan is in use. The Fire Service College Moreton-in-Marsh recommends that the outlet is one half to two thirds the size of the inlet. However, Dr. Stefan Svensson of Lund University Sweden informs us[194] that to optimise the available air flow through the structure the ratio should be at least 1:1 which provides 75 percent efficiency, or the outlet should be twice the size of the inlet to achieve 90 percent efficiency. In the author's test set-up which demonstrates a typical configuration of PPV for a residential building, the efficiency of the air-flow through the outlet window was only just over 40 percent.

---

194 Svensson. S, Fire Ventilation; Swedish Rescue Services Agency; Radnings Verket 2000

$$OI_r = \frac{\frac{A_f}{A_t}}{\sqrt{1+\left(\frac{A_f}{A_t}\right)^2}} = \text{Effectiveness of outlet to inlet ratio (percentage)} \qquad \text{Eq.18.3}$$

Where:

OI_r = Air-flow factor (the effectiveness of air-flow according to the outlet to inlet ratio (percentage))

A_f = Area of vent outlet (m²)

A_t = Area of vent inlet (m²)

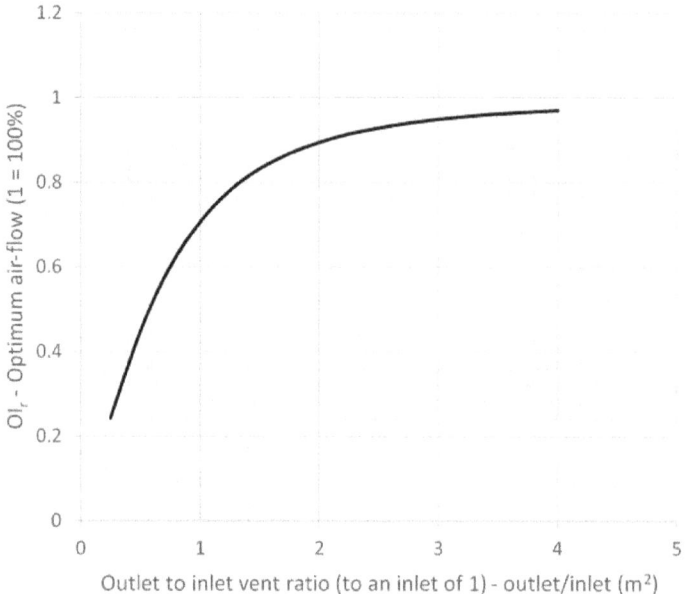

Figure 18.1: *If optimising the PPV air-flow through the structure, the outlet should be twice the size of the inlet to achieve 90 percent efficiency (Svensson).*

A tactical objective when using PPA would be to quickly force an under-ventilated fire into a well-ventilated state, removing dangerous smoke and fire gases from the building and reducing compartment temperatures in order for firefighters to undertake fire suppression, search and rescue operations in a safer and more controlled environment. This of course has its own risks but by following strict guidance and protocols based on scientific research the exposure to risk can be reduced. When assessing the exhaust location(s), the impact of PPA will be noticeable within seconds. When the exhaust is first created the buoyant flows will result in a neutral plane (smoke interface) located somewhere in the window depending on the location of the fire and stage of fire growth. The high pressure hot gases (smoke) will flow out the top of the window above the smoke interface. A gravity flow of cooler ambient air will flow in the bottom, below the smoke interface. Once the fan is turned on, the smoke interface should drop to the window sill and the exhaust should become a unidirectional flow indicating the fan flow path has been established. A smoke interface above the window sill on the exhaust opening while conducting PPA indicates more air flow is required or an obstruction exists between the inlet and the exhaust. This suggests additional actions such as increasing the fan flow by adding a fan or increasing fan throttle are required while ensuring that no obstruction exists in the intended fan flow path. If increased exhaust vent flow cannot be established within a short period of time, crews should stop the fan and consider implementing a different tactic. The 7213 report from NIST[195] demonstrates how PPV was used to overcome a fast developing room fire in ventilation controlled conditions and again suggested that a critical flow-rate to overcome high fire pressures from a very intense 12MW room fire was in the region of 4.65m³/s (16740m³/hr).

An overall analysis of appropriate test data suggests that recommended critical air-flow rates to overcome intense compartment fire pressures are somewhere between 3 and 10 m³/s as follows –

VTT Finland Report 419	Recommended critical air-flow rates to overcome compartment fire pressures are somewhere between 6.66m³/s and 8.33m³/s (24000m³/hr to 29988 m³/hr) where venting intense room fires using PPV	6.66m³/s and 8.33m³/s (24000m³/hr to 29988 m³/hr)	
NIST USA Report 7213	NIST report 7213 suggested that a critical flow-rate to overcome high fire pressures from a very intense 12MW room fire was in the region of 4.65m³/s (16740m³/hr). In fact, the actual test identified this air-flow-rate potentially as a minimum critical limit against a very intense room fire, where it took over a minute to completely reverse smoke flows issuing into the hallway at the top of the fire compartment door.	4.65m³/s (16740m³/hr)	
Based on Thomas's correlation (1970) to hold back pressurised smoke flow	Intensity of Room Fire	Minimum Air Volume flow-rate required to Overcome Fire Pressure (Pa) at 1.4m² internal door	
	1 MW	4.6m³/s	
	3 MW	6.4m³/s	
	5 MW	7.3m³/s	
	7 MW	8.1m³/s	
	10 MW	9.1m³/s	
	15 MW	10.5m³/s	
VTT Finland Report 326	A critical air-flow rate is recommended as 96m³/hr with optimum performance achieved at 144m³/hr. This is total airflow applied to the available air flow-path. In the test house the air flow-path was estimated at 110m³ so 96 x 110 = 10560m³/hr or 2.9m³/s is the critical flow.	96-144m³/hr per m³ in the flow-path (from inlet to outlet) to be vented (4.4m³/s is the optimum flow and 2.9m³/s is the minimum critical air flow).	

Continued overleaf.

195 Kerber. S and Walton. W; Effect of Positive Pressure Ventilation on a Room Fire; NISTIR 7213; 2005

Travis House Study USA 2007	The Travis House study* in the USA was the subject of an Interflam technical research paper in 2007 and provided some detailed data from 43 non-fire room experiments in a three storey brick and wood joist building. The research was quite specific in its aims of investigating the effects of distance between fan and inlet, the size and number of outlets, as well as what effects the volume of the flow-path has on vent flow rates when using positive pressure ventilation.	In the tests Fan #3 appears closest to the author's test rated fan air-flow and demonstrated a 1.06m^3/s best air-flow through the 130 m^3 flow-path in the research study. This falls far short of critical air-flow requirements according to most research recommendations discussed above.
Underwiters Laboratories (UL) fire test data, USA 2016**	For PPA to be effective the pressure created by the fan must be greater than the pressure created by the fire. Although fan size does play a role in the effectiveness of PPA, exhaust size plays a greater role. Providing enough exhaust to reduce the pressure in the fire room to a pressure below what the fan is capable of producing in the remainder of the structure is essential for safe PPA operations. A fire in post flashover state, venting to the exterior was seen to produce between 9Pa and 11Pa of pressure in the upper layer 1ft from the ceiling. This means for the fan to prevent flow from the fire compartment to an adjacent compartment, the adjacent compartment [hallway] needs to be at least 9Pa, preferably 11Pa or higher.	The most effective way to ensure that the pressure from the PPA in the adjacent compartments is higher than the pressure in the fire room is to have the exhaust openings in the fire room be larger than the inlet of the opening to the fire room. The inlet size was thought to be the opening where the fan was placed. However, according to the UL research, the true inlet is the opening to the fire compartment. UL's testing demonstrated a 2:1 exhaust to inlet ratio was much more effective than a ratio of 1:1 or less. Although under non-fire conditions, the pressure in the bedroom with one window open is less than in the remainder of the structure, when fire is introduced it creates additional pressure. As the heat release rate of the fire increases, the pressure in the fire room increases. At the point where the fire room pressure matches the remainder of the structure, combustion products will flow from the fire room into the structure again. This increases temperatures, and transfers smoke and toxic gases from the fire compartment to the remainder of the structure.
Author's Test Data	>2.45m^3/s (>8820m^3/hr) is the potential airflow in a residential house using the test ventilator	>2.45m^3/s (>8820m^3/hr) is the potential airflow in a residential house with a 110m^3 flow path using the test PPV ventilator (non-fire situation) cold-flow data.

* Ezekoye, Svensson, Nicks; Investigating Positive Pressure Ventilation (Travis House Study) Interflam 2007
** Fire Service Summary Report: Study of the Effectiveness of Fire Service Positive Pressure Ventilation During Fire Attack in Single Family Homes Incorporating Modern Construction Practices; Firefighter Safety Institute (UL) 2016

Table 18.5: International research recommendations of the critical air-flow rates required for PPV to reverse the flow-path and direct combustion products away from advancing firefighters.

What does this mean to the Firefighter?

The use of positive pressure fans to clear smoke and direct heat away from advancing firefighters before the fire has been extinguished is termed by many as positive pressure attack (PPA), or offensive PPV (ventilation). The ability to create a smoke free path and reduce temperatures in the approach and search routes may enable firefighters to advance on the fire and locate trapped occupants with speed and safety. However, as with any fire-ground strategy it brings its own risks and firefighters must be effectively trained and equipped to deploy PPA under a strict regime of tactical awareness and understanding.

There should be clear protocols in guidance notes that inform on the minimum number of trained firefighters needed to deploy PPA safely and effectively and roles such as command, fan operator, attack hose team, safety hose-lines, and safety observer should be considered. It is essential to have and maintain a communication link with interior crews as fire dynamics can change very quickly where PPA becomes ineffective. It is also important to ensure the fire compartment has been located and an outlet vent has been created before a fan's air-flow is directed into the building. The importance of creating an adequate flow-path cannot be emphasised enough for if the outlet vent is too small, the combustion products and possibly the fire itself may reverse back towards advancing firefighters. It is important therefore to allow a 30-60 second time delay before deployment into the building after PPA is started and a fan operator should stay with the ventilator and maintain communication with interior crews throughout. It may be necessary to reduce or increase the airflow, depending on fire conditions and the ventilation profile as presenting, or possibly even turn the fan away from the inlet vent opening. The impact of an exterior wind on the use of PPA must be a prime consideration. As such, a wind heading into the inlet vent may over-pressure the fire compartment when added to the ventilator's air-flow and cause the flow-path to reverse. Alternatively, a head wind into the outlet vent may counter any positive effects of the ventilator's air-flow.

PPA ventilators should not be operated after a crew has entered the building unless they are in a position to effectively size-up fire conditions and request it themselves. If an urgent evacuation of firefighters suddenly becomes necessary at any point due to rapidly deteriorating fire conditions, an immediate decision should be made to either maintain the ventilator in position with air flowing into the inlet vent, or to turn it away and stop the air-flow entering. This could be an extremely critical decision and in general, the air-flow should be maintained unless fire is coming out of the inlet vent (doorway). This is a decision to be made based on the fire conditions presenting and the location of firefighters. If a situation arose where the firefighters had strayed into the flow-path hot zone, between the fire and the outlet vent, the ventilator airflow should immediately be turned away from the inlet vent. It is clear that a good knowledge of fire behaviour is critical if firefighters are to utilise PPA safely and effectively. A well-placed ventilator and a knowledgeable crew may certainly save many lives when using this strategy to good effect.

18.6 'AIR TRACK' OR 'FLOW PATH'?

In fire behavior language, the term 'air track' had come into common use in the UK and across parts of Europe and Australia in the 1990s, describing how air would flow into an opening below the smoke interface and combustion products would flow out. This term was reinvented and replaced with a new term 'Flow Path' by NIST researchers in their reports[196] in 2009, although the two terms are interchangeable. The fire service has always been good at renaming well-used and established terms. Remember, Braidwood had originally assigned the oldest term reference 'draught' to this form of fire behavior and air movement (above).

- *The 'air-track' is the 'point to point' route that is taken by air flowing into a structure and combustion products leaving the structure'*[197] *(Grimwood 2008)*
- An early scientific definition of 'air-track' as referenced; **Gravity Current** – *also termed gravity wave – An opposing flow of two fluids caused by a density difference. In firefighting terms this is basically referring to the under pressure area where air enters a building or compartment and the over-pressure area where smoke, flame or hot gases leave*[198]. *(Grimwood 2003)*
- *The flow path is the space through which fire, heat and smoke progress, moving from an inlet and the high-pressure fire area to where the fire wants to go – toward lower pressure oxygen sources, outlet areas such as door and window openings'*[199] *(UL-NIST 2012)*
- *'Flow paths can be defined as the movement of heat and smoke from the higher air pressure within the fire area to all other lower air pressure areas both inside and outside of a fire building' (Kerber)*

The <u>term</u> 'flow path' will be used throughout the book to describe what is, in reality, serving the same function as air-track. This term 'Flow path' is now becoming more internationally recognised and used although the term 'air-track' still remains in use in some texts.

Lt. John Cierello, a firefighter in Brooklyn, New York rightly points out that there are two parts to the flow path and each part has its own tactical relevance. The first two definitions above seem to describe this best.

1. The natural flow of air coming in to feed the fire
2. The natural flow of smoke, heat and flaming combustion heading away from the fire, towards and out of vent openings

For the purposes of teaching and defining these two distinct parts of the flow-path we will term these the 'cool zone' (air coming in to feed the fire) and the hot zone (smoke, heat and flame heading towards vent openings).

1. **Cool zone** (we can control at the door or vent inlets)
2. **Hot zone** (we can control at the vent outlets)

---

[196] Kerber.S & Madrzykowski. D; Fire Fighting Tactics Under Wind Driven Fire Conditions: 7-Story Building Experiments: NIST Technical Note 1629; National Institute of Standards and Technology, USA 2009

[197] Grimwood. P; Eurofirefighter; Jeremy Mills Publishing 2008

[198] Grimwood. P and Desmet. K; Tactical Firefighting CEMAC 2003; http://www.olerdola.org/documentos/cemac-kd-pg-2003.pdf

[199] Examination of the Impact of Ventilation and Exterior Suppression Tactics on Residential Fires, UL, National Institute of Standards and Technology (NIST) and the Fire Department of the City of New York, July 2012. Presentation, May 2013

Figure 18.2: Flow-path cool zone between the inlet and the fire and hot zone between the fire and the outlet

When windows are vented and openings are made you can be certain that the fire, in its exigent search for oxygen, will nearly always head towards the point where the opening is made. There is nothing new in that and experienced/well trained firefighters have always understood it. In practice, a *ventilation profile* may relate to the quantity of air/oxygen feeding into a fire compartment and influencing fire development, just as an opening serving as an air inlet or a smoke outlet, elsewhere in the building, can create a flow-path and cause the fire to spread and travel in a variety of directions. If we create an opening at the street door into a fire building this may initiate a primary flow-path, impacting on fire development, intensity and directional fire spread. If we create a vent opening elsewhere, say an upper storey window, this again can change all of those factors and reverse the flow-path, even possibly worsening conditions and intensifying the fire further still. On the other hand, sometimes a flow-path reversal can work to our advantage.

The ability to alter the direction of the flow-path and cause it to head away from firefighters working inside a building has been at the very root of tactical venting operations for decades. If we open a door to enter the fire compartment it may head right at us, but if we vent a window on the opposite side of the room, chances are it will head there if we 'push' it a little with a hose stream or if we create an opening at high level. The Swedish fire service have even used 'stable door' venting tactics where the compartment entry door has been cut below the lock to open the lower half of the door whilst the room window has been vented at high level. The natural air movement caused by the difference in air density (hot air rises) causes cool air to flow in at low level and heat to exit at high level in the room. This effect may also be achieved by the use of smoke-blocking devices[200]. If we utilize the air-flow from a positive pressure fan, we can force it completely out of our path, if effectively applied according to established protocols. The *ventilation profile* of a fire compartment refers to the amount of ventilation (percentage) available through openings, as a ratio to the floor space – (area of ventilation openings/area of floor space) as A_v/A_f. So for 6m^2 of vent opening in a compartment with a 16m^2 floor space; the vent profile is 38 percent. As a rule of thumb guide, vent profiles above 25% in a small compartment (as above) will be required to achieve fuel controlled burning. At this point the heat release from the fire is at its highest intensity. In larger compartments, a much lesser percentage in the ventilation profile will achieve fuel-controlled burning at maximum fire intensity.

200 http://www.rauchverschluss.de/index_e.htm

18.7 SMOKE BLOCKING DEVICES

There are a number of portable smoke (or exterior wind) blocking devices used by firefighters and these are gaining global popularity. These units are used to provide greater control over door openings whilst entry is being made to fight the fire and also to reduce/prevent exterior wind entering windows and causing a wind driven fire to send searing heat and smoke into the path of interior firefighters. Such tools can be used to prevent smoke from entering stair-shafts or particularly in extended length corridors >10 metres in residential buildings where evacuating entire floors before gaining entry to the fire apartment can be resource heavy operations. By siting a smoke curtain blocking device, the corridor remains relatively smoke free should occupants attempt to exit their apartments further along the corridor whilst firefighting is underway. In some cases, there can be 10-20 apartments per floor and to raise occupants and evacuate prior to fire attack can be a time-consuming operation. Therefore, consistent with 'defend in place' strategies commonly used in apartment blocks (building specific) the smoke blocker can be used to maintain the escape corridor on the fire floor for occupants who decide to leave of their own accord, once firefighters have opened the door to the fire involved apartment. This tool allows greater control over the flow path and supports the principal of automated smoke ventilation in the corridor.

18.8 FLOW-PATHS ARE DRIVEN BY PRESSURE DIFFERENTIALS

The movement of air, smoke, fire gases and heat (or flaming combustion) are driven by differences in pressure between the inlet/s and outlet/s. The biggest pressure influence is from the heat of the fire which causes air and gases to expand and head towards any point of escape (outlet). In terms of hot air/gases, they will also rise on convection currents through the building, heading for stairways, shafts and any vertical openings. In most cases these currents are caused by the hot smoke but in very tall buildings the stack effect caused by natural air movements related to external and internal temperatures (pressures) may influence the movement of smoke (flow-path).

Higher smoke production rates and heat release rates have decreased the desired smoke lifting of the interface that is traditionally the intent of ventilation. If visibility is increased, then search and fire attack are easier to accomplish. As the fire receives air it produces more smoke than can be ventilated out of the house and therefore may not lift the smoke layer. It is possible to get an idea of what may be happening inside the structure by observing the flows at the front door after it is opened. In the UL horizontal ventilation experiments there was an obvious air flow in through the front door toward the fire location. The average velocity through the bottom of the front door for room fire experiments where the door was opened was 1.5 m/s (3.4 mph) although this can push as high as 10m/s (22 mph). There may also be an air tunnelling effect that resulted from the in rush of air through a front street door. Window vent outlets can reach smoke velocities to 20 m/s (44 mph) and roof vent outlets even higher. The actual speed of air/smoke movement in the flow-path can be anything from 1-2 m/s to 15-30 mph.

Wind Driven Flow-Paths

The effect of pressure differentials caused by wind velocity and direction can have devastating effects in the flow-path. If the wind is heading into an opening in the hot zone it will most likely cause a flow-path reversal (Fig.18.3). This may direct searing flame and heat into the path of advancing firefighters with temperatures increasing instantly tenfold, placing firefighters who are located in the cool zone into a totally untenable environment where survival times are just a few short seconds. If the wind is heading into a cool zone opening (entry door), it may still cause the fire and heat to head back towards the firefighters on the interior unless adequate vent openings have been made on the opposite side of the fire, as is required when using positive pressure attack ventilation (PPA) fans. NIST research into wind driven fires used full-scale live fires that were ignited in furnished rooms of an apartment. Due to excess fuel generation (or lack of ventilation) the room of fire origin could not transition to flashover until windows self-vented and introduced additional fresh air with oxygen to burn. Without a wind imposed on the vented window, the fire did not spread from the room of origin and never left the apartment of origin. Even with no externally applied wind, creating a flow path from the outside, through the fire apartment into the corridor and up the stairs to the bulkhead door on the roof increased the temperatures and velocities in the corridors and in the stairwell to create hazardous conditions for fire fighters and untenable conditions for occupants on the fire floor and above in the stairwell.

With an imposed wind of 9 m/s to 11 m/s (20 mph to 25 mph) and flow path through the fire floor and exiting out of the bulkhead, temperatures in excess of 400°C (752 °F) and velocities on the order of 10 m/s (22 mph) were measured in the corridor and <u>stairwell above the fire floor</u>. These extreme thermal conditions are not tenable, even for a firefighter in full protective gear.

These experiments demonstrated the 'extreme' thermal conditions that can be generated by a 'simple room and contents' fire and how these conditions can be extended along a flow path within a real structure when wind and an open vent are present. The wind effects on the fire were so strong that two 27-inch PPV fans in unison could not overcome the effects of a wind driven condition.

Flow-path creation

A flow-path will generally occur as soon as a fire reaches the ceiling. At this point it will start to spread out and flow along the ceiling towards any opening that exists. This may be a door, a window or an open shaft of some sort. Where openings do not exist, the heat from the fire may create them naturally, or occupants/firefighters may do so. Whilst this is happening, air is rushing in to feed the base of the fire at low level and if this air supply becomes depleted, the fire will begin to decline, producing a large quantity of combustion products in the smoke. If an opening is created, the fire will likely burst back into life and head for that opening, with additional ait feeding in at the base of the opening. The path of fire travel will depend on the availability of further openings or vertical routes to upper levels, or the exterior.

> Taking a look at fire videos of simulated or training corridor fires, one might ask; what makes a fire turn right along a corridor when leaving a fire apartment instead of left (or vice versa)? Think about that!

18.9 FLOW-PATH REVERSALS

A flow-path reversal is slightly more complex and can be a very dangerous occurrence, or if planned for by firefighting actions, it may be very advantageous. Imagine it like this – you are driving fast along a three-lane motorway in heavy traffic when suddenly, all the cars have turned completely around and are now heading back towards you at the same speed! That sounds pretty scary right?! That analogy demonstrates just how flow-path reversal can suddenly appear! The phenomenon that is referred to as *flow-path reversal* has been responsible for multiple firefighter deaths and is something to be anticipated and controlled (or avoided). In EuroFirefighter[201] (published 2008) a flow-path reversal that occurred at a fire in Illinois, causing firefighters to jump from upper floor windows, was described in detail (p52).

> 'At a three-storey house fire (fire at ground level) in Illinois, the firefighters working the interior (second floor) were calling for some ventilation. The street door to the structure was open, serving as the air inlet point [and smoke outlet]. As [firefighters] in a tower ladder broke out the windows on the third floor there was a sudden and massive air movement up the stair-well. It was this sudden release of high-pressure [hot] smoke that caused the massive air movement up the stair, pulling fire upwards from the lower floor and forcing several firefighters to jump from upper levels'.

Another example of a high-rise flow-path reversal appears on p51:

> 'They exited the apartment and headed down the hall, but a nasty thing happened when they opened the stairwell door, sources say. The stairwell door acted like a ferocious maw, sucking heat and smoke down from the burning apartment. For Jahnke and Green (Houston Fire Department) the effect was overwhelming. The smoke grew thick as a blindfold; a torrent hot air whirred past. The two firefighters tried to beat a retreat ... The violent shift in the air current created high confusion between the escaping firefighters by sucking the heat away from the fire. To Jahnke it seemed like they were heading towards the fire, not away from it ... Captain Jahnke lost his life in the process'.

If the fire is heading out of an opening, say a window to the exterior, on a lower floor but an opening is made on a higher floor or at the roof, the initial window outlet may then become an air inlet and the flow of hot gases and smoke will reverse in direction and head for the higher outlet. A flow-path reversal could occur in other ways, for example if a window fails, or is vented, and a wind heads into the opening causing the fire to head for lower pressure areas in the building. Also, if automated smoke control (extract) systems are incorrectly configured to pull hot smoke and gases towards the access stair instead of away from, this can reverse the flow-path towards advancing firefighters in a corridor.

Figure 18.3: Flow-path reversal occurs where the original inlet now becomes the outlet.

201 Grimwood. P; EuroFirefighter p52; Jeremy Mills Publishing UK; (jeremymillspublishing.co.uk) 2008

There are other occasions where the flow-path can be reversed in situations where air can actually travel down a stair-shaft as openings occur through fire spread, or are created. In fact, some smoke control systems are designed to do this, directing combustion products towards a smoke shaft in the building. The dangers of flow-path reversal are clear to be seen where firefighters may be occupying the cool-side of the flow-path, only for it to be reversed. However, some venting actions may even pull smoke and heat away from firefighters and place them in the cool zone.

Here are just a handful of many examples of flow-path reversal that have occurred in recent years where multiples of firefighters have lost their lives (there are many more as well as numerous near misses):

Watts Street Fire, New York City, 1994

At 1936 hours on March 28, 1994, FDNY responded to a report of heavy smoke and sparks from a chimney of a three-story apartment building at 62 Watts Street in Manhattan. The initial response to the fire call was 3 engines, 2 ladders, and a battalion chief. On arrival they saw smoke and sparks issuing from a chimney but no other signs of fire. The engine companies were assigned to ventilate the roof above the stairs by opening the scuttle and skylight, and two three-person teams advanced hose-lines through the main entrance to the ground and first floor apartment doors.

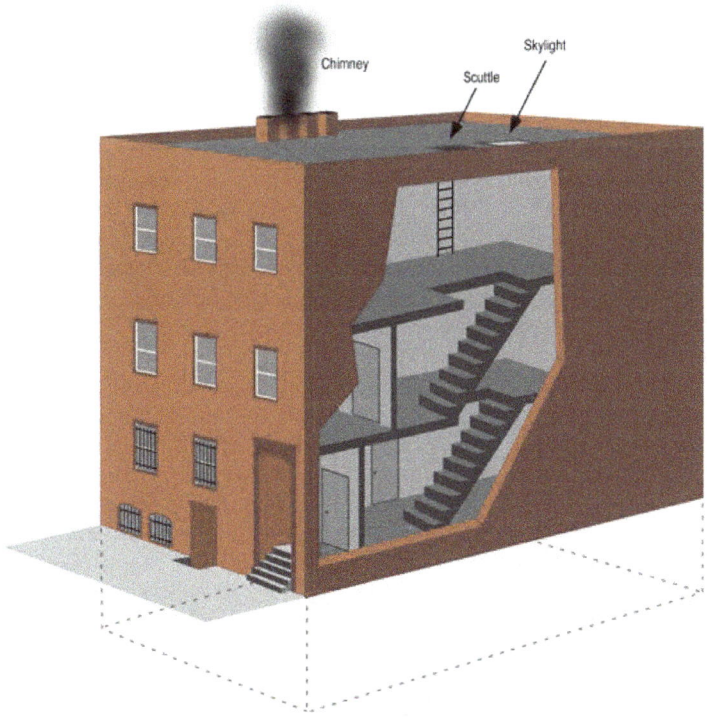

*Figure 18.4: The Watts Street fire building in New York where 3 firefighters were killed by a **flow-path reversal** event in 1994 (see chapter 19 for CFAST zone model of the fire)*

The ground floor team forced the apartment door and reported a momentary rush of air into the apartment, followed by a warm (but not hot) exhaust, followed by large flame issuing from the upper part of the door extending up the stairway. *'The fire blew out a door on the first [ground] floor and moved along the ceiling to the stairway and up to the second [first] floor,'* said Lieut. David Deering, who was at the scene. *'They (3 firefighters) were caught in the hallway.' 'They were going up to make a search for any occupants,* said a department spokesman. *In fact, all the residents had already safely left the building,* officials said[202].

The ground floor team was able to duck down under the flame and retreat, but the three firefighters on the upper floors were engulfed by the flame which now filled the stairway.

An amateur video was being taken from across the street and became an important source of information when later reviewed by the fire department. This showed the flame filling the stairway and venting out the open scuttle and skylight, extending well above the roof of the building. Further, the video showed that the flame persisted at least 6½ minutes (the tape had several pauses of unknown duration, but there was 6½ minutes of tape showing the flame).

The fire itself demonstrated an initial flow-path out of the chimney but was unable to obtain an adequate air supply on return and became grossly under-ventilated. As openings were made at high level (roof) and the apartment door was opened up by the hose-team, the flow of air reversed down the chimney and, assisted by additional air entering from the street door, the ventilated fire exited via the apartment door and travelled into the stairwell. All three firefighters on the upper levels were tragically killed by this event.

Cherry Road Fire, Washington DC, 1999

A fire in a Washington DC townhouse resulted in tragedy as two firefighters were caught and killed by a reverse flow-path. At approximately 00:24:00, firefighters began entering the ground floor via the front door. Conditions on the ground floor were described as 'heavy smoke,' with thick black smoke coming from the doorway. Within two minutes, the front window at ground level was taken out by firefighters to provide ventilation. The window was removed from the inside, due to obstructions from security bars on the outside. Firefighters were also opening the second story windows on the front of the house. The occupants had left the second story windows on the backside of the house open.

Firefighters positioned by the sliding glass doors at the rear basement (split) level, reported that the basement was fully charged with smoke and that upon arrival a few flames appeared briefly. The sliding glass door was broken out in two stages. First the right half was taken out at approximately 00:26:20, then the left side was removed approximately 20 seconds later, due to obstructions from security bars. After the sliding glass doors were broken out, firefighters entered the basement to conduct a search. They reported that there were a number of small fires on the floor of the basement, and that the fires began to increase in size after the sliding glass doors were opened. The basement firefighters reported that a tunnel or path was open in the smoke that enabled them to find their way out of the basement to the exterior, just prior to the basement becoming fully involved with fire.

As the flow-path reversed upwards at high velocity (19 mph), taking fire up into the ground level, two firefighters were killed and a third was seriously burned but survived.

---

202 Fireman Dies Battling Blaze in SoHo; **New York Times; March 29th 1994**

Tactical Ventilation – survive in the 'Flow-path' • 381

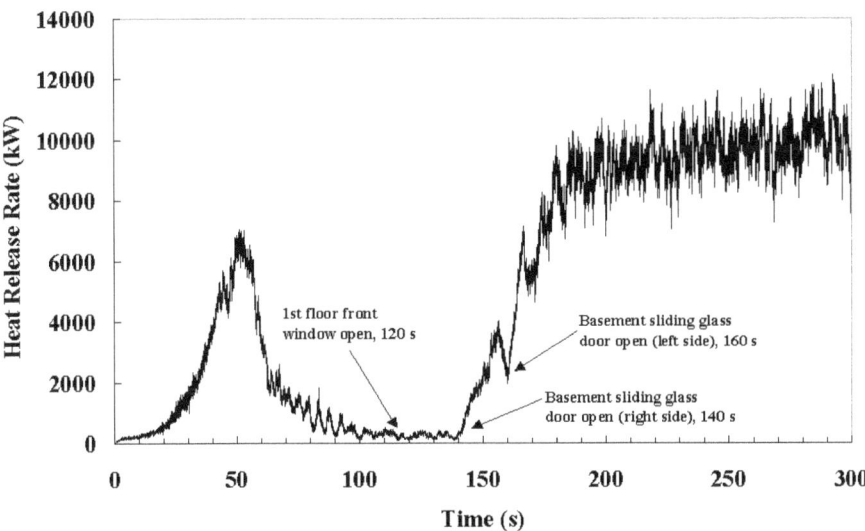

Figure 18.4: Estimated heat release rates over time in the Cherry Road fire Washington D.C 1999 **Image courtesy of the National Institute of Standards and Technology (NIST) USA**

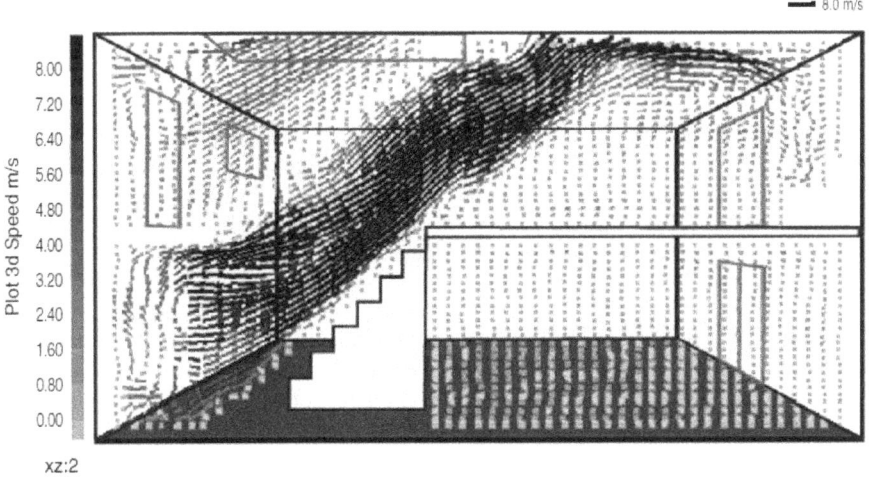

*Figure 18.5: Velocity of air, fire and hot smoke movements in the **flow-path reversal** at the Cherry Road fire, Washington D.C 1999 reached >8.5m/s. The direction of the flow-path from the vented basement sliding windows to the rear of the structure, and up the stairs to the ground floor street door at the front of the building, is simulated by NISTs computer aided analysis of the fire development.* **Image courtesy of the National Institute of Standards and Technology (NIST) USA**

Shirley Towers Fire, Southampton UK 2010

Two Hampshire FRS (UK) firefighters were tragically killed after they were unable to locate a small fire involving window curtains and nearby furniture (candle originated). In the smoke logged split level apartment they carried on upwards, past the fire, to upper levels before opening windows above the fire floor to vent the smoke. The fire, that had by now caused the window to partially fail, suddenly flow-reversed up the apartment stairs, trapping and killing them

18.10 DOOR CONTROL AND REVERSING THE FLOW-PATH TO TACTICAL ADVANTAGE

In 1996 I worked with a team of international firefighters assembled in Valencia, Spain, for a series of live fire experiments. One particular experience has always lived with me to this day when Chief Fire Officer Jose Miguel Basset demonstrated exactly how the flow path can be manipulated to tactical advantage:

> As we crawled into the room, the fire's roar was somewhat disconcerting. The thick smoke from the fire's plume was banking down setting an 'interface' at about one metre from floor level and the heat radiating downwards from the ceiling could clearly be felt through the substantial layers of our protective clothing. I looked directly above our position, into the darkness of the smoke, and noted some yellow tongues of flame rolling the ceiling, detaching themselves from the main body of fire that blazed in the furthest corner of the compartment. We had advanced about a metre into the room as I reached for the nozzle of the high-pressure booster-line and discharged the briefest 'pulsation' of water-fog into the overhead.
>
> There was no drop-back in terms of water particles and the series of 'popping' sounds suggested that the fog was 'doing its thing' in the super-heated gas layers. The tongues of flame dispersed for a few brief seconds before resuming their eerie 'snake-like' dance towards the open access point (doorway) situated behind us. 'Hold the water' shouted Miguel over the SCBA comms' radio. As we crawled further into the room I realized then that I was placing my deepest trust in the man.
>
> The smoke continued to bank down around us and I watched in awe as several 'balloon-like pockets of fire gases ignited, each for a brief second, in front of my eyes about a metre from the floor. I could sense the moment of compartmental flashover' was fast approaching and I instinctively reached for the nozzle again. 'WAIT', shouted Miguel – he laughed as he reached back and kicked the access door almost shut. I felt extremely vulnerable but then, as if turned off by a tap, the fire suddenly lost its 'roar' and the rolling flames in the plume above dispersed completely. Everything went dark as the fire 'crackled' and the smoke banked right down to the floor. There was an eerie silence within this blinding experience that seemed all too familiar to the 'firefighter' in me. Miguel took the nozzle out of my hands and discharged several brief 'pulsations' of water-fog, on a wide setting, into the upper portions of the room.
>
> Again, there was no 'drop-back' [of water] and you could almost sense the minute particles of water suspending themselves within the super-heated flammable gas layers. The steam 'over-pressure' and humidity was negligible and any air movement went unnoticed. More importantly, the thermal radiation from above had lessened considerably reducing the likelihood of a flashover. Then I heard Miguel's voice over the fire-ground comms' calling for an exterior tactical venting action and almost instantly

the smoke layer began to rise as firefighters in the street vented the window serving the room. The fire in the corner of the room became visibly active again as it increased in intensity, however this time the tongues of flame in the ceiling layer were heading towards the open window and away from our position.

Jose Miguel Basset was the Chief Fire Officer of the Valencia (County) Fire Brigade in Spain. He was a practical man who had learned much about fire and its behaviour under various conditions. He had 'played' with fire over a number of years, experimenting alongside his trusty team of firefighters, pushing ventilation parameters to their limits in an attempt at gauging the effect of venting on fire growth. Within the fiery depths of this acquired structure training situation He had demonstrated quite clearly how firefighters may utilize tactical venting actions to attack a fire's progress and that simply by closing the access door or opening a window at its highest level you can avert or delay a backdraft or flashover situation. He also showed how firefighters can reduce thermal radiation from above by reversing the direction of a fire's plume away from the access point, as described.

Despite London Fire Brigade's expertise in holding off ventilation until the fire was under control, I had first seen the use of door control techniques in New York City in 1975 where FDNY ladder crews were well practised in using such methods along in combination with the flow from a 10-litre fire extinguisher to reduce the heat release in a room fire and gain some time whilst they searched for victims. The FDNY firefighters were always pretty adept at using the first few seconds of a door's opening (prior to closing it) to get a view under any smoke layer; to use the light from the fire to enable some quick layout assessment, and then to sweep with a tool along the floor and behind the door to feel for obstructions, occupants or holes in the floor ahead of them. Most firefighters elsewhere would perhaps just enter the fire compartment with the door wide open behind them and direct water onto the fire, possibly with flames and heat heading straight at them. They are actually in the flow-path and in this case, the wrong side of it. However, Chief Baset's demonstration reminded me how we can turn the flow-path around and take control of it, simply by coordinating compartment door control with external ventilation. No water was needed at that stage, we were simply experimenting with the flow-path in a training scenario. It showed me how important door control is and how the flow-path can be effectively managed. However, imagine if the vent opening was made by firefighters working on the interior. This scenario has happened many times on the fire-ground and sometimes with tragic outcomes. The firefighter goes to the window and ventilates at that point causing fire to head directly towards his/her position. The firefighter is now the wrong side of the flow-path and may end up with severe injuries cause by flaming combustion or super-heated smoke. On occasions firefighters are forced to exit the window itself, which may be at some height from the ground.

Door Control –

'The FDNY firefighters were always pretty adept at using the first few seconds of a door's opening (prior to closing it) to get a view under any smoke layer; to use the light from the fire to enable some quick layout assessment, and then to sweep with a tool along the floor and behind the door to feel for obstructions, occupants or holes in the floor ahead of them'.

Where 'point to point' openings exist, or are created, in a fire-involved structure, the velocity or momentum of the flow-path will increase. Where a forced draft (wind) is involved the velocity of the air-track may increase further still. What this means is that where the first opening is created in a structure, a second opening may increase the air-track's velocity and this may lead to rapid fire development, or some sort of fire phenomena. In effect the second opening is often the catalyst, although a single opening is sometimes all that is needed. If we vent a window with an open entry point behind us then we may set the air-track in motion. If we close the door behind us, the single opening of a window may not be enough to increase velocity in the air-track and air-inflow is more controlled. However, where fires are isolated in this way, by closing entry (or other) doors, there are distinct advantages and disadvantages:

Advantages of Fire isolation –
- Asserts greater control over fire growth & development
- Reduced Heat Release Rate from the fire
- Reduced thermal radiation where close to the fire
- Limited flow-rate is more effective
- Less likely to experience Flashover or Backdraft
- Reduces likelihood of windows failing (unplanned ventilation)

Disadvantages of Fire isolation –
- Thermal balance is destroyed and;
- The air-track is depleted and visibility decreases
- The heat in the overhead is brought down to the floor
- May lead to a rich-mix of fire gases accumulating
- CO levels increase and O2 levels decrease
- Remaining occupants may suffer because of these effects

So there is a conflict here as by restricting air-flow into the compartment/structure, we may assert greater control over the fire's development but cause carbon monoxide levels to rise and oxygen levels to drop, as well as increasing heat levels as thermal balance is destroyed, causing smoke and heat to drop down and hit the floor. This is a call that only the nozzle or search team can make – with an officer sited at the entry doorway, feeding hose in and filling the **door control** assignment.

18.11 DOOR CONTROL ASSIGNMENT

The 'door control assignment' (DCA) refers to the deployment of a firefighter with the primary attack crew who has the specific role of controlling the fire compartment door opening to control the flow-path and reduce flashover potential. There are a number of points for consideration here before door control should become part of the SOP for interior attack.

Staffing	The assignment of a door control firefighter (DCA) means at least one additional person is deployed internally. In some cases this might mean two if working alone in SCBA is considered an issue here.
	Technically the DCA can be part of a crew of three, with two entering the fire compartment and the DCA remaining at the door to observe fire conditions and smoke interface (or neutral pressure plane), communicate with the crew and control the door as necessary. In some cases a crew commander is used for this assignment. In effect, this is a safety control measure.

Fire development	The closing of the fire compartment door will reduce the fire's rate of heat release but will also cause the smoke layer to drop leading to a thermal layer that is equally distributed in general, throughout the compartment. However, if the window to the compartment has not already vented then the heat build-up and flow-path reduction will delay this process. However, without water applied to cool the gas layers the fire can still progress towards flashover in some cases, particularly in larger compartments.
Flashover control	In Flashover & Nozzle Techniques (1999)* the author described computer modelling by US fire investigator David Birk that demonstrated the varying effects that different access door openings have on fire growth & development in a hotel room fire. With the ventilation controlled fire initially restricted to a burning chair he reports time to flashover as being greatly affected by such openings as follows – 36 inch door opening: Flashover in 2.38 minutes 2 inch door opening: Flashover in 2.82 minutes 6 inch door opening: Flashover in 4.28 minutes 3 inch door opening: Flashover in 6.97 minutes Door closed: Flashover not achieved
Communications	The change in fire conditions as a fire compartment door is closed will dictate how tactical advantage might be gained. It might be detrimental to what the interior crew are trying to achieve and in some cases they will request that the smoke layer remains high to assist the search. As the door is closed down the light from the fire will no longer be there to assist a quick visual sweep of the floor layout, possibly showing victims on the floor. Therefore, a close communication triangle between the interior crew, the DCA and the IC outside the building is critical. The interior crew and the DCA must communicate effectively to enable effective flow-path control.
Training	The DCA should be planned for, staffed for, equipped and trained for. It is a role that requires an in-depth understanding of compartment fire behaviour.
Equipment	When a fire compartment or street entry door is partially closed onto a hose-line it is essential that some form of door control device is used to ensure the door closes no further to jam closed on the hose-line. This can occur where a sudden pressure release occurs inside the fire compartment if a brief but intense ignition of fire gases occurs. This has happened before where firefighters have become trapped inside the fire building as the door closes tightly on the hose and restricts or closes its flow down to zero. The use of smoke-blocking devices can also be used to control the flow path where compartment doors are breached for entry by firefighters.

Table 18.6: Requirements for an effective door control assignment.

---

*   Grimwood. P; Flashover & Nozzle Techniques; 1999; www.eurofirefighter.com

18.12 HAZARDOUS CONDITIONS IN THE FLOW PATH

In 2008 the author described this phenomenon (air-track/flow-path) in the book 'Eurofirefighter', as being created by 'momentum and inertia forces', where pressure continues to build in a confined room fire compared to exterior atmospheric pressure. If the door is opened or the window is vented the room pressure normally reduces. The higher pressure (fire room) will flow towards the lower pressure (exterior) and sudden and massive movements of smoke, heat and flaming combustion could occur any time this happens. An analogy was made between the front and back door of a house left open on a windy day with wind entering at the front door. You may note that nothing will happen if a door separating the flow of air between these two location points remains closed. However, if that interior door is opened there will be a massive air movement towards the rear door that may slam closed. Compare this to a fire situation where pressures between openings can be influenced by:

- Exterior wind
- Buoyancy driven air movements within buildings caused by density differences between hot smoke and cool air (convective heat transfer)
- Stack effect
- Fire room pressure
- Automated mechanical ventilation configurations, in hallways, corridors or stairs

---

What does this mean to the Firefighter?

A room or building fire will seek ways to discharge its fast-expanding hot gases, smoke and flames to the exterior. Sometimes it will take the obvious routes such as windows nearest the fire or head up stair-shafts, but in other instances it will redirect these combustion products back into the structure and exit elsewhere.

This movement of combustion products, in exchange for air to further feed fire growth, takes place in a 'flow-path'. Therefore firefighters require a basic understanding of how to influence such effects in a fire to their tactical advantage and at ALL COSTS, avoid placing themselves, or others, in the wrong side of the flow-path.

Typical examples are:

1. Wind effects that can drive combustion products along a flow-path at speeds of 20-40 mph if an exterior window fails through heat or is inadvertently vented by firefighters
2. Firefighters might advance to a level above the fire without extinguishing the fire first; or isolating the fire or closing the door to the fire room below; or a siting a second crew with a hose-line to protect their escape route
3. Firefighters enter above a basement fire as a window breaks, or is inadvertently vented in the basement area, creating a flow-path up the stairs
4. Firefighters enter a large area to fight a fire and a vent opening is created or occurs above their heads that is not directly over the fire, causing the fire to head across the ceiling to reach the vent opening.

5. The stack effect in a tall building can work to tactical advantage or disadvantage, depending on exterior temperatures and the height (floor level) of the fire in the building. On a very cold day with a fire in the lower third of a tall building its likely that the stack effect will create a flow-path that will 'push' the fire straight at firefighters as it heads for stairs or interior smoke and service shafts.
6. An automated venting system configured incorrectly may create a flow-path that directs air towards firefighters as they gain entry to the fire floor in long corridors.

Most certainly, prior to the fire being located and the flow-path determined – any venting actions without clear directives or justifiable purpose should be strongly discouraged.

Bidirectional Flow-path

Where a flow path is bidirectional (moving in two, usually opposite, directions) two flows are competing at the same opening

Fig.18.6; A bidirectional flow-path exists as there is movements in two directions at one opening (doorway)

Unidirectional flow-path

There are at least two openings in the fire room or building, where one is primarily serving as an air inlet and the other serving as an outlet for hot gases, smoke and flaming combustion, where heavy smoke or flaming combustion **fills the entire outlet**.

Fig.18.7; A unidirectional flow-path exists as there is flow-path movement in one direction – from inlet to outlet

There is only one opening in the fire room or building, where air normally enters the lower part of the opening and smoke issues from the upper portion. The point where the pressure differences exist at the doorway is known as the neutral pressure plane (NPP), with the more visible smoke-interface with cool air existing slightly above this point.

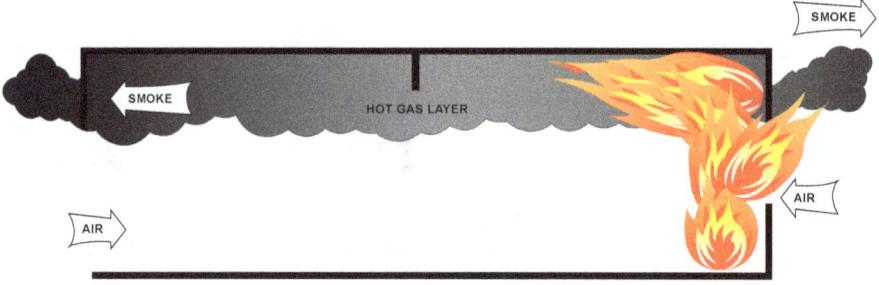

Fig.18.8; A bidirectional flow-path exists as there is movements in two directions at two openings (doorway and window)

18.13 TAKING CONTROL OF THE FLOW-PATH

An in-depth understanding of the *existence, creation, prevention and control* of flow paths is key to ensuring firefighters are working in the safest and most comfortable conditions in a fire building. In all instances we should try to avoid placing ourselves between the fire and the exhaust vent.

There are three main *control measures* we can use to take control of a flow-path and use it to our advantage on the fire-ground. These are critical:

1. Isolate the flow-path
2. Ventilate the flow-path
3. Cool and inert the fire gas layer existing near the ceiling

Each of these tactical options may be applied either to advantage, or disadvantage, and therefore training in these control measures is vital.

- **Isolate the flow-path;**
  This is a simple approach to limiting air flowing into the fire zone that will reduce the fire's heat release and burning rate and delay rapid fire spread throughout the building. This can be achieved in several ways, for example by first closing the entry door that is feeding air in until an interior hose-line deployment is ready to go with water at the nozzle. As crews advance in towards the fire, closing doors as they are passed is an excellent strategy. This will sectorise compartmentation and until the fire is located, protect the exit routes for firefighters. As a classroom exercise, apply this principle to previous case histories where firefighters have become trapped or lost their lives and then consider if such an approach might have saved them. Finally, having located the fire and arriving at the fire compartment, a door control assignment can position and hold at the doorway to control/reduce air flow into the fire room as crews advance in towards the fire. This tactic can reduce the fires heat release and enable water available at the nozzle to optimise effect. If vehicle tank water was in use at the time, less water would be needed to combat less heat release from the room fire and the tank supply would last longer. However, a constant augmented water supply is, of course, a priority. Another typical situation where isolation of the flow-path is critical is during a **VEIS** operation (**Vent-Enter-Isolate-Search**), using exterior entry points such as bedroom windows to undertake rapid entry, room search and snatch rescues. In this case, closing the room door prior to search to isolate the flow path will generally increase the chances of a rapid and safe occupant extraction.

- **Ventilate the flow-path;**
  Actually, the opposite to the above approach but can be used effectively if carefully coordinated with interior crew location and firefighting actions. Therefore, a high level of communication between the incident commander and all interior crews is vital. If the fire has been located and is confined and easily accessible, an exterior venting action may be used to extract smoke, steam hot fire gases and other combustion products ahead of the fire stream being applied. PPA is another method that can be used to vent and control/direct a flow path to tactical advantage (reverse the flow-path).

- **Cool and inert the fire gas layer existing near ceiling;**
  It is important to maintain the height of thermal layering in hot gases near the ceiling and any visible smoke interface (near the neutral pressure plane) for as long as practical to ensure the base fire can be easily located. If using isolation tactics, the thermal layering may well drop to the floor whilst the heat release is reducing. This is good for safety but may delay the approach or search. The cooling of the hot gas layer is a good thing but must be applied in short controlled bursts if the layer and interface are to remain high. The production of water vapour in the smoke will also have some inerting effect in the flammable fire gases and lessen the chances of a fire gas ignition. When approaching rooms or compartments displaying grossly under-ventilated conditions with heavy dark smoke sometimes pushing out under great

pressure, a controlled entry into the environment is required using CFBT door entry procedures, behind a sixty second burst and pause cycle into the ceiling, via a partially opened door (just wide enough to apply the stream and get a brief view of the interior). The use of PPA can be introduced at this stage, prior to entry into the fire compartment, to direct the flow-path away from firefighters and discharge a hot steamy gas layer out via the exhaust vent. In this case, as always, communication between the interior crews and the incident commander are vital as it would be the internal firefighting crew that would call for this application of PPA.

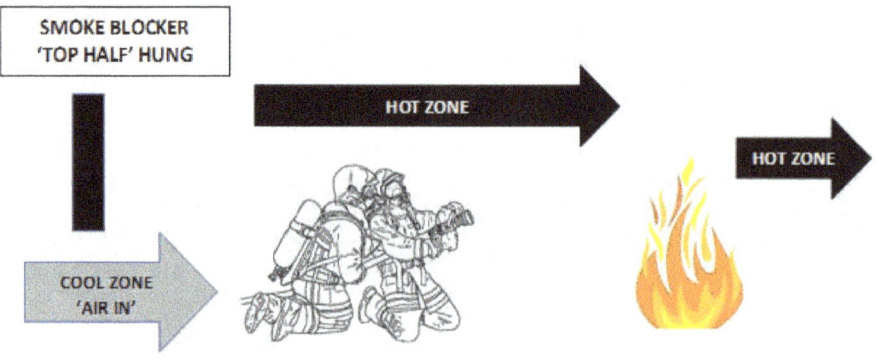

Figure 18.9: The placement of a smoke-blocker, hung to cover the top half of the entry doorway, creates an optimal flow-path with air flowing into the door at low level and smoke leaving at the higher level from the window opening. The ceiling reservoir created prevents heavy smoke flowing into the entrance hall or corridor and changes a bidirectional flow-path (Fig.18.8) to unidirectional advantage.

What does this mean to the Firefighter?

The flow-path is the route taken as smoke leaves the building and air feeds in to the fire. This ventilation flow-path through the building will be different at every fire and it is critical we identify where it is and where we are when entering the building and placing ourselves directly inside the flow-path. In some cases there is only one opening, which is the door we enter and leave by. In such a case the flow path starts and ends there but visits the fire also.

The most dangerous place we can be is the wrong side of the flow path – that is between the fire and the exhaust vent opening (probably a window). Even if that vent is inactive, meaning it is yet to open! Should that vent open for any reason, let's say a firefighter ventilates it from the outside; firefighters located in the red zone (some call this the 'kill zone') may be in immediate danger.

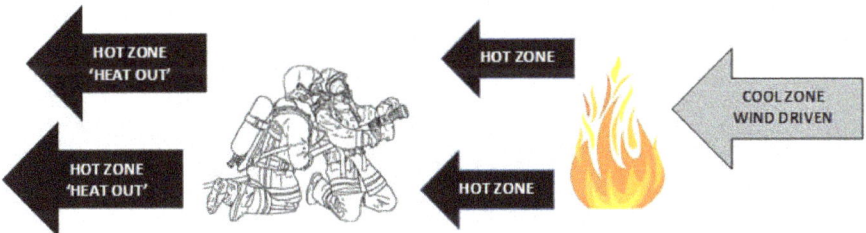

Figure 18.10: A unidirectional 'wind-driven' flow-path reversal with equal temperatures spread at both ceiling and floor levels. This is the most hazardous of all flow-paths and places firefighters very much in the hot zone. This dangerous event may also occur where automatic opening vents (AOVs) activate behind the firefighters, combined with pressure differential forces such as stack effect in tall buildings (see Fig. 19.15).

This is one of the most important sections in the book and all firefighters should be trained in methods and techniques used to 'reverse the flow-path' away from their position.

18.14 BASEMENT FLOW-PATH REVERSALS AND TACTICS

The true dangers of a basement flow-path being created as firefighters are advancing near or down stairs into the basement are rarely contemplated or little understood. There is much anecdotal evidence of firefighters being 'sucked' down stairs as a window fails in the basement area where a grossly under-ventilated fire exists. As the fire receives its first charge of air in through the basement window, the sudden fire development can actually cause a surge of high velocity air moving down into the basement. Very soon this flow-path from top to bottom (ground floor to basement) will reverse and send vast quantities of heat, smoke and possibly flaming combustion back up the stairs. This may all occur within a few short seconds and therefore any tactical venting of basement windows should be carefully coordinated with the interior crew and the IC.

Excerpt from the book 'Fog Attack' (Paul Grimwood) p117, 1992 –

Case No. 12: Two-storey Brick House – Baltimore City, MD, USA
(Fire Engineering – October 1986, p.50)

Heavy smoke was pushing from around first floor windows and doors of the two-storey brick 'row-house' as Baltimore City firefighters arrived at the scene. Despite the heavy smoke, no fire was visible from outside. Lieutenant Nelson Taylor of Engine Co. 8, and two firefighters with a preconnected handline, crawled into the kitchen through a rear door in an effort to locate the fire. They had made their way approximately five feet into the smoke-filled room and were nearing an interior stairway leading to the basement, when an 'explosion' occurred, causing the basement to erupt into flames. A typical backdraft situation one might have thought, but this was definitely not the typical backdraft. Lt. Taylor was pulled or 'sucked' down into the flaming basement by the sudden events, not blown outward! Lt. Taylor was severely burned over 65 per cent of his body, and suffered a head injury when he fell downstairs. He succumbed to his injuries about 12 hours later.

Analysis provided in Fog Attack – The firefighters in the kitchen, having determined that the seat of the fire was in the basement, approached the door to the basement stairs, fully anticipating that they would be met by heavy smoke and superheated gases venting upward toward them when the door was opened. Instead, there was an immediate flow of air from the kitchen, down the stairway to the fire in the basement. It seems probable that the fire had vented itself, probably through one or more of the basement windows, just as the firefighters were opening that fateful door upstairs. An outward flow of smoke and heated gases from the windows was thus established as the door was being opened. Then the sudden opening of the door in the kitchen provided the channel for cool air to enter, and a replenishing oxygen supply to flow into the fire area. It was this flow reversal of high velocity air that knocked the firefighters off balance, causing Lt. Taylor to fall or to be swept into the burning basement.

In effect, then, this incident was the reverse of the 'typical' backdraft. The primary ventilation was probably accomplished through a basement window, the lowest point, when the fire self-ventilated. The interior stairway from the kitchen above then became the channel for air – and oxygen – to flow into the fire.

London basement fire 2005

At another basement fire in London in 2004 two firefighters were to lose their lives where one firefighter was apparently 'sucked' down the stairs as a flow-path was created. The event culminated in extreme fire behaviour, either a backdraft or fire gas ignition. What set the flow-path in motion was a fire involving high level racked storage in the basement, burning through the ceiling on the far side of the basement. The fire burned through the ceiling into an adjacent compartment, in fact this was a stair-shaft. As firefighters on the roof noticed smoke issuing from a doorway serving the stairs they opened the door to access what they believed was a fire. As the flow-path exited at the top of the stair it released high-pressure from the basement fire into the stairs which in turn, drew air in from the street side entry door, the two deceased firefighters had used to access the basement. The author presented this theory when approached by the official FBU[203] investigation into the specific fire phenomena involved in the firefighter's deaths.

Firefighting tactics at basement fires

- Always deploy adequate water for the risk, this means a mainline hose and not a hose-reel
- Increase your resources on-scene as soon as a working fire is confirmed
- Always support the attack team with a safety line *whilst not crowding the first line* in
- Never ventilate a basement fire whilst firefighters are in the basement unless the crew have located the fire, can assure they are not in a position between the fire and the outlet vent, have water on the fire and call for any venting action themselves
- Place hose-lines to cut off vertical fire spread

203 Fire Brigades Union (FBU) UK 2004

- Plan for early (10-15 minute) reliefs at the nozzle due to heat build-up – basement fires can be punishing on firefighters
- Be aware that the hose-line is at ceiling height as it enters the basement and may be subjected to high temperatures – protect it
- Establish the building layout and structural stability in case of extended duration firefighting operations

Chapter 19

Smoke Control – System Categories and Configurations

19.1 MECHANICAL SMOKE VENTILATION SYSTEMS (MSVS)

'As the fire and smoke entered the corridor we knew we had to get back to the safety of the stairs quickly. The heat was unbearable and visibility was zero. We followed the hose-line back to the riser outlet in the corridor, just a few short metres. However, we couldn't find the stair door and in our blind panic, we went past it. It seemed we had just a few seconds in which to find our way out of this and then by a miracle we found the stair door. However, the heat had caused the door to expand and jam shut in the frame. We had to break the vision panel out to gain some leverage so that the door would open but finally we managed to get into the stairway. The fire behind us was spreading into an adjacent flat by now and the safety of people above the fire was severely compromised. This was a code compliant single stair building with a 7.5 metre corridor travel distance from the furthest flat door to the stair. Had this distance have been longer, I don't think the four of us would have made it out'.

UK Firefighter – UK SE Region Fire

'Eurofire' Fire Engineering Conference 2011 – Paris

Several corridor flashovers had previously resulted in the deaths of multiple firefighters in St. Petersburg, Russia (9 firefighters) 1991; Vandalia Avenue, New York (3 firefighters) 1998; followed by *Neuilly*-sur-*Seine*, just north of Paris 2002 (5 firefighters). There are many other recorded fatal incidents where firefighters (not in multiples) have been overcome by fire, heat and smoke whilst attempting to reach the stair in long corridors.

In 2011 the author's three-year research project[204] into the impact of mechanical smoke ventilation on corridor fires, supported by Kent Fire and Rescue Service, was presented to the 'Eurofire' conference in Paris that was attended by over 500 eminent fire engineers from around the world. The study looked at how various automated ventilation systems were currently being configured in the UK to remove smoke emitting from a fire involved

---

[204] Grimwood. P, Access Design, FRM Journal, July/August 2011, Fire Protection Association/Institute of Fire Engineers

apartment, via an extended length service corridor (15-30m). The smoke was then directed out through an AOV (automatic opening vent) in the exterior wall, or via a smoke shaft exiting at roof level. This type of system was also finding its way into hotel and office buildings. To enable direct comparisons across a broad range of system configurations in both 15 and 30-metre-long corridors, an under-ventilated fire was utilised to model the various venting configurations that created hazards for firefighters occupying the flow-path using the desktop NIST CFAST zone model[205].

The study used data that the National Institute of Standards and Technology (NIST) had produced for a computer model of an apartment fire in New York City. The objectives of the author's research were primarily to compare the impact of various smoke control systems on flow-path effects that might expose firefighters to a wide range of fire conditions in the access corridors/hallways, and establish some design parameters and firefighting tactics to counter each possible scenario.

In March 1994, the New York City Fire Department (FDNY) responded to a fire in a three-storey apartment building in Watts Street, Manhattan. On arrival, there were little signs of fire. Firefighters were deployed to the first floor and into the stairs above the fire apartment. When the door to the first-floor apartment was forced open, a flow of warm air (<100°C) issued from the apartment, which very quickly turned into a large door flame transporting into the hall and up the unprotected stairway (flow-path), engulfing three firefighters at the second-floor landing. The flame persisted for some 6½ minutes, resulting in their deaths.

FDNY requested the assistance of NIST to model the incident in order to understand the factors which caused such an event. The NIST CFAST model[206] was able to reproduce the observed conditions, and supported a theory of the accumulation of significant quantities of unburned fuel from a vitiated fire in an apartment which had been insulated and sealed for energy efficiency.

What does this mean to the Fire Engineer/Architect?

There is a critical fire behavioural function of every building fire termed the 'flow-path'. It is referred to throughout this book and is of great relevance to both the firefighter and fire design engineer, for either profession can influence how an ensuing flow-path will likely improve or worsen firefighting conditions for firefighters, during their approaches to search and rescue and firefighting operations. The installation of most automated smoke control systems are configured solely to enhance means of escape routes for building occupants, keeping them relatively smoke-free and tenable for a period of time. The direction of air-flow is normally determined by building geometry and natural vent openings. However, where corridor ventilation is provided, particularly in extended lengths (beyond code compliance) the configuration of the smoke control system should be designed so that it supports firefighting operations and meets tenability criteria that assists firefighters. The SCA residential design manual and BS 9991 now provide up-to-date detailed design guidance.

---

205 The Consolidated Model of Fire and Smoke Transport; https://www.nist.gov/el/fire-research-division-73300/product-services/consolidated-fire-and-smoke-transport-model-cfast
206 http://www.fire.nist.gov/bfrlpubs/fire95/PDF/f95090.pdf retrieved 26.7.2016

Author's Research Paper – Data inputs

For the purposes of this study of *under-ventilated* fire conditions, comparing 14 different ventilation systems and configurations in access corridors, the core data inputs used by NIST in the Watts Street apartment fire were repeated for consistency in all models, except for the fire behaviour training unit, where inputs were more relevant to a much smaller fire compartment and a controlled fire load with an earlier door entry at 800 seconds. The fire area in the Watts Street apartment measured 85.4m^2 with a 2.5m high ceiling. This is close to the 'average' sized apartment found in the UK.

In the NIST model, the apartment fire grew to about 500kW over 5-minutes simulation time, then rapidly throttled back as the oxygen concentration dropped below 10%. Temperatures in the apartment peaked briefly at about 300°C and then rapidly dropped below 100°C as the burning rate fell. The concentration of carbon monoxide rose to about 3,000ppm and a large amount of unburned fuel accumulated within the apartment volume during this stage of vitiated combustion. On opening the apartment door at 2,250 seconds, the door flame grew within a few seconds to 5MW, with stair temperatures in excess of 1,200°C, while conditions in the apartment remained relatively cool (lower layer temperature <160°C).

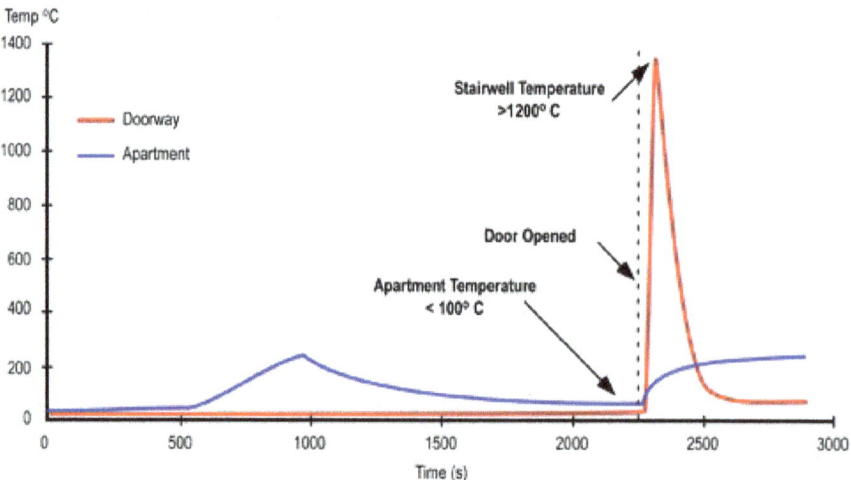

Figure 19.1: The time-temperature gradient of apartment and stairwell temperatures at the Watts Street fire in New York where three firefighters lost their lives in 1994.

The author's study used original data inputs from the NIST Watts Street simulation representing a grossly under-ventilated fire compartment, but a 15m long corridor was also added between the fire apartment and the stairs to enable a range of venting solutions and smoke shaft arrangements to be modelled under the same fire conditions. In all cases, the automated ventilation was activated at 1,800 seconds into the simulations, with door entry to the fire compartment timed at 2,250 seconds, based on the Watts Street fire itself.

Smoke Control – System Categories and Configurations • 397

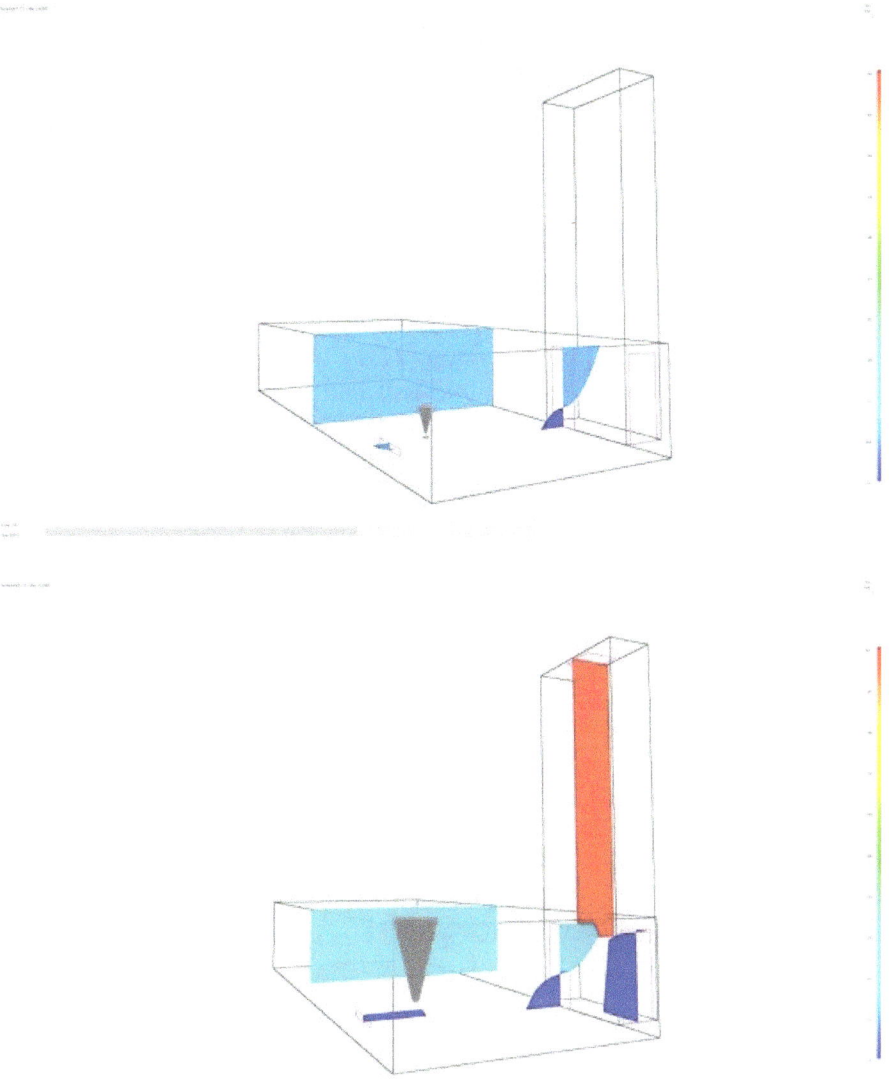

Figures 19.2 and 19.3; The NIST model of the Watts Street fire (reproduced by the author with identical data input) where 3 FDNY firefighters died as a result of a flow-path reversal in 1994 – closed apartment door (left) open apartment door (right).

Fire behaviour

An understanding of practical fire behaviour supports a view that the CFAST computer zone models produced in the author's study represent viable corridor 'flashover' conditions (ignition of smoke in the corridor), where ventilation design configurations may serve to influence fire development. In most situations, the remote corridor ignitions of unburned combustion products are made worse by forced or natural air-flow paths that cause depressurisation of the corridor, or a substantial movement of air through and beyond the space. In each situation as modelled, the under-ventilated combustion products would generally ignite in the corridor; although the duration and intensity of the gas burn-offs may vary. In some situations, the corridor is seen to depressurise extensively, causing more combustion products to be drawn out of the apartment to mix with air in the corridor.

Such ignitions in the gas layer may demonstrate one or more of the following scenarios:

- low-intensity high-level flaming at the ceiling (may be hidden in the smoke)
- flaming at the smoke/air interface
- backdraught
- smoke explosion or 'flash-fire'
- full corridor 'flashover'

The output data suggests that corridor heat flux to the floor reached 21kW/m² in some situations (note that 20kW/m² at floor level is normally accepted as one indicator of flashover[207]) It has been established that a maximum heat flux at firefighter locations of <5kW/m² is an acceptable operational limit for firefighter exposure under extreme circumstances for less than one minute, while 10kW/m² is considered critical and within the life-threatening range. No account has been made for any firefighting water applications, because it may be the case that such conditions dictate an immediate evacuation or extraction of firefighters from the space is needed.

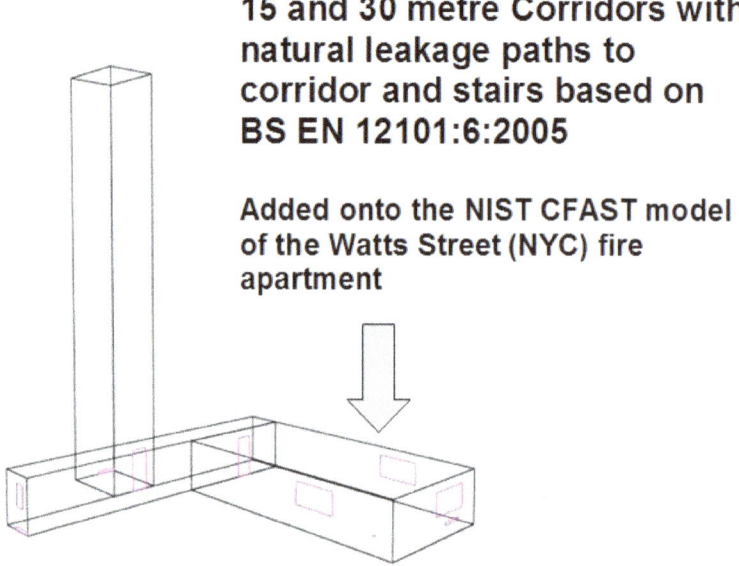

Figure 19.4; The author's CFAST design model was used in his research to evaluate the effects of corridor depressurization/pressurization in a range of designs and automated ventilation configurations, where the fire compartment is grossly under-ventilated.

207 Drysdale. D; An Introduction to Fire Dynamics; John Wiley & sons Ltd; 1985, 1998, 2011

Study findings

The various methods used to provide natural ventilation to common area stairs, corridors and lobbies have evolved over several decades. We are now seeing a move to smaller vents and smoke shafts, in conjunction with fan-assisted systems that work on the principle of creating pressure differentials between the stairs and the corridor. In some cases, a powerful air movement is forced through a corridor and pressure differentials are not dominant, particularly with smoke flushing or dispersal systems based on the 'push-pull' principle. However, what is important to the designer may conflict with what is important to the firefighter. In general, smoke shafts will assist firefighters in the vast majority of situations, but in a narrow range of circumstances, the operation of a smoke shaft may cause tactical disadvantage to the firefighter and, in rare circumstances, particularly where a compartment fire is grossly under-ventilated, it may enhance or cause unusually rapid fire development.

> **What does this mean to the Fire Engineer/Architect?**
>
> *'What is important to the designer is sometimes at conflict with what is important to the firefighter'*

This study identified that where compartment fires develop slowly, leading to low-temperature under-ventilated conditions, a remote ignition of the fire gases may occur in the corridor as firefighters gain access to the fire compartment to begin the firefighting phase. This event may create untenable conditions in the corridor for the firefighters, which may be inescapable. In particular, the naturally vented smoke shafts and exterior wall vents serving corridors may initiate unfavourable air-flow paths and excessive depressurisation, 'pulling' flaming combustion into the corridor where firefighters are advancing their hose-line. In the case of the automatic opening wall vent located in the corridor (simulation 5), it was notable that a smaller $0.5m^2$ vent opening created far less severe conditions in the corridor than the $1.5m^2$ opening.

Where fan-assisted systems were modelled (simulations 8-14), it was observed that, while there may be less likelihood of untenable conditions being created in the corridor in comparison to naturally vented scenarios, the systems creating depressurisation of the corridor were less effective in protecting firefighters. The modelling of vent configurations that seemed to offer better protection to firefighters when gaining access to these under-ventilated fires were those that created neutral or slight positive pressures in the corridor, both prior to and after the fire compartment door was opened fully. The systems that performed most effectively (simulations 13-14), by protecting firefighters in the corridor and ensuring smoke infiltration into the stairs was least likely, was that which provided +50Pa pressurisation to the stairs.

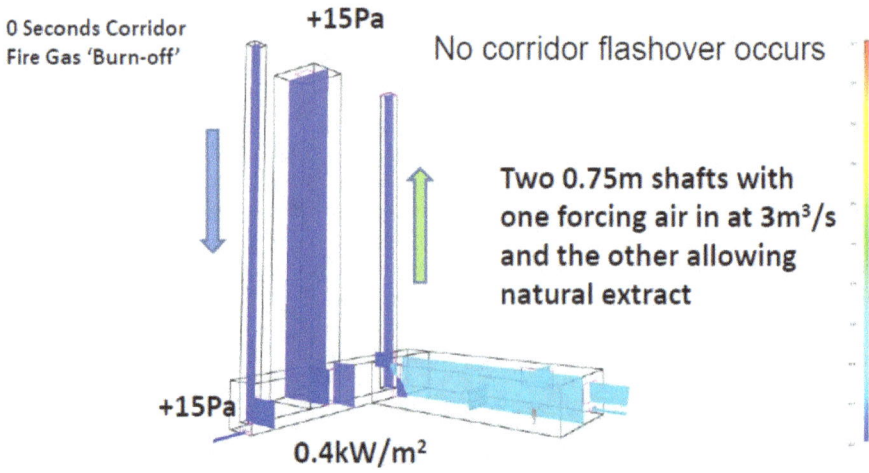

Figure 19.5; One example of the 14 CFAST models where a flow-path is created to tactical advantage for firefighters where a smoke flushing (dilution) system is used in an extended corridor situation (30-metres end to end). Although in a single stair building ideally the wings should be separated by fire resisting doors, this configuration demonstrated no corridor flashover and effective protection to the stair.

What does this mean to the Firefighter?

You may find that some buildings, constructed since 2001, have extended the corridor lengths beyond normally accepted code compliant lengths (maybe as far as 30 metres from the stair to the furthest flat door). In order to do this the corridors are normally protected by fan-driven automated smoke control systems activated by smoke detection in the corridor. It is likely that in the vast majority of cases pre-2015, the AOV outlet (exterior wall or into a smoke shaft) is located adjacent to the stairs. This may cause smoke, heat and possibly fire to leave the flat, if the door is open as a flow-path has been created, and head towards the stair. This configuration can be overridden by a smoke control panel at the main entrance and deactivated if the incident commander considers it is placing firefighters in untenable conditions. New guidance and design standards have been introduced since 2015 that include measures to protect the firefighter. These include sprinklers in the accommodation where corridors extend beyond 15 metres one-way travel to a stair and smoke control configurations that cause smoke, heat and combustion products to head away from the stair access.

As a result, a key finding of the study is that, for design purposes, the firefighting phase should be addressed more accurately using two worst-credible scenarios, at the point firefighters are opening the fire compartment to gain access. These should account for a relevant post-flashover fire, as well as a grossly under-ventilated fire.

Smoke Control – System Categories and Configurations • 401

Natural	1	Steel container fire behaviour training unit (under-ventilated door entries)
	2	Baseline unventilated 15m x 1.8m x 2.4m high
	3	Unventilated corridor with stair 1.0m^2 AOV and door to stairs fully open
	4	Corridor with stair 1.0m^2 AOV and door to stairs fully open – and 0.5m^2 window AOV in Corridor open
	5	Single 1.5m^2 window AOV in Corridor open
	6	ADB 1.5m^2 Natural Shaft
	7	2 x 75m^2 natural shafts (non-code compliant)
Fan Extract	8	Mechanical Extract 0.65m^2 Shaft to 2m^3/s – with Natural make-up shaft
	9	Mechanical Extract 0.5m^2 Shaft to 4m^3/s – Natural make-up air from stairs
Smoke Flushing	10	2 x 0.75m^2 3m^3/s Mechanical Flushing Shafts (one serves as natural extract)
	11	1 x 0.75m^2 3m^3/s Mechanical Flushing Shaft to 1m^2 corridor window AOV
	12	French system – 2 fan assisted 'push-pull' 0.18m^2 shafts (1.5m^3/s extract shaft and 0.9m^3/s inlet shaft)
Pressurization	13	50Pa Pressurization to stairs with stair door into corridor open to 0.1 (hose-line)
	14	50Pa Pressurization to stairs with stair door into corridor fully open (hose-lines)

Table 19.1: *The fourteen different design configurations used for evaluating and comparing automated smoke control systems protecting corridors.*

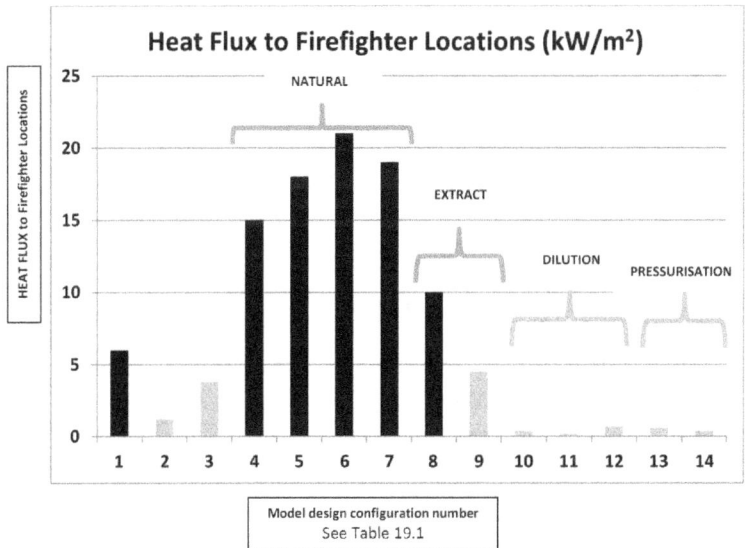

Figure 19.6 *The columns denote Heat Flux to the corridor floor (vertical axis) with darker zones demonstrating heat flux at or in excess of 10Kw/m^2 which represents dangerous and untenable conditions for firefighters.*

19.2 SMOKE CONTROL IN A LONDON RESIDENTIAL TOWER FIRE

The fire apartment was located at the end of a 20-metre corridor (serving other unaffected apartments) with a central stair and a window AOV at the opposite end. There was a 3m² natural venting (BRE) smoke shaft protecting the stair. The design layout was modern (2009). The 7 MW fire in the flat entered the corridor prior to firefighters arriving at the fire floor and spread along to the AOV. The flow path was directed mainly inwards through all windows in the post-flashover fire flat and AOV, creating an unfavourable flow path directed at firefighters as they entered the fire floor. The smoke shaft located within a lobby effectively prevented smoke from entering the stair. However, a positive pressurisation system in the stair would have avoided the negative flow-paths that created extremely difficult conditions for firefighters in dealing with the fire.

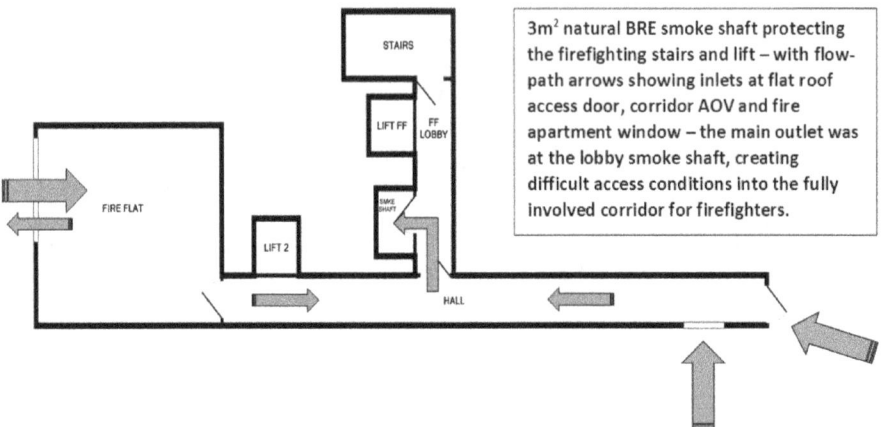

Figure 19.7; Floor plan of fire floor demonstrating flow-path inlets and outlets in a London tower block fire with natural ventilation provided by a corridor AOV and a BRE smoke shaft protecting the firefighting shaft (stairs and lift lobby).

The main objective of smoke control is to protect the stair, which in this case it undoubtedly did. However, there is also the consideration of assisting the firefighting approach and such a design configuration did nothing to achieve this. A stair pressurisation system

19.3 EXTENDED LENGTH CORRIDORS IN UK BUILDINGS

In early 2013 the author worked closely with the smoke control association (SCA) review panel in the UK to develop an integrated design strategy[208] for fire service intervention in new residential apartment blocks. It was here where smoke control systems were being introduced to justify extended dead end common area corridor travel distances up to 30-metres in length. Where buildings used to have two stairs under prescriptive building codes, a stair was now commonly being removed to provide single stair access and egress, with escape routes founded on performance based principles.

---

208 Guidance on Smoke Control to Common Escape Routes in Apartment Buildings (Flats and Maisonettes); Smoke Control Association (SCA), 2015

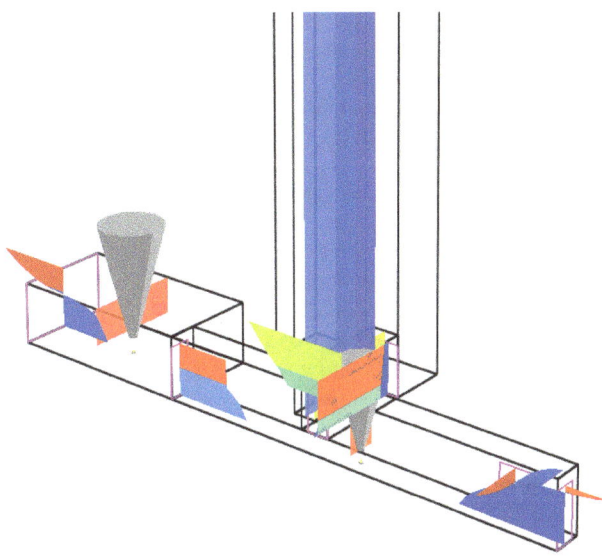

Figure 19.8; CFAST representation of the fire floor in the London tower block fire showing flow-path routes compromising the entire corridor, making firefighting access dangerous and difficult.

What does this mean to the Fire Engineer/Architect?

Fire engineering approach (extract in italics taken from Building Regulations ADB)

There is no obligation to adopt any particular solution contained in an Approved Document (Building Regulations – ADB) if you prefer to meet the relevant requirement in some other way.

***0.30** Fire safety engineering can provide an alternative approach to fire safety. It may be the only practical way to achieve a satisfactory standard of fire safety in some large and complex buildings and in buildings containing different uses, e.g. airport terminals. Fire safety engineering may also be suitable for solving a problem with an aspect of the building design which otherwise follows the provisions in this [Building Regulations ADB] document.*

***0.31** British Standard **BS 7974** (Fire safety engineering in buildings) and supporting published documents (7974 PDs) provide a framework and guidance on the design and assessment of fire safety measures in buildings. Following the discipline of BS 7974 should enable designers and Building Control Bodies to be aware of the relevant issues, the need to consider the complete fire-safety system and to follow a disciplined analytical framework.*

There is also alternative design guidance available in documents such as BS9991, the residential SCA guide (apartments), BS 9999 and the single stair office SCA Guide. The fire engineer, building developer, architect and smoke control system designer may all decide to use such alternative guidance where appropriate.

Such an approach enabled greater use of commercial floor space and was generally supported using fire growth and occupant escape computer models. The author wrote the fire service intervention section at s5.4 of the SCA residential guide (it should be noted that the SCA design guidance is referenced for use in BS 9991:2015 and designers of smoke control systems have a responsibility to take this design guide into consideration) –

s 5.4 *The national guidance for the design and construction of apartment buildings, flats and maisonettes has historically encouraged the principles of compartmentation and ventilation to protect occupants and assist the firefighting operation, particularly where a 'stay put' policy is in place. Such design principles are based on a reliance that occupants not affected by fire or smoke will remain within their flats during the firefighting intervention phase. However, occupants not at immediate risk are free to evacuate at any stage and enter common escape routes despite the principle in design and advice to 'stay put'. Consideration should be given to the likelihood of uncontrolled occupant movements within common areas and both designers and the Fire and Rescue Service (FRS) should establish and align their strategies and areas of responsibility, taking such scenarios into account.*

There are, however, some key aspects that should be considered when designing smoke control systems as compensation for extended travel distances:

- In the design and construction of extended length dead-end corridors beyond 7.5 metres served by a single stair, BS 9991 recommends that, in single stair buildings above 11m in height with accommodation on two (or more) sides of the common stair, the various wings of the building should be isolated by fire doors in order to prevent corridors from becoming contaminated by smoke. Where corridors are provided with smoke control systems as compensation for extended travel distances these cross-corridor doors may be omitted, providing the designer is able to demonstrate that the ventilation provisions can provide acceptable tenability for occupants using escape routes to exit the building.
- Despite the absence of suitable research in this area, designers should be aware that single direction travel distances over 30m in length (measured from the staircase door to the furthest flat entrance door) in common escape routes are considered to present onerous conditions for fire fighters. It is therefore recommended that they are not proposed.
- It is acknowledged that it is unlikely a mechanical smoke ventilation system can be designed to maintain tenable conditions in corridors for escaping occupants once firefighting has begun or if the door to the flat of fire origin remains open for any other reason.
- It should be noted that where 'stay put' policies are incorporated into the building's fire strategy, particularly where single direction travel exists, the approving authorities may place a much greater emphasis on the need for sprinklers or fire suppression systems in flats and/or the provision of cross corridor suitably fire resisting doors to separate extended corridors into manageable sections.
- As outlined in section 6.4.1 of this [SCA] document, it is recommended that smoke ventilation systems extract smoke away from the stairs where single direction travel distances exceed 7.5m. This will help to ensure that the staircase is adequately protected from the products of combustion, as well as assist with fire fighter access (ideally fire fighters will be approaching the flat of fire origin with the flow of inlet air). The FRS have expressed their concerns in relation to the lengths

of undivided corridor that fire fighters may have to navigate to carry out rescues or reach safety in situations where fire conditions deteriorate during intervention. Therefore, the designer should consider that corridor sub-divisions may increase fire fighter safety.
- Where cross corridor doors are provided, the ventilation strategy should take these into account when demonstrating performance capability.
- There needs to be an understanding and acceptance of responsibility between approving authorities, smoke control system designers and the Fire and Rescue Service that the risk to occupants should not be increased beyond that provided by a recognised prescriptive code compliant layout (ADB). Where extended common escape route smoke ventilation solutions are proposed, it is recommended that designers discuss all relevant fire service intervention aspects with the local fire and rescue service.

Design guidance for 'extended corridors'

The primary design objective[209] of any system should be to maintain relatively smoke free conditions within the staircase such that it can be used for evacuation and fire service access/egress at all times. Where the travel distance from the furthest apartment to the door to the staircase or the door to a sterile lobby does not exceed 7.5m (4.5m in buildings <11m to the highest floor), this is considered to be the only design objective. Therefore, where a mechanical ventilation system is provided to the common escape routes, it is not considered necessary to assess conditions within the corridor or the sterile lobby against specific performance objectives. The system provided should have at least equivalent performance to a compliant (ADB) natural ventilation system and conditions should not be made worse. Where a compliant natural ventilation or pressurisation system is provided then generally no further consideration is required.

Where the building design follows BS9991 or, where the regulator is willing to accept a proposal that travel distances from the furthest apartment entrance door to the staircase door do not exceed 15m, and the building is provided with residential sprinklers, then irrespective of building height it is considered appropriate for the smoke ventilation system to meet the same performance criteria as a system provided for a building where the travel distance from the apartment of fire origin to the door to the staircase does not exceed 7.5m.

Where sprinklers are not provided and where the travel distances from the apartment to staircase or sterile lobby are over 7.5m but do not exceed 15m, the performance objectives of the system are to maintain the staircase relatively free of smoke and to ensure the designer's specified tenable limits for means of escape are met within the corridor. Additional performance objectives for protection of firefighters (tenable limits for firefighting[210]) are also required. Further provisions for firefighting access, such as firefighting shafts, may be required. Note: generally, natural ventilation is not appropriate for corridors of this length, so the system provided should be mechanical. As an alternative to a mechanical smoke ventilation system, sprinklers to the apartments and natural ventilation may compensate for the extended travel distance to 15m.

Where the travel distances from the apartment to staircase or sterile lobby exceed 15m, the performance objectives of the system are to maintain the staircase relatively free of

209 SCA Residential Design Guide 2015
210 BS PD 7974-5:2014 and SCA Residential design guide 2015

smoke and to ensure the designer's specified tenable limits for means of escape and Fire Service operations are provided within the corridor. This may necessitate discussions with the Fire Service and regulatory authority having jurisdiction (AHJ), prior to design or installation of the smoke control system, to ensure the system performance allows the operational requirements for fire fighters to be met, irrespective of building height.

Any ventilation design will form part of an overall fire safety strategy and should not be designed in isolation. The designer of a smoke ventilation system should define how it fits into the fire safety strategy.

It is recommended in the guidance that extracting (or directing) smoke away from the stairs should be the default position where travel distances are in excess of 7.5m. This will normally result in the extract shaft(s) or outlet vents being positioned remote from the stair. Any proposed deviation from this important design objective should be discussed with the regulatory authorities at the earliest instance.

	Design options for extended corridors in single stair buildings – (max 4.5m TD in buildings <11m high)	Ventilation	Sprinklers Always required in buildings >30m high	Design Fire	Firefighter Tenability Analysis
1	Travel distance <4.5m	Natural	Not required	None	Not required
2	Travel distance <7.5m	Natural	Not required	None	Not required
3	Travel distance 7.5-15m	Natural	Required	>2.5 MW steady	Required
4	Travel distance 7.5-15m	Mechanical	Not required	4-6 MW growing	Required
5	Travel distance 15-30m	Mechanical	Not required	4-6 MW growing	Required
6	Travel distance >30m	Not allowed			

Table 19.8: Design strategy options for extended corridors (BS 9991 and SCA Guidance)

Note: 7.4 (BS 9991:2015) Escape routes from flats and maisonettes with corridor or lobby approach[211]

To prevent exposure of escaping occupants to smoke and heat in the internal corridor or lobby, either:

a) *the travel distance between the exit doors from the dwellings and a smoke-free area should be limited, and the amount of smoke and other combustion products in the internal corridor or lobby kept to a minimum by providing either cross-corridor fire doors and ventilation, or a mechanical smoke ventilation system; or*

b) *an independent alternative escape route should be provided from each dwelling either by way of a corridor at another level or through an external common balcony.*

Where travel distances from the furthest dwelling entrance door to the means of escape staircase cannot be met then a mechanical smoke ventilation system and/or sprinkler and [natural] smoke ventilation system may be used to extend the travel distances (see Clause 11, Table 2, 14.1.3 and Annex A of BS 9991:2015).

For buildings not provided with a mechanical smoke ventilation system, escape routes

---

211 BS 9991:2015; s.7.4; Escape routes from flats and maisonettes with corridor or lobby approach, British Standards Institute

should be provided. For buildings provided with a mechanical smoke ventilation system, escape routes should also be provided in accordance with Figure 6 and Figure 7, with the exception that cross-corridor fire doors and openable or automatically opening vents (AOVs) may be omitted.

Firefighter Tenability

It took several years before the author's research[212] into *firefighter tenability and safety in extended corridors* was finally addressed in the relevant design guidance and during this period, hundreds of 'fire engineered' extended corridors formed part of the design for residential apartment blocks and some commercial buildings throughout the UK. Many of the smoke control systems protecting these corridors are now considered to be configured in a way that may actually worsen fire conditions for firefighters, by extracting or directing smoke and heat towards a smoke shaft or automated vent opening, located adjacent to the firefighting access stair. The possibility may still arise in these buildings with such venting configurations (generally pre-2016) where firefighters are caught in deteriorating fire conditions that extend into the corridor, creating a situation where they may have to evacuate themselves back along >30 metres of untenably hot and smoky conditions to the safety of the firefighting shaft, or stair. It has been demonstrated both by computer aided simulation and fire-ground experience that firefighter tenability in parts of these long stretches of corridor may compromise the ability of firefighters to escape safely. Over 20 firefighters are known to have died in corridor 'flashovers' or forced draft fires over the past 30 years. Therefore, it is absolutely critical that future design specifications, where normally compliant occupant travel distances are greatly extended, meet with the SCA design guidance as well as relevant codes and standards requiring accommodation sprinklers or correctly configured mechanical smoke ventilation systems (MSVS), where corridor lengths exceed 7.5 metres. Wherever MSVS is used to justify extended corridors, firefighter tenability must be addressed in accordance with design guidance.

Exposure Condition	Maximum exposure time (minutes)	Maximum air temperature (°C)**	Maximum radiated heat flux (kW/m^2)	Remarks	Recommended distance from apartment door
Routine	25	100	1	General fire-fighting	15-30m
Hazardous	10	120	3	Short exposure with thermal radiation	4-15m
Extreme	1	160	4 – 4.5	For example, snatch rescue scenario	2-4m
Critical	<1	>235	>10	Considered life threatening	0-2m

** Measured at a height of 1500mm from FFL

Table 19.9: SCA guidance on firefighter tenability using CFD computer-aided analysis to take spot-temperatures readings to demonstrate the corridor temperatures on the corridor approach side (from the stair) of a flat door.

212 Grimwood. P; The hazards of mechanical smoke ventilation systems (MSVS) impacting on under-ventilated fires in residential buildings with extended corridors; EuroFire, Paris, 2011

The requirement is now for smoke design engineers to demonstrate how their system will meet the requirements in Table 19.9 in terms of spot temperatures recorded, using graphic CFD slice images (Fig.19.11). The guidance recommends how these maximum temperatures and heat fluxes should be shown at 0-2 and 2-4 metres on the firefighting approach side of the compartment door (Fig.19.10). In meeting these maximum temperatures and fluxes, it may be necessary to increase air flow (m³/s) in the firefighting mode, particularly if extracting towards the firefighting approach. A typical system configuration might see a 5m³/s extract rate in firefighting mode, based on a 4 MW time dependant fire.

*(Note: a growing **time dependent** design fire design of 4-6 MW should be used, according to SCA design guidance, where the corridor travel distance is in excess of 15 metres in length (from furthest flat door to stair door – A **steady state** design fire <2.5 MW may be applicable to corridors with less than 15 metres travel distance).*

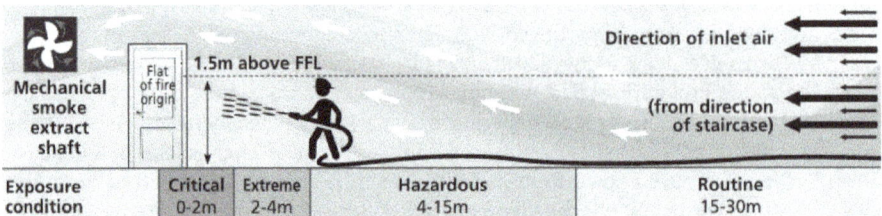

Figure 19.10: s5.3.3 in the SCA design guidance on firefighter tenability recommends using CFD computer-aided analysis to take spot-temperatures readings to demonstrate the corridor temperatures on the corridor approach side (from the stair) of a flat door.

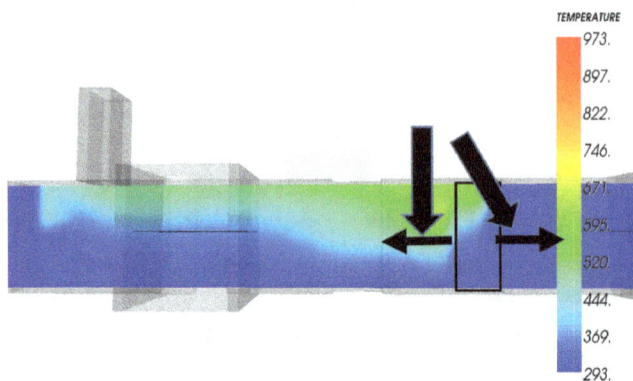

Figure 19.11: A typical CFD slice of a 30-metre dead end corridor with smoke being extracted via the smoke-shaft on the left at a rate of 5m³/s (firefighting mode), causing 'make up' air to flow into the corridor from the firefighting stair at the right-hand side of the image. It is demonstrated that temperatures along the firefighting approach side of the flat door (at 1.5 metres from the floor) are only 23 deg. C (The scale is in Kelvin) right up to within 2 metres of the door, whilst corridor temperatures on the extract side of the door are at 322 deg. C within 2 metres of the flat door reducing to 227 deg. C at 4 metres. Therefore, if airflows were configured towards the stair, firefighters would be approaching untenable conditions. **Image courtesy of FDS Consult, Fire Engineering, Dartford in Kent UK**

Smoke Control – System Categories and Configurations • 409

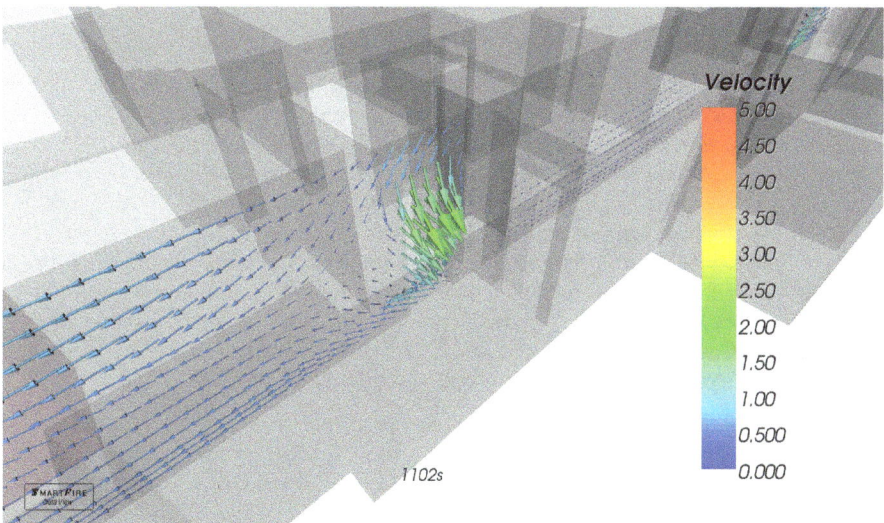

Figure 19.12: A typical slice from a CFD analysis with velocity arrows denoting the speed and direction of air-flow in the corridor as smoke exits from the flat door.

Note: At the time of going to print with this book, the Smoke Control Association (SCA) are producing similar guidance for the application of smoke management proposals that may enable single stair offices to surpass building height, occupancy and travel distance limitations existing in current prescriptive guidance. However, the regulatory authorities are advised to ensure that such measures are carefully reviewed in accordance with documents such as BS9999 and BS 7974, as well as ADB.

19.4 SMOKE CONTROL USING PRESSURE DIFFERENTIALS

Smoke control using pressure differentials is implemented through several different classifications of systems, with differing requirements and design conditions. The design conditions have been placed in separate system classes which may be used to implement a design using pressure differentials for any given type of building.

The two basic requirements for a pressure differential system are:

- A mechanically driven supply of fresh air into each protected space, to maintain the pressure at a level higher than that in the fire zone;
- A leakage path from the accommodation to the external air, to maintain a continuous flow out of the protected space, preventing pressure equalization. Depending upon the design it may also be necessary to provide for overpressure relief from the protected space

> **BS EN 12101 – Design objectives for pressurisation systems**
>
> *According to the BS EN standard (12101) a pressure differential system is one that uses a system of fans, ducts, vents, and other features provided for the purpose of creating a lower pressure in the fire zone than in the protected space – It is specifically intended to cover the protection of stairwells, lobbies and corridors that form part of a protected escape route or firefighting shaft. The aim is to establish a pressure differential across any leakage paths that will ensure that smoke moves away from the protected space. This is achieved by maintaining the protected space at a pressure higher than that of the fire zone. It is essential that adequate air release shall be provided from the accommodation to ensure that a pressure differential is maintained. The velocity of hot smoke and gases from a fully developed fire could reach 5 m/s and under these conditions.*
>
> *It would be impractical to provide sufficient through-flow of air wholly to prevent ingress of smoke into the lobby. It is assumed that firefighting operations, such as the use of spray, contribute significantly to the holding back of hot smoky gases. It is, however, essential that the staircase itself be kept clear of serious smoke contamination.*

The 12101 standard specifies pressure differential systems designed to hold back smoke at a leaky physical barrier in a building, such as a door (either open or closed) or other similarly restricted openings. It covers methods for calculating the parameters of pressure differential smoke control systems as part of the design procedure. It gives test procedures for the systems used, as well as describing relevant, and critical, features of the installation and commissioning procedures needed to implement the calculated design in a building. It covers systems intended to protect means of escape such as stairwells, corridors and lobbies, as well as systems intended to provide a protected firefighting bridgehead for the Fire Services.

> **What does this mean to the Firefighter?**
>
> Smoke control by pressurisation is designed to ensure that, according to design standards, the smoke (flow-path) moves away from the 'protected space'. Therefore, if the protected space is a stair then smoke should move away from the stair. However, some firefighting lobbies or corridors may configure a natural or mechanical ventilation system where smoke (flow-path) moves towards the stair or lobby. However, as long as the smoke does not move into the stair but leaves the lobby/corridor via smoke shaft or external wall opening, this is regulatory compliant. This type of design can still create problems for firefighters as the forced flow-path can place firefighters on the wrong side of the flow path and directly in the hot zone. The 2015 SCA guide has considered this problem of firefighting access and deals with directly through recommended design guidance at section 5.4 of their document (see 19.3 above).

SYSTEM CLASS	EXAMPLES OF USE	SPECIFICATIONS
CLASS A SYSTEM	Means of Escape Defend in place Apartment building	• The airflow through the doorway between the pressurised stair and the lobby/corridor shall not be less than 0.75 m/s. • The pressure difference across a closed door between the pressurised stair and the lobby/corridor shall not be less than 50 ± 10 Pa. • The open door can indicate an open flow path through a simple lobby.
CLASS B SYSTEM	Means of escape & firefighting **Firefighting shaft**	• The air supply shall be sufficient to maintain a minimum airflow of 2 m/s. • The pressure difference across a closed door between the pressurised stair and the lobby/corridor shall not be less than 50 ± 10 Pa. • The open door can indicate an open flow path through a simple lobby.
CLASS C SYSTEM	Means of escape Simultaneous evacuation Offices	The airflow velocity through the doorway between the pressurised space and the accommodation shall be not less than 0.75 m/s. • The pressure difference across a closed door between the pressurised space and the accommodation area shall not be less than 50 ± 10 Pa. • In the event of simultaneous evacuation it is assumed that in the stairways, some smoke leakage can be tolerated. The airflow due to the pressurisation system shall clear the stairway of this smoke.
CLASS D SYSTEM	Means of escape Sleeping risk Hotels	• The airflow velocity through the doorway between the pressurised space and the accommodation shall be not less than 0.75 m/s. • The pressure difference across the door between the pressurised space and the accommodation area on the fire storey shall not be less than 50 ± 10 Pa.
CLASS E SYSTEM	Means of escape Phased evacuation Hospitals	The airflow velocity through the doorway between the pressurised space and the accommodation shall be not less than 0.75 m/s. • The pressure difference across the door between the pressurised space and the accommodation area on the fire storey shall not be less than 50 ± 10 Pa.

Table 19.10: BS EN 12101 system options

19.5 PRESSURISATION OF ESCAPE ROUTES, STAIRS AND STACK EFFECT

Stack effect and Neutral Pressure Plane (NPP)

The natural buoyancy of air density causes an upward flow within the building such as in stairwells and elevator shafts, if average indoor temperatures are a few degrees higher than the exterior. In case of a lower temperature outside the building in relation to the inside of the building, the air inside the building has a lower density than the air outside the building. This density difference results in a pressure differential between inside and outside. This is known as (normal) ***stack effect*** or chimney effect. However, if the air is hot outside, a downward airflow occurs in an air-conditioned building, which is called reverse stack effect. For both normal and reverse stack effect situations, a horizontal plane normally exists where the pressure inside a building is equal to that of the exterior; which is regarded as the neutral pressure plane (NPP). Furthermore, no horizontal flow occurs at the neutral plane due to zero pressure difference between inside and outside of a building. Determination of the location of the neutral plane is critical in order to evaluate the flow due to stack effect. Analytic equations have been developed to locate the neutral plane for certain simple cases, e.g., leakage openings. However, the combinations of leakage openings in a buildings shafts and compartments are certainly more complex and require in-depth attention using computer models.

The ***Neutral Pressure Plane (NPP)*** is defined as the locus of points on the building envelope at which the indoor-to-outdoor pressure differential is zero. Since the stack effect is often the dominant driving force for air leakage, the NPP is typically assumed to be located along a horizontal plane at the mid-height of the structure – at least under winter conditions and in the absence of wind and mechanically induced pressure differentials. In practice, the NPP does not have to be horizontal (since the driving forces vary over the building's surface), nor at the mid-height of the building (since this assumes an equal distribution of leakage pathways over the building's height), nor is its location static (since the driving forces can vary on a second-by-second basis). In fact, the location of the NPP is highly variable and a building can even have several Neutral Pressure Planes.

Swedish fire engineer **Anders Dagneryd** is one of a growing number of fire engineers who is of the opinion that smoke control in tall buildings may be impacted so greatly by stack effect that pressure differential systems based on smoke extraction from the stair lobby or corridor serving the fire apartment, using a smoke shaft, may not be at all effective. He proposes that positive pressurisation, may be necessary for buildings in excess of 30 metres high.

19.6 PRESSURE DIFFERENTIAL SYSTEMS IN VERY TALL BUILDINGS >30 METRES

Anders Dagneryd writes ....

> 'In the UK, the commonly referenced BS standards (for example BS 9999) suggests a height limit where a *more capable* smoke control system shall replace the *less capable* smoke ventilation system. This limit is 30 m for commercial buildings. For residential buildings, there is no such limit defined in the code and accordingly it is the responsibility to develop these limitations in relation to the functional requirements in the Building Regulations 2010.

Such commercial buildings in general demonstrate a high occupant load and accordingly a higher risk. Therefore, a BS EN 12101-6:2005 standard compliant system is recommended for a building higher than 30 metres. This is a robust system with the capability to mitigate any risk of combustion products entering the staircase enclosure. The longer persons have to remain in a building affected by fire the higher is the risk of being exposed to hazardous conditions. The performance of systems and compartmentation tends to deteriorate when exposed to fire and smoke tends to migrate to other areas of the building, outside of the fire compartment. There is research that suggests it is therefore advisable the building shall be evacuated before such systems reach their maximum resistance to fire by a safety factor of 2, in relation to the expected evacuation time. When the evacuation time is calculated, it is also suggested to use a conservative travel speed, research has indicated the travel speed will significantly reduce when required to negotiate a high stair.

The price for floor-plate area and office space is continually increasing. Accordingly, the buildings are growing taller in order to generate a higher return per square metre of the floor plate, and the requirements to efficiently use every square meter for leasable space is increasing. We are now creating vertical communities with five figure occupant loads and evacuation times from one hour up to several hours are predicted. During evacuation, there could be thousands of persons in the stair case enclosure above a fire floor totally reliant on the stair to be protected from smoke entrainment.

For these very tall buildings, when a number of persons equal to a small city may be required to evacuate a building, we have to understand that the occupants may be exposed to high levels of physical exertion where people are likely to get injured. A malfunctioning or poorly designed smoke control system has the potential to expose persons to harm or death. As with the stair case the lift shaft is also a vertical pathway that has significant capacity to transport and spread smoke and combustion products in a building. The lift shaft is also a system that may be used to evacuate occupants and to effectively transport fire fighters and their heavy equipment nearer to the fire floor. With this the importance to protect the lift shaft is justified. The lift shaft has significantly less friction than a stair case and the sliding doors have a significantly higher leakage rate than the smoke proof doors protecting a stair shaft enclosure.

This makes the lift shaft a highway for the transport of dangerous combustion products. In such tall buildings evacuation lifts and firefighting lifts must be pressurised and this requires significantly higher air-flows to pressurise the shafts than for a stair case enclosure.

In a performance-based design of a smoke control system the following parameters should be considered:

- Building height and geometry
- All flow-paths such as leakages, façade openings, doors, shafts, rooms, ducts, void dampers etc.
- Minimum and maximum design pressure differentials
- Maximum and minimum building temperatures
- Stairwell temperature
- Pressure differential per floor in the stairs
- Design maximum and minimum outdoor temperatures
- Wind effect
- Smoke feedback via external intakes to the interior of the building

The design shall consider the protection of the stair case enclosure and lift shaft intended for fire service access and elevators intended for the evacuation of people in the building.

Where necessary the system must be able to mitigate the transfer of smoke from areas of the building affected by fire to non-affected areas. For buildings that could not be easily evacuated within a short time frame the lifts could be used for evacuation. This will require all lift shafts intended to be used for evacuation to be protected by a smoke control system.

Pressurization Systems

The intention with a shaft pressurization system is to maintain an overpressure in the shaft, for example a stair shaft or lift shaft, in relation to a firefighting lobby to mitigate smoke entrainment. In the UK, there are two methods open to the designer; (a) design the system in accordance with BS EN 12101-6, or; (b) design the system in accordance with the functional requirements of Document B of the Building Regulations 2010.

Whenever a pressure difference exists over an *air flow path* occurring at any leakage path, a flow is induced which is exponentially proportional to the pressure difference. Consequently, the stack effect results in an inflow into the shaft/building below the neutral pressure level and an outflow above'.

In the design of a pressure differential smoke control system to protect a fire fighting shaft the system can either be designed in accordance with the criteria provided in BS EN 12101-6:2005 or designed in accordance with the functional requirements of the Building Regulations 2010.

For smoke control in staircase enclosures there are in general three types of pressure differential systems:

- Pressurization
- Depressurization
- Combined pressurization and natural ventilation (smoke flushing, or dilution)

Effect of stack and wind effect on a Vertical Shaft

1. Stack effect

The temperature of air is proportional to the density of the air. In case of lower temperature outside the building in relation to the inside of the building, the air inside the building has a lower density than the air outside the building. This density difference between a shaft and the outside ambient air generates a pressure differential between the shafts and outside, refer to Fig.19.13.

In cases of summer conditions, the outside air can be warmer than the inside air, resulting in an inside air density higher than the outside air density, refer to Fig.19.13. This situation is commonly referred to as the reverse stack effect. Reversed stack effect has less impact than the normal stack effect in buildings in the UK, but should still be considered in the design.

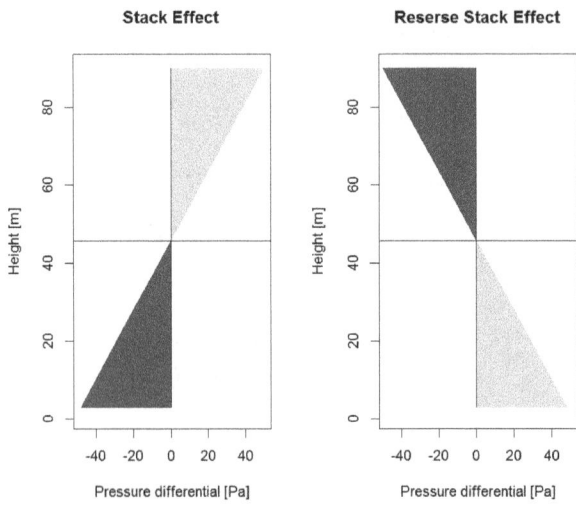

Figure 19.13 – Stack effect and reverse stack effect

The pressure distribution above, with the neutral plane in the middle of the building, only occurs when the shaft leakages are evenly distributed over the height of the building. The neutral plane represents the level at which the shaft the pressure will be the same as outside ambient pressure, and is represented by the horizontal line in the middle of the chart. An increased pressure differential across a flow-path generates an increased flow, refer to equation 19.1. By definition, the flow is always from a zone with higher pressure to a zone with lower pressure. Accordingly, air will normally flow from the outside of the building into the shaft on all levels below the neutral plane. On all levels above the neutral plane there will be an outwards flow.

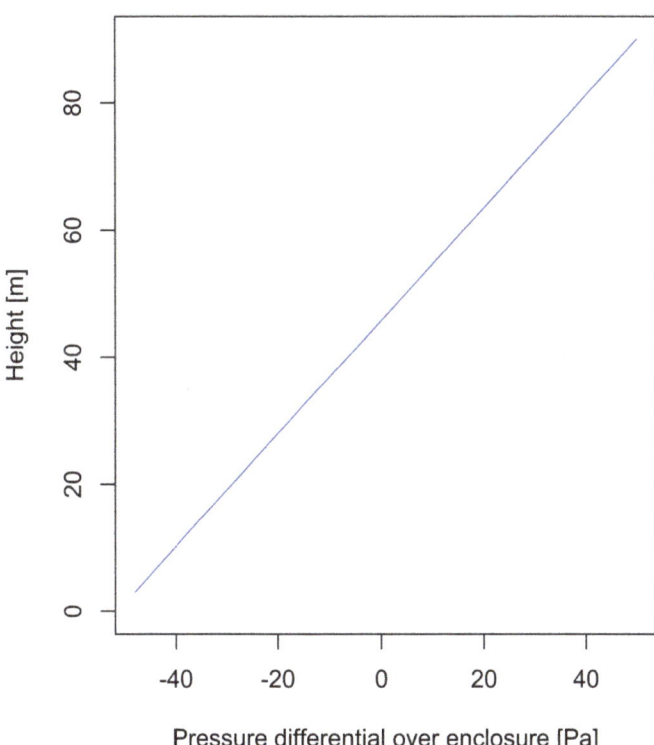

Figure 19.14: A representation of the output from a CONTAM[213] simulation describing the pressure differential in relation to outside ambient air occurring in a vertical shaft with a temperature of 23 degree Celsius and with an outdoor temperature of minus 3 degree Celsius. The shaft is 90 m in height and has 30 storeys connecting to the shaft. There are no openings in the shaft.

213 CONTAM is a free download of multi-zone airflow and contaminant transport analysis software produced by NIST USA

If a vent or other opening is created at the top of the shaft, the neutral layer will move upwards towards the opening at the head of the shaft. In this case, with no leakages except for the opening at the head of the shaft, the neutral layer will move all the way up to the top of the stair, refer to 19.15. Accordingly, there will be an inflow of air on every level below the neutral plane and an outflow at the opening at the top of the building, refer to figure 19.15. For stair-cases up to 30 m this generates a negative pressure well suited for ventilating any smoke that has entered the staircase enclosure, though caution shall be taken not to extract the fire out in the staircase enclosure due to the negative pressure in the stair shaft enclosure.

It is not recommended to ventilate a stair via an opening at the head of the stair for buildings higher than 30 meter as this could generate a significant risk. With the increased pressure differential between the staircase enclosure and the fire compartment the risk to extract smoke and fire out in the stair shaft enclosure increases. Extracting the fire out into the stair shaft enclosure is a significantly dangerous scenario, and accordingly it is suggested not to install a vent at the top of a staircase enclosure higher than 30 m above ground.

Figure 19.15 shows the output of a CONTAM simulation of a 90m high building with 30 storeys, with an opening at the head of the stair, minus 3 degree Celsius outside temperature and 23 degree indoor temperatures. Without influence from any wind the simulation indicates a negative pressure of 100 Pa in the lower parts of the shaft. A negative pressure of 100 P can easily pull unlatched doors open, such as fire doors into a stairway or firefighting shaft. It may also cause negative pressures high enough to make apartment doors difficult to open, which could obstruct the means of escape.

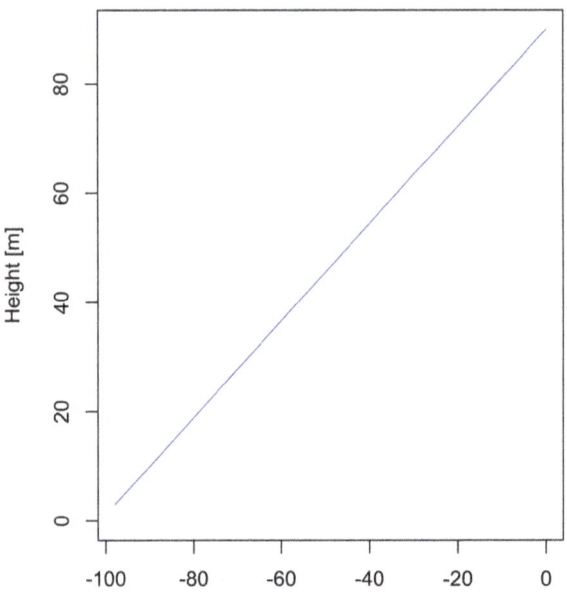

Figure 19.15: Stack effect with opening located at the top of the shaft enclosure

If a vent or other opening is opened at the lowest part of a shaft without any other openings, the neutral layer will move all the way down to the lowest level of the shaft, refer 19.16. The CONTAM simulation that is presented in Figure 19.16 indicates a positive pressure differential over the stair shaft enclosure of approximately 100 Pa which is above the level when a door in is considered to be easily opened, which could obstruct the means of escape.

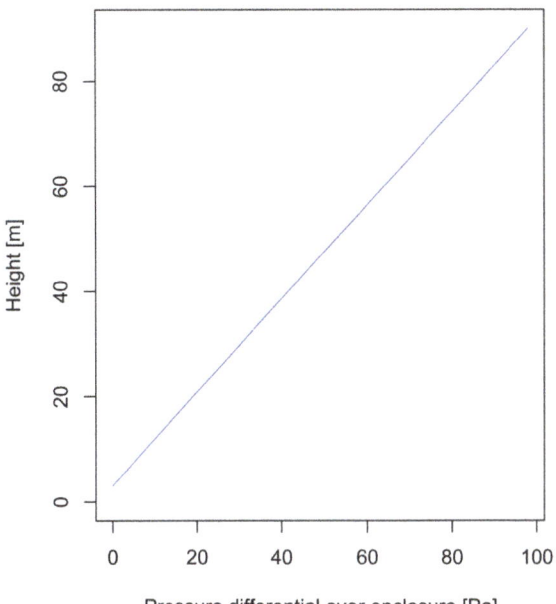

Figure 19.16 – Stack effect with an opening located at the base of the shaft enclosure

2. The effect of wind

If a vent or other opening is created at the top of the shaft, this opening will be affected by the wind. If the opening is horizontal only negative pressures will occur due to wind over the opening. If the opening is vertical, both a negative pressure could occur due to wind past the opening or a positive pressure due to wind towards the opening.

It must also be considered that wind could also affect other openings in the building[214], for example an open or broken window in the fire compartment. This could generate a positive or negative pressure differential over the opening that must also be accounted for in the design of the smoke control system.

Figure 19.17 represents a 90m high building with 30 storeys, no stack effect is assumed and the horizontal ventilation opening at the top of the stair is affected by the wind. In this case a 9.4 m/s design wind was assumed on a horizontal opening. The horizontal opening was assumed to be exposed to a pressure coefficient of -1. Normally a pressure coefficient of -0.7 to -2.5 can be assumed depending on the design of the roof. With the suggested input the wind will generated a negative pressure on the ventilation opening of -56 Pa.

214 BS EN 1991-1-4:2005

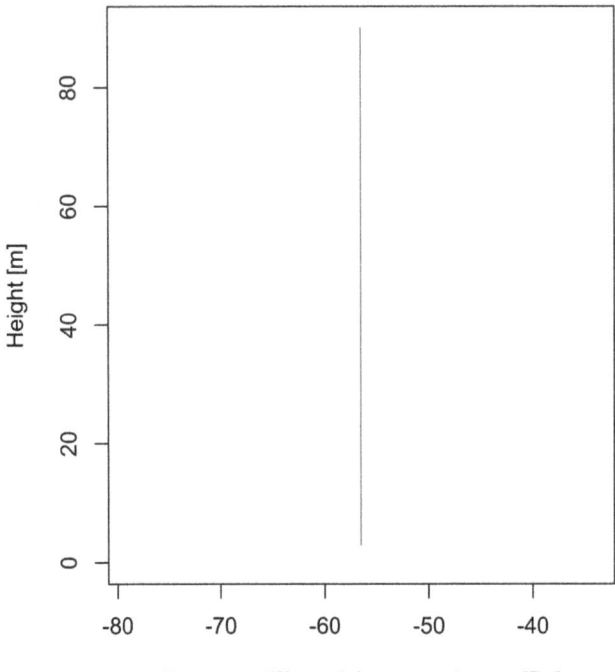

Figure 19.17: Wind effect with opening located at the top of the shaft enclosure

3. The effect of a combination of wind and stack effect

If a vent or other opening is opened at the top of the shaft and the shaft is exposed to both the effect of wind and stack the neutral layer will no longer be located inside the building. Figure 19.18 shows the result of a combination of wind and stack effect in accordance with the previous scenarios. A negative pressure of approximately 150 Pa is generated in the staircase even with a moderate pressure coefficient.

The conclusion from the exercise above is that it is counter-productive, and could become seriously dangerous to the occupants of the building, to install openable vents at the top of a staircase in a building with a height exceeding 30 m above ground.

There will always be leakages in a stair shaft enclosure and it is likely doors will open and close. When there is more than one opening in a shaft it becomes more difficult to predict the location of the neutral plane. As soon as more than one leakage point is introduced a flow will occur that which will generate pressure differentials due to friction.

Figures 19.15 and 19.16 describe a scenario with both a high level and low level opening in the shaft.

Wind and Stack effect, High level opening

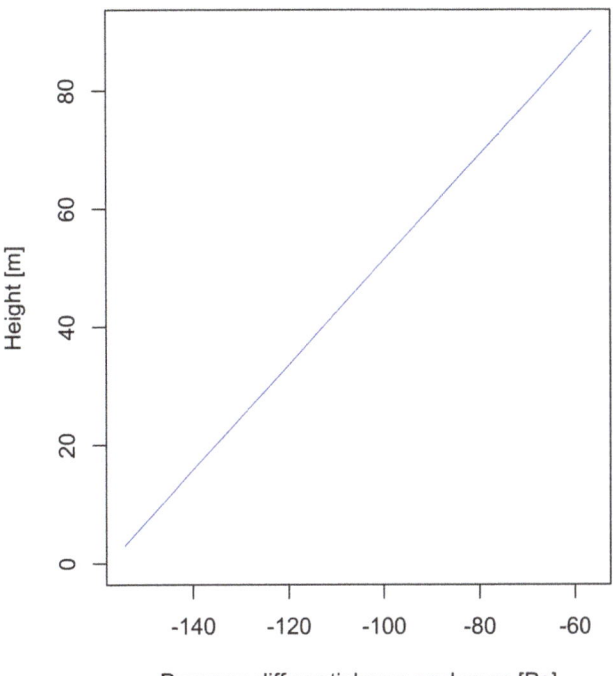

Figure 19.18: Wind and stack effect with opening located at the top of the shaft enclosure

> A general discussion on stack effect assumes the temperature to be uniform throughout the entire structural space. By this assumption, Klote deduced a formula (ASHRAE Handbook of Smoke Control Engineering) for calculating the neutral plane, demonstrating that the height of the neutral plane (NPP) is less than half of the building's space height, and that the vertical temperature is not important to the neutral plane. However, due to strong air-entrainment and fire source, the location of the neutral plane could be changed and even possibly raised above the mid-height of the building. Similarly, smoke that should find its way into a stair or lift shaft may stratify before reaching the top of any shaft and collect at lower levels. Therefore, a detailed evaluation on the location of any neutral plane under a building fire scenario may also be of particular design relevance.

In Figure 19.19 the opening at the top of the shaft has area that is twice as large as the area of the opening at the lower level of the shaft. In 19.20 the opening at the lower part of the shaft has an area that is twice as large as the area of the opening at the top of the shaft.

In a simulation of the pressure differentials a building it is assumed the mass of the inflow and out flow must be equal. The pressure differential over an opening can be described by the orifice equation from ASHRAE, Equation 3.3, refer to Equation 19.1.

$$m = CA\sqrt{2\rho\Delta p} \qquad \text{Eq.19.1}$$

Where:

m = Mass flow [kg/s]

A = Geometric area [m²]

C = Flow coefficient for opening

ρ = Density of at the specific temperature [kg/m³]

Δp = Pressure differential over opening [Pa]

In accordance with **Eq. 19.1** the pressure differential increases exponentially with increased air velocity. In this case with the same mass flow over both openings the higher-pressure differential will occur over the smallest opening, which will move the neutral plane in the direction of the largest opening. Accordingly, the highest-pressure differential will occur over the smallest opening.

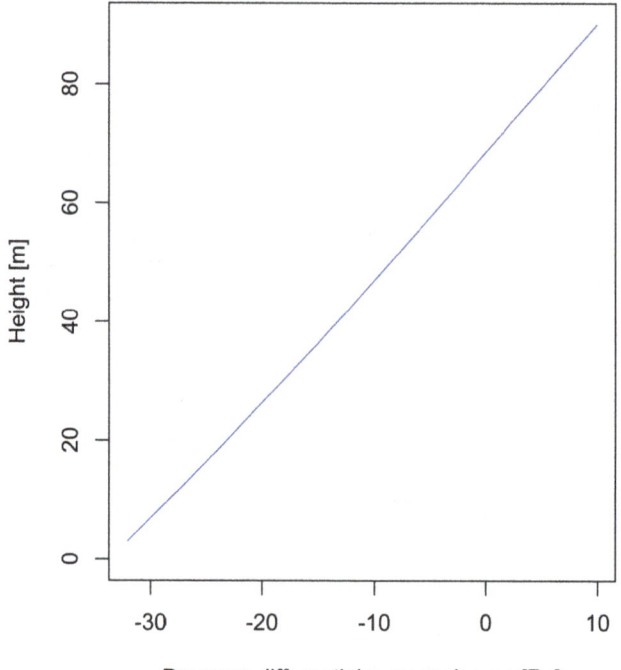

Figure 19.19: Stack effect with larger opening at high level and smaller opening at lower level

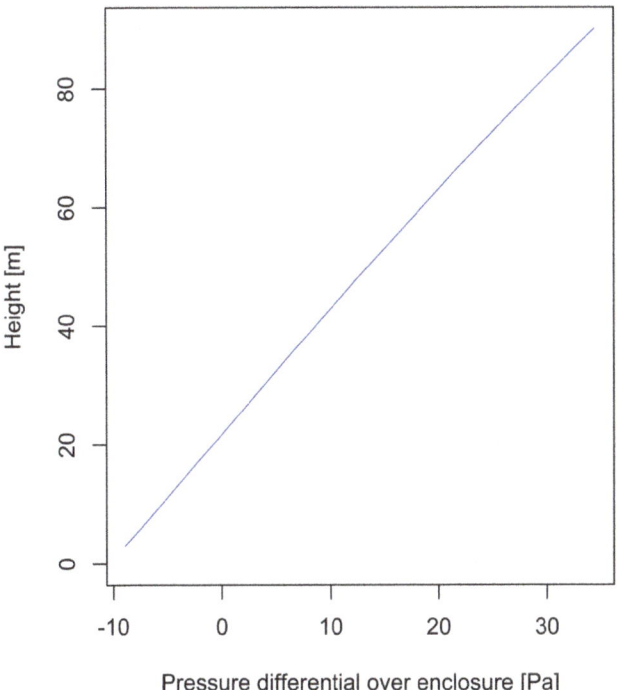

Figure 19.20: Stack effect with smaller opening at high level and larger opening at lower level

Pressurization of a shaft in accordance with BS EN 12101-6:2005

The intention with shaft pressurization is to create a positive pressure differential such that there will always be, at all levels in the building and under all likely conditions, a positive pressure. This is to mitigate the over pressures generated by local stack effect in the fire enclosure or firefighting lobby. If an over pressure (that is higher than the local over pressure generated by the fire or wind towards the facade opening of the fire enclosure) exists at all levels of the building in relation to the lobby the risk to extract the fire out in the staircase enclosure is eliminated. The shaft will no longer perform as a chimney that could extract the fire out into the staircase enclosure, with the risks associated with this. The system will also be able to mitigate smoke entrainment out in the staircase.

A pressurised shaft does not have any direct openings to the outside of the building and the shaft is accordingly not directly exposed to the effect of wind. Effect of wind will still occur as a direct pressure differential over openings in the façade. Most critical are openings in the fire compartment that are intentionally opened or have been created due to the effect of fire. A positive pressure on the façade has the potential to push smoke into shafts. The effect of stack effect will always affect the staircase as long as there is a temperature difference between the shaft and outside the building.

Figure 19.21 shows the output from a simulation of a 90-meter high building with a pressurization system in accordance with BS EN 12101-6:2005. The building in the

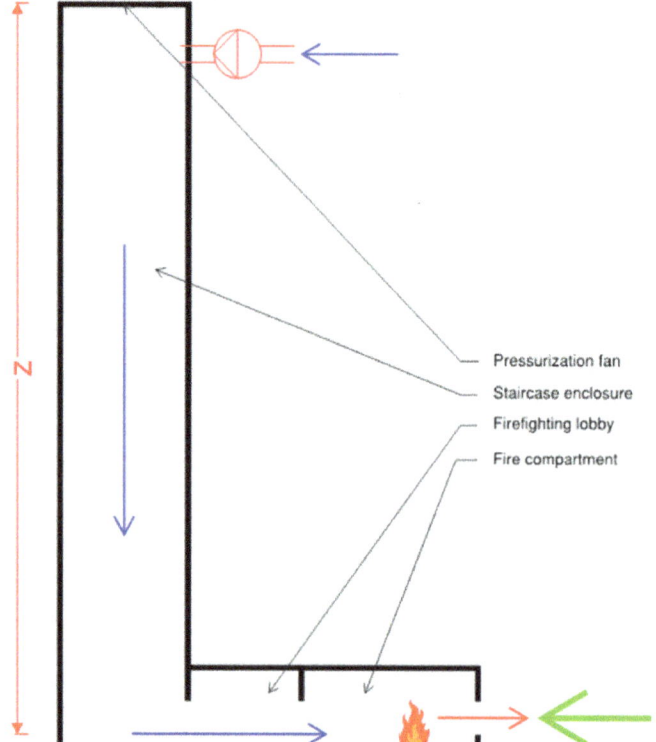

Figure 19.21: Pressurization system – BS EN 12101 – 30 storey building

simulation has 30-storeys, the shaft temperature is 23 degree Celsius and the outside temperature is minus 3 degree. On the lowest level of the building a minimum 50 Pa overpressure will be generated in relation to the outside of the building. With this design the pressure differentials increases with the height of the building.

From the output of the simulation, at the top of the stair shaft enclosure the pressure is approximately 150 Pa which is significantly above the level where a door could be easily opened. Accordingly, the design does not meet the 100 Newton opening force of a door due to the stack effect. This indicates there is a maximum shaft height for a stair-shaft pressurised strictly in accordance with BS EN 12101-6:2005, without any pressure relief at the top of the building.

Limitations

Pressurization systems are designed to operate within a pressure differential range. The limitations for the pressure differential system is the minimum and maximum pressure differential over the shaft enclosure. This defines the limitations for the height of the staircase. As the building is exposed to stack effect due to the temperature difference between the shaft and the outside ambient air of the building, the pressure will vary over the height of the building. In ASHRAE – Handbook of Smoke Control Engineering 2012[215]

---

215 American Society of Heating, Refrigerating, and Air Conditioning Engineers (**ASHRAE**)

Stack effect, Pressurization

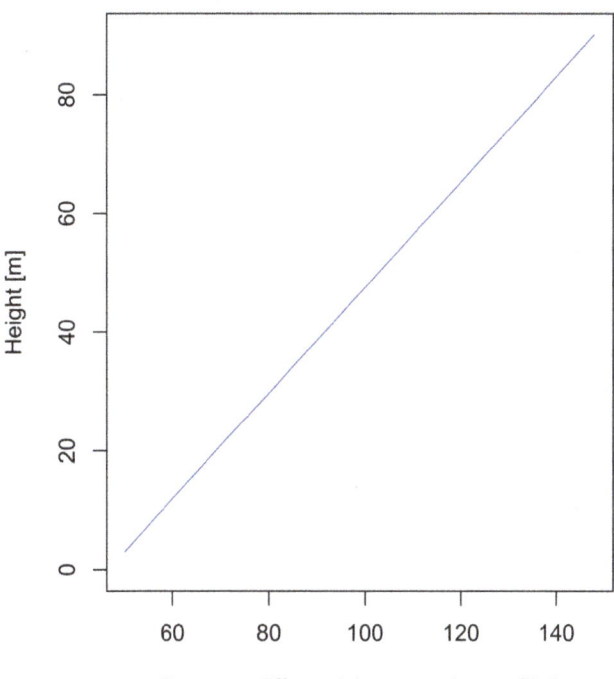

Figure 19.22: Stack effect in pressurised shaft enclosure

an outside design temperature of -3.1 degree Celsius is suggested for the lowest likely condition and 28.3 degree Celsius for the highest likely conditions. These temperatures represent the dry bulb temperatures and cover 99.6 % of all occurrences on the reference location.

With this information, the maximal pressure difference from shaft to the outdoors can be calculated by the use of Eq. 19.2, *refer to ASHRAE - Handbook of Smoke Control Engineering 2012, equation 3.34.*

$$\Delta p_{so} = 3460 \left(\frac{1}{T_o} - \frac{1}{T_s} \right) z \qquad \text{Eq.19.2}$$

Where

Δp_{so} = The pressure differential from shaft to the outdoors (Pa)

T_s = The absolute temperature in the shaft (K)

T_o = The absolute outdoor temperature (K)

z = The distance above the neutral layer (m)

In the UK, the maximum temperature difference is generated by the winter temperature and accordingly the summer temperature will not be the limiting factor for the maximum height of the shaft enclosure. This equation represents worst conditions with no leakages between the staircase and corridor in a residential building, and significant leakages from the corridor or lobby to the outside ambient air of the building. The pressure differential assumes one open door on ground floor. This pressure differential will have the most significant effect on the highest located door in the stair shaft enclosure. Doors located in the means of escape paths must be openable at all material times in case of fire. BS EN 12101-6: 2005 suggests a maximum force to open the door applied at the door handle shall not exceed 100 Newtons (N) when the door is shut. NFPA 101-2015 require a maximum door opening force of 133 N applied to the door handle.

The requirements for a 133 N door opening force in NFPA 101-2015 are less restrictive in relation to maximum allowed pressure differential and door opening force than the requirements in BS EN 12101-6:2005. As a consequence, a higher stair shaft can be accepted with NFPA with normal pressurization and the design margins increases. On the other hand, 33 % more force could be required to open a door if the building is designed in accordance with the recommendations in NFPA 101:2015 compared to BS EN 12101-6:2005. As NFPA 101-2015 is a well-recognised code applied in a number of countries and updated on a regular basis it is reasonable to argue the 133 N from NFPA 101 could be used as the design criteria but this require approval as this is a deviation from the requirements in the standard.

For the design, it has been assumed that:

- The width of the door leaf is 900 mm and the height is 2100 mm
- The door handle is located 825 mm from the vertical lines that intersects the hinges
- The force to overcome the door closing device and friction is 25 N applied at the handle
- No friction is assumed for the door
- The minimum pressure differential in the staircase shall be 50 Pa

A moment equation generates a maximum allowed pressure differential over the door of 23 Pa with the 100 N in accordance with BS EN 12101-6 and 55 Pa with the 133 N in accordance with NFPA 101-2015.

This indicates that a building located in London (using Heathrow airport reference temperatures) with a stair shaft temperature of 23 degree Celsius, an outdoor ambient temperature of -3 degrees and with the door input as indicated above, a pressurization system designed in accordance with the prescriptive methodology in BS EN 12101-6:2005 –

- Can have a maximum shaft height of 20 m with a maximum allowed door opening force of 100 N, or;
- Can have a maximum shaft height of 49 m with a maximum allowed door opening force of 133 N, according to NFPA 101-2015.

If a higher shaft is required, a performance based design using a node network model (for example PFS or CONTAM) is required. With a node network model, it is possible to design systems with an unlimited theoretical height. For further information refer to BS EN 12101-6:2005.

Performance based Design of a Pressurization Shaft

With a performance-based design of a pressurization system the low-pressure sections of the staircase enclosure are pressurised with a pressure control fan to 50 to 75 Pa and in the high-pressure parts of the staircase, the pressure is relieved via a pressure sensor controlled damper to the outside, or a pressure controlled extract fan reduces the pressure to 50 to 75 Pa. If the building is of such height that the capacity of the supply and extract system cannot maintain the pressure in the shaft between 50 to 100 Pa by simply providing air supply at the top and bottom of the shaft, additional pressure controlled intermediate fans must be provided in the shaft. The difficulty with more than one independently pressure controlled fans is the risk of chasing which requires sophisticated control systems to control the fans. It must therefore be understood these systems are qualified life safety systems requiring significant amount of maintenance from qualified personnel.

Figure 19.23 indicates the intention with a performance based shaft pressurization system. The idea with the system is to mitigate the effect of stack by inducing a flow in the shaft such the friction in the staircase will be equal to the effect of stack. In theory, this system can handle an unlimited height of the shaft, but in practice environmental effects are likely to limit the reasonable height of the open shaft. As this type of system to a significant extent is depending on the friction of the shaft, the pressurization of a lift shaft will require significantly higher flow than is required for the same type of pressurization of a stair case enclosure.

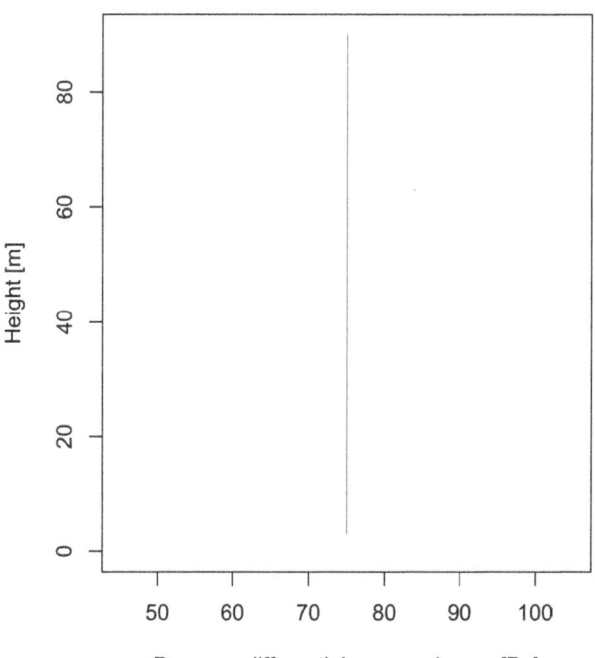

Figure 19.23: Stack effect in pressurised shaft enclosure designed with performance based methodology

Depressurization Systems

With a depressurization system, the fire compartment is depressurized in relation to the staircase enclosure to maintain a pressure differential in relation to the staircase (Figure 19.24). The problem with the depressurisation system is that only limited leakages are allowed for the system to perform properly. This requires the façade to be fire rated or for the fire enclosure to be located underground, this to mitigate inflow of ambient air, refer to BS EN 12101-6:2005, Section 9.1. This limits the use of a depressurization system.

It is also required to have an extract system that can handle likely temperatures in the fire compartment. As the extract system is a life safety system and not a ventilation system it is required to be able to handle all likely temperatures even a fully developed fire in the fire enclosure following a sprinkler failure (refer to PD 7974). The benefit with the system is it is only required to maintain a design pressure differential between the fire enclosure and the staircase and accordingly only experiences limited influence of stack effect and wind.

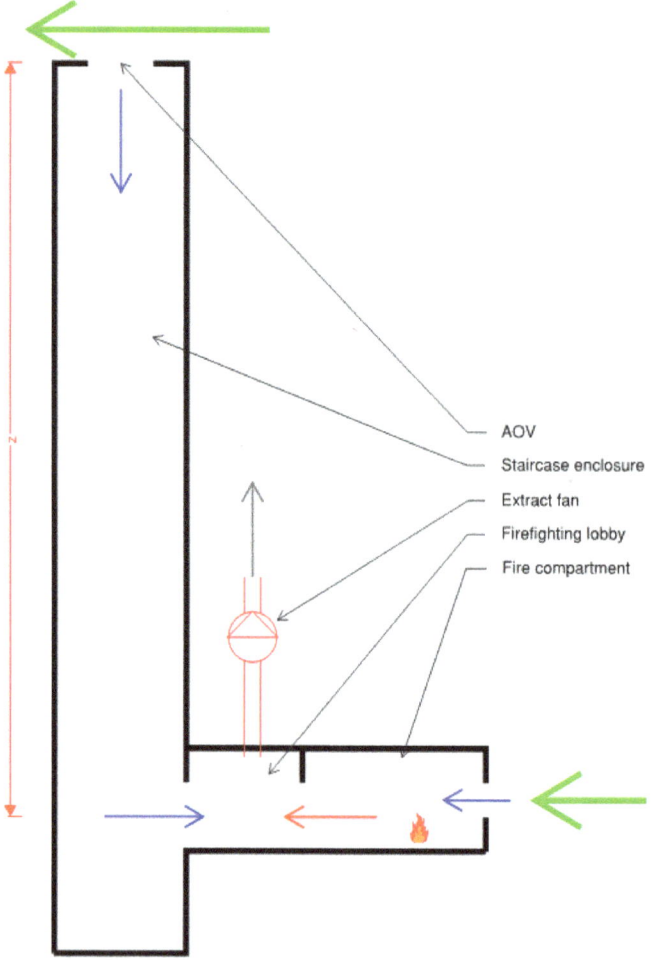

Figure 19.24: Depressurisation system

Lobby Ventilation System

Lobby ventilation systems according to BS 9999:2008 are allowed in lieu of a smoke shaft up to a building height of 30 m. A lobby ventilation system with makeup air via an AOV at the top of the staircase exposes the staircase enclosure to the effect of wind, and stack effect. From Figures 19.15, 16 and 17, it has been clearly indicated that makeup air via an AOV at the top of the building has clear disadvantages. A lobby ventilation system is just a ventilation system and not a smoke control system able to sufficiently protect a shaft in higher building. The standards BS 9999:2008 and BS 9991:2015 suggest the system should not be used for buildings higher than 30 m, but the technical maximum height of the protected shaft depends on the building design, minimum outdoor temperatures, indoor temperatures, and the buildings exposure to wind.

The following example simulated in CONTAM is intended to explain the limitation with this system. The example represents a 90-metre high building, significantly above the 30 metres when a lobby ventilation system is recommended in the standards. This demonstrates the risk when attempting to extrapolate calculations for a lobby ventilation system above the recommended 30 m height. It should be noted; this calculation is only an example with the intention to highlight problems that could occur and could not be used as a general statement. The simulation for any other example will have different output as even minor alternations can have significant impact on the output of the simulation. Information and justification of input, acceptance criteria and methodology has been limited in this chapter. For further information refer to ASHRAE – Handbook of Smoke Control Engineering 2012 and BS EN 1991-1-4 2005.

Building description:

- Building height – 90 m
- Shaft height – 90 m
- 30 story building with the fire floor on Level 1
- A horizontal AOV is located on the roof, with an area of 1.5 m^2 and a flow coefficient of 0.6
- The area of each door is 2m^2 with a flow coefficient of 0.8
- Façade opening between the fire room and ambient has an area of 10 m^2
- For the environmental design, it is assumed the building is located in central London
- Without further justification one door with the leakage area of 0.01 m^2 is located at each level and has a flow coefficient of 0.6

Conditions

- The shaft temperature is 23 degree Celsius
- Ambient temperature is -3 degree Celsius
- A mean temperature in the fire enclosure of 600 degree Celsius
- The reference wind is 10.1 m/s, with a velocity of 9.35 m/s at 90 m above ground
- A pressure coefficient of -1 is assumed for the wind pressure over the AOV
- No wind pressure is assumed on the façade
- All doors are closed except the doors on the fire floor from the staircase enclosure to the corridor and from the corridor to the fire enclosure
- The AOV at the top of the building is open
- No other leakage areas are assumed

Assumptions

- Without further justifications, it is a assumed a 2 m/s flow velocity over the door in the staircase enclosure between the staircase and the lobby/corridor is required to mitigate smoke entrainment from corridor/firefighting lobby to the staircase enclosure
- Without further justification, the following is assumed for the friction of the staircase enclosure:
  - The stair in this simulation is assumed to have the following input data:
    - Stair shaft dimensions 2.5 m x 5 m
    - Closed threads
    - Flow – Low population density
    - Length – 3 m per storey
  - The above information is used to translate the friction per floor to a relative orifice, in accordance with ASHRAE handbook of Smoke Control Engineering, 2012, Equation 3.25.

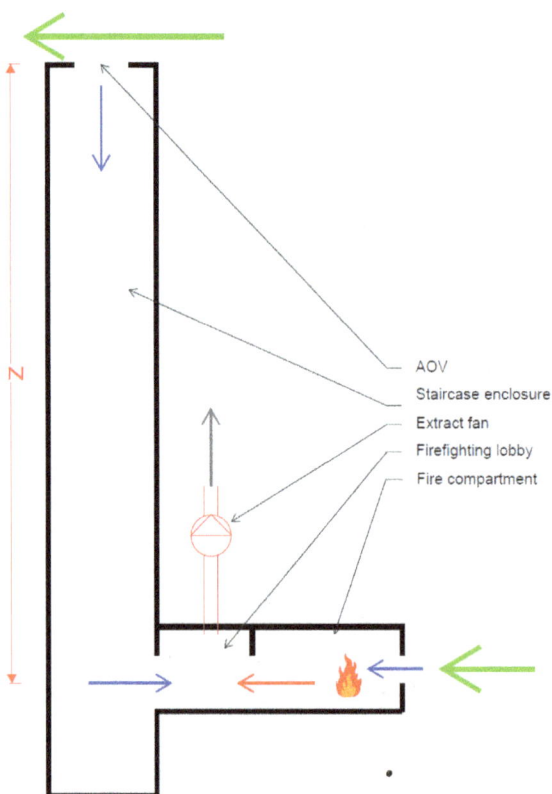

Figure 19.25 – Lobby ventilation system with AOV located at roof level

For this type of performance based design is should be noted that for all smoke control systems the effect of wind and stack must be considered in accordance with the definition of a performance base design. This is also further clarified in PD 7674-2:2002, Section 6. If the system is justified in comparison to the so called BRE-shaft, the BRE Project Report No. 79028, Section 5.4.4 clearly identifies that the effects of wind also must be considered in the design.

The extract fan will extract the required volume from the firefighting corridor/lobby. Accordingly, the design pressure differential will occur in the lobby/corridor as an inflow to this area is occurring both from the staircase enclosure and the fire compartment (refer to Fig.**19.25**). As the required inflow over the door from the staircase enclosure is defined by the input, this flow will define the design pressure differential between the lobby and outside ambient air.

The sum of the following effects will define the pressure in the lobby:

- Wind generated pressure over the AOV
- Stack effect due to temperature difference between the shaft and outside ambient air
- The friction in the staircase/Pressure differential per floor
- The pressure differential over the open door between the stair enclosure and the corridor/lobby

The summarized pressure differentials in this 90-metre high building generate a pressure differential of 162 Pa in the firefighting lobby on the fire floor. This pressure differential occurs between the lobby and ambient pressure. By definition the same pressure differential of 162 Pa that occurs from ambient air via the AOV and the staircase to the lobby will also occur from the lobby, via the fire compartment to outside ambient air, refer Fig. 19.25. As the flow-path is open this pressure differential can only be generated by a pressure differential over the opening. To achieve a pressure differential of 162 Pa over the door between the lobby/corridor to the fire compartment and from the fire compartment to outside ambient air a flow of more than 34 m^3/s of combustion gases from the fire compartment to the lobby/corridor is occurring.

The inflow to the lobby from the staircase and from the fire compartment must be balanced by an extracted flow by the extract fan connected to the lobby/corridor. With the suggested input the extract fan must extract between 15 to 46 m^3/s depending on the temperature of the extracted gases. The 15 m^3/s represents 23-degree Celsius temperature of the fire gases and 46 m^3/s represents 600-degree Celsius temperature of the gases. As combustion is likely to occur in the corridor if the extracted gases are at 600 degree Celsius, the required extract rate is likely to be closer to extract rate required for the 600 degree Celsius gases than the extract required for the 23 degree Celsius. It is not difficult to understand that an inflow of 15 to 46 m^3/s 600 degree burning gases at a velocity of up to 23 m/s into the corridor/lobby will generate significantly hazardous conditions in the lobby/corridor, even for fire fighters. The extracted flow to the lobby/corridor will generate an inflow to the fire compartment of 13.7 kg/s. In accordance with Thornton's Rule this additional high-velocity flow will generate an additional heat release rate of approximately 41 MW in the fire enclosure (equivalent to a wind driven fire) due to the increased ventilation of the fire compartment.

It should also be noted, with this suggested design the door of the base of the stair will be exposed to a pressure differential of 162 Pa. This will generate a door opening force which will be significantly higher than the 100 N suggested as the maximum recommended door opening force.

Lobby Ventilation System with insufficient capacity

The next question is, what happens if the extract fan has insufficient extract capacity and cannot sufficiently protect the staircase enclosure to the expected degree? The main difference between a pressurization system and a lobby extract system is that for a pressurization system, even if the pressurization fan fails the stair shaft will not perform as a chimney and extract the fire out in the staircase.

Assuming a 6 kg/s extract rate instead of the 18.5 kg/s that was required for the scenario above. If the extract system has insufficient capacity, very hot fire gases could be extracted out from the fire compartment to the lobby and the staircase that will act as a chimney. A 6 kg/s extract rate equals to an extract rate of 5 m^3/s at 23 degree Celsius or 15 m^3/s at 600 degree Celsius.

This scenario will generate an inflow to the staircase of between 4.4 m^3/s at 23 degree Celsius and 30 m^3/s at 600 degree Celsius in the staircase and lobby enclosure, refer to Figure 19.26. The simulation indicates a system with insufficient capacity will not be able to mitigate the effect of wind, stack effect and friction and combustion products may enter the staircase enclosure. By extracting the fire out in the staircase enclosure, 'hazardous' to' lethal' conditions are likely to be generated for persons and fire fighters in the staircase enclosure.

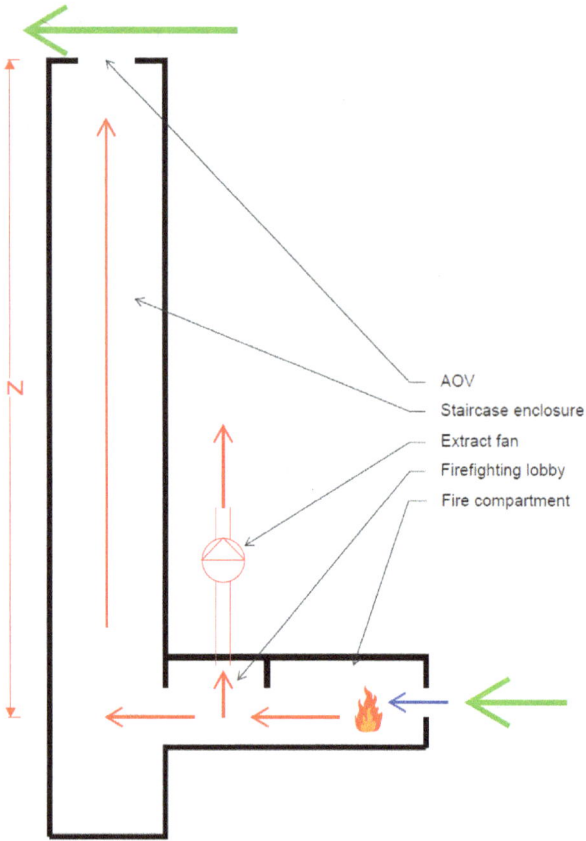

Figure 19.26 – Lobby ventilation system with insufficient capacity

Design Tools

- *Hand calculations and excel calculations*

By following the methodology for pressurization in BS EN 12101-6:2005 and the ASHRAE – Handbook of Smoke Control Engineering 2012, pressurization of a building up to the maximum height in relation to stack effect could be performed with basic tools, such as for example excel or hand calculations.

- *Node systems*

CONTAM developed by NIST – National Institute of Standards and Technology and PFS, developed by Professor Emeritus Lars Jensen at Lund University are flow network models, used for example to develop pressurization system. CONTAM is the most commonly used network model and is considered a well validated model. The program is also the program suggested to be used to design pressurization system in the ASHRAE handbook of Smoke Control Engineering, 2012. It is worth pointing out that NIST is also the developer of the most commonly used CFD program – FDS. The two programs complement each other and there are no technical conflicts between the two programs. When designing a pressurization systems all air flow paths, such as open doors, ducts, façade openings, shafts, cracks and leakages must be analysed as one system, refer to ASHRAE – Handbook of Smoke Control Engineering 2012, page 227. This system must be analysed in relation to the effect of stack and wind and the simulation program must be able to provide output including pressure differentials over all doors and other parts with design limitation for pressure differentials.

- *CFD*

It has been suggested to use CFD models for analysing the airflow conditions in tall buildings. In the fire engineering industry, FDS is among the most commonly used CFD models. While CFD is well-suited for analysing the airflow and contaminant transport characteristics on a small-scale level, i.e., within a single zone of a building, the computational resources required to perform a CFD analysis for an entire high-rise building are currently prohibitive (refer to NIST Technical Note 1887). The high computational requirements result amongst others from the small leakage paths and cracks that need to be modelled. Furthermore, it is not acceptable to limit a CFD calculation to the corridor or lobby. All interfaces interacting with the flow such as wind, stack effect and flow from, or via all other communication enclosures, must be included in the simulation. Pressure differentials created by wind over an AOV at top of the stair, against the façade or stack effect must be accounted for as these are systems with a capacity to create significant pressure differentials and high flow volumes (refer to PD 7974-2:2002, Section 6).

Consequently, a full CFD model would be computationally very demanding and effort to develop flow paths for all small leakages challenging. Results from a CONTAM model could be used as boundary conditions for a limited CFD model, combining the strengths of both approaches, while maintaining the computational feasibility and taking into account all necessary boundary conditions.

19.7 SMOKE-FLUSHING (DILUTION) SYSTEMS

Simon Lay is an experienced fire engineer, now a Director at Olsson Fire and Risk, who offers a view[216] that pressurization systems can be fraught with design problems where the environmental impact in particular, creates too many variables that cannot be addressed effectively in very tall buildings. Lay informs; 'Results from commissioning can be highly sensitive to the wind and temperature conditions on the day of testing. This is well recognized within design codes for pressurization systems. For example, the BS EN 12101-6-2005 includes protocols to normalize the system test against the climate conditions on the day of the test. However, no account is taken of what might happen on a different day under different wind or temperature conditions. The impact of this will vary with the height and location of the building, but in cities which see large variations in temperature throughout the year, and in the case of very tall buildings where wind effects are a continuous feature, the significance of setting a system to work under a single climate condition is likely to result in system performance problems under other conditions'.

Mr Lay provides many other examples in his paper surrounding the important challenges and design aspects that need to be considered for pressurization systems in very tall buildings. Air leakage paths can include stairway doors, windows, gaps in walls, natural leakage through wall materials, elevator doors, service shafts, facades and raised floor systems. There are standard estimates which are recommended in pressurization design guides. However, if the calculations are to be correct, such that the system performs as intended, the engineer designing the system will be reliant on these variables being static from the point of design, through the rest of the design process, subcontractor design, construction, commissioning, fit-out, refurbishment and so on for the lifetime of the building.

Whilst it is common to include tolerances in the design process, these cannot guarantee sufficient design flex to accommodate significant changes in leakage paths. Also, the need for integrated building services design to be efficient means that it is very difficult to achieve changes to components such as ducts, fan sizes, power supplies and relief dampers after the initial design process. It can very difficult to balance the different air flow requirements for creating positive pressure in a core with closed doors and the large volume required when doors are open. This problem becomes exaggerated in very tall buildings. There are design solutions intended to overcome the problems of balancing air flows and door opening forces. These range from simple weighted dampers to complex variable drive fans or damper arrangements linked to pressure sensors within the shaft. However, even with these arrangements there are still practical limits to the height of shaft that can be pressurised. There is an inherent expectation by some designers that following the simple calculation processes in design standards will lead to an acceptable design, without tackling some of the trickier design challenges. There is also little value in statements in codes which require the designer to make the contractor or building owner aware of the restrictions arising from the design of the pressurization systems. This approach fundamentally fails to recognize the practicalities of the building design and construction process, and does little but attempt to indemnify designers against inevitable changes in buildings.

Mr Lay does provide an alternative solution to such design for smoke control in tall buildings and offers the Beetham Tower in Manchester as an example where a smoke

---

216 Lay. S; Pressurization systems do not work & present a risk to life safety; Case Studies in Fire Safety 1 (2014) 13–17; Elsevier

flushing (dilution) system is installed. The Beetham Tower consists of a multi-level Podium block containing; conferencing facilities, bars, restaurants, reception, open atrium and back of house areas. Rising from the podium is a single high rise tower which includes hotel accommodation (up to level 22), a 'Sky Bar' (level 23), plant areas (level 24), residential accommodation (levels 25 to 46) with a single penthouse apartment at the top floor level (level 47).

Lay further informs that the natural stack in the building works with the system, wind effects enhance the system performance and the design does not have an adverse impact on door opening forces. In tests, the system at the Beetham Tower was found to deliver approximately 70% of its performance through natural stack action without any mechanical fan assistance at all. This provides a very robust design as potential maintenance failures are mitigated by a reduced, but still adequate, performance. Similarly, the system was found to need approximately 1/3 of the air supply plant that a pressurization system would require. A key benefit of the system was that opening doors to the stair did not significantly impact on performance and the system could be applied on multiple floors simultaneously, giving fire fighters much greater flexibility in firefighting operations.

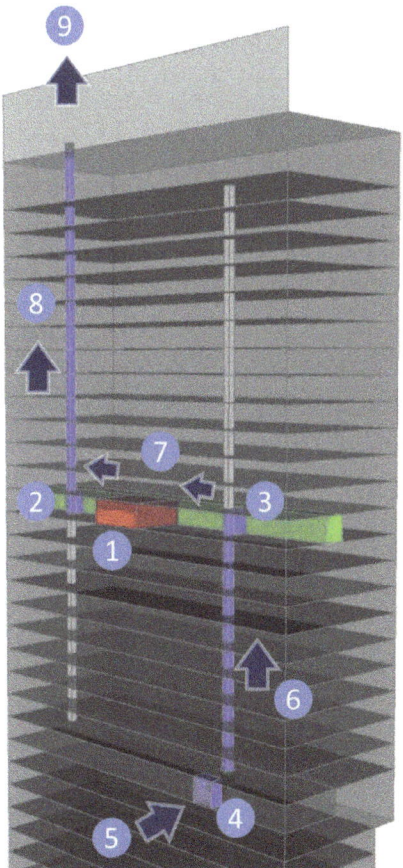

In the event of a fire in an apartment, (1) smoke will enter the common corridor, activating the system. With system activation, a damper will open into the outlet smoke shaft on the fire floor only. Fans in the plant room (4) start up and draw fresh air from the outside (5) into the inlet shaft.

On the fire floor, this air is forced into the common corridor (3) and then flows along this common corridor (7) to the outlet smoke shaft (2) collecting smoke from the fire along the way.

The diluted smoke passes into the outlet smoke shaft (2) and flows upwards (8) to the outlet vent (9).

The natural stack effect in the building is such that with the fans (4) shut down, there is still a flow of air through the system, even with cool smoke from a suppressed fire.

Under fire service control, subsequent floors can be connected to the system by manually activating the inlet and outlet dampers on affected floors.

Figure 19.27: Beetham Tower, Manchester UK. **Image courtesy of CTBUH 2004 October 10~13, Seoul, Korea**

In a paper[217] by McGrail, Lay and Chow (2004) the authors explained that It was considered that an all-natural solution could not readily provide smoke control on the fire floor, although it could provide an effective smoke clearance provision. A system based on drawing air up through a duct with a fan (smoke extract), could be effective, but also has the potential to draw additional smoke into the protected corridor space if not correctly balanced. Therefore, a smoke flushing system was deemed the most viable solution, whereby air is supplied to the common corridors creating a positive pressure, with natural relief. A flushing system has the added advantage that where the fire door to an apartment is closed, the effectiveness of the smoke sealing is significantly enhanced by the corridor being at positive pressure with respect to the apartments. Also, when persons are escaping from the apartment of fire origin and the apartment door is open, air is encouraged to flow through the apartment door opening, reducing the quantity of smoke, which enters the common corridor. Of course, a disadvantage is that the additional air could feed into the fire compartment and cause the fire to reach flashover more quickly, particularly without any initial pressure relief in the compartment.

When creating a CFD model (NIST FDS v3.1) to assess the variables involved in smoke control in a 60m tall building, the engineers used inputs replicating an apartment fire with heat release peaking at 14 MW. The typical fire size considered to apply to hotel bedrooms and apartments is a heat release rate of no more than 2,500 kW. The fire in this case peaks at up to 14,000 kW. This very high heat release rate arises because of the high level of ventilation available in this fire once the window breaks. Failure of the window occurs at c. 380s. The failure of the window leads to an initial increase in fire size (ending the incipient fire scenario which was occurring up to that point), this then leads to a flashover in the apartment and a sudden, very rapid growth in the fire. The severe fire curve predicted by the room fire model has been applied as this ensures a conservative basis to the design and adds extra confidence to the findings.

Six scenarios were considered to investigate different aspects of the proposed system. These cases are summarized at Table 19.11:

Following consultation with the regulatory authority and the Fire Service, an additional scenario (Scenario 6) was added to the study. This scenario included the addition of a fire fighting shaft within the model. The door to the firefighting shaft was propped open to form an opening approximately 200mm wide (to represent the obstruction of the door opening by a hose connected from the floor below). In practice, the smoke control system on the corridors should maintain clear conditions in the corridor of the fire floor, allowing the fire service to connect to the dry riser on the fire floor and hence leave the door to the firefighting shaft closed. However, as standard firefighting practice would be based on approaching from the floor below the fire, this more onerous condition was investigated. In addition to the opening to the firefighting stair, a vent at high level in the firefighting shaft (the 1.0m² vent at the head of the shaft) was incorporated, allowing air and smoke to flow into this shaft. A limited amount of smoke did in fact enter the firefighting shaft and although this was acceptable to the fire service, at a visibility of >10m, it was recommended that fire main outlets would benefit this type of system if installed in the corridors and not within the stair.

---

217 McGrail.D;, Lay. S; and W.K. Chow. W; CTBUH (Council for Tall Buildings and Urban Habitat) 2004 October 10~13, fire safety Conference Seoul, Korea

Scenario 1	• Fire compartment located adjacent to the lift landing area. • Air supply rate to corridor 1.0 m³/s. • Air input and vent operated simultaneously upon activation of smoke detector in common corridor.
Scenario 2	• Fire compartment located adjacent to the lift landing area. • Air supply rate to corridor 1.0 m³/s. • Air input and vent operated independently, the vent being activated only when smoke reaches a detector close to the vent.
Scenario 3	• Fire compartment located adjacent to the lift landing area. • Air supply rate to corridor 1.5 m³/s. • Air input and vent operated simultaneously upon activation of smoke detector in common corridor.
Scenario 4	• Fire compartment located adjacent to the lift landing area. • Air supply rate to corridor 2.0 m³/s. • Air input and vent operated simultaneously upon activation of smoke detector in common corridor.
Scenario 5	• Fire compartment located at the end of the common corridor. • Air supply rate to corridor 2.0 m³/s. • Air input and vent operated simultaneously upon activation of smoke detector in common corridor.
Scenario 6	• Fire compartment located adjacent to the lift landing area. • Air supply rate to corridor 2.0 m³/s. • Air input and vent operated simultaneously upon activation of smoke detector in common corridor. • Firefighting shaft included in model with leakage to firefighting shaft assessed.

Table 19.11: Scenarios were considered to investigate different aspects of the proposed smoke-flushing ventilation system for the Beetham Tower.

Key Design Specifications

- An inlet shaft is provided serving all residential levels (except the penthouse level). The inlet shaft has a fire rated, motorised damper at each level (which is normally in the closed position).
- Smoke detectors on the common corridors (2 per level are proposed) that will (via a common fire alarm system) instruct the damper on the fire floor to open. The fire alarm system will also start up an air supply fan, attached to the inlet shaft at the Level 24 plant room level.
- A back-up fan is provided and a minimum 2 power supplies are taken to the fans and dampers
- The air supply fans provide a *minimum* of 2.0 m³/s at any given residential floor although a design rate of at least 2.5 m³/s is recommended as the target for the system.
- The smoke detectors which activate the supply air system also instruct the damper to the exhaust shaft on the fire floor and roof to open.
- All dampers are motorised and provided with a back-up power provision.
- The exhaust shaft (and inlets to the exhaust shaft) have a minimum clear area of 0.5m² and are provided with a cowl at roof level to prevent adverse wind effects on the shaft.

19.8 FIRE ENGINEER'S SMOKE CONTROL CALCULATIONS (SIMPLIFIED APPROACH)

This section outlines a simple approach to calculating either fan capacities or vent areas for single-storey, large-undivided-volume buildings where the fire is directly below the smoke reservoir. It is not intended for use with more complicated building geometries. It may be of particular use to Building Control Authorities, Fire Authorities, students and others, who have to check smoke ventilation proposals or to give approval.

The mass of smoky gases entering the layer per second (*m*) is given by Equation 19.3 where P_f is the perimeter of the fire in metres; and $y^{1.5}$ is the height from the fire (usually but not always the floor) to the chosen height of the smoke-layer's base, in metres (m).

Using Equation 19.3:

$M = 0.19 \times P_f \times Y^{1.5}$ Large-area rooms such as auditoria, stadia, large open-plan offices, and atrium floors, etc. where the ceiling is well above the fire.

$M = 0.21 \times P_f \times Y^{1.5}$ Large-area rooms, such as open plan offices, where the ceiling is close to the fire. *(Note: it is not known how and under what conditions one should regard a ceiling as being close to the fire. Until better evidence appears, it is hereby suggested that C_e should take the value 0.21 whenever Y is < three times $\sqrt{A_f}$).*

$M = 0.34 \times P_f \times Y^{1.5}$ Small rooms such as unit shops, cellular offices, hotel bedrooms (prior to flashover or full involvement), etc. with ventilation openings predominantly to one side of the fire (eg from an office window in one wall only). Thus most small rooms will take this value.

Design Parameters –

- Fire perimeter size (P_f) = metres
- Heat output of fire (Q_f) = kW
- Height from floor level to ceiling = metres
- Clear layer height (Y) = metres
- Smoke depth (d_b) = metres
- Sprinkler fusing temperature (T_s) = °C
- Ambient temperature (T_o) = °C (or Kelvin)

Design Fire –

Fire perimeter size (P_f)	=	9 metres
Heat output of fire (Q_f)	=	2500 kW
Height from floor level to ceiling	=	2.6 metres
Clear layer height (Y)	=	2.2 metres
Smoke depth (d_b)	=	0.4 metres
Sprinkler fusing temperature (T_s)	=	68°C
Ambient temperature (T_o)	=	15°C (or 288 Kelvin)

Calculation of the temperature rise of the smoke above the ambient temperature of the smoke layer, along with the resultant absolute temperature of the smoke layer:

θ_{sm} = temperature rise of the smoke layer above ambient temperature, ignoring the cooling effect of sprinklers (K)

C = specific heat capacity of air = 1.01 (kJ/kgK)

$$\theta_{sm} = \frac{Q_f}{mC} \qquad \text{Eq. 19.4}$$

So, for a design fire on a department store floor (for example) of 1,200 m² floor area and a 2.6 m high ceiling, the following design parameters are agreed between the designer and the regulatory body –

Therefore –

$M = 0.21 \times P_f \times Y^{1.5}$ =

$M = 0.21 \times 9 \times 2.2^{1.5}$

= 6.17 kg/s

So –

$$\theta_{sm} = \frac{Q_f}{mC} =$$

$$\theta_{sm} = \frac{2500}{6.17 * 1.01}$$

= 402 K *(ignoring sprinkler cooling)*

So –

$$\theta_{sp} = T_S - T_O \qquad \text{Eq. 19.5}$$

=

$\theta_{sp} = 68 - 15$

= 53 K temperature rise of sprinklers

Following on, the actual rise in temperature of the smoke layer as an average of θ_{sm} and θ_{sp}

θ = the temperature rise of the smoke layer above ambient temperature (including the effect of sprinklers)

$\theta = \theta_{sm} + \theta_{sp}$ =

$$\theta_{sm} = \frac{402 + 53}{2}$$

= 227.5 K *(including sprinkler cooling)*

At this point we can calculate the absolute temperature of the smoke layer as –

$T_c = T_o + \theta =$
$T_c = 288 + 227.5$

= **515.5 K** *(including sprinkler cooling)*

Now we can calculate the extract rate required from the space -

V_{fr} = volume flow-rate of smoke extraction required (m³/s)

$\dot{p}$ = density of ambient air (1.2 kg/m³)

$$V_{fr} = \frac{M\, T_c}{\dot{p}\, T_o} \quad = \text{Eq. 19.6}$$

=

$$V_{fr} = \frac{6.17 * 515.5}{1.2 * 288}$$

= **9.2 m³/s**

Chapter 20

1-5-12 Fire Command Concept

'Keep in mind that when we limit our exposure to information, or when information itself is scarce, our picture of reality suffers. We become oblivious to both opportunities and hazards. Trends become invisible. History disappears.'

John Salka FDNY 18th Battalion

Over many years of experience accumulated through attending, fighting, investigating and researching building fires, one thing that stands out is the commonality of time-lined impacts of fire command decision making. It has been said earlier that Incident response data demonstrates that a large percentage of fires become worse following fire service arrival, before control is ever achieved. In the author's study of 5,401 UK building fires representing all building fires over a three-year period (2009-2012), where firefighters required breathing apparatus and deployed water hose-lines to deal with developing fires, 47% of fires spread beyond the compartment of origin and 30% of these spread to involve upper floor levels before control was achieved. In another study in London, 25% of building fires became worse following fire service arrival prior to the fires being extinguished. Some situations demonstrated key strategic and tactical omissions or errors occurring in the earliest stages of firefighting where command decision making had the greatest impact on incident outcomes. It was in the first sixty seconds where critical actions or decisions could have made the biggest difference in locating and saving trapped occupants. It was in the first five minutes where the deployment plan established a path of 'no immediate return' that may have impacted greatly on the success of the entire firefighting operation.

The fire-ground is extremely dynamic and decisions are often made under great duress with each specific situation presenting a diverse range of challenges and outcomes. It was noted that in some situations firefighters were deployed to an apparent point of 'no return', for fear of losing ground on the spread of fire through a building, even though their resources were not supported and their exposure to risk was rapidly escalating. It is here that the 'error chain' can be broken by a second arriving commander fresh onto the scene. It may be at this critical point within the first twelve minutes of an initial response arriving on-scene that a command decision to 'pull out' and regroup for re-deployment might have saved multiples of firefighter lives.

On the subject of decision-making processes, Dr Gary Klein has investigated the subject of recognition primed decisions (RPD). According to Klein, 'fire-ground commanders will make 80 per cent of their decisions in less than one minute'[218]. It was discovered that emergency-scene decision-making relies heavily on experience, especially when the fire-ground commander is faced with a time-pressure situation. The RPD decision-making model combines two ways of developing a decision; the first is recognizing which course of action makes sense and the second is evaluating the course of action through imagination to evaluate if the actions of the decision make sense. The RPD paradigm of decision-making applies to fire-ground command because decisions on the fire-ground are under time pressure conditions and experience of the fire commander plays a large part in determining if the appropriate decision will be made.

However, some key points of guidance here from Arthur Perlini, Professor of Psychology at Algoma University, Canada, when discussing his work in fire commander's decision making, based on 'gut feeling' and a strong sense of past experience:

- *Less informed leaders fall victim to 'pattern matching' ...*
- *There are tendencies to compare each new situation with old patterns, without paying attention to the actual individual needed cognitive efforts.*
- *There is also a tendency for people to over emphasise recent events rather than a combination of all experience in total.*
- *There is greater recollection of instances that confirm beliefs, and therefore an over-estimation of the likelihood of an event repeating itself.*
- *Many are prone to see patterns that don't actually exist.*

It should be our objective to provide buildings that support 'safe' occupation for both firefighters and occupants in those early stages of a building fire. It must also be a primary aim of firefighters that we implement actions and tactics that reduce a fires heat release or rate of fire spread as soon as we arrive on-scene. This strategic objective alone may save even more lives.

20.1 DECISIONS MADE IN THE FIRST SIXTY SECONDS CAN SAVE OCCUPANTS' LIVES

As a first-arriving fire commander steps off the fire engine at a building fire there are so many key indicators that will determine 'where', 'how' and 'if' to deploy internally is the correct decision at this early stage.

The immediate **and critical** tactical considerations in that primary 60-second period are:

- **Occupancy type?**
- Firefighting resources and **staffing available?**
- Persons reported **trapped or missing?**
- Occupants located at windows requiring **immediate rescue**
- **Fire location (undertake as near a 360 deg. exterior survey as possible)?**
- Is building **sprinklered?**
- **Wind direction and speed (and likely impact on fire)?**
- **Isolate the flow-path** at the open front or rear door (if relevant).

---

218 Klein,G; Sources of Power; How People Make Decisions; Cambridge, Massachusetts; MIT Press 1998.

20.2 DECISIONS MADE IN THE FIRST FIVE MINUTES CAN SAVE PROPERTY

*The first 'sixty second' primary survey should dictate the tactical options for the next five minutes and beyond. At this point the 'SLICERS' options (chapter one) come into play.

- Exterior holding position?
- Consider water-fog injection into a confined fire compartment?
- Hi-flow exterior attack?
- Transitional attack?
- Protect surrounding exposures first?
- Deploy an interior attack?
- Undertake interior search and rescue?
- Implement additional hose-line management, a safety hose-line or interior door control assignments?

During this initial phase of firefighting operations, it is important to determine and communicate with great clarity the tactical objectives to all firefighters on-scene.

*Here are some key points for discussion with firefighters … Another 60-minute classroom debate! It might, at first, appear simple and straightforward to request an exterior stream be applied into a window discharging fire and smoke. However, what is the strategy? This will determine how the water is to be used! Is it a 'holding' strategy whereby the water may be from an appliance tank supply of around 1800 litres with the intention to slow or control fire spread whilst awaiting additional resources to arrive? Or, has a constant flow supply been sourced? Are we preparing to actually extinguish the fire and if so, is a high flow-rate required because of the potential/involved energy in the fire load? If we empty our water tank in a rapid hi-flow attack (not a holding attempt), do we have secondary water available within 60-120 seconds? Or, are we preparing for a transitional attack where following a 30-60 second exterior stream in through the window to reduce the fire's heat release (fire reset), we immediately follow-up with an interior deployment to extinguish the fire? If this is the case are we on tank water or constant augmented supply? Are we able to deploy the first (or a secondary laid) hose-line with breathing apparatus already started? The key to all this is that the secondary interior line must apply water within thirty seconds (or near) of the primary exterior line shutting down. If they are the same hose-line then this may be a real challenge, relevant to on-scene staffing. Clarification in the brief is critical here and everyone needs to understand the tactical objective.

20.3 DECISIONS MADE IN THE FIRST TWELVE MINUTES CAN SAVE FIREFIGHTER LIVES

The author undertook a review some years ago, of several major incidents where multiple firefighter life losses occurred. It was noted that the 12-minute period after arrival on-scene was commonly the most hazardous on a time-line where a chain of events had set in, leading to firefighter's lives being lost. A fire commander arriving on-scene within the first 12 minutes of a developing fire may have the advantage of coming in as a fresh pair of eyes. However, it takes courage and experience to be able to make that call to either continue with the current strategic deployment or change direction, particularly if that means withdrawing crews to the exterior and starting again. It has been the case that crews responded to alternative sides of a large structure and self-deployed believing they were solely in charge, or an element of freelancing has seen an uncontrolled number of firefighters enter with accountability. In other cases, the deployment may potentially be in towards a headwind, or with a strong wind into the entry door and towards the fire, but without an outlet to relieve internal pressures. In any such situation, it may be hard to determine how many firefighters are working in the risk zone and exactly what their roles are. A quick determination of the building fire indicators coupled with what is a necessary deployment at the time of this secondary arrival should dictate if any evacuation of firefighters, either in part or as a whole, is required. If it's an unoccupied retail unit as opposed to a fully occupied hotel, this will clearly also impact on any such decision in respect of firefighter safety and exposure to risk.

Take a look at some NIOSH[219] and other[220] fire reports and debate with colleagues where tactical decisions made at this early stage (12 minutes) to withdraw, regroup and re-deploy may have worsened firefighter safety or improved it. In some cases, the building may have been lost but all firefighters went home safe. In other circumstances the exposure to risk may have been increased. With hindsight, decision making is always much easier!

Bill Gough is a senior fire commander who has served over four decades with the West Midlands Fire Service in the UK. A major part of his academic work (to PhD) in firefighter safety and incident command is directed at identifying cultural differences and psychological influences that directly impact how some firefighters expose themselves, and their crews, to unnecessary levels of risk without any reasonable justification according to a logical risk profile.

20.4 BILL GOUGH WRITES ON SAFETY CULTURE

Just after 1.30 pm on March 29, 2015 Captain Peter Dern of Fresno Fire Department arrived at the scene of a single storey family residence on fire. A little over a minute later, whilst undertaking roof ventilation, he stepped onto the weakened garage roof and fell into the seat of a fully developed fire. Within hours, the event was being viewed globally. Courtesy of social media, firefighters, their commanders, managers and their communities across the globe could watch the scene unfold on You Tube and at the time of writing there has been almost 240,000 views. A week after the accident the Fresno Fire Chief assembled a Serious Accident Review Team (SART). Their detailed and very comprehensive report was published in January the following year.

The report called for a refocused approach to risk management and risk assessment but the dominant finding was the need for cultural change. Creating an organisation wide culture that gives safety a priority is a classic journey and never a destination. A journey that starts with a challenge that is notoriously difficult – defining the meaning of culture. Academics, scholars and Scientists have a mixed bag of socio/psycho versions. The Cortland Incident report defines culture as:

'The sum of attitudes, customs, and beliefs that distinguishes one group of people from another. Culture is transmitted, through language, material objects, ritual, instructions, and art, from one generation to the next'.

---

219 https://www.cdc.gov/niosh/fire/
220 http://www.everyonegoeshome.com/

A generalised version is offered by the Centre for Safety Research: '... *who and what we are, what we find important, and how we go about doing things around here*' (Hudson 2001)[221].

A www search for 'safety culture' will reveal 51 definitions but many will offer an abbreviated form that captures the generalisation: '... the way we do things around here ...'

In this simple statement lies the real challenge to a safety culture, which we, who?

When Fire Service commanders and managers develop their top down 'cultural change' programmes they should ask themselves when dealing with ownership of cultural safety, when demonstrating safety leadership and holding tool box talks, whilst walking the walk and talking the talk, who is the we doing the things that cause injury?

UK Practice differs from that in the US. For a Crew Chief, especially the equivalent to a Captain to be injured at an incident in the way Pete Dern was is not only a rarity, it is most unlikely. A study of operational injuries undertaken in 2015/16 identified that in the UK on average 1,100 firefighters sustain operational injuries every year 75% of them will be in the role of firefighters, 15% Crew Commanders (Lieutenant) and 10% Watch Commanders (Captain). Closer analysis revealed that the vast majority of injuries did not involve a high stakes activity where risk-v-gain assessment could result in physical injury. 87% of all operational injuries are sustained before reaching the hazard zone. Or when hazards are under control, risk is reduced and 'critical' decision making is no longer a safety challenge (Gough 2016).[222]

Creating a safety culture that may impact firefighter injury is not going to be an easy journey. But behavioural change is not possible without understanding how or what is influencing the 'moment of choice' of the people being injured, without in some way visiting some of those factors that influence their moment of choice. There are many factors that influence behaviour but here the focus is placed on when is the way we do things around here actually concealing cultural behaviour based on 'we've always done it this way'? The question cultural change managers need to consider is, is there really any difference between cultural practice and habit, which has the strongest influence, which is easier to re-shape or re-train, the habits of a lifetime or the influence of the workplace. The manager who is walking the walk and talking the talk of cultural change will eventually get to that workplace that is the Fire Station. The cascade of re-branding safety, the new message which is portrayed as 'ours' will be transmitted by that manager. At that workplace the challenge they should give each individual is one of self-reflection: which is mine, (my way of doing things) and which is theirs, (their way of doing things)? Which do I own and which is influenced by others? Which do I do because I want to and which do I do because others expect it? This start point explores the possibility that a personal habit can influence group practice, and as a natural progression, over time group practice can become embedded culture.

There is an inextricable relationship between habit and the moment of choice. From at least one perspective it would appear that habit is a good thing because so much of what we do demands quite some thought (cognitive processing) and takes up head space to get the nerves working the muscles, moving the limbs that get things done. This is when a frequently performed habit quite simply takes up less head space, it gets filed away as part of our working memory. Arguably, this is what skill and expertise is all about, well-practiced habits enabling us to work on auto-pilot. This brings us back to the need for understanding the factors that influence the moment of choice. In the study mentioned above for the first time, injury

---

221 Hudson, P. (2001) *Aviation Safety Culture*. Safe Skies Magazine
222 Gough, B (2016) *The Moment of Choice,* Poster Presentation in the Proceedings of the 2016 Fire Service Research Event, November 2016. West Midlands Fire Service Headquarters, Birmingham, England

causing accidents were analysed for their 'Active Error' types. This revealed that the majority of critical injuries resulted from skill based errors (35%).

James Reason describes how skill represents routine, highly practiced tasks that people are very good at 'most of the time'. He also points out that 'strong habit intrusions are amongst the most common of all error forms', and that 'they are most likely to occur during the performance of highly automated tasks'. (Reason 1997). Skill based errors leading to injury represent the down side of habit, we could be so used to doing things on auto-pilot that we can miss subtle changes when they occur in the work space, when things are almost but not quite the same. When more than expected head space is needed to deal with them it turns out things can be missed. These are the 'cues' that inform situation awareness. Now this can mean that, just like Pete Dern we could actually enter a hazardous environment without waiting for the head space to respond to the nuances of contextual difference. Routine practice becomes habitual, so it requires less head space. This means there are circumstances that we pay less attention to. When subtle changes occur, we are challenged to notice them particularly if the strength of the habitual practice outdoes the 'cues' in the environment. This means that the head space will respond to the strongest cue – the habitually practiced one when it might not be the right action and … someone could get hurt!

But what is the 'cultural' dimension of habit?

The relationship between culture and habit is revealed when work as planned isn't quite the same as work as it gets done, when through habit, a step in the standard operating procedure is itself habitually overlooked or set aside for all good intentions. Reason characterises this error type as a violation and offers a word of caution to the cultural change manager who attempts to control behaviour through the design of SOP's: 'one of the effects of continually tightening up safe working practices is to increase the likelihood of violations being committed' (Reason 1997:51)[223]. The Human Factors paradigm describes violations as either being routine or exceptional. When behaviours are controlled by decision rules such as 'must' or 'under no circumstances', procedures will be routinely violated or when an incident simply does not fit the SOP model an exceptional 'work around' or adaptation is used. In the operational injury study 4% of injuries occurred during routine violation and 8% exceptional violations. The dilemma this creates at the moment of choice is when to comply and when to violate and this is where the link with habit and culture is possibly its strongest, when routine violation becomes the alternative routine, ingrained within 'the way we do things around here'. When these habitual nuances aren't re-aligned in the workplace habitual behaviours become routine violations and firefighters get hurt.

This is where the role of the supervisor is important because they are in the middle of the mix. The first point of contact, they're right there at the action end of the system, the so called 'sharp end'. This supervisor prominence is principal at the heart of classical accident theory 'The supervisor or foreman is the key man in industrial accident prevention. His application of the art of supervision to the control of worker performance is the factor of greatest influence in successful accident prevention' (Heinrich 1959). As the custodian of SOP's, they can also be the enabler or maybe even the practitioner of habituation, of routine violations. The Fire Station is that place where after the cultural change message has told them where and what risk is, after SOPs have been re-aligned to their new branding and structure for dealing with hazards and risks; that place where work as designed and imagined isn't exactly the same as 'work as it gets done'.

223 Reason, J. (1997) *Managing the Risk of Organisational Accidents*. Ashgate Publishing, Aldershot

The Fire Station, the crews, shifts or watches that populate it are often described as a band of brothers and sisters, a family, a clan. This too is what the cultural change manager will at times be challenged by. After the publication of the Cortland Incident report the mood of the clan was mixed but one particular discussion thread stood out as an exemplar of the mood, accredited to an author identified as Nick:

'I guess we will just hire anyone now and inform them there is NO risk with firefighting, everything will be done from outside ... put soccer moms as IC's, never make an interior attack where roof venting is critical, never ask weaklings to throw ladders, JUST throw away the gear- set up master streams from the outside or wait, how about an aerial stream on all house fires ... what (censored) is wrong with people??? (censored) happens, it's a DANGEROUS JOB, and RISKS and DANGER plays a part. Man, these safety whores are overdoing it. Because of 1 or 2 incidents, we have these assholes with their 500 page reports on this and that ... shut up already' (Statter911 2016)[224].

This may represent an individual view, it may have endemically become the clan view and reflect the clan culture. Supervisors are members of the CLAN – they're CLAN CHIEF's – sometimes NICK is the SUPERVISOR. Rather than 'what' is the most difficult challenge to cultural change it can be more a case of 'who' which raises a number of questions for the cultural change manager:

- How far does NICK's influence go?
- Does it spread throughout the CLAN?
- How many CLAN's have a NICK?
- What if NICK is a supervisor?

There can be no doubt that the Fire Station is a significant cultural pinch point, its where the way things get done is shaped. The influence of the clan can be so embedded that cultural change can take a generation. To start the journey contemporary research now argues for finding something that's a winner, maybe even go for the quick win. Then why not use the culture of the clan to claim a quick win?

Contemporary research also tells us that reacting to the lessons of failure is the old way of safety. That we should instead be taking a positive view, why things go well, why when the majority of roof venting operations are achieved without incident are some not? Would the adoption of a KILL ZONE critique lead to a wiser risk taker? Why not get Nick and the Clan to take an after-action review of successful operations in the roof venting KILL ZONE? It was Heinrich that told us that for every fatal accident there are 29 that cause varying degrees of injury but that there are also 300 close calls or near misses. Does the secret to preventing 330 accidents that result in varying degrees of injury from hurt pride to fatality lie in every clan taking time to critique uneventful KILL ZONE operations to identify why, on that occasion everything went as planned (Heinrich 1959)[225].

The Cultural change plan of any Fire Service should seek understand that cultural change takes time, maybe even a generation. In defining culture understand who owns the 'way we do things round here', the change plan should seek out the cultural pinch points, the workplaces with a dominant clan culture, use the language of the clan, seek out and use the safe supervisor, weaken and hopefully realign the dominant macho voice of Nick.

---

224 Statter, D. (2016) *Report calls for Fresno FD 'culture change' after garage fire that critically burned Capt. Pete Dern*. Discussion Blog at: statter911.com/2016/01/20/report-calls-for-fresno-fd-culture-change-after-garage-fire-that-critically-burned-capt-pete-dern/

225 Heinrich, H. (1959) *Industrial Accident Prevention*, 4th Edition. McGraw Hill, London

Look for a 'winnable' route to change, could KILL ZONE critique be a useful tool in the box, what role could Nick have in shaping such a programme?

20.5 'DESCRIBE YOURSELF IN ONE WORD'

As final words in this book I would ask the reader this one question – *describe yourself in one word?* Now that's a great interview question and if you ever get asked it, well maybe s20.5 might go some way in preparing you for it! There are many words I would expect you can use to describe yourself and all may be very descriptive of your demeanor or personality. The fact that you have got this far in the book means you either read the sports pages in the newspaper first, or you like to spoil a story by reading the end before trawling through the entire book; or it might mean you are a real student of your profession and you've arrived here through some scholarly process. You might describe yourself as *conscientious, enthusiastic, prompt, studious, tenacious, inquisitive, courageous, proactive* or any one of a hundred other descriptors. These are all good traits that represent the kind of person who will generally do well in life. However, be prepared to answer the next question in the interview! *Give me several examples of how you consider that you actually meet that descriptor?*

Probabilistic analysis ('Flapper's' story)

I remember clearly a conversation I had once with a young crew commander at a fire station I worked in London's west-end. He appeared a reasonably clever guy and always put a lot of thought into his own strategic command approach. However, he had an unspoken nickname amongst the team – we called him 'flapper' (now that's another story), but never to his face! Now 'flapper' and I never saw 'eye to eye' over his personal view that everything we did in life, especially at work, was heavily influenced by a conscious *probability decision based process.* In fact, we had deep and lengthy discussions over his command concept. Now I have never been a big fan of probability analyses, in firefighting or in fire engineering methodology. I do appreciate that statistically, we might get one fire in a particular building once in every hundred years or so. However, my own approach to firefighting response and deployment was founded upon a strong experience-based desire for *caution.* This wasn't taught, it was learned the hard way through the many little things we could and should have done better. That one fire in a hundred years could be tonight! The fire engineering profession once told us that the tunnel under the sea linking England with France was so well protected from fire that probability analyses suggested even one fire in a hundred years was highly unlikely. Well guess what – we had four working fires in the tunnel in twenty years!

Well as a fire commander, 'flapper's' operational and tactical approach was quite simply based upon a reactive response. In fact, the probability of a fire alarm in one of our tall high-rise buildings actually turning out to be a fire led to what I would call, a complacent approach. Now that's fine for ninety-nine false fire alarms. It was clear, 'flapper' and I just didn't see things on the same level. I was *proactive* and he was *reactive*, so much so that each journey up in a 30-storey hotel lift to yet another fire alarm 'reset' caused conflict when I insisted we followed procedure, myself taking the lift to the floor below the reported fire floor, whilst he carried on up in the elevator to the fire floor itself. There were many conflicts between us and all of them were founded upon his 'probability' concepts and my strict adherence to procedure and proactive belief that 'every call could be the one'! Poor 'flapper's' constant desire to calculate and estimate everything on the basis that an event (or fire) was statistically improbable, meant that his command approach was flawed and forever tinged with complacency. I wonder how the four inner-city firefighters (two without breathing apparatus) who since became trapped in a lift that stalled *at the fire floor*, with flames roaring along the corridor in a blazing tower block, might view such an approach now.

Well 'flapper's' probabilistic analysis of every tactical approach saw him rise to the highest levels of command in the London Fire Brigade, in charge of the entire operational response, so he must have done something right. I always look back and think of a single word to describe our own demeanor and I think 'flapper' and I could easily stand for *'complacent'* and *'safe'* between the two of us! I'll let you decide which approach you feel would describe me best and of course, which of any single word you would use to describe yourself!

Always be alert, proactive and remain in control.
'Stay safe' *and good luck.*

Paul Grimwood

Post Script

Group Commander Rob Lawson
Kent Fire and Rescue Service (ret)

Think of where we have come from; to where we are now; to where we need to be in future – it's for you to make the changes needed!

The built environment is a changing place, new construction methods and materials being tried as environmentally and cost effective solutions are sought.

The traditional construction methods of our predecessors were well known and every good firefighter would know what to expect on their patch. The characteristics and danger signs were shown in simple manuals and remained the core of instruction for the recruit and a timely reminder for the old hand.

Modern construction however requires a more investigative approach from incident commanders and the firefighters who take the greatest risks as part of their duties. Many systems can only refer to laboratory fire test data, the actual conditions in use and fire loadings vary considerably and may never be revealed till the first real life incident. Firefighters must remain forever vigilant to safely do their work, for it is this that will set sound tactics and save life and property.

There can be no greater source of learning than through specialist works and the dedicated fire community where experience and discussion promotes the objectivity and knowledge which contributes to safe operations.

Paul Grimwood's Eurofirefighter 2 is one more step on that path but by no means is the end. It is for all those in the fire industry to do their best to keep the built environment and those that take the risks as safe as possible. It is the role of all to continue this work and share life protecting knowledge.

CPSIA information can be obtained
at www.ICGtesting.com
Printed in the USA
BVHW060851130919
558158BV00040B/8/P